MODERN POWER STATION PRACTICE

IN 8 VOLUMES

VOLUME 1

PLANNING AND LAYOUT

CONTRIBUTORS TO VOLUME 1

R. Briggs

J. L. Davies, B.Sc.

W. L. Martin, C.Eng., M.I.Mech.E.

D. L. McKie, C.Eng., M.I.C.E.

A. Meredith, A.M.C.T.

C. J. Smith

E. F. F. W. Usher, C.Eng., I.Mech.E., F.I.E.E.

MODERN POWER STATION PRACTICE

SECOND REVISED AND ENLARGED EDITION

VOLUME 1

PLANNING AND LAYOUT

Published for and on behalf of the

CENTRAL ELECTRICITY GENERATING BOARD

BY

PERGAMON PRESS

OXFORD · NEW YORK · TORONTO

SYDNEY · BRAUNSCHWEIG

Pergamon Press Ltd., Headington Hill Hall, Oxford
Pergamon Press Inc., Maxwell House, Fairview Park, Elmsford, New York 10523
Pergamon of Canada Ltd., 207 Queen's Quay West, Toronto 1
Pergamon Press (Aust.) Pty. Ltd., 19a Boundary Street,
Rushcutters Bay, N.S.W. 2011, Australia
Vieweg & Sohn GmbH, Burgplatz 1, Braunschweig

First published in five volumes during 1963-1964

Second revised and enlarged edition 1971

Library of Congress Catalog Card No. 75-86200

Printed in Hungary

08 006454 X

08 016436 6 (set of 8 volumes)

CONTENTS

LIST OF ILLUSTRATIONS

PREFACE

In 1963 the C.E.G.B. first published *Modern Power Station Practice*, a revised version in book form of a correspondence course prepared by engineers, scientists and managers within the Board. This course was intended to benefit mainly engineers and other technical staff with at least H.N.C. standard of education in their particular disciplines. A further edition (the third) of the correspondence course was completed early in 1969.

The first two editions were somewhat limited in range and depth of treatment of the subject, particularly with regard to the auxiliary technologies: this fact and the continuing acceleration of technical development made it necessary for the editors to make a fresh approach in preparing for the third edition. As a result this edition was substantially enlarged to give a wider, deeper, and more up-to-date coverage of subject matter, and it was re-arranged into nine sections, A to J, each of which comprises five lessons. The complete course now gives a choice of instruction in practically all aspects of power plant technology, from design to management, in subjects as diverse as nuclear power engineering and water chemistry. A further section, K, is in the course of preparation and concentrates on the commissioning of new plant. Sections A to H are published sequentially as Volumes 1–8, but the Management Section J has not so far been prepared.

The main plant conditions described in the text are based upon the 500 MW unit which is at the present time being commissioned in considerable numbers in this country and from which early operating experience has been and is being gained. In addition, some material is provided on 660 MW units, the first of which is expected to be commissioned at Drax Power Station by 1971. This concentration on modern plant has meant the virtual elimination of items such as stoker fired boilers, etc., but as the age of nuclear power gets under way, this omission seems particularly appropriate.

As a correspondence course of instruction the third edition may be considered to be difficult although rewarding to the student, and many of the Board's staff have expressed the view that the lessons should be made available as reference information. It is also known that there is considerable interest outside the Board's organisation, in having this information made more generally available. Mainly for these reasons the present set of books has been produced, but in order to minimise costs and delay in publication, the composition and arrangement are closely similar to those of the third edition of the correspondence course, no further major revision having been made. This arrangement is not entirely satisfactory in a work for general publication, particularly when one considers the enormous editorial task faced by the editorial panel of dealing with over $1\frac{1}{4}$ million words supplied from the pens of a hundred authors; but the Generating Board hope that the reader will be well compensated by the unique assemblage of information which these volumes represent.

A work of this size would not have been possible without the whole-hearted support of numerous manufacturers in the supply industry who have provided many of the drawings and photographs which so liberally illustrate the text.

I would like to thank the authors who have contributed such a fund of knowledge to provide this encyclopaedic range of power station work and have done this essentially by spare time effort.

Much spare time work has also been contributed by the editorial panel who gave devoted service not only to draw up the syllabus and edit the work, but in many cases to provide much of the original material. They did this in face of some provocation by the Chairman, who started off so dissatisfied and ended, if not entirely satisfied, enormously impressed by the prowess of their work.

Finally I would thank the Personnel Department of Board Headquarters who have worked hard to administer the correspondence scheme.

ALAN SHERRY
Chairman Editorial Panel

CHAPTER 1

SITING AND DESIGN OF POWER STATIONS

1.1. INTRODUCTION

The first supply of electricity for general public use was made available at Holborn Viaduct, London, in 1882.

The demand expanded swiftly and by 1948 there were over 550 separate electrical undertakings with an aggregate generating capacity of 11,600 MW which were vested in the British Electricity Authority on Nationalisation.

The British Electricity Authority, besides being responsible for the generation and main distribution of electricity, had powers to exercise general control over the fourteen Area Boards formed for the regional distribution of electricity.

Since 1948 the structure of the industry has been changed, first by the Electricity Reorganisation (Scotland) Act of 1954, which established the South of Scotland Electricity Board to take over the functions of the South-East and South-West Scottish Electricity Boards and the two generating divisions, and the Electricity Act, 1957, which came into operation on the 1st of April 1958, dissolving the Central Authority and establishing in its place the Electricity Council which is responsible for the overall guidance of the industry, and on which the Chairmen of the Central Electricity Generating Board and the twelve Area Boards sit.

The 1957 Act transferred responsibility for the generation and main transmission of electricity to the C.E.G.B. The Board's principal duty is to develop and maintain an efficient, co-ordinated and economic supply of electricity in bulk for all parts of England and Wales and to provide bulk supplies for the Area Boards to distribute. The Board is empowered to provide bulk supplies for the Scottish Boards, the railway undertakings and undertakings outside Great Britain.

The Board's undertaking is organised in five Regions, each in the charge of a Regional Director, for the management and operation of power stations and the transmission system in their areas.

The planning and development of the system, which includes both power stations and supergrid transmission lines, is the responsibility of the CEGB headquarters departments with the Project Groups responsible for the construction of power stations and the main transmission system throughout England and Wales. Now once again the Government have decided the central direction of the supply industry needs strengthening and are introducing new legislation to reconstitute the Electricity Council as the Electricity Authority with new powers to plan and control the policy of the industry as a whole.

This chapter explains the selection and location of new power stations, outlines the major design features, and the procedures followed to secure the consents necessary for their construction.

1.2. DEVELOPMENT OF THE SYSTEM

The pattern of future development of the system, including the location of new generating plant, must provide for the future demands for electrical energy in the most economic manner taking into account the availability and relative cost of fuel. For this purpose a development plan, based on estimated future maximum demand and unit requirements, must be prepared.

1.2.1. Historical Growth

In the twenty-year period from 1948/49 to 1968/69 the growth in demand has increased about fourfold from 9,382 MW to 37,738 MW, an annual growth rate of 7·3%. If the growth rate is examined over a four/five year period, it is found that this has not been uniform but was 8·7%, 7·9%, 7·5% and 4·7%. During the same twenty years, electricity supplied has risen from 40,314 to 173,418 million units, an increase of $4\frac{1}{4}$ times, and the system load factor increased from 49·4% to 52·7%. Whilst such figures show past experience and can be a useful guide when forecasting future trends, they require most careful analysis in comparison with such factors as economic growth, market trends, industrial, commercial and domestic and competitiveness of alternative fuels.

To enable a strict comparison to be made and trends to be estimated, all demands are adjusted to standard weather conditions.

With a given state of electrical development it is possible to estimate the relationship between weather conditions and electricity demand and to calculate the yearly peaks that would correspond to the range of weather likely to be experienced in a span of a hundred years. The median of the yearly peaks so calculated for the hundred years is described as the demand in "average cold spell conditions"; thus, in half of the year the peak demand would be higher and in half the year it would be lower than the demand in average cold spell weather. The temperature conditions employed are close to freezing, not the average daily temperatures.

1.2.2. Estimates of Demand

Unlike the gas industry, the electricity supply industry is unable to store electrical energy during periods of low demand to provide for periods of peak demand. Therefore, it is necessary to provide sufficient plant to meet the peak demand in the system which usually falls within the months of December to February.

For the estimates, Area Boards provide for each point of supply within their system the maximum demand and the total units required for each winter for eight years ahead of that in which the estimates are prepared, based on knowledge of local developments and business trends. The Area Board estimates are adjusted to allow for the difference

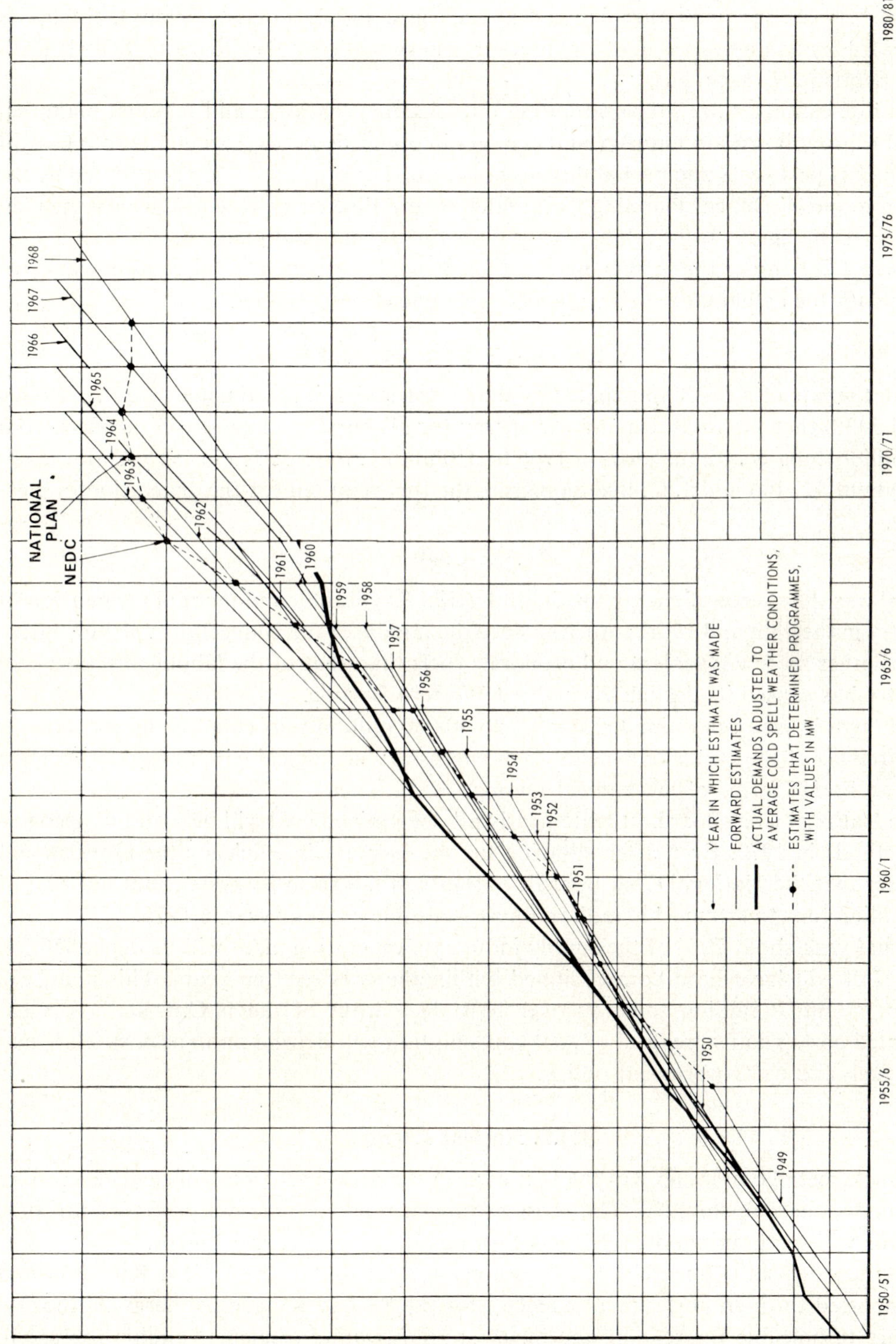

FIG. 1.2.2. Forward estimates and actual peak demands in ACS conditions.

in time between the maximum demand at each grid supply point (diversity), the losses on the transmission system and for direct transfers made by the Board to British Railways, Scotland, France, etc.

Separate estimates are prepared taking into account national and regional planning, probable growth rates in industry and changes in domestic needs, together with trends of growth over past years and probability in the future. These estimates are considered by the Research and Technical Planning Committee of the Electricity Council, which then recommends the figures to be adopted as a basis for the national plan.

Figure 1.2.2 shows the estimates of future demand corrected in successive years, together with the actual corrected demands experienced until 1968/9.

1.3. ENERGY RESOURCES

With the rate of growth anticipated by the Government, it is estimated (see Fuel Policy Cmnd 3438) that the total demand for energy for all purposes inland will increase from 298 million tons coal equivalent in 1966 to 310 m in 1970 to 350 m in 1975. Of these figures, about 92, 106 and 136 mtce represents the fuel required for the generation of electricity.

1.3.1. Coal

Coal was the source of energy for the Industrial Revolution and the rapid expansion of industry in the 18th and 19th centuries. Since the middle of the 20th century, the introduction of other more convenient and economic fuels than that of the labour intensive coal industry has caused its decline throughout Western Europe.

Although it has been estimated that some 40,000 million tons could be mined, only in limited areas of the British coalfields do the necessary geological conditions exist for the most economic forms of mechanical mining.

The National Coal Board estimate nearly all deep mined coal will be won by mechanical methods in 1970. Even after allowing for the increase in efficiency the Government consider that the National Coal Board's output will fall from its present output level of 160 million tons to around 120 million tons of saleable coal per year in 1975.

In that year about 70% of the supply industry's generating plant will be dependent on coal. Half will have been commissioned within the previous ten years. This plant will require about 70 million tons, or over half the output of the N.C.B. at the present estimated cost. The future level of coal consumption will depend upon its cost relative to other fuels and on Government policy.

1.3.2. Nuclear Energy

Natural metallic uranium was the fuel used in the early gas-cooled nuclear reactors to be commissioned up to 1970. The more advanced types of reactors proposed for 1970 onwards will use uranium dioxide fuels with possibly 1 or 2% enrichment.

Uranium ores in commercial qualities are not available in the United Kingdom and are imported through government agencies for the United Kingdom Energy Authority, who are responsible for refining the metal and fabricating elements suitable for use in commercial power stations.

In the Magnox Reactor, 1 ton of natural uranium produces the heat equivalent of 10,000 tons of coal; the heat cost is very low, but the high capital cost of plant necessary to produce and utilise the heat elevates the total cost of generation to above that of the best modern coal-fired power stations being commissioned on the coalfields in the early 1970's.

In the Advanced Gas-cooled reactor (A.G.R.) which will come into commission in the early 1970's, 1 ton of uranium oxide fuel will produce heat equivalent to over 75,000 tons of coal. This type of reactor will produce electrical energy more than competitive with the best coal and oil-fired stations.

1.3.3. Oil

Since 1945 the British petroleum industry has increased the quantity of crude oil imported for processing from 2 million tons to 83 million tons in 1969. New refineries are under construction on a number of estuaries to meet the increasing demands for petroleum products which will have a capacity of 100 million tons by 1970. This policy is resulting in increasing quantities of "home-produced" fuel oil which in the immediate vicinity of the major refineries can provide heat at a lower cost than coal.

The supplies and costs of the oil are subject to international relationships and political pressures outside normal commercial considerations. Fuel oil carries a tax of 2·4 *d.* per gallon, representing an increase of over 40% on the cost, and if this were removed fuel oil would be competitive over the whole of the country.

Although the sulphur content of oil is often alleged to be the cause of heavy atmospheric pollution, based on the same quantities of heat from normal oil and coal, supplies from some northern coalfields produce similar quantities of sulphur dioxide. Low sulphur oil stocks are becoming available from the new developing fields such as North Africa and may be used at sites where consideration of atmospheric pollution from sulphur dioxide is of special importance, such as in the London conurbation.

1.3.4. Natural Gas

Since natural gas was first discovered under the North Sea off the Humber in the autumn of 1965 at least four important gas fields have been identified. From the wells already established by 1969 a steady flow is coming ashore and is being distributed over the pipe system (gas grid) which is rapidly being constructed to supply all the Gas Area Boards by 1970. The policy is to make the fullest use of this new "nature" fuel as quickly as possible and to this end the Gas Council have entered into a number of contracts to provide bulk supplies to major chemical works and other industries.

The recoverable reserves are estimated as 25×10^{12} cubic feet. Production is expected to rise from 2000×10^{6} cubic feet per day in 1972 to about $4{,}000\times10^{6}$ cubic feet in 1975.

A trial conversion of one boiler to gas firing at Hams Hall Power Station has proved most successful; studies for the seasonal use of gas in power stations have been carried out and proposals for the conversion to dual firing have been made to the Minister of Power who, because of difficulties in the coal industry, has not approved further conversions.

Although in the Minister's paper "Fuel Policy" presented to Parliament in November 1967 it was visualised 9 million tons of coal equivalent of gas would be used in power stations by 1970 rising to 14 million by 1975, no agreement had been reached two years later and the extent of the future use of gas for electricity generation remains in doubt.

1.3.5. Water Power

The water resources of England and Wales are limited and the increasingly high standard of living and intensified industrialism of the country requires that more and more potable or high-quality water be made available for those purposes. The only areas where hydroelectric power might be developed economically are Northern and Central Wales, the Lake District and the Upper Tees, but development in parts of these areas would be the subject of very severe amenity objection. Hydroelectric stations with an aggregate capacity of 470 MW constitute just over 1·0% of the total installed generating plant capacity. Of this 360 MW is pumped storage. The use of water for pumped storage schemes to provide peaking generation offers the main hope for future development in the hydraulic field.

1.3.6. Summary

The Electricity Supply Industry exists to supply electricity at the lowest possible cost, the biggest single item of which is the cost of fuel.

The Minister of Power's policy states "the Government have decided the Generating Boards should base their choice of fuel on an economic assessment of the method of generation which will enable them to supply electricity at the lowest system cost.... In deciding to give consent to new power stations the Minister ... will also take into account such wider economic considerations as may be relevant" and "proposals for oil fired stations would be the subject of special scrutiny."

The application of this policy takes into account support for coal, balance of payments and similar problems. Power stations are therefore not necessarily sanctioned in their order of simple economic merit.

1.4. ECONOMICS OF FUEL TRANSPORT

When the distribution of the estimated loads and the sources of the fuel supply are known, the most economical means of getting the energy in the fuel to the load centres must be considered.

The three main physical means employed by the Generating Board for the transport of coal from the colliery to the power station are rail, road and water.

1.4.1. Rail

The Generating Board, in co-operation with British Rail and the National Coal Board, has introduced modern methods into the transport of coal which is drastically reducing freight rates over hauls longer than forty miles or so. Trains consist of 32-ton steel hopper-bottomed wagons equipped with disc type air braking so they can be operated as fast mineral traffic marshalled in trains carrying 1000 tons or more.

Where possible the wagon loading facilities at the major collieries are being converted to load complete trains which, when they arrive at the latest power stations, equipped with a continuous loop or run-round siding system, can discharge the coal into hoppers beneath the tracks whilst the trains are still in motion. This system of working, described in detail in Section 1.12.6.1, because it utilises wagon, engine and track time more efficiently, has made rail hauling the most economic means of transport in most localities.

The amount of rail-borne coal has increased from 21 million tons, 48% of the Board's total supplies in 1958, to 40 million tons, or 61% in 1968.

1.4.2. Road

For short and medium hauls, road transport offers a simple, flexible and cheap means of transporting coal. Unfortunately, the large road wagons necessary for economy cannot be loaded at all pits, and the already overcrowded road system is unsuitable for the large amount of additional mineral traffic which would result if increasing quantities of the fuel were moved by this means. Because of the problems of road transport and the economy of rail transport the road transported quantities will be reduced; over the last year the amount of coal carried by road was to 10·6 million tons or 16·0% of the total supplies.

1.4.3. Water

The transport of fuel by water falls into two categories, canal and sea.

(a) *Canal.* Prior to vesting day, considerable use was made of canals to convey fuels to the numerous smaller stations situated on canals and rivers. The increasing coal consumption of the large new stations, which are often not located to receive barge traffic direct, and the limited size of the barges which can use the existing system of shallow canals, in conjunction with the increase in cost, has led to a decrease in the percentage of fuel handled by this method to 2·4 million tons, or 3·6% in 1964.

(b) *Sea.* Certain coalfields—Scottish, Northern, North-Eastern and South Wales—are favourably situated near suitable ports for the transportation of coal by collier to power stations situated on the coast and the major estuaries in the southern part of the country. This mode of transport was formerly the most economic for long hauls, but unless the hauling distance is similar to that from Northumberland to the River Thames, rail is more economic. The Board charter 7,000 ton colliers, the largest available for the "home trade", which show the lowest transport cost to bring coal from the northern and north-eastern coalfields to stations on the Lower Thames Estuary. Most of the Board's fleet of smaller colliers are retained to service the older stations which the larger vessels cannot reach. Colliers of 2000 to 2500 tons deadweight known as the "up-river" class were specially constructed to negotiate the fixed bridges of the River Thames, and are used to supply the Battersea and Fulham power stations. Since the supply and transport of fuel has been planned on a national basis the average distance fuel is transported by inland methods has fallen from 37 miles to 24 miles and by all methods from 113 to 72 miles.

In the year 1958 the amount of fuel conveyed by sea was 9·7 million tons or 22·5% of the coal supplies; in 1968 the quantity was 10·0 million tons or 15·0% of the coal supplies.

1.4.4. Other Methods

The Generating Board has a small number of stations near collieries which take coal direct from the pithead by conveyor belt. Some 3·0 million tons or 4·5% of the fuel was conveyed in this manner in 1968.

1.4.5. Transport of Oil

The major oil-fired stations at present under construction are located near oil refineries and take their supplies direct by pipeline, which is the cheapest means of conveying the fuel where large quantities are required.

Whilst the stations on the estuaries and coast converted to oil-firing in the 1950's use coastal tankers up to 5000 tons deadweight, for large stations tankers of about 40,000 tons provide the lowest fuel cost. The lighter grades of fuel oil used at some small stations for peak load operation and for lighting up pulverised fuel boilers are often delivered by road. For large quantities such as are required at the new peak load gas turbine stations, rail transport is most practical and economic.

Where a station site is adjacent to a refinery which is its sole source of supply, very careful consideration has to be given to the resources of the supplying company, their sources of crude oil, and alternative supplies in an emergency, as well as the economic and technical factors, before such a scheme is adopted.

1.5. TRANSMISSION OF ELECTRICITY

The construction, from 1926, of the 132 kV overhead transmission network by the Central Electricity Board, interconnecting the main centres of generation each of which up to that time had its full complement of generating plant to meet its local loads, brought about a major change in system planning; from purely local considerations those of national and regional loads and fuel economics became paramount. The 275 kV supergrid transmission system which was put in hand after Nationalisation, has now been replaced by 400 kV on the major trunk routes. These heavy duty quad-conductor lines have a carrying capacity of 1800 MVA per circuit—thirteen times that of the standard 132 kV circuits and over three times that of 275 kV lines.

The primary purpose of the grid system was to provide interconnecting capacity between areas for security against all normal operating hazards with a smaller margin of spare plan than formerly, thereby saving capital expenditure on generating plant. It now also provides means for the bulk transmission of electrical energy.

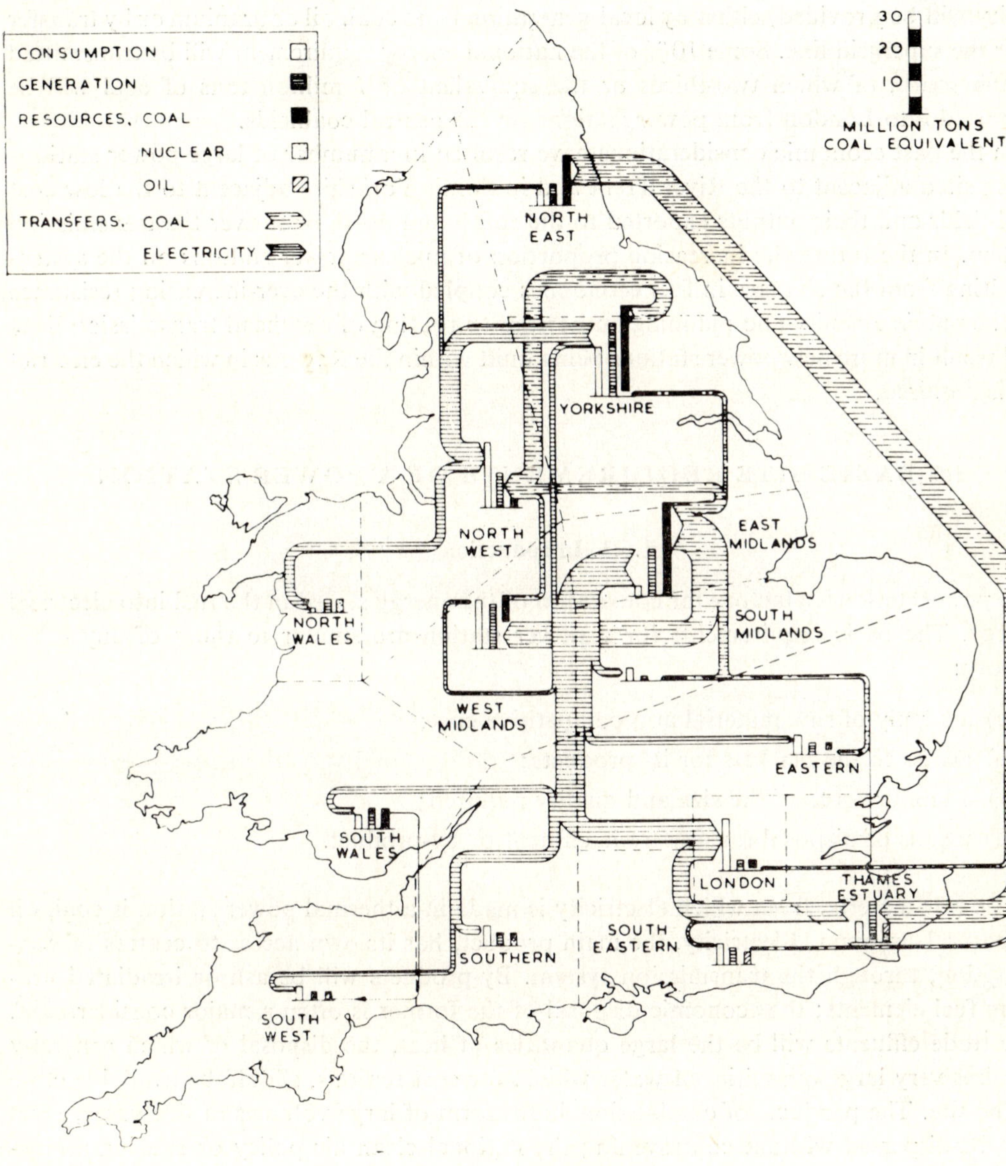

FIG. 1.5. Energy pattern. Consumption generation resources and transfers, 1969/70.

The interconnecting capacity allows large units to be used as the system can cover their loss if a set comes off load, and also accommodates the output of the 2000 to 4000 MW stations at present under construction.

Figure 1.5 shows the energy requirement in each area for 1969/70, and how the electricity will be provided, either by local generation from coal, oil or uranium or by transfer over the supergrid line. Some 10% of the national energy requirement will be transmitted to the south, of which two-thirds or the equivalent of 7 million tons of coal will be imported into London from power stations on the central coalfields.

In the past economic considerations have resulted in a number of large power stations being sited adjacent to the River Trent and in South Yorkshire adjacent to the low cost coal-fields and their outputs exported to the south and north-west over the transmission system. In the future the increasing proportion of nuclear power stations on the system resulting from the changes in fuel economics coupled with the ever-increasing resistance by the public amenity and planning authorities to routing of overhead transmission lines will result in more new power stations being built within the Regions in which the electricity is required.

1.6. BASIC SITE REQUIREMENTS OF A POWER STATION

1.6.1. Introduction

A power station is a factory for conversion of the energy stored in the fuel into electrical energy. The basic requirements for a power station are similar to those of any other factory:

(a) a supply of raw material at a competitive cost;
(b) access to the markets for its products;
(c) a labour force of the size and quality required;
(d) means of disposal for any trade effluent or by-product.

The raw material from which electricity is made in a thermal power station is coal, oil or natural uranium. Electricity, the main product, has its own access to centres of consumption, through the transmission system. By-products will be ash or irradiated uranium fuel elements; the economic disposal of the former is often a major consideration. The trade effluents will be the large quantities of heat, the disposal of which generally requires very large quantities of water which, for cost reasons, should be available close to the site. The products of combustion, in the form of large volumes of flue gases, must also be dispersed without contravening the national clean air policy or causing atmospheric pollution.

These considerations form the basis of site selection and location and apply to the requirements of the site area and also to the facilities required in the surrounding district.

1.6.2. Site Requirements

When an area of *search* has been defined, the broad site requirements must be settled before detailed investigations begin. For this purpose it is necessary to know the capacity of the plant that will be required in the particular location:

(a) for immediate development (five to seven years ahead);
(b) over the next ten to twenty years.

(a) provides the information to enable the initial development to be planned and (b) enables site area to be selected sufficiently large for the ultimate development and the services, railway sidings, jetties, water supplies, access and transmission connections to be developed in the most economical manner to allow for future requirements and for the full utilisation of the natural resources. It also ensures that even with the electricity demand doubling every ten years, the number of sites required can be kept within a reasonable limit and planning authorities can be informed of the new site requirements for some years ahead.

The information required for (a) demands that a decision be taken on the size of set and associated plant to be installed. (b) is a matter for intelligent forecasting (planning) on the basis of past load growth, anticipated future fuel availability and considerations of probable technical development.

For the purpose of this chapter the plant considered is:

(1) coal- and oil-fired stations of between 2000 MW and 4000 MW using 500 MW sets and boilers;
(2) nuclear stations of 1000 MW and over;
(3) pumped storage;
(4) peak load gas turbines.

1.6.3. Environment

A modern 2000 MW coal-fired station may exceed 900 ft in length with a boiler house 200 ft, cooling towers 375 ft and a single chimney 800 ft or more high. This bulk, introduced into the landscape, becomes a major feature for distances often exceeding 5 miles from the site in a level landscape and the discharge of flue gases and the saturated plume from cooling towers, together with fuel handling, present visual and loss of amenity problems. The necessary juxtaposition of the demanding physical requirements is most often found in rural areas, less often in zones of major industry and, on planning grounds, residential areas are usually avoided. Usually, therefore, there is a most limited choice from the point of view of environment.

Coal

The second generation of nuclear reactors (A.G.R.s) are less bulky than the equivalent capacity conventional stations and their compact visual intrusion into the landscape, their lack of tall chimneys and external plant on the ground ensures they are more acceptable from the amenity aspect.

To ensure that in the event of an accidental release of radioactivity, action can be taken to fully safeguard local population the first nuclear power stations had less than 500 people within one mile.

In 1968 the Minister of Power announced that as a result of advances in technology the safety of a gas-cooled reactor in a pre-stressed concrete pressure vessel was such that it now could be constructed and operated much nearer built-up areas than had so far been permitted.

The bases for assessing a site's suitability are:

No population living within $\frac{2}{3}$ mile unless it is practicable to take effective emergency action such as evacuation of all population within a given time and preventive medication.

Detailed investigation and surveillance extended to 2 miles with a view to the possibility of more extensive emergency action in less likely circumstances.

Massive populations beyond 2 miles may be taken into account and each site must be considered individually.

For the purposes of emergency action and control of development once a site is licensed any proposal to increase the population within $\frac{2}{3}$ mile and any groups of 50 people or more up to 2 miles are carefully considered.

As new designs for reactors are evolved with new fuels, better containment and greater safety margins they will be accepted for operation even closer to centres of population.

Gas turbines, with alternators designed to operate as synchronous compensators, are economic to build but expensive to operate. They are more suited for peak load operation on load centres, preferably where their installation can defer major cable reinforcements as this can considerably enhance their economic justification. Provided noise attenuators are fitted in their exhaust system they can be installed on old power station sites, often in the older industrial and residential areas of major cities, for peak load operation without risk to amenity.

1.6.4. **Nature of Site**

The area of land required will depend upon the following factors.

1.6.4.1. Fuel Storage

The ideal fuel storage area should provide space for about a quarter of the annual burn and in a 2000 MW station this requires approximately 50 acres. Such a store enables fuel to be brought into the station at a fairly uniform rate throughout the year, matching the output and deliveries from the collieries, so fully utilising the specialised rolling stock required. Because of the more uniform rate of consumption, fluctuation of stocks under normal conditions is least at base load stations. In the later years of operation, when a station is on peak load duties, the stocking capacity is fully utilised, building up over the year at a uniform rate the stocks which are consumed in a few winter weeks (see Section 1.12.6.5). Over the country as a whole the Board aims to have 6 weeks' stock in hand at the beginning of winter.

The storage of fuel oil in steel tanks above ground surrounded by earth banks or bunds for containment in the event of tank failure, together with the necessary facilities to heat the viscous fuel, is much more expensive per ton than coal. Because of the large quantities of fuel oil in store or being processed in the major refineries which are often connected to the station by a pipeline, and the ease with which oil can be handled and transported in bulk, storage for about three weeks' winter consumption is provided.

The storage of uranium fuel which provides 10,000 to 70,000 times the amount of heat per 1 ton than coal, does not require major space provision. At nuclear power stations, stocks of new fuel elements are held but the numbers are usually associated with the most economic size of the batch which can be manufactured (each station has its own fuel element design) rather than a particular amount of energy held in stock.

1.6.4.2. Station Buildings

The area occupied by the station buildings will be a small portion of the site which, in the main, will be decided by the needs of the fuel store, railway sidings, cooling towers, switch compound construction area and store-yard. A modern, 2000 MW station with cooling towers will occupy an area of at least 200 acres.

The occupied areas of a typical station which receives coal by rail and using cooling towers, would be:

Station buildings	15 acres
Coal stores and sidings	55 acres
Cooling towers	45 acres
Switch compound or buildings	10 acres
Surrounding area and approaches	40 acres

A considerable area will be required in addition for the storage yard, workshops and offices foı construction purposes, although in some cases where land is difficult to acquire, such land may be rented outside the station site area.

A nuclear station of 1200–2000 MW capacity will require an area of about 70–80 acres of which about 15 acres will be occupied with permanent buildings. Because the reactor pressure vessel and the heat exchangers are often supplied in sections and fabricated and tested on site, the balance of the site is required as a construction area.

For a gas turbine station for peak load operation a site area of some 5 acres is adequate for a two-set station of 110 MW capacity with oil storage for 70–80 h continuous operation.

1.6.4.3. Site Levels

A site should be reasonably level, not liable to flooding and not so high above the source of cooling wateı that excessive pumping power is required to supply water for cooling purposes.

A site requiring extensive filling to bring the level above the general flood level or excessive excavation for the purposes of siting buildings, greatly increases the site preparation costs, but in order to provide for the major capacities now required such additional cost can often be justified. For base load stations, where subsoil conditions are suitable, it can be economic to excavate deeper for the turbine house foundations to save on pumping power (see Section 1.12.4).

1.6.4.4. Foundation Loads

Modern boilers when filled with water are heavy structures, weighing possibly 16,000 tons. Magnox type nuclear reactors with their thick concrete biological shields can weigh about 60,000 tons. These structures impose loads of up to 6 tons/ft^2 on the sub-soil which must be able to support such loads with suitable foundations.

1.6.5. **Cooling Water**

The most difficult problem in siting a 2000 MW thermal power station is the disposal of 10,000 million Btu of heat per hour from the latent heat of the steam condensed (see Section 1.12.5).

For this purpose, large quantities of cooling water are required to circulate through the condenser for direct cooling: at least 53 million gallons of water per hour is required when operating on full load. If the source, a river or estuary, is inadequate to disperse this quantity of heated water, then cooling towers are used. These work on a closed circuit and require only make-up water which averages about 2% of the quantity circulated (see Section 1.12.5).

A station using cooling towers is about 0·5% less efficient than a similar station which is direct-cooled. This is because the cooling tower station needs extra pumping power and the higher temperature of the re-cooled circulating water results in a lower condenser vacuum.

1.6.6. **Access**

Access is required to bring on to site:

(a) the constructional materials and plant;

(b) fuel supplies;

(c) operational staff, etc.

Good road access is essential for construction, and rail and sea facilities are useful advantages. Direct access to a main trunk road to bring in heavy loads is desirable. Transformers, alternator stators and other items of main plant often cannot be accommodated over much of the road system, especially in rural areas, because of their weight or dimensions; these items are often shipped from the port nearest to the place of manufacture to the point nearest to the site which the special shallow-draft, end-loading vessels built for this purpose can reach.

Because traffic during the construction period is greatly increased, local roads adjacent to the site are often reconstructed to avoid undue inconvenience or risk to other road-users.

When the station is operating the overriding consideration is access for fuel supplies, and the site must be conveniently situated either close to a main railway line to accept rail-borne fuel, or in areas remote from the coalfields on an estuary or the sea coast, to take its fuel from ships. For the economic transport of coal by sea, provision should be made for the largest colliers in the British home trade. These carry 6500 tons of fuel and require 24 ft depth of water when loaded.

Where new oil-burning stations are not within economic pumping distance of refineries, tankers of 40,000 tons deadweight are the most economic. These are 700 ft long and require 40 ft of water when loaded.

Only small quantities of fuel are required at nuclear power stations and road transport is used. The irradiated fuel is moved in 50 ton flasks on special transporters to the nearest rail head.

1.6.7. Fresh Water Supplies

For domestic use, boiler make-up and cooling tower make-up, fresh water is desirable although potable water is only essential for the domestic supplies and possibly construction purposes. Even good quality sewage works effluent can be used in cooling towers.

Because of the cost of treatment of sea and estuarine water, supplies for the boilers amounting to about 1·5 million gallons per day when on full load should be provided from the potable supplies where possible. The 150,000 gallons per day required for construction purposes usually presents little difficulty, because it is not required in the same period as the major operational supplies.

The increasing demands from industry and domestic consumers for fresh water make the quantities required for electricity generation increasingly difficult to obtain, especially in south-eastern England, and in future, although expensive, more boiler water supplies will have to be evaporated from river or sea water.

1.6.8. Ash Disposal

The electricity supply industry is the major consumer of the lower grades of coal which are being produced with an increasing ash content.

During the early years of operation of a 2000 MW station, when operating on base load something approaching a million tons of ash each year may be produced, which if consolidated would cover about 28 acres of land to a depth of 30 ft in one year. When selecting a site, therefore, very careful consideration must be given to the provision of suitable economic ash disposal, either on low-lying ground which can be filled, worked-out mineral workings or by the creation of landscaped hills.

In extreme cases schemes have to be carried out such as that for taking ash in special rail wagons from the East Midlands to Peterborough, where it is sluiced into the vast excavation left by the brick-workings. Such schemes can be very expensive, increasing the effective cost of fuel by about 4 to 5% (see Section 1.12.7).

1.6.9. Transmission

A route must be available for the transmission lines from the site to the nearest point on the 400 kV supergrid system or major load point on the Area Board system which can accept the station's output. Increasing opposition from the public, amenity societies and planners to overhead lines makes line routes increasingly difficult to obtain, and sometimes the only solution is to underground sections of the line. The use of underground cabling, however, can have a major effect on the selection of sites when considered from economic merit because cables cost sixteen times more than overhead lines of equivalent capacity, i.e. over £ 900,000 per mile.

1.7. LOCATION AND INVESTIGATION OF SITES

1.7.1. General

From the study of the load estimates, probable fuel availability and cost, a first approximation can be made of the area, possibly some hundreds of square miles in extent, in which the new station should be located. Because of the size of the system and the flexibility provided by the supergrid transmission, the biggest units can be accommodated and site requirements can be clearly defined in the terms explained in the previous section.

The first step is to locate and study all possible sources of information on water resources, including navigation facilities, planning, including present and future development of the area, amenity considerations, such as recreational areas, historical features and areas of natural beauty, and the probable subsoil conditions in the area. Such information may be obtained from the River Board reports, Admiralty Charts and "Pilots", County Plans, Geological and Ordnance Surveys, local libraries and Government Aerial Photographic Library.

1.7.2. Circulating Water

The approximate quantity of cooling water required is estimated. An examination of records such as found in *The Water Year Book* will indicate the flows in the non-tidal section of rivers, while from Admiralty Charts cross-sections of the tidal channels can be plotted and with the aid of tide tables tidal flows can be calculated. From the results a number of small areas located on suitable rivers and the coast can be marked within the "area of search" on a map. The suitability of the rivers must be considered, not only for the quantity of water available but also for the possible effect on fish life of discharge of heat to the river and the possible effect on the interests of other users.

Where the rivers are tidal or the site is on the coast, investigations must be carried out to confirm calculated water flows by such means as measuring the speed and direction of floats, or by following patches of dye in the water. Such detailed studies are also necessary to determine the probable dissipation of heat by mixing surface radiation and loss by mass flow. Such studies also determine the location of the cooling water intake and discharge points to avoid recirculating heated water through the station. This work, which may take some months, must be done before the final selection is made.

On the coast, the movement of beach material under the influence of winds and tides must be studied to ensure that neither the circulating water intake, discharge, harbour works, nor site will be affected. Usually little information of this nature is available and observations of the coast line and soundings of the sea-bed may have to be taken for at least a year. Hydraulic models in which conditions such as tides, local water and silt movements can be reproduced, are often used when planning and designing the circulating water works and jetties for power stations.

For a typical example of river cooling capacity exercise see Appendix 1.

1.7.3. Initial Selection of Sites

On a map of the area of search the position of all suitable water sources can now be marked, together with the areas of land thought to be suitable from study of the development plans, Ordnance maps and similar sources of information.

For sites using rail-borne fuel supplies the land areas so marked near the main rail routes to the source of fuel are studied and noted for detailed consideration.

Where water-borne fuels are to be used, the areas near deep water or suitable for the construction of ship unloading facilities are studied. Information on deep water channel approaches, navigation hazards and port facilities can be obtained from Admiralty Charts and "Pilot" instructions.

The correlation of these features provides a number of sites which appear to have suitable land area, fuel access and cooling water. The study of large-scale Ordnance maps will provide information on levels, gradients and drainage. From the information obtained, possibly a dozen site areas will be found justifying ground inspection. After inspection a number will be considered unsuitable on amenity or engineering grounds, leaving possibly half a dozen for detailed investigation.

1.7.4. Geology

1.7.4.1. Thermal Stations

The heavy plant to be built on the site will impose heavy loads on the ground under the foundations of the buildings. The general nature of the soil can usually be obtained from records and maps of the Geological Survey. Before making a final decision on a site, a detailed survey of the subsoil conditions must be carried out to determine the ability to carry the loads and to estimate the costs of suitable foundations which can vary widely. Where cooling water works require tunnels to and from the sea or rivers, usually a major item, subsoil investigations including permeability tests and ground water tests are necessary to enable the feasibility and cost of building the tunnels to be determined. These investigations usually require a number of boreholes to be sunk, some perhaps over 120 ft deep. To determine the thickness, strength and other physical properties of the strata under the site samples are tested both *in situ* and at soil mechanics' laboratories. In industrial areas it is essential to know the previous uses of the land to locate old foundations, mine workings, etc. (see Section 1.12.3).

1.7.4.2. Pumped Storage

Sites for pumped storage schemes require exacting geological conditions. They must be located in hill country or on cliffs where sites for two reservoirs preferably with 1000 ft or more difference in height can be found. Usually the reservoirs are built where the hill contours form a basin of suitable capacity or in a valley across which a dam can be constructed. In certain cases the sea can be used as the lower reservoir or it may be necessary to form an entirely artificial basin by means of an embankment. The selection of the site of a dam and the site proving tests to determine the suitability of the rock for dam and foundation conditions require specialised expert advice, and because of the consequences of failure of such structures the design, construction and periodic inspection have to be carried out by an engineer or an approved panel of experts. The reservoir subsoil should be naturally impervious or easily sealed in order to avoid leakage of water, with consequent financial loss, deterioration of the foundation rock and, in the case of coastal schemes, possible contamination of fresh water sources with salt water and injury to surrounding vegetation. A supply of water is required for the initial filling and for makeup to compensate for evaporation losses. In coastal schemes or where a river flows through a valley forming the lower reservoir this presents no difficulty. In the second case elaborate precautions may be needed for releasing flood water from the lower reservoir to prevent the power discharge from over-topping the dam (see Section 1.11.3).

The Ffestiniog pumped storage scheme which has a capacity of 360 MW has an active upper reservoir volume of 60 million cubic feet with a daily rise and fall in water level of 65·5 ft.

1.7.5. **Airfields**

In considering the area, careful search must be made for airfields, both commercial and military. Regulations to safeguard aircraft impose stringent restrictions on the height to which the buildings and chimneys can be built, and therefore on the location of modern generating stations where boiler houses often exceed 200 ft, cooling towers 375 ft, and chimneys 800 ft in height.

1.7.6. **Acceptability**

As a result of these investigations a list of sites of varying degrees of acceptability and cost will be prepared on which to base the economic comparison required for the Annual Development Plan.

The acceptability of the station to the Board, the local authorities and other guardians of amenity is a much more difficult matter, but just as vital for assessment. From the time the Board approves investigations on site and announces its intention, close liaison is maintained with the local planning authorities and other interested parties to enable the "acceptability" to be accurately determined so that a final decision on the construction of a station can be taken.

1.8. ANNUAL DEVELOPMENT PLAN

1.8.1. Basic Information

Each year a "Development Plan" is prepared which covers the generation and main transmission requirements for the fifth, sixth and seventh years ahead. For the preparation of the Plan the following data is obtained:

(a) the quantities of fossil fuels available;
(b) estimate of electricity demands within each Region, both at peak and in units per annum;
(c) the cost in pence per therm of supplying the fuels available to each site by various means of transport;
(d) a list of power stations and sites suitable for extension, together with estimates of capital costs;
(e) the cost of the transmission works necessary for the station output to be fed into the National Grid system.

Estimates of the maximum demands and unit demand are prepared for each year and approved by the Electricity Council (see Section 1.1). As experience shows that an allowance of 17% must be made for plant not available at time of peak demand the load estimates are increased by 17% to establish the capacity of plant required to avoid load shedding at the time of winter peak, assuming weather of average severity. The capacity of the new plant required is provided by deducting the capacity of the plant planned to be available at the beginning of the year.

1.8.2. Selection of Projects

When selecting the projects to be included in a Development Plan, all aspects must be given due weight. Broadly, when the practicability of the site has been determined the two major considerations are:

(a) the economic merit of the proposed project compared with the others available;
(b) amenity considerations of the site vicinity and associated transmission line routes.

1.8.2.1. Economic Comparison

Table 5 gives typical examples of the economic comparisons between coal, oil, nuclear and pumped storage stations based on the price levels and interest rates etc. prevailing in 1965. Realistic comparisons have to take into consideration many factors besides capital cost, including the effect on the overall system cost over each project's life. Broadly, each project is considered as being installed to meet an increment of load growth; it operates, however, as an integral part of the system at a load factor varying over its life according to its relative fuel cost (*d.*/kWh).

As plant is operated in "order of merit" a low fuel cost station would operate at a high load factor in the early years of its life but would be relegated down the order as new lower cost plant is installed in subsequent years. To take account of these effects it is necessary to forecast the operating costs of the whole system over the life of the plant (say 30 years). Then by simulating the operation of the whole system, the role of the proposed plant may be established and its actual operating costs, and effect on whole system costs, may be determined.

The total effective cost of the plant may now be built up. All items of cost are expressed as effective annual costs in £/kW, which is obtained by applying an annuing rate to the initial costs, or where applicable to the present value of the varying annual costs arising over the lifetime of the plant. Firstly, to the annual charges on the capital cost of the plant are added other fixed charges largely independent of the operating régime. The running charges arising from merit order operation as found above are then added. The total cost of installing and operating the plant is then adjusted to take account of differences in operating régime. If the new plant operates at a load factor below that corresponding to the average system load factor, some extra generation by other plant is needed

TABLE 1. ECONOMIC COMPARISON OF PROJECTS

1. Station, MW s.o.	2000	2000	1200	800
2. Fuel	Coal	Oil	Nuclear	Pumped storage
3. Thermal efficiency, %	37·0	36·5	41·5	(72·0)‡
4. Specific fuel cost, *d.*/kWh	0·390	0·350	0·115	—
5. Economic life, years	30	30	20	55
6. Estimated average lifetime load factor, %	32	65	75	11
Capital costs, £/kW s.o. Including interest during construction				
7. Generating plant	43·1	41·0	91·3*	34·9
8. Transmission plant	28·2	28·2	29·8	30·1
Annual charges on generating plant, £/kW/annum				
9. Capital charges and other fixed costs	4·42	4·24	10·27	3·21
10. Fuel and other running costs	6·26	9·92	3·26	[illegible]
11. Gross cost of installing and operating new plant in merit order	10·68	14·16	13·53	5·57
12. Adjustment to operating cost to level of system energy increment	−0·11	−4·43	−4·85	4·41†
13. Net cost of meeting system power and energy increment	10·57	9·73	8·68	9·98
Annual charges on transmission plant, £/kW/annum				
14. Capital charges and other costs	2·66	2·66	2·88	2·84
15. *Total standardised annual cost to system*, £/kW/annum	13·23	12·39	11·56	12·82

* Includes initial fuel charge.
† Includes credit for higher availability at peak.
‡ Conversion efficiency.

to meet the energy increment for the system. Conversely, if it operates at a load factor above that corresponding to the average system load factor, then there is a reduction of generation by other plant. In either case an appropriate adjustment to costs is required in arriving at the total costs of generation for economic comparison (Items 1, 2 and 3, Table 1).

Transmission plant charges are the final item. The sum of all costs then provides the standardised annual costs for comparison of alternative projects, within the general pattern of future development of the system. In practice, in arriving at the final figure a vast amount of detailed analysis is necessary to determine the net results of the interaction of operating costs, plant life, capital costs, etc.; and because of the many assumptions about the future that have to be fed into the calculations it is necessary to exercise cautious judgement in interpreting the conclusion. It should be borne in mind that this section deals briefly with a very complex subject.

1.8.2.2. Amenity

Those aspects of siting which cannot be assessed in economic terms but affect the area surrounding the sites and the Board's public image must also be taken into account. These are the effect on local amenity, including both visual impact and level of atmospheric pollution, and opposition likely to the associated transmission.

This appraisal must also take into consideration the conservation of flora and fauna, which is specifically mentioned in the 1957 Act.

The possibility of strong opposition to a project on these grounds, the true worth of the site to the national interest in the light of merits claimed for the area by outside parties, must all be most carefully weighed before a site is finally adopted. Such an assessment is mainly based on long and sometimes difficult experience with the public and planning bodies' reaction to various sites and environments; for similar types of environment the reactions vary widely in different parts of the country (see Section 1.12.2).

1.8.3. Procedure and Finance

A provisional list of the power station extensions proposed for each year under review is then drawn up by the Chief Planning Engineer at Headquarters and embodied in a draft Development Plan together with the transmission proposals prepared by the Chief Transmission Engineer in consultation with the Regions.

The draft Plan is then reviewed by the Regions and the Headquarters departments concerned and any modifications considered necessary are embodied. The final Plan is considered by the Generating Board and approved and the programme for the development of the system for the fifth year ahead is released for construction.

The Board's capital financial requirements are based on the development Plan and the capital allocation is agreed with the Government's Capital Sanctions' Committee.

1.9. PLANNING AND AMENITY

1.9.1. Introduction

Section 37 of the Electricity Act, 1957, requires the Board, the Electricity Council and the Minister to take into account any effect that any proposed works would have on the natural beauty of the countryside or on any flora, fauna, features, buildings and objects of special scientific, architectural or historic interest. In addition to requiring regard to be given to preservation of natural beauty and conservation of flora and fauna, Section 26 of the North Wales Hydro Electric Power Act, 1955, provides that the Generating Board must appoint a landscape architect and consult with the Countryside Commission and the Local Planning Authority on the situation, design and external appearance of any dams, buildings, etc., constructed under the Act.

For all new projects apart from consultation with the Planning Authority, the views of the Countryside Commission are invited together with those of other parties with special interests, and close consultation continues through all stages of development. The following gives an idea of the procedure, the nature of the problems arising and the bodies most concerned with the amenity aspects of the Board's work.

1.9.2. Planning Considerations

In formulating proposals for power station sites and transmission line routes, planning interests are considered and regional standing conferences, Economic Planning Bodies and the Planning Authorities at County and local levels in areas of search are consulted informally during the preparation of the schemes. In particular, the probable impact on the local environment, the existing quality and type of development in the locality are studied together with the available information on recreational and amenity interests and on the Planning Authorities' proposals for future development. Several sources of information are used, the most comprehensive and valuable of these being Local Authority development Plans and the reports of Regional Economic Planning Bodies.

1.9.3. Development Plans

The 1968 Town and Country Planning Act provides for the replacement of County Development Plans drawn up under previous Acts by Structure and Local Plans.

The Structure Plans will cover large areas and define broad policy taking account of regional economics, land use, population movements, communications and major services. The first such plan is the Greater London Development Plan at present in draft. A working party of Greater London Council/C.E.G.B. officers prepared a report on the future power station and substation needs and as a result the plan includes three areas for consideration for future power stations.

Local Plans must generally conform with the principles in the appropriate Structure Plan and will show in detail how the local area is to be developed.

The Local Plans for London each cover a London Borough who is responsible for its preparation.

1.9.4. **Amenity**

The main amenity considerations dealt with under the Planning Act may be called the visual effects of development. Other amenity features are given statutory consideration by the Clean Air Act and the Rivers (Prevention of Pollution) Act, and others; the matters covered by these Acts, although not subjects for planning control insofar as they affect the Board's developments, are often the subject of grave local concern, and are given critical examination at the detailed investigation stage.

On amenity and architecture, the Board has available the advice of a member who is an eminent architect. Landscape architects are employed to advise and to prepare schemes to help the integration of the power stations and transmission lines into their surroundings.

There are few areas zoned in development plans for industrial use or even for urban development which provide the technical requirements for power station siting and there is a risk in such surroundings that the large scale of the buildings may alter the balance and character of the locality (see Section 1.12.12). However, industrial plant can provide exciting components in otherwise dull landscape, for example the steelworks at Margam or the massive blocks of cold storage depots in the Fenland, and Dungeness power station over the surrounding shingle. Care must be taken to make the interruptions with the existing scenery clean, positive and compact, with the countryside reaching right up to the plant, and to avoid the scattering of auxiliary buildings into the surrounding countryside.

1.9.5. **Nature Conservancy**

Following the Dower and the Hobhouse Committee reports on National Parks and the Special Committee report on Wild Life Conservation, in March 1949 the Nature Conservancy was constituted by Royal Charter to "provide scientific advice on the conservation and control of the flora and fauna of Great Britain, to establish, maintain and manage nature reserves in Great Britain including the maintenance of physical features of scientific interest and to organise and develop the scientific services related thereto". For the proper discharge of the Conservancy's responsibilities, other provisions were incorporated in the National Parks and Access to the Countryside Act, 1949.

Apart from the provision of Nature Reserves a duty is placed on the Conservancy to notify the Local Planning Authority concerned of any site which is considered to be of special scientific interest (S.S.S.I.) by reason of its flora, fauna, geological or physiographical features. This purely consultative procedure ensures that scientific interests are brought into planning considerations at the appropriate time and does not empower the Conservancy to prevent development.

In all cases the views of the Conservancy are sought on new power station proposals, and where the Conservancy interests are affected close liaison takes place.

1.9.6. **Countryside Commission**

Under the Countryside Act 1968 the functions of the National Parks Commission were transferred to the Countryside Commission who were also given additional powers to promote open air recreation.

The Commission has designated extensive areas as National Parks and is continuing to designate areas of outstanding natural beauty. It will also assist Local Authorities in developing smaller Country Parks especially for recreation. By 1968 9% of the land in England and Wales was covered by National Parks with 7% confirmed as A.O.N.B's.

The Commission must make representations to any responsible Minister and for Local Planning Authority on any proposed development likely to be prejudical to the natural beauty of the area or to public access for recreation.

The Board has close liaison with the Commission who are consulted on all new proposals.

1.9.7. National Trust for Places of Historic Interest or Natural Beauty

The Trust was founded in 1895 with the object of preserving as much as possible of the history and beauty of the country for the people. It has since become an organisation incorporated by an Act of Parliament to ensure the preservation of lands and buildings of historic interest or natural beauty for public access and benefit. It has also under its care gardens of national importance, bird sanctuaries, nature reserves and several hundred farms. In all, the National Trust now administers more than 250,000 acres of land in England, Wales and Northern Ireland. Lately the Trust have been specially concerned in "operation Neptune" and acquiring attractive stretches of coastline.

1.9.8. Royal Fine Art Commission

The Commission was appointed by a Royal Warrant in 1924 to "enquire into such questions of public amenity or artistic importance as may be referred to them by any of our Departments of State, and to report and give such advice on similar questions when so requested by public or quasi-public bodies where it appears to the Commission that their assistance would be advantageous". Subsequently this was extended to "call the attention of our Departments of State, or of the appropriate public or quasi-public bodies to any project or development which, in the opinion of the Commission, may appear to affect amenities of a national or public character" and "to call before them such persons as they shall judge likely to afford them any information, and to visit and inspect such places as they may deem it expedient".

The Generating Board invite the observations of the Commission on architectural designs for power stations.

1.9.9. Council for the Preservation of Rural England

The Council, which is a private society with no statutory responsibility, was set up in 1926 and its sister body, the Council for the Preservation of Rural Wales, was founded in 1928.

The objects of the Council are stated to be:

(a) the protection of rural scenery and the amenities of country towns and villages;

(b) to act as a centre for advice and information upon any matters affecting the protection of such amenities;
(c) to arouse, form and educate public opinion.

The Council has a wide range of interests with some 45 constituent bodies and nearly 200 affiliated bodies, including the Royal Institute of British Architects, Town Planning Institute, County Councils Association, National Trust, the County Land Owners Association, Royal Society for the Protection of Birds and the Pure Rivers Society.

The Council has promoted or supported legislation to increase the protection given to rural scenery and has been responsible for forming the Parliamentary Amenities Group, consisting of Members of Parliament in all parties of both Houses. It has provided witnesses and evidence at innumerable public inquiries, including those into power station siting proposals, and has given evidence before many Royal Commissions and Committees, and is often consulted by local planning authorities when considering applications for permission for development.

1.9.10. Fishery and River Interests

The intake of large volumes of water, its chlorination to prevent growth of marine organisms in the cooling water works, and heating before discharge may affect life in rivers and to a much lesser extent in the sea.

In law, river water is polluted by heating, quite apart from the increase in solids given to the river when cooling towers are purged.

The statutory bodies are as follows.

(a) *Fresh water and tidal estuaries.* Under the Water Resources Act 1963, River Authorities were set up to replace River Boards and have taken over all fishery protection functions. They have powers to make by-laws and under the River Pollution Acts, 1951, etc., are the authorities responsible for the control of new discharges. Before a discharge can be made into the river or estuary a licence, to which conditions covering the allowable temperature rise, increase in solids and biological oxygen demand of the water after use may be attached, must be obtained. The River Authorities also have powers under the Salmon and Freshwater Fisheries Act, 1923, extending to the sea adjacent to the coast, under which they can take action only when salmon or other fish have been killed by an effluent. Under the Water Resources Act, 1963, a licence to abstract water must be obtained and payment made for the use of the water.

(b) *At sea.* In certain cases Sea Fishery District Committees may be able to take action to effect the cessation of the discharge of effluents harmful to fishing interests. The burden of proof rests on the Committee. They have objected to the discharge of pulverised fuel ash in certain sea areas.

1.9.11. Land Utilisation

In Great Britain, and more especially in England and Wales, land is a very scarce natural resource; with about 45,000,000 people on 37,000,000 acres, the total land per head is just over 0·8 acre. The world, as a whole, is almost twenty times better off. The land actually used for farming in England and Wales is 0·55 acre per head of population and the potential usable land is about 0·6 acre per head. For the world, the figures are 4 to 5 acres and 1·125 acres respectively.

The other requirements for land are housing, recreation, transport and communications, defence purposes and agriculture. There are various estimates of the annual loss of agricultural land. It is probably 40,000 acres a year for urban and industrial development, of which total less than 1000 acres are taken for power station purposes.

The current rate of production of pulverised fuel ash is approaching 10 million tons a year and about 35% is sold for block making and as bulk filling. The Board has large-scale schemes to recover land for useful purposes such as that at Peterborough where some 30 million tons of ash will be used to fill 1,000 acres of derelict pit working and Gale Common, Yorkshire, where 750 acres of land affected by mining subsidence will be filled, and a landscaped hill some 120 ft high will be formed.

1.9.12. The Committee on Air Pollution (Beaver)

In their final report in 1954 the Beaver Committee, which was set up to examine the cause and effect of air pollution, stated that the serious immediate problem was the visible pollution by smoke, grit and dust and the biggest culprit was declared to be domestic fires discharging at relatively low level. The Committee made recommendations for new legislation to prohibit or regulate atmospheric pollution.

1.9.13. Clean Air Act, 1956

This Act, giving legislative effect to the recommendations of the Beaver Committee, excluded power stations, which are now scheduled under the Alkali Works Regulations Act. A certificate of registration is needed from the Alkali Inspectorate of the Ministry of Housing and Local Government before chimney emissions are allowed. The height and number of chimneys to be provided at each new station are agreed with the Chief Alkali Inspector.

1.9.14. Emission Problems

Although from the earliest days of the industry there has been the problem of discharging the products of combustion from power station chimneys, the advent of large pulverised fuel-fired boilers aggravated the problem of dust and grit discharge and oil-firing, whilst reducing dust emission, increases the risk of pollution from sulphur dioxide and the possibility of the emission of acid smuts.

The Ministries of Power and Housing and Local Government, for normal industrial chimneys, still apply the recommendation of the Committee set up in 1937 by the Electricity Commissioners, that a chimney should be $2\frac{1}{2}$ times the height of the tallest local building. The Board considered this rule inadequate for installations the size of modern power stations, and has carried out considerable laboratory and field research to determine the best method of taking chimney discharges clear of the turbulent air zone around buildings, so as to keep the pollution at ground level to a minimum.

Research and experience have produced the single chimney with multi-flues 650 ft to 800 ft in height with a discharge velocity of not less than 60 ft/sec. This arrangement provides the mass flow and thermal buoyancy which under normal atmospheric conditions at least doubles the effective chimney height at which the gases disperse. Electrostatic precipitators remove 99·3% of the fly ash and field tests show that the effect on the atmosphere at ground level is so low that in most cases it can be ignored.

When investigating a new site, consideration is given to the effect of the chimney discharge on the emission level in the light of the local topography, development, and the existing atmospheric pollution. Calculations are made to determine the ground level concentrations of dust and gas and by comparing these figures with those calculated for existing stations where atmospheric pollution is measured, it is possible after allowing for increase in size of plant to assess the levels of pollution likely to result.

Although Government Medical Assessors have stated that they are satisfied that 2000 MW stations would not add appreciably to the hazards of health, the ideal solution is the removal of sulphur dioxide from the flue gas.

Neither of the Board's existing flue gas washing plants at the coal-fired Battersea and the oil-fired Bankside power stations, both of which use the river water for the purpose, can be regarded as entirely satisfactory as the maintenance and operating costs are extremely heavy and local concentrations of washed gas which has little buoyancy, are not unknown. Research into dry methods of removing sulphur dioxide gas from flue gases has not yet provided an economic system.

Details of the recorded results of all atmospheric pollution gauges in the country, including those of the Generating Board, are summarised in the D.S.I.R. Pollution Bulletins.

1.10. GOVERNMENT CONSULTATIONS AND THE CONSENT PROCEDURE

1.10.1. **Introduction**

Any plan to increase national productivity and living standards depends for its fulfilment on sufficient electricity being available, for which a steady supply of new sites is necessary. Consultations with the Ministries concerned therefore take place as soon as areas for detailed investigation are known. The procedure ensures that no unnecessary effort or money is spent on an unacceptable area, and that by the time the Ministry receives the Board's formal application for consent any fundamental objections from other government departments have been considered.

1.10.2. Aircraft Obstruction

Tall structures present a hazard to aircraft in flight and the need for safeguards has to be agreed with the Ministry of Defence for each site. For sites in the vicinity of airfields any possible height limitation must be determined at the time of the ground inspection and when the Ministry are informed of the approximate height and position of the structures, they inform the Board of Trade who may consult private aerodrome owners.

Under the Civil Aviation Act 1949, the Board of Trade can require obstructions to be lit for civil aircraft and the Ministry of Defence for service aircraft.

Generally the position may be summarised as:

> Any structure 300 ft and above is an obstruction and may require lighting. Each case is considered on its merits. Any structure 500 ft or over requires lighting.

In the immediate vicinity of large international airports such as London (Heath Row), buildings may be limited to 150 ft in height and some hundreds of square miles in extent can have a height limitation of 500 ft. Such regulations severely limit the siting of power stations in the vicinity of major airports.

1.10.3. Inter-departmental Committee

The principal consultation with government departments takes place when the Board agrees to adopt the site, but in advance of the time of Section 2 Application. Details of each proposal are submitted initially to the Ministry of Power who are responsible for notifying them to all other government departments. Departments with interests are represented at a meeting of the interdepartmental committee or submit their views in writing. The Ministries mainly concerned are Power; Housing and Local Government; Agriculture, Fisheries and Food; Ministry of Defence; Department of Employment & Productivity; Ministry of Transport; and Board of Trade. By this procedure the Generating Board is informed of any fundamental objection to the proposed development at an early stage.

The Ministries of Housing and Local Government, and Agriculture, Fisheries and Food provide general guidance on areas of search at the time of the initial investigations.

1.10.4. The Consent Procedure

The Generating Board is responsible to the Minister of Power and his consent has to be obtained before a power station can be established or extended, the procedure to be followed being that laid down under Section 2 of the Electric Lighting Act, 1909, modified by the Electricity Act, 1957. Under Section 35 of the Town and Country Planning Act, 1947, the Minister's consent also embodies deemed planning permission for the proposed development.

The procedure for obtaining Section 2 consent and the deemed planning permission is, therefore, a combined one and both aspects have to be considered together.

The application to the Minister and the planning authority is accompanied by a plan showing the extent of the site in colour and technical particulars giving approximate information on station capacity, cooling water requirements and a general description of the plant for the power station. The design of the station has to be settled before the details can be provided. Once supplied, they are legally binding on the Generating Board and the Minister's consent will usually stipulate that detailed plans with complete technical particulars are to be supplied at a later date if required.

The Minister usually gives directions to the Generating Board to give publicity to their proposals. Arrangements also have to be made with the local authority for the deposition of the plans and particulars in a place where they are open for public inspection. Notice is given in the national and local press of the Generating Board's intentions, and the public given a period of thirty days for lodging objections. At the end of that time the Board is in a position to appraise the nature and weight of the opposition. Arrangements are made to see or write to any objectors if there appears a likelihood of reaching agreement and obtaining the withdrawal of their objections. At the same time as the particulars are supplied to the Minister, an application is made for deemed planning permission to the Local Planning Authority and copies of the drawings and particulars are supplied.

Upon receipt of the application for deemed planning permission the Local Planning Authority consults certain other bodies (River Authorities, Ministries, Yachting Associations, Council for the Preservation of Rural England, etc.) who they consider are interested parties. If, after consideration and consultation, they wish to suggest modifications to the proposal or conditions, these are discussed with the Generating Board and, if agreed, are entered on the planning permission. If the Local Planning Authority does not approve the project, or wishes to impose conditions unacceptable to the Generating Board, they send the deemed planning application to the Ministry of Housing and Local Government with a copy to the Generating Board.

After the Minister of Power has had further consultations with other interested Ministries which have been consulted in the Inter-Departmental Committee and has considered the observations of the Local Authorities, public bodies and private individuals, he decides whether the proposal affects the interests of other parties sufficiently to justify a public inquiry, and if so orders one. If the Local Planning Authority objects, the Minister must hold a public inquiry.

1.10.5. **Public Inquiries**

The Ministry appoints an inspector to conduct the Inquiry. The reasons for objections from individuals or bodies claiming to have an interest are heard and the range of subjects that may be discussed is not prescribed in any way.

The prodecure is that the Generating Board has to state its case to justify the proposed development and put forward witnesses as to its nature and probable consequences, each witness being subjected to cross-examination by opponents of the scheme or their legal representatives. The objectors then successively state their case and if they put forward expert witnesses they, in turn, are subjected to cross-examination. The objectors' legal representatives summarise their grounds for objection, the Generating Board's representatives reply and the Inquiry closes.

Preparation for an Inquiry entails the assembly and presentation of a great volume of technical information, facts justifying the installation, technical reasons for the selection of the site and design details of the station when completed, together with an estimate of its effect on rivers, local industries, etc.

If the Generating Board has been unable to purchase the requisite land by negotiation then at the same time as the application to the Minister of Power for his consent under Section 2, an application is made for confirmation of a Compulsory Purchase Order under the Acquisition of Lands Act, 1946. Should the owner object to the enforcement of the Order a Public Inquiry is held at which the Generating Board has to justify the use of the area required, in detail. Whenever possible this inquiry is held with that for the Section 2 Consent.

On the completion of the Inquiry, the Inspectors and Assessors prepare their reports and make their recommendations to the Minister. The Minister then decides if the application should be granted having regard to the interests of all concerned and, if he is satisfied, issues his consent. The Minister may attach conditions in the consent; these may be those proposed by the Local Authorities, other departments or objectors, if considered reasonable, or the Minister may impose conditions he considers suit the particular circumstances.

The Ministry has not considered it necessary to hold a Public Inquiry into a Section 2 application since 1963. In the previous ten years sixteen inquiries were held and consent was twice refused. Where held the Public Inquiry procedure added more than six months on average to the period between application and the Minister's decision.

1.10.5.1. Other Statutory Consents

The Minister of Power's consent does not absolve the Generating Board from obtaining other statutory consent which may be required to build or operate plant. As well as those mentioned specifically in the text, some of the more important are as follows.

1.10.5.2. Water Resources Act, 1963

The River Authorities established under this Act control abstractions from surface and underground water and have powers to develop the water resources in their areas. A licence to abstract water gives a statutory right to do so and the water may be used away from the source. This gives some scope for siting power stations away from the actual riverside.

1.10.5.3. Nuclear Installations Act, 1959

Section 1 requires the Board to obtain a licence from the Minister of Power to build and operate nuclear installations including power reactors. The licence gives the Minister's Nuclear Inspectorate wide control over the design, construction and operation of such installations.

1.10.5.4. Harbours Act, 1964

This Act requires the Board of Trade's consent to all harbour and associated works of £500,000 or more in value.

1.10.5.5. Coast Protection Act, 1949

Section 34 requires the Minister of Transport's consent to the construction of works below high water mark.

1.10.5.6. Summary

A considerable number of other Acts such as those appertaining to the Port of London and similar Authorities, affect the Board's works. Most of these provide for public inquiries and appeal procedures against refusals or the terms imposed in the licence or consent.

The Electricity Supply Industry is the only one subject to such prolonged consent and planning procedures. Besides requiring considerable money, time and manpower to provide the technical and legal information needed for negotiations and enquiries, such procedures extend the period of uncertainty between application and authorisation of construction to two years or more.

The additional operating cost of one day's delay in commissioning a 500 MW unit in the mid 1970s will be of the order of £25,000 in addition to the capital charges on the unremunerative capital of possibly £8,000.

1.11. INTRODUCTION TO POWER STATION OUTLINE DESIGN

The first part of this chapter has given some account of the considerations which must be applied in the search for suitable sites for modern power stations and the factors which have changed these requirements over the last few years. The final part of this chapter will show broadly how the site is developed once it has been approved by the Board and the Minister of Power has given his consent to the construction of the power station.

1.11.1. Definition of Outline Design

A more accurate phrase would be "initial basic station design", using the word "station" in the widest sense to include all the equipment and buildings which are normally found within the confines of a power station site. Basic power station design for the purpose of this lesson has been taken to mean the disposition and grouping of the various individual buildings and external plant items on the site, thereby taking fullest possible advantage of the site's resources, and the disposition of the major plant or groups of plant in the main station building leading up to a determination of the shape of building and its leading dimensions.

1.11.2. Design Objectives

The objectives of any aspect of power station design are to achieve the lowest capital cost and ease of construction, together with simplicity and efficiency in the operation and maintenance of the station. In attempting to reach these objectives, a number of features have to be considered. These features are listed below although, all being equally important, the order is not intended to denote priority:

Efficient operation

(a) Reliability of operation.
(b) Simplicity of operation.
(c) Safety in operation.
(d) Good working conditions.
(e) Ease of maintenance.

Minimum expenditure

(f) Low capital cost.
(g) Minimum operating cost.

With both of these it is necessary to take into account:

(h) Simplicity in design.
(j) A good integrated design.
(k) A pleasing appearance.

The objective of the initial design study is to evolve a completely integrated design conceiving the station as a single entity and not a collection of individual pieces of plant thrown together in haphazard fashion.

1.11.3. Types of Stations (see Fig. 1.11.3A)

Stations at present being built for the C.E.G.B. can be broadly classified under the following headings:

(a) Coal-fired.
(b) Oil-fired.
(c) Dual-fired (combination of coal and oil).
(d) Nuclear.
(e) Pumped storage.
(f) Gas turbines.

Types (a), (b) and (c) are very similar in design, the main difference being the type of fuel used to generate steam in the boiler. Coal-firing requires a more extensive plant for discharging, conveying and preparing the fuel for the boilers than oil-firing which is, consequently, cheaper.

A nuclear station is similar to the first three types in all respects other than the source of heat for steam generation, which is provided by a nuclear reactor.

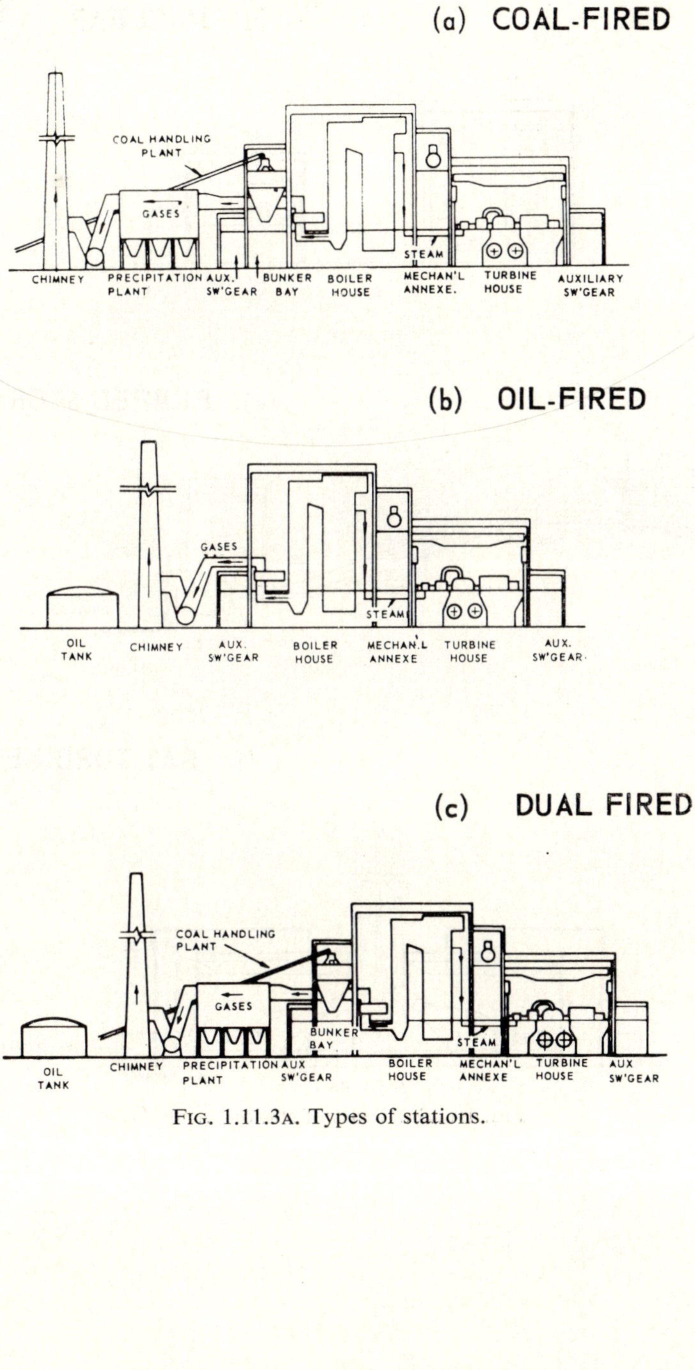

FIG. 1.11.3A. Types of stations.

(d) NUCLEAR

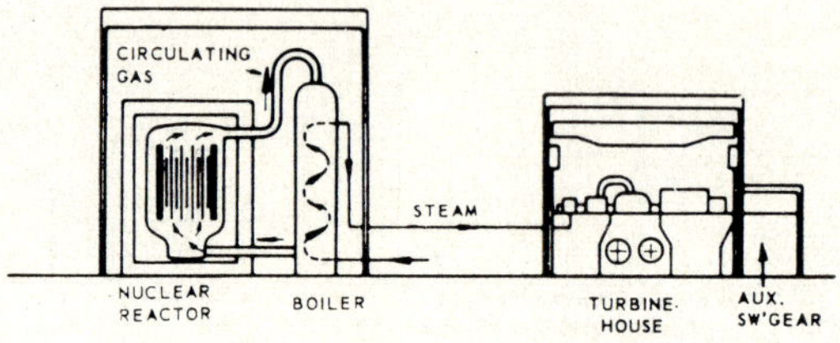

(e) PUMPED STORAGE

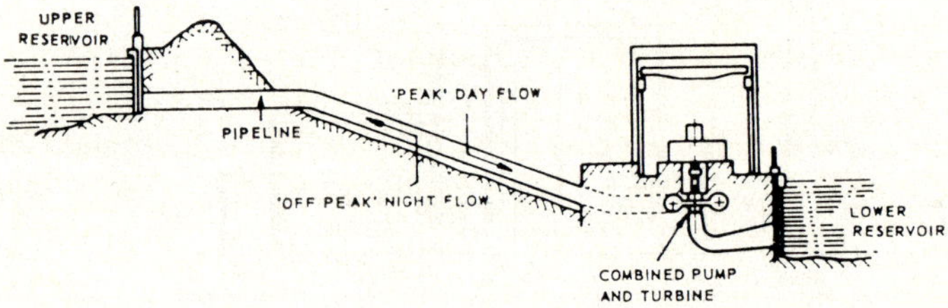

(f) GAS TURBINE

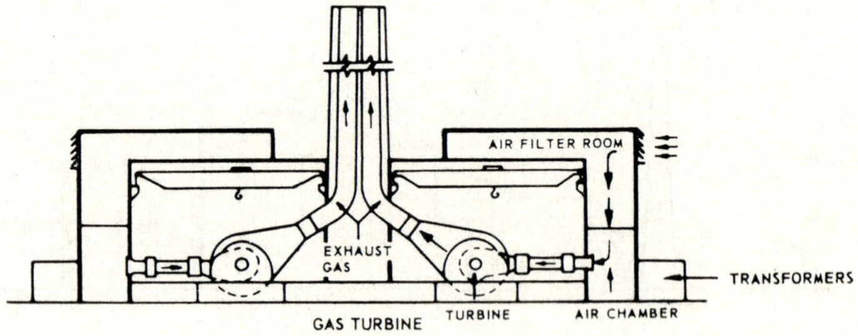

FIG. 1.11.3A. Types of stations (*cont.*)

The most difficult feature to provide in a nuclear power station site is an adequate supply of cooling water. Because of the limitation imposed by the uranium metal and canning material, the operating temperature is low and to utilise the heat in the most efficient manner a dual pressure steam cycle, 692 lb/in^2 389°C, 283 lb/in^2 389°C is used in the latest Magnox station, which provides a thermal efficiency of the order of 30–33%.

For a 650 MW station using this steam cycle some 27 million gallons per hour of water is required for cooling, a quantity only available on the major estuaries or sea coast.

In future designs the more efficient Advanced Gas-cooled Reactor, known as the A.G.R., will be built; the steam cycle adopted for this design is the same as for a contemporary conventional station using 500 MW units, the thermal efficiency being approximately 41·5%.

Electricity cannot be stored on a large scale, but the same result can be achieved indirectly by pumped storage. Britain has a high peak load with an average load factor of about 50% and the water resources are very limited. In such conditions pumped storage schemes offer economic advantages where sites exist with suitable rainfall and geological strata and the topography allows the construction of two reservoirs at different levels (see Fig. 1.11.3A(e)).

At off-peak periods cheap power can be provided from highly efficient base-load plant to pump water from the lower reservoir into the storage reservoir. At times of peak demand the water flow is reversed and its hydraulic head is used to generate power. By this means, improved use is made of the best thermal stations in the system and power is provided at peak periods at a lower cost than that otherwise provided by less efficient stations. The fuel so saved will often more than offset the additional fuel consumed in providing pumping energy. The re-use of water allows a much greater generating capacity without material increase in the civil engineering costs which make up the bulk of the capital cost.

In addition to providing a peak supply the plant can also be used for standby duty, since it can be ready for operation at very short notice to cover unexpected load increases, or breakdowns elsewhere on the system.

Considerable flexibility is possible in planning storage schemes. The variables of pumped capacity, pumping periods and reservoir capacity must be considered together to give the most economical scheme when integrated with the system loading and the other plant in the system.

Because of its essential simplicity of design, pumped storage can be built at a lower cost than conventional thermal stations, and needs only a small staff to operate it.

Gas turbines are also available for peak load generation. In this type of plant modified aircraft jet engines are used to generate high-temperature gases at low pressure which are then expanded through a two-stage power turbine directly connected to an electrical generator.

Dependent upon the electrical power output required and the type of aero engines used, machines have been built to give outputs up to 70 MW by incorporating up to four jet engines in the design. Other designs are being developed that would allow a maximum output of 110 MW. A particularly useful characteristic of this type of turbo generator is its capacity to take up full load within 1–2 min of starting from cold. Because of the simple cycle involved with this type of machine, auxiliary equipment is minimal. The demand for cooling water is limited to the amount required for hydrogen and lubricating oil cooling and if necessary this heat can be dissipated by air-cooled coolers.

Because of the high noise level produced by the jet engines, measures must be taken to attenuate the noise to an acceptable level in the vicinity of the station. This is done by housing the jet engines in special muffs and the main machinery in a building of heavy brick construction without windows and with double doors. Special attenuation devices are provided in the air intake and in the exhaust duct work.

This type of generation plant is comparatively cheap to construct, but is fairly expensive to run. In consequence the duties for which it is used are confined to peak lopping and emergencies. The annual load factor of this plant when used for peak lopping is about 3% and storage of 70–80 h supply of fuel oil usually in three or four tanks is provided. Provision is made to operate the generator for system synchronous compensation whenever required (see Fig. 1.11.3B).

Gas turbine sets of up to 25 MW and of a design similar to that installed in main gas turbine stations have been introduced at the modern stations to provide the auxiliary power for each 500 MW unit. In emergency or times of peak load these generators supply the auxiliary power requirements of the main generating plant thereby making the auxiliaries independent of system frequency fluctuations and ensuring that the auxiliary plant performance remains unaffected. They also enable plant to be restarted when a power station is cut off from the National Grid.

These machines can either be accommodated within the main building or in a separate building situated conveniently on site.

1.11.3.1. Thermal Stations

Of the fourteen stations which have been designed by the Board since 1960 only two are planned to burn oil with one dual-fired station, burning both coal or oil to take advantage of the short-term economic fluctuation in market prices of fuel. The remaining eleven are pulverised coal-fired stations.

The predominance of rail delivery of coal and the use of cooling towers for these stations is shown by Tables 2 and 3 respectively.

The majority of stations designed since 1960 contain 500 MW sets with unit pulverised fuel-fired boilers, are tower cooled and have their coal supplied mainly by rail. The greater part of this lesson, therefore, deals with stations of this type, but the field is so wide that it is only possible to touch on the salient features of "station outline design".

Table 2.

Fuel delivery	Number of stations
Rail-borne	10
Water-borne	1
Combination of rail- and water-borne	1
Oil pipeline	2

Table 3.

Type of cooling	Number of stations
Closed cooling tower system	10
Direct-cooled system	4

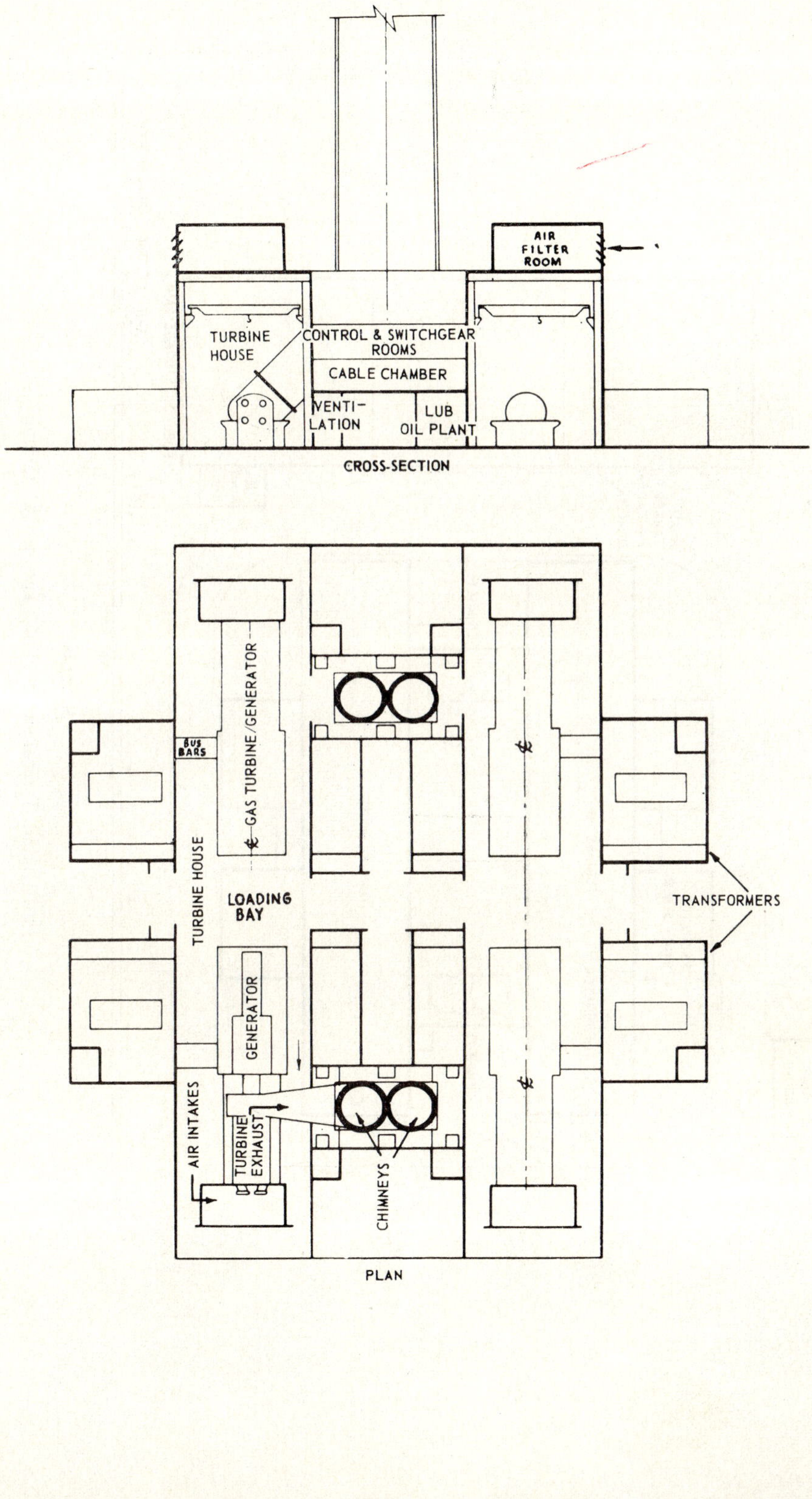

FIG. 1.11.3B. Typical plan and cross-section of a gas turbine station.

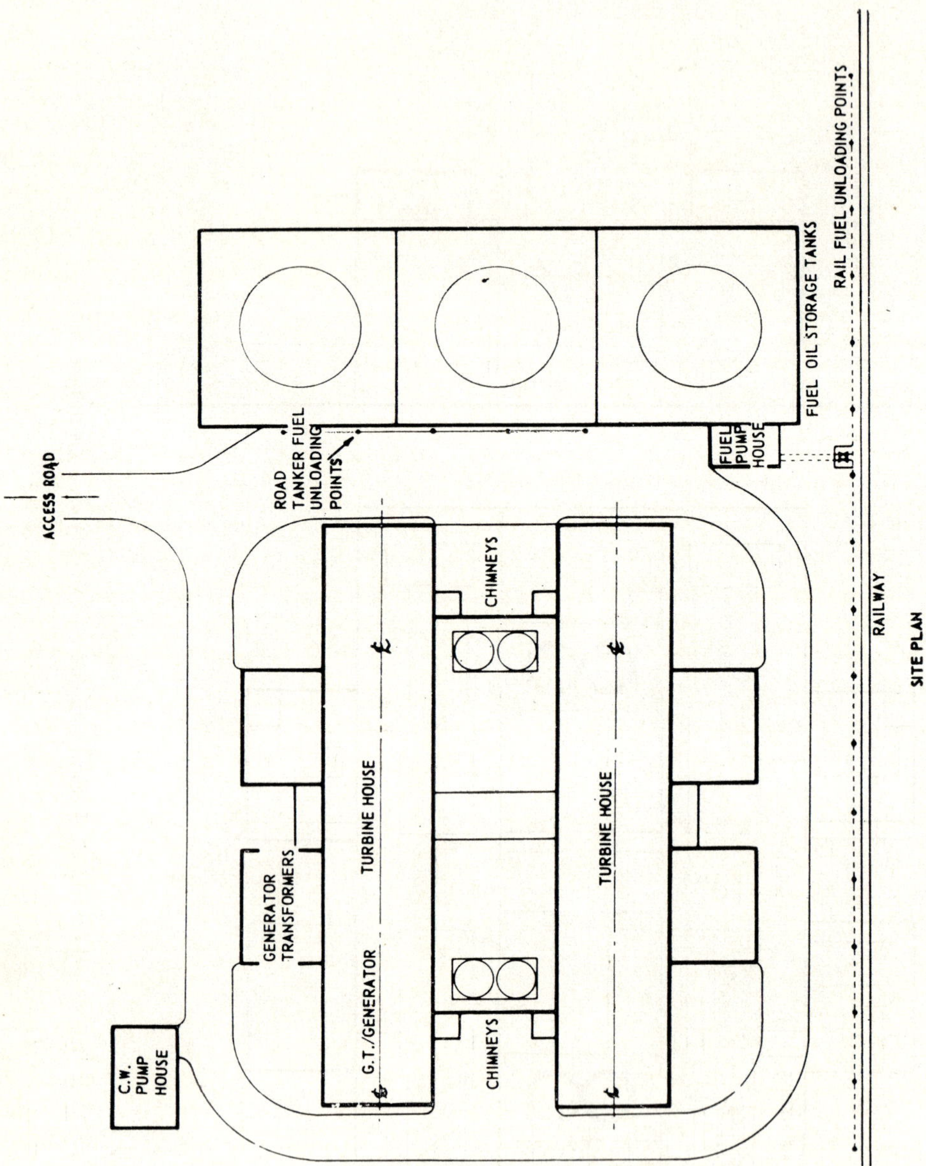

FIG, 1.11.3C. Typical site plan of a gas turbine station.

1.12. SITE LAYOUT

1.12.1. General

It is not possible to say that there is a recognisable site layout pattern which is adopted at most stations, but like the design of the station itself the problem of site layout is to find an optimum arrangement which results in minimum overall cost consistent with ease of erection, ease of operation and good appearance within the limits of the available site (see Fig. 1.12.1). There are always conflicting requirements to be reconciled and alteration of one feature often has repercussions on others so that every major aspect of the design must be considered in relation to the others. The main factors influencing site layout are discussed briefly in the following paragraphs.

1.12.2. Foundations

When selecting a site the general foundation conditions are determined by sinking exploratory boreholes. These must be extended to cover the whole site in detail before site layout decisions are made. Such tests determine the most economic location of the main buildings and indicate any geological features which might limit the area in which the station can be built. All other things being equal (which they rarely are) a situation would be chosen for the main station buildings where foundation costs would be a minimum.

A piled foundation is usually adopted where the subsoil is inadequate to carry the required loading, and it is necessary to penetrate to some depth to reach a load-bearing strata.

A raft foundation is adopted where ground conditions are suitable to carry the required loading. The raft which can be designed of solid concrete, or of cellular construction, is used to distribute the point loadings equally over the ground.

Assuming that there is no marked preference on the score of cost, the cellular raft has the advantage in that some accommodation is available for running pipes and cables below basement level.

1.12.3. Mining Support

Where underground coal mining is expected to be carried out in the vicinity of a station during its lifetime, it is necessary to prevent subsidence of the main components on site by refraining from mining within a closely defined volume of ground below the station. The shape of this volume is known as a pillar of support and varies according to the surface area to be supported and the geological characteristics of the ground. This shape is formed by extending, from the extremities of the foundations on the pillar, outward sloping lines below ground level at approximately 35° to the vertical. The Generating Board have an agreement with the National Coal Board to restrict this supported area and thereby reduce the quantity of any coal so sterilised. The station designer should aim to keep this area to a practicable minimum; since under the agreement some compensa-

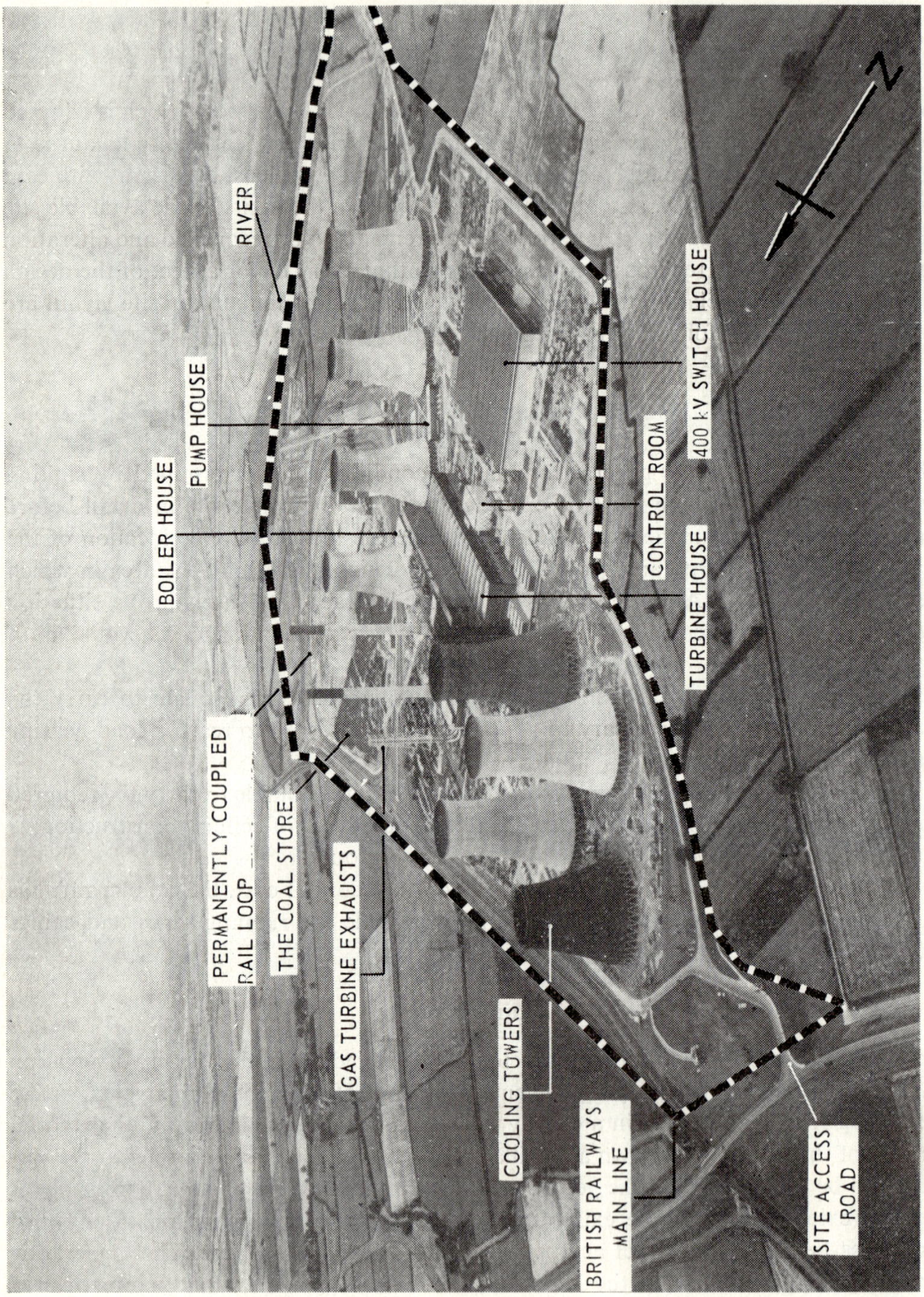

FIG. 1.12.1. Aerial view of the 500 acre site for West Burton power station nearing completion.

tion to the Coal Board is payable, the area is usually limited to the main buildings and cooling towers (if any) and sometimes the switch compound but not the coal store and sidings.

1.12.4. Site and Station Levels

1.12.4.1. General

The main factors governing the choice of station and site levels are:

(a) the need to protect the station against the risk of flooding;
(b) capital cost of civil works;
(c) circulating water pump running cost;
(d) ease and speed of construction.

For economy and general convenience on the site, it is in most cases desirable to construct the station basement, roads and rail sidings at the existing ground level (see Fig. 1.12.4.1(d)). This avoids the necessity of extensive excavation and removal of spoil, or the importation of filling material. If a site is above the predicted maximum sea or river level the necessity for protecting against flooding, factor (a), does not arise. If the site is below flood level there are two main methods of affording protection. The surest way is to lift the station basement, and all other installations essential to the safe operation of the station, above flood level. This method is adopted where the cost of required filling material is cheap and cost of pumping cooling water is not materially increased (see Fig. 1.12.4.1(b)).

Where the filling material and pumping costs are expensive, flood banks are relied upon, along those sides of the site below flood level (see Fig. 1.12.4.1 (c) and (d)).

With closed cooling tower systems (see Section 1.12.5.3) the circulating water has to be pumped through the condenser and then up to the distribution headers inside the cooling towers. The level of a station with this system is not dependent on the absolute sea or river level and, in consequence, the choice of station basement level is little influenced by pumping costs.

However, with direct cooling systems (see Section 1.12.5.2) the water has to be pumped from the adjacent sea or river so that the height between the station level and water level is very important. This water level is itself constantly changing, due to tidal variations, and the maximum pumping head necessary is associated with the lowest water level.

Because of high pumping costs, advantage is taken at direct-cooled stations of the principle of siphonic action. Atmospheric pressure is about 14·7 lb/in^2 and is capable of supporting a column of mercury 30 in high or a 34 ft column of water. When the cooling water side of a condenser is exhausted of air by vacuum pumps, the atmospheric pressure acting on the surface of the sea or river is sufficient to lift the water to a height of about 32 ft after which it will flow through the condenser and then into the outfall system (see Fig. 1.12.4.1(e)). In practice it is necessary to limit the siphon to a suction head of about 27 ft. At greater heads depending upon water temperature, liberation of dissolved air could take place which would expand and break the siphonic action. It is equally important to design and maintain an air-tight circulating water system in the siphonic section

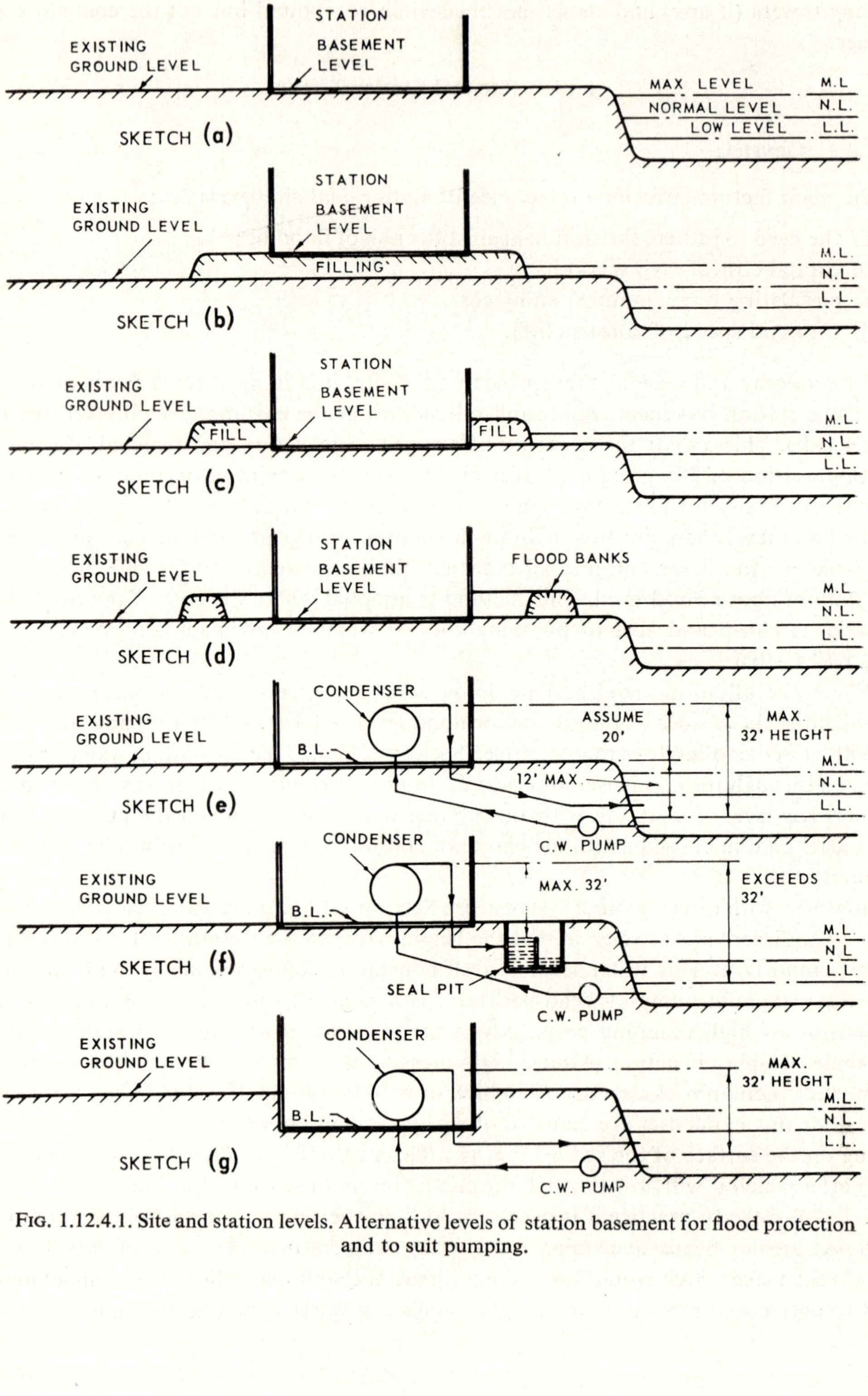

FIG. 1.12.4.1. Site and station levels. Alternative levels of station basement for flood protection and to suit pumping.

to prevent the entry of air and the attendant risk of breaking the siphonic action which would result in cessation of flow.

There are sites, however, where with the station basement at site level, the top of the condenser would be considerably more than 27 ft above the lowest level from which the water has to be pumped. If this is the case one of two alternative arrangements is usually adopted. Either the condenser is lowered relative to the site level to reduce the height (see Fig. 1.12.4.1(g)), or it becomes necessary to construct a seal pit or weir to restrict the length of the siphon leg (see Fig. 1.12.4.1(f)). The choice is usually made by examining the advantages and costs of both schemes. It may be that if the station is founded on rock the cost of excavation to lower the condenser and basement level would be very high in which case a seal pit would be adopted. On the other hand, if the ground on site is sloping it may be more convenient to have the turbine basement lower than the boiler basement, and advantage could be taken of this slope to locate the condenser at a lower level.

In direct-cooled systems where the siphon is equal to or less than 27 ft the circulating water pumps have to generate a head sufficient to overcome culvert and condenser friction only. At stations where the height of the top of the condenser is greater than 27 ft above the lowest sea level, the CW pumps are designed to generate a head that will overcome both friction and the difference in height between the level of the condenser above the sea level and the siphon leg.

1.12.5. **Circulating Water Systems**

1.12.5.1. General

The circulating water system is perhaps the most important of the common station services and its design requires very careful consideration. Although there are exceptions, it has become common practice in stations to provide one constant speed circulating water pump for each turbo generator, fully bussed on both the suction and discharge sides. This arrangement is adopted for the pumps supplying the condensers in direct cooled and closed cooling tower systems, and has been considered in most cases to provide the requisite degree of control over water quantity (by varying the number of pumps in use) in the cheapest and most reliable manner. Not only does the fully bussed system permit satisfactory control of water quantity but it ensures a supply of water to all condensers in all but the most extreme and unlikely emergencies, such as failure of electricity supply to all pumps.

The basic aims of the designer are to provide:

(a) a guaranteed supply of water to all running sets at all times;
(b) ready and efficient control of water quantity to give optimum station efficiency under all conditions of station load and water temperature;
(c) a balanced supply of water to all sets without recourse to unnecessary "throttling";
(d) minimum maintenance needs easily carried out;
(e) minimum overall capital and operating costs consistent with the foregoing requirements.

Types of CW Systems. The main types of CW system used by the Board are as follows.

(a) Direct-cooled systems—in which the cooling water is abstracted from the source (which may be river, estuary or sea), pumped directly through the condensers and returned to the source. A hydraulic gradient for a direct-cooled CW system is shown in Appendix 2.

(b) Closed cooling tower systems—in which the condenser cooling water is pumped round a closed system incorporating cooling towers wherein the heat rejected by the condenser is dissipated; an external source of water (river, canal, estuary, etc.) is required to make good tower evaporation losses and purge losses. A hydraulic gradient for a closed cooling tower CW system is shown in Appendix 3.

1.12.5.2. Direct-cooled System

This system is usually adopted when the station is sited on the coast or on a river estuary where there is an adequate supply of water available at all times (see Fig. 1.12.5.2).

A cooling range of between 8° C and 10° C approx. is the average temperature rise between condenser water box inlet and outlet. The heat rejection rate of 660 MW and 500 MW turbines for this cooling range is about 2625×10^6 Btu/h and 2065×10^6 Btu/h respectively.

It is necessary to establish the correct relative position of the intake and outfall, these points often being decided from a hydrographic survey or hydraulic model tests (see Section 1.7.2). If they are too close together, recirculation will take place and this will reduce plant efficiency and increase running costs. If they are moved further apart the capital cost of the connecting mains may be increased. They should, therefore, be in the position relative to each other that produces a balance between these two costs.

For the connecting mains from the intake to the station and the return to the outfall, reinforced concrete culverts, spun concrete pipes, cast iron, or steel pipes are used depending on the quantity of the water, the subsoil conditions, the availability of materials and the relative costs based on the capital and running costs. The velocities in these culverts or pipes can vary from 5 ft/sec to 20 ft/sec, normal flow velocities being about 10–15 ft/sec throughout with all sets running. The lower velocities are associated with culverts or pipes of larger cross-sectional area and consequently more expensive to construct. The higher velocities, for an equivalent quantity of water, require a smaller and hence cheaper culvert or pipe but pumping costs are increased. Some optimisation is therefore essential at the design stage to ensure that the correct balance between capital and running costs is achieved.

Suction velocities usually require to be less than the discharge velocities, to permit ebb and flow tides and general currents to sweep away accumulated debris and to eliminate air-entraining vortices which could lead to cavitation in the CW pump. For this reason it is often considered cheaper to reduce the pump suction culvert length to a minimum at the expense of the pump discharge pipework, and therefore the main circulating water pumps are usually located near the intake. The connections from the pump house to the

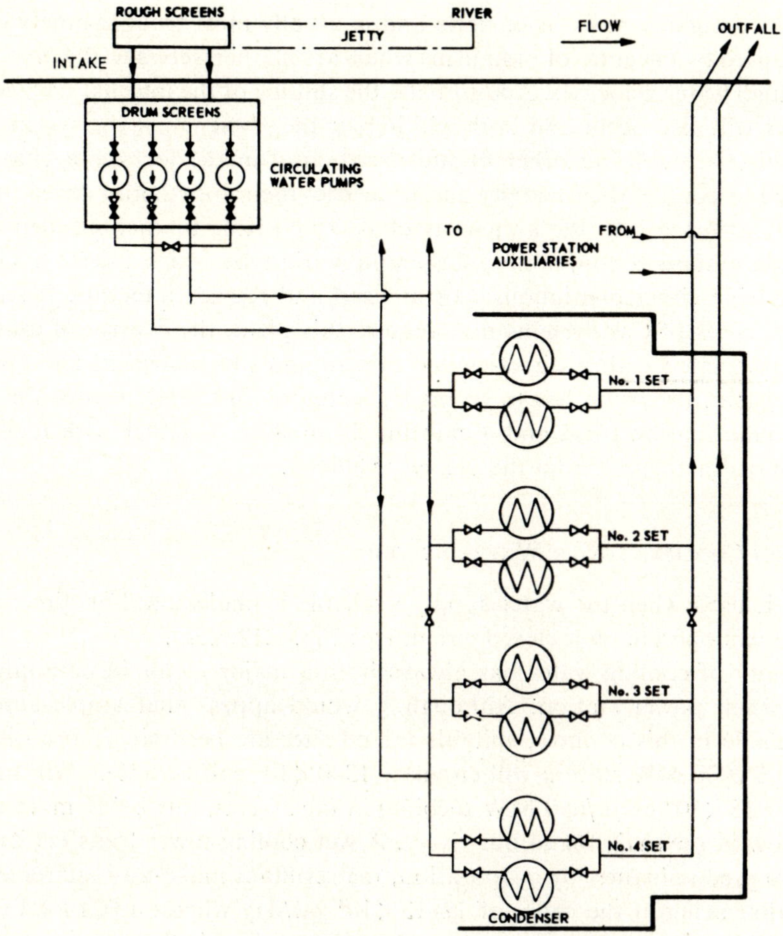

FIG. 1.12.5.2. Typical diagrammatic arrangement of a direct cooling water system.

turbine house should then be kept as short and as straight as possible after the suction culvert lengths and the location of the CW pumps has been established.

Care should be taken in routing the culverts or pipes so as to avoid the crossing of culverts or obstacles that could create air pockets in the system. Their design should take into account the maximum pressures on the culvert both external, imposed by the earth, and internal imposed by the internal water pressure. Other conditions such as the effect of external loads or pressures on empty or near vacuum condition culverts or pipes should also be considered.

Where pumps are taking their supply from a natural source such as the sea or a river, screening plant is provided before the pumps. The screens may be of the cup or disc type, but due to the variations in water level, direct cooled stations generally use band screens or drum screens. These fine screens are self-cleaning, the debris being continuously removed. Coarse bar screens are usually placed before the fine screens to prevent the ingress of large debris such as baulks of timber and tree trunks, usually brought down river by flood.

Fouling often occurs in pipes or culverts and is usually most troublesome when they are new; it is caused by the entry of organisms which attach themselves to the pipe surfaces in spite of the high water velocities. Additionally the sliming of the internal surfaces of the condenser tubes will also occur and both will impair the efficiency of the system. Where crustaceans (mussels, etc.) and other organisms abound in the locality a chlorination plant is installed to control their activity and to restrict their growth and development by the injection of chlorine into the system as close as possible to each condenser inlet. The chlorination system is commonly of the automatic type which injects a measured quantity of chlorine either continuously or at fixed and regular intervals. It should be noted that both types of CW systems may require this plant, the control of crustaceans usually being associated with a direct-cooled system and the control of sliming with a tower-cooled system. Provision has to be made, therefore, in the site layout for a house to contain the chlorination plant and a chlorine drum store, together with facilities for loading and unloading drums from the delivery vehicles.

1.12.5.3. Closed Cooling Tower Water System

This system is used when the water supply available is inadequate for direct cooling, and the condensers operate on a closed circuit (see Fig. 1.12.5.3A).

The availability of cooling water has always been a major factor in determining the siting of a thermal power station. Although it would appear that ample supplies of water are available in this country, suitable inland sites are becoming more difficult to find. Typically a 2000 MW station will circulate 1300×10^6 gal/day ($13{\cdot}5 \times 10^6$ gal/h/500 MW turbine or 15×10^6 gal/h/660 MW turbine) in the system, this being more than the dry weather flow in the majority of our rivers. A wet cooling tower transfers its heat to the atmosphere predominantly by evaporation, the resultant make-up requirements of a 2000 MW station being in the order of $12\text{–}13 \times 10^6$ gal/day when on full load (see Fig. 1.12.5.3B). In such stations the only supply of water required is that necessary to make up the amount used in purging the system of dissolved solids, i.e. salts concentrated in the system due to continuous recirculation and to replace the amount lost by evaporation from the cooling towers.

The amount of make-up to replace purge depends on the quality of the water supply, but is usually about 2% of the quantity circulated. The water required for evaporation loss is usually about 1% so that the total quantity of water required for make-up purposes is approximately 3% of the quantity circulated in the system. Although this quantity is taken into the station the 2% used is purged and is returned, so the loss to the source of supply is 1% of the quantity circulated. The 2 : 1 ratio of purge to evaporation will ensure that the concentration of solids in the water purged to the river will never rise above 50% of the original concentration.

This system permits a large measure of flexibility in the availability of water, or a temporary failure in the make-up pumping equipment. Provided the cooling tower ponds are of sufficient capacity the station can operate for an appreciable time without make-up.

The most common problem with a closed cooling tower system is siltation. The usual

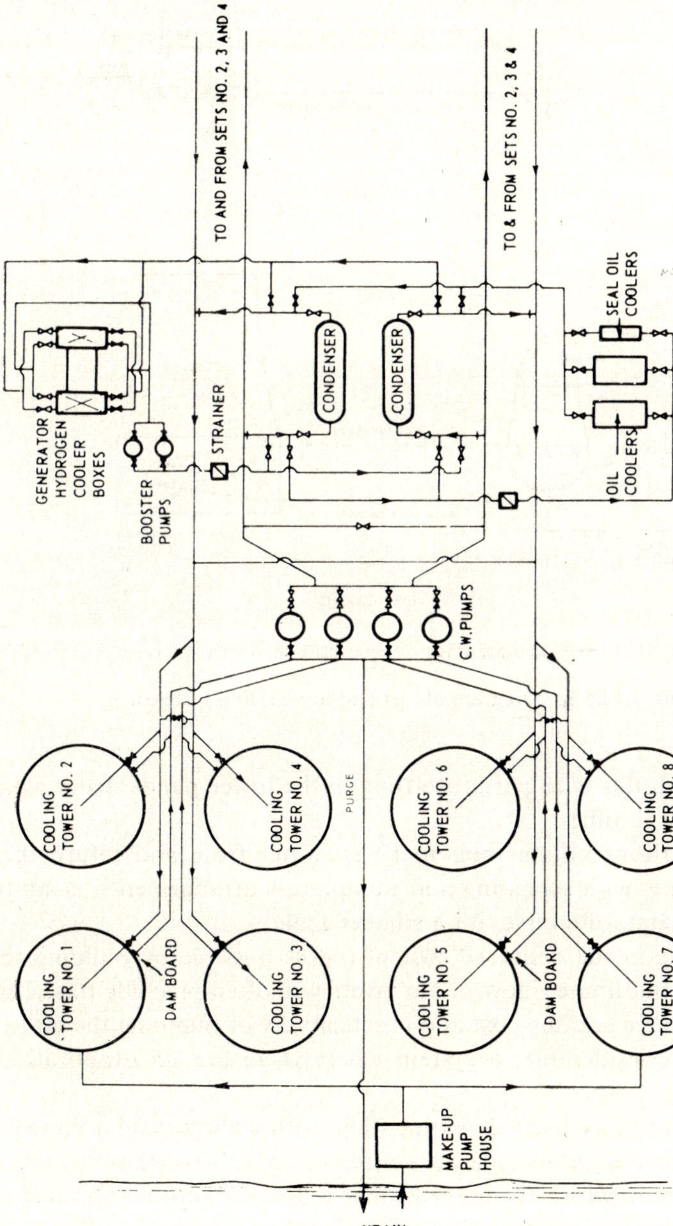

FIG. 1.12.5.3A. Typical diagrammatic arrangement of a closed cooling tower water system.

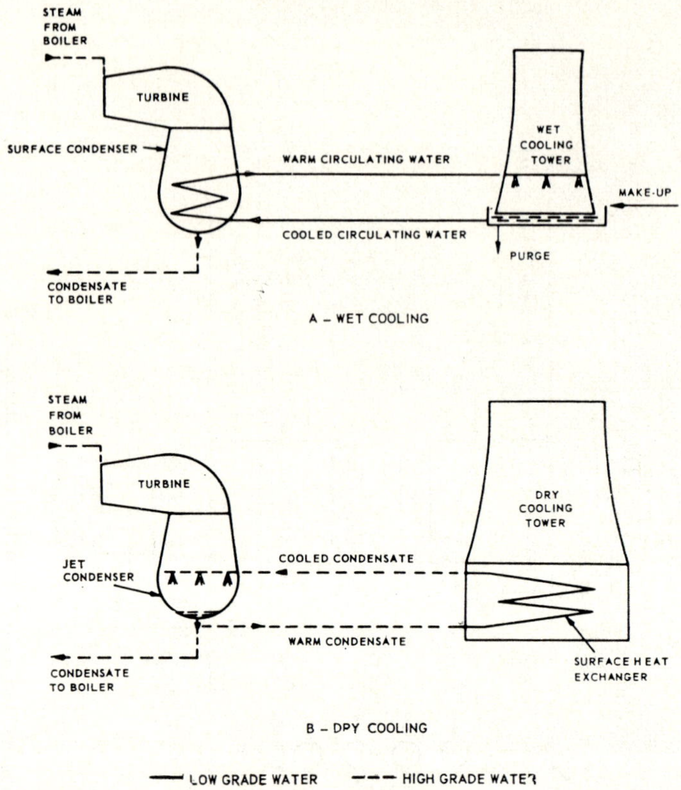

FIG. 1.12.5.3B. Diagram of wet and dry cooling systems.

method of dealing with this is to permit settling in the tower ponds, from which it must be removed from time to time.

The majority of stations obtain their make-up water from and return the purge to rivers. A similar intake, with screening and pumphouse arrangements, is adopted as for direct-cooled systems but, of course, on a smaller scale.

A substantial saving in the delivered cost of fuel is possible by building the stations close to the coalfield, but limited flow in our rivers will often preclude this. Therefore on sites where water is scarce and the cost and practicability of pumping the large quantities of water required is considerable, a system where there are no evaporation losses is attractive.

The Generating Board has been experimenting with such a system known as a "dry cooling tower". In this design no separate source of cooling water is required, the condenser is of the jet type in which the cooled water is sprayed into the exhaust steam, and the resulting condensate is circulated through the tower as part of a closed circuit from which 2% of the water is bled continuously as boiler feed water. The net rejection of heat to the atmosphere is effected by finned tubular coolers of a special design, mounted in the base of a hyperbolic concrete shell.

The dry cooling tower is considerably larger than a conventional wet tower, the design base diameter for a 250 MW tower being in excess of 400 ft. A diagrammatic comparison of the "Wet" and "Dry" cooling systems is shown on Fig. 1.12.5.3B.

Cooling towers are usually grouped and sited so that the CW system as a whole is as compact as possible, maintaining a minimum clearance of half the tower base diameter between adjacent towers, and between the towers and any object which might obstruct the air flow into them. Towers should be so positioned that any spray at base of tower does not blow on to public or station roads in frequent use, or onto the station sidings, and it is desirable to reduce the risk of coal and ash dust blowing into the tower ponds.

1.12.6. **Coal and Fuel Oil Handling**

1.12.6.1. Railborne Reception and Discharging

(a) *Coal.* All but two of the twelve coal-fired stations designed since 1960 have all their coal intake delivered by rail, and over 60% of stations already in commission are also coaled by rail, as Table 4 shows.

Table 4. Transport of Solid Fuels

Method of transport	1958		1968	
	Millions of tons	Percentage	Millions of tons	Percentage
Rail	20·8	48·4	40·4	60·9
Sea	9·7	22·5	9·9	15·0
Road	10·2	23·7	10·6	16·0
Inland waterway	1·7	3·9	2·4	3·6
Conveyor belt	0·6	1·5	3·0	4·5
Totals	43·0	100·0	66·8	100·0

The annual coal consumption of a 2000 MW station, depending on its load factor, amounts to approximately 5 million tons, i.e. 20,000 tons daily over 250 days. (A calculation for deriving this figure will be found at the end of this section.)

To handle this tonnage in conventional wagons through tipplers would require so much time to operate as to be almost impracticable, and a sidings layout comprising standage tracks to accommodate two days' mean delivery of coal (i.e. one day's fulls, one day's empties) would be required and this would amount to approximately 16 miles of rail track.

To speed the delivery of coal trains to large modern power stations, agreement was reached between the British Railways Board and the Generating Board to handle hopper bottom wagons on a "permanently coupled" basis. Permanently coupled working means the keeping together of the main line locomotive and its train of wagons as a single unit, from entering to leaving the site, including discharge over a ground unloading hopper.

By 1970 the railborne coal intake is expected to amount to 60 million tons of which over two-thirds will be moved by the permanently coupled train system. The advantages of this system are the faster turn round of trains at the power station, which normally will not exceed one hour. It is expected that trains will achieve at least $3\frac{1}{2}$ trips per week between colliery and power station thus giving greater wagon utilisation together with capital economy in track work (3–5 miles only required) plus the associated savings in operational costs.

A design of hopper wagon has now been adopted having rapid self-discharging characteristics and it is the Board's intention to use this wagon for the transport of coal to all new power stations. This wagon is the 32-ton capacity two-axle wagon with six bottom doors operated automatically by lineside equipment. It discharges the coal whilst travelling at a speed of approximately $\frac{1}{2}$ m.p.h. It has been designed with a separate top canopy to facilitate its introduction to collieries where there is a height limitation. In this form its height will be 10 ft 6 in (i.e. 1 ft $5\frac{1}{2}$ in below its full height) and its capacity will be approximately 26 tons. The length of the wagon is the same for both capacities, i.e. 29 ft 8 in over buffers.

The largest train at present envisaged by British Railways is 1500 tons gross load and 1000 tons payload. With the 32-ton wagon the length of the train for this payload will be 32 wagons (950 ft) plus locomotive and brake (100 ft) = 1050 ft. Accommodation in the sidings layout should therefore be 1200–1300 ft to provide working clearance. Where possible it would be prudent to allow for greater lengths, if site space allows, to provide for longer trains should B.R. wish to introduce them at some later date.

Sidings layout. Depending on the space available and the volume of traffic to be handled, the choice lies between the "loop" system or the "run-round" system. The latter, although necessitating the main line locomotive running round its train, is accepted by B.R. as permanently coupled working.

(i) *Loop system.* This system permits trains to enter the site, run over the unloading hopper in succession and depart, without stopping. The minimum radius of curve that can be used is 9 chains (594 ft) and a length of straight track is required before and after the hopper. In consequence of this the smallest loop with parallel sides will encompass an area approximately 1500 ft wide by 2000 ft long. The loop can, however, be "pear"-shaped to suit site space (see Fig. 1.12.6A).

(ii) *Run-round system.* This system is acceptable where the space available is long and narrow. The trains are hauled successively over the unloading hopper, the main line locomotive uncoupling to run round its train, either before or after discharge, prior to departure. This introduces additional time and complication into the train movements, and from choice should only be adopted at stations having a comparatively small coal intake (see Fig. 1.12.6A).

(iii) *Weighing.* To provide for check weighing of all coal supplies, weighbridge platforms for gross and tare weighing are located respectively before and after the unloading

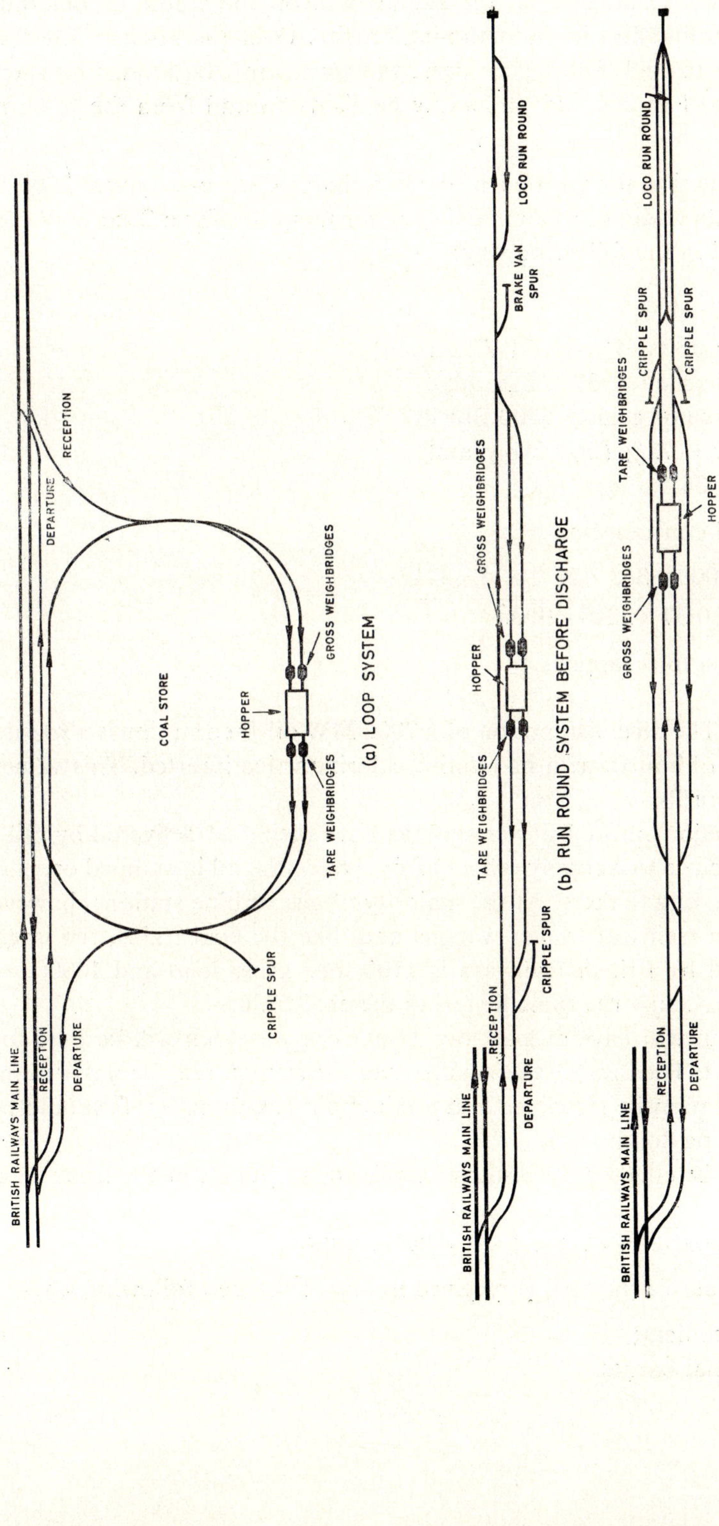

FIG. 1.12.6A. Typical schematic siding layouts.

hopper on each unloading track. As in the discharging cycle, weighing is automatic whilst the train is travelling at a speed of approximately $\frac{1}{2}$ m.p.h.

(iv) *Gradients*. Gradients on the sidings will on the whole be determined by the level of the main line and levels obtaining on site. Over the hopper, and for a distance before and after to include weighbridges, the track must be level. Elsewhere on site it should not exceed 1 in 260 and preferably be rising to and from the track hopper and weighbridges.

A timing calculation for a railborne coal discharge cycle at a typical 2000 MW station is as shown in Appendix 4. The annual coal consumption for a 2000 MW coal-fired station is calculated in the following way:

Data

4×500 MW units (4×465 MW sent out)
Thermal efficiency = 37% base load
Calorific value of coal = 9900 Btu/lb
Load factor = 75% (first five years)
Btu/kWh = 3412

Annual coal consumption =

$$\frac{4 \times 465{,}000 \times 365 \times 24 \times 0{\cdot}75 \times 3412}{0{\cdot}37 \times 2240 \times 9900}$$

= 5·1 million tons (approx.)

(b) *Fuel oil*. The oil consumption of a 2000 MW oil-fired station is also calculated in a similar way, but of course with its relative calorific value inserted. This value is approximately 18,500 Btu/lb.

None of the major oil-fired power stations have their fuel delivered by rail, since they are usually sited close to a supply refinery from which the oil is pumped or, in some cases, delivered by sea-going tankers. At the majority of gas turbine stations, however, the fuel oil is delivered in trains of tanker wagons and, like the coal trains, the largest train at present envisaged by British Railways is 1500 tons gross load and 1000 tons payload. Such a train will occupy the same length as the coal train.

A sidings layout will have at least two tracks one of which will be used for the locomotive run-round. Because of the smaller site areas required for gas turbine stations, siding space will often be restricted and a minimum length of 500 ft should be provided for a 500-ton oil payload.

Discharge will be effected by positive displacement pumps at the track side.

1.12.6.2. Waterborne Reception and Discharging

(a) *Coal*. Waterborne coal is received in one of the two following ways:

(i) sea-going colliers;
(ii) river or canal barges.

Sea-going colliers are used at stations located on the coast, or on navigable river estuaries. Vessels of varying tonnages are in service at different stations, the largest of these colliers being capable of carrying about 6500 tons of coal. These are approximately 370 ft long and have a draft of 23 ft at 6500 tons.

In locating the quay from which the colliers will be discharged, it is necessary to avoid interference with other shipping, and have a sufficient depth of water for berthing. To qermit berthing at all states of the tide, it may be necessary to construct the quay well offshore, or to carry out extensive dredging.

Of the new 2000 MW inland stations only one is designed to accept part of its coal intake by canal barge and for unloading a barge tippler has been installed.

(b) *Fuel oil.* Waterborne oil is received in one of the two following ways:

(i) sea-going tankers;
(ii) river or canal barges.

Tankers are used at stations located on the coast, or on navigable river estuaries and the same requirements as for berthing the colliers apply. These tankers, the largest of which is capable of carrying approximately 45,000 tons of oil, are 740 ft long with a draft of 37 ft.

River or canal barges can be used at gas turbine stations where the oil burn is relatively small. These barges can carry up to approximately 500 tons.

1.12.6.3. Roadborne Reception and Discharging

(a) *Coal.* Road vehicle unloading facilities are arranged for standby or supplementary purposes and, where provided, normally comprise no more than a single ground unloading hopper into which road vehicles can discharge and from which coal is conveyed by belt to some convenient point in the main coal-handling system.

(b) *Fuel oil.* Again, because of the large quantities of oil required for the large 2000 MW and above oil-fired power stations, road tankers are only used for supplying some of the gas turbine plants. Discharge from these is by positive displacement pumps at the roadside and usually alongside the oil storage tanks.

1.12.6.4. Conveyance from Unloading Point to Station Bunkers or Coal Store

The "main line" of the coal-handling plant runs between the coal unloading point (be it wagon hoppers or jetty) and the boiler house, and for maximum economy this line should be as short as possible. The "branch line" used for stocking out and reclaiming to and from the coal store should also be kept to a minimum length. For a station of 2000 MW deep troughed twin conveyors of 1500 tons/h each are usually provided for the "main line", one being a standby for the other. The "branch line" conveyor is a single 3000 tons/h reversible belt.

It is common practice to provide screening and crushing plant (should oversize coal arrive from the colliery) in the bunker conveyor run and it is usually at this point, after the crushers, that the supply to and the return from the coal store is introduced.

In designing a system, a conveyor inclination should not exceed 18° to the horizontal, and a height of approximately 10 ft should be allowed at transfer points for the inclusion of a chute. For a typical diagrammatic arrangement see Fig. 1.12.6C.

1.12.6.5. Coal and Fuel Oil Storage

Coal storage. Although it is now common to provide for a store of capacity equal to a quarter of the annual station burn, over half of this capacity may be regarded as a strategic reserve for use in emergency and it is consequently unnecessary to incur heavy expense in providing fixed handling equipment for the more remote part of the store which is usually served by mobile plant. Tests have shown that coal of any quality now in use may be stocked to a considerable depth provided it is properly consolidated and it is now assumed for planning purposes that nearly all coal stores can be at least 30 ft deep. The store is usually accommodated inside the rail sidings loop described earlier in this section.

The Generating Board have had to accept, at the new stations, the weekly mean delivery of coal over 5 days, this being due to the shorter working week adopted by the N.C.B. Consequently two-sevenths of the daily coal intake will be put out to store for reclamation at weekends, when coal is not being delivered. Normally the stations will be coaled over two shifts (i.e. 16 h); however, British Railways have the right to request the Board to accept trains "round the clock" for which they will, in return, offer a reduction in freight rates.

Fuel oil storage. Modern oil-fired stations are usually located adjacent, or fairly close to, an oil refinery or where their needs can be easily supplied by sea-going tankers. Therefore oil storage arrangements vary according to the station's position since there is not a definite policy on oil storage by the Board.

If sited adjacent to the refinery often only a surge tank will be required holding about a day's supply. If the refinery is a little distance away then oil storage will amount to between 4 to 7 days' supply, depending on whether the refinery has adequate tankage for holding the fuel for the station's requirements. In both these cases the oil is pumped from the refinery to the power station.

On the coastal or estuarial power station it is usual to hold at least three weeks fuel burn. Storage tanks for holding the oil should be carefully sited to preserve visual amenity and to protect other plant and buildings from the attendant fire risk. Bund walls, or containment areas, should be sufficient to contain the volume of oil stored, should a tank burst.

1.12.6.6. Hydraulic Transport of Coal

The technology of the hydraulic transport of coal by pipeline has been considered as an alternative method to the other forms of transport. The Board has one experimental line linking a power station to a colliery and is carrying out detailed experimental and design investigations.

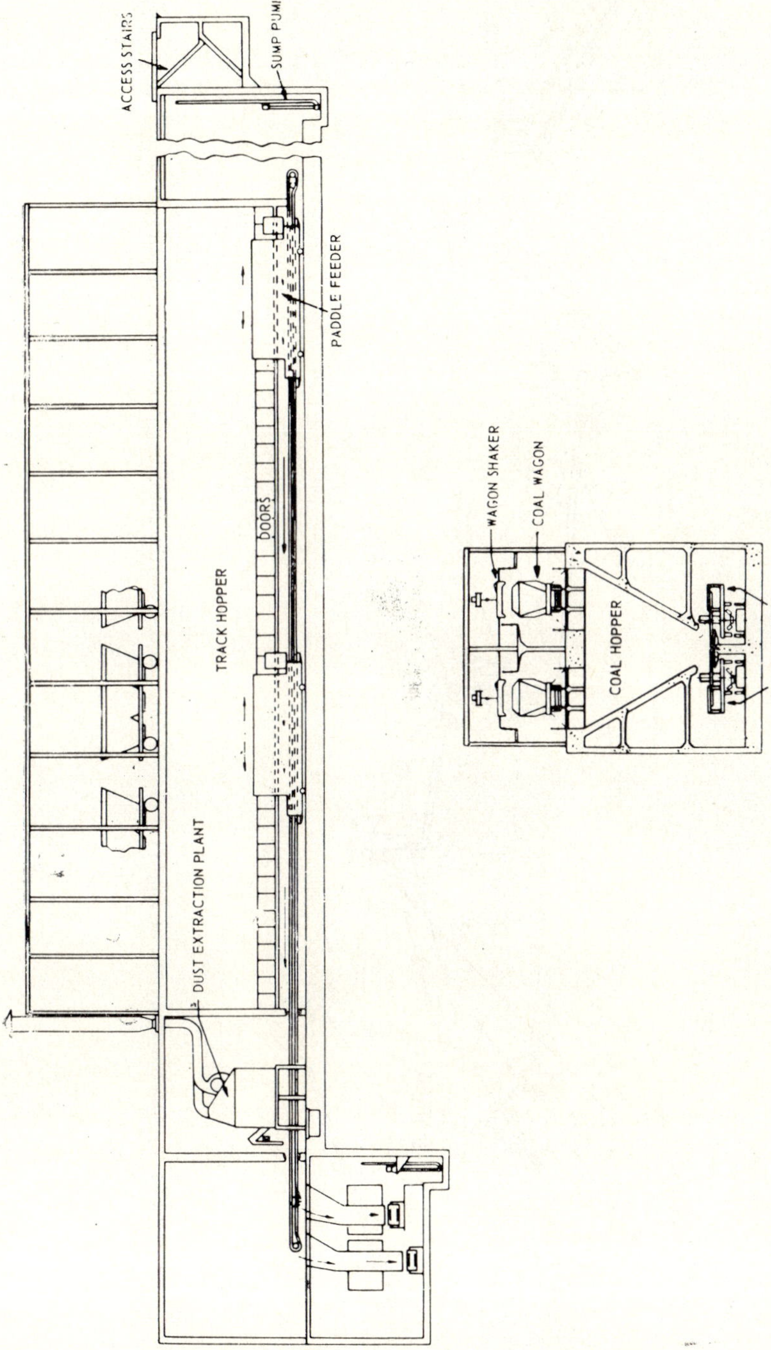

FIG. 1.12.6B. Arrangement of track hopper for permanently coupled rail working at railborne coal-fired power stations.

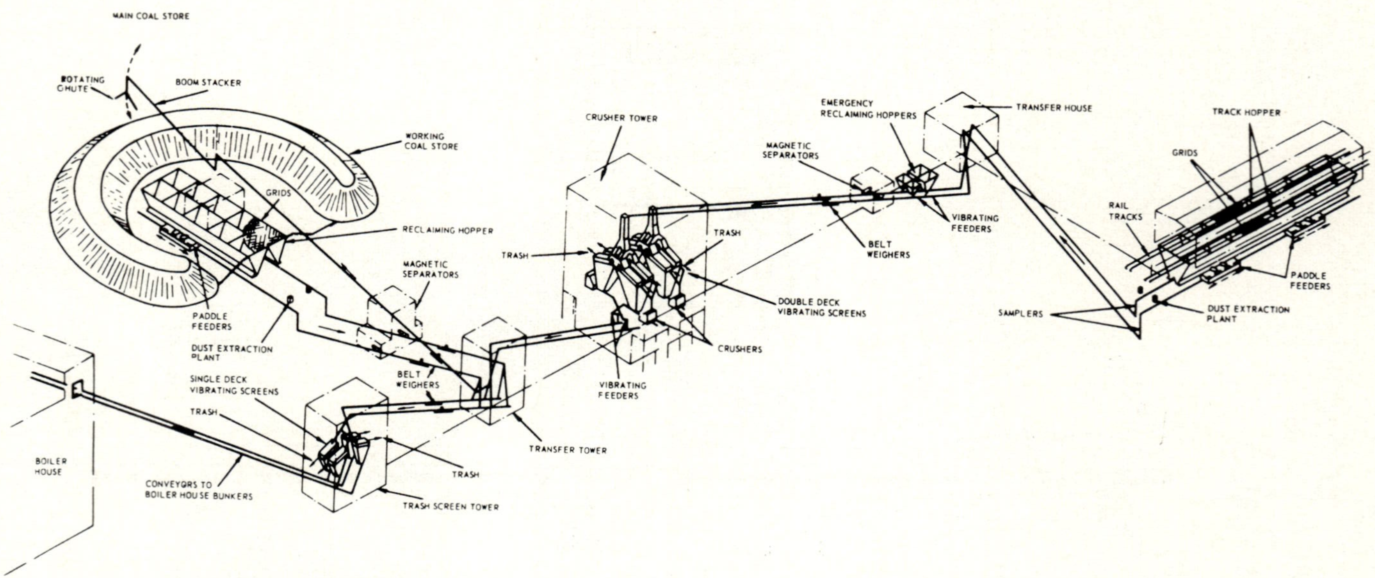

FIG. 1.12.6C. Typical diagrammatic arrangement of coal conveyance from the unloading point to station bunkers or coal store.

1.12.7. **Ash and Dust Disposal**

Various systems are in use for removing the furnace bottom ash from the boilers, but the one principally used is the sluicing system. With this, the ash that collects in the hopper below the boiler furnace is washed into a sluiceway by high-pressure water jets. This sluiceway leads to an ash sump usually located close to the boiler house and from this the ash can be taken away by lorry or in trains of wagons.

The dust from the boiler (now known as P.F.A.) is collected in hoppers beneath the precipitation plant, from which it is removed by pipeline to the dust bunkers.

The location of these ash sumps and dust bunkers should be so chosen that pipeline and sluice lengths are kept to a minimum. P.F.A. is very similar to the volcanic ashes used by the Romans, in combination with calcined limestone, as a cementing material. This combination continued to be widely used until the discovery of Portland Cement less than 150 years ago. Unlike the volcanic ashes P.F.A. is already finely divided and does not require crushing or pretreatment.

In recent years the possibilities of utilising this material have begun to be systematically investigated and its commercial use has increased. The most promising avenues for the economic utilisation of P.F.A. in quantity appear to be:

(a) mixed with clay to produce bricks;
(b) as a constituent of concrete;
(c) to make aerated concrete blocks;
(d) to make lightweight aggregate by sintering;
(e) as a structural fill material for embankments, etc.

On some sites the Generating Board have allowed contractors to erect these block-making processes or sintering plants (c) and (d), and for this purpose some consideration should be given to the provision of space for the buildings associated with these processes. This area should be close to the dust bunkers and great care taken to ensure that such an installation will not create a noise or dust nuisance nor must it detract from the efforts already applied towards preserving visual amenity.

Such commercial plants only account for a small proportion of the station's ash and the remainder must be disposed of from the site. If because of suitable local disposal areas the pumping of the pulverised fuel dust from the dust bunkers is uneconomic, then it becomes necessary to transport the dust by rail in specially adapted pressurised wagons. These wagons are modified "Presflo" cement wagons, capable of carrying up to 21 tons of dry P.F.A. which are loaded from the station dust bunkers. An additional sidings scheme in parallel with that used to deliver the coal and operating on the same permanently coupled basis is required both at the power station and discharge terminal for handling these wagons. Since P.F.A. is, on average, 20% of the station coal burn then for a 2000 MW station about a million tons of dust per annum must be disposed of — this representing some 4 or 5×1000 ton payload trains daily.

1.12.8. **Transmission Requirements**

The cost of providing outdoor switchgear is less than that of indoor gear so that except in special circumstances, such as exposed sites near the coast or areas of heavy atmospheric pollution or where space or amenity considerations are paramount, outdoor gear is used. Costs of connections are at a minimum when the compound is adjacent to and parallel with the turbine house. Furthermore, the compound should be so disposed on the site that the outgoing feeders may be arranged in an economical manner. To avoid the risk of spray blowing from the base of any cooling towers on to outdoor switchgear, cooling towers should be well removed (the minimum distance being half the tower base diameter) from a switch compound and preferably on the leeward side of the prevailing wind; the relative position of coal store and compound should also be carefully chosen.

For a 4×500 MW×400 kV switch compound or switch house the areas required are approximately 800 ft long × 700 ft wide and 800 ft long wide × 750 ft wide respectively. This includes space for transformers and banking connections, 400 kV sealing ends, roads and hardstands considered essential for normal equipment handling and access facilities. The switchgear is usually arranged in 70 ft wide bays and maximum height of structures or main hall is also 70 ft. A strip of land, along the length of the compound or switch house, 120 ft wide is required for the transmission tower reserve.

Because of the large ground areas required for the 400 kV switch compounds, in many cases amounting to more than the area occupied by the main buildings themselves, an attractive alternative for reducing the area appears to be the use of "double-deck" or "high-level" switch houses. In such a design the switchgear would be arranged in "tiers" and not side by side in bays as at present. Obviously this arrangement would be acceptable where site area is limited and the increased height, possibly up to 150 ft, would not raise objection.

1.12.9. **Road and Rail Access**

The station should be so arranged on the site that if possible the access road from the nearest main public road is as short, direct and as level as possible.

Rail access into the site is not essential for construction purposes but when provided the spur to the turbine house loading bay and the workshop is often retained for servicing purposes when the station is fully commissioned.

1.12.10. **Ancillary Buildings**

It is common practice nowadays to arrange the administration offices, the canteen, the station workshops and the stores in a building or buildings which are separate from the main station building. In some cases a bridge connects the offices with the station at operating floor level. It is often considered convenient and architecturally desirable to arrange the administration block adjacent to the station on the main access road.

1.12.10.1. Offices

Usually there should be a central office building, housing the majority of the personnel who require office accommodation. A small number of personnel will be housed in offices adjacent to working areas (e.g. stores, control room (see Section 1.13.6), etc.) The location of the central office building will depend on the overall station layout and on any special condition which may affect the disposition of the ancillary buildings. However, although it is impossible to say where on the site the office building will be located, certain conditions must be fulfilled:

(a) freedom from noise;
(b) freedom from dust (e.g. precipitators);
(c) easy access to other ancillary functions and to the main building;
(d) easy access between the maintenance engineers and the maintenance workshops;
(e) easy access between the personnel concerned with operation and the control room;
(f) reasonable aspect.

1.12.10.2. Laboratory

The chemist and his assistant provide a comprehensive chemical service to the station. Samples dealt with in the laboratory come from all over the station, consequently the siting relative to the flow of materials is arbitrary. However, the work carried out is of a delicate nature requiring a high standard of cleanliness. Station laboratories should be situated to avoid dust, vibration and noise. Hence they should be well away from the plant buildings, and preferably on the ground floor. The administration block is normally satisfactory, but a separate building is preferred. The laboratory should include a separate office for the station chemist and in modern stations an office for his deputy. The laboratory should be easily accessible from other buildings, have adequate working space and good natural lighting preferably from windows facing north.

1.12.10.3. Workshops

The workshops are provided in order to carry out maintenance work on the plant, although the extent of this is decided by the Region in which the power station is located and is affected by (a) the regional policy about the use of central workshops, (b) the distance from the central workshops or other large workshops at an existing station, (c) the type of plant installed, (d) the distance from the manufacturer's factory or workshop.

With the increase in size of plant installed, the problem of handling in the workshops has become of great importance. Where previously the mechanical workshop was the only area served by an electric overhead crane, it is now considered desirable to provide cranage over the major parts of (a) the mechanical workshop, (b) the electrical workshop, and (c) the welding shop.

Special care should be taken to ensure that the various workshops are sited to facilitate the duties they perform and they are usually associated with the main buildings and with the stores building which provides the materials necessary for effecting repairs.

It is frequently found satisfactory on the new large stations to provide an island of ancillary accommodation which contains administration, welfare, workshops and stores functions and has easy access to the main building. Recently, however, stations have been planned with the workshops attached to the end of the turbine hall and have the use of one of the turbine hall cranes.

Wheeled vehicle access is necessary to all workshops from the turbine house, boiler house and other plant buildings.

1.12.10.4. Stores

The stores building on a power station is provided for keeping (a) spare parts for the items of plant installed in the station, (b) raw materials for use in the workshops and on plant, (c) binnage items, e.g. nuts and bolts, (d) protective clothing, (e) miscellaneous items used generally throughout the station, e.g. cleaning materials.

The greatest demand for materials from the stores comes from the workshops and therefore these two functions should be planned in close association. There is also a considerable flow of material from the stores to the main building, but it is best that the stores building is kept separate from the main building so that room for expansion is not limited.

All incoming goods will arrive at the reception area or go direct to the heavy stores. Road access to the stores must be direct, unimpeded and suitable for long wheel-based vehicles.

1.12.11. **Construction Requirements**

Where the construction of other units is expected to continue after the first unit has been commissioned it is usually desirable to arrange for the coal feed to the bunkers, the CW supply to the turbine condensers and the ash discharge from the boilers to enter the "starting" end of the station (i.e. that end which will be completed first). This avoids any inconvenience which may be caused by having to arrange temporary support for future bunker conveyors and having to work over running plant, CW works and ash plant, etc. which is in operation.

Considerable areas, often amounting in total to 50–100 acres, are required by the contractors during erection for storage of parts and for site offices. These storage areas should be so disposed that access from them to the station is as short and convenient as possible for each contractor.

It is often necessary to provide some permanent "labour camps" for the men working on construction.

1.12.12. **Amenity Considerations**

Whilst recognising that production of a reliable supply of electricity at the lowest possible cost is the paramount consideration, it is the Board's statutory duty to pay attention to the appearance of new power stations, both in detailed architecture and in its suitability for the environmental amenity.

Very often the architect may suggest a number of arrangements of buildings or cooling towers in order to achieve the correct massing in the landscape and to improve appearance. This work is done in close collaboration with the engineering design staff to ensure that the optimum construction and operational design is still achieved at minimum cost.

Landscape architects are also engaged with a view to integrating the station and transmission equipment in the immediate locale so far as is practicable with the surrounding countryside.

1.12.13. **Typical Site Layouts**

It will be appreciated from the foregoing sections that it is almost always impossible to satisfy every requirement perfectly. If the circulating water system is as short as possible then the connections between the generator transformers and the switchgear may be increased in cost; if the coal conveyors are reduced to minimum length then some other services may suffer, and so on. The weight to be credited to each factor in terms of money, operational convenience and amenity value is dependent on the type of station, the site and its environment. Three different solutions to site layout problems are illustrated in Figs. 1.12.13A, B and C, and it is only necessary to supplement the sketches with the following brief descriptions.

Figure 1.12.13A shows the site layout of a 2000 MW station comprising 4×500 MW units, utilising a closed cooling tower system and with railborne coal supply. The disposition of the main components on the site is such that a second station could be accommodated at a later date within the existing boundary.

A balance between the engineering and architectural requirements is achieved. The 400 kV switch-house is placed "indoors" and situated adjacent and parallel to the turbine house, while the cooling towers are grouped in two sets of four at either end of the station; this is an architectural requirement which though not detracting much from operational convenience requires an additional pump house. However, views of the station from the surrounding country are greatly improved.

A loop system of sidings is adopted for coal delivery and P.F. dust despatch. The workshop and stores are located close to the turbine house and the boiler make-up water treatment plant is central at the rear of the boiler house. The administration block, canteen and welfare services are adjacent to the access road at the "finishing" end of the station (i.e. that end of the station that is completed last).

Figure 1.12.13B shows the site layout of a 2000 MW station comprising 4×500 MW units with a direct-cooled circulating water system and a pipeline oil supply.

Advantage is taken of the local terrain to screen the station which is located mostly on reclaimed saltings. Fuel oil storage tanks (for one week's full load operation) are situated at the rear and close to the boiler house. Fuel is delivered by pipeline from an adjacent oil refinery.

The CW intake culverts are short and straight between the pump house and turbine house. Alongside the pump house a jetty has been provided for off-loading any item of plant arriving by sea, the geographical location of this particular site being such that to bring in heavy and bulky loads by road would be extremely difficult.

Because of salt spray and not because of architectural requirements the 400 kV switchgear is "indoors" and is situated adjacent and parallel to both the turbine house and the control room. The heavy workshops are placed at the starting end of the station and are a continuation of the turbine house thus allowing the main crane to travel into the workshop. The light workshops and stores are adjacent to and at the rear of the heavy workshops.

The administration, welfare, and canteen block is separate from the turbine house and

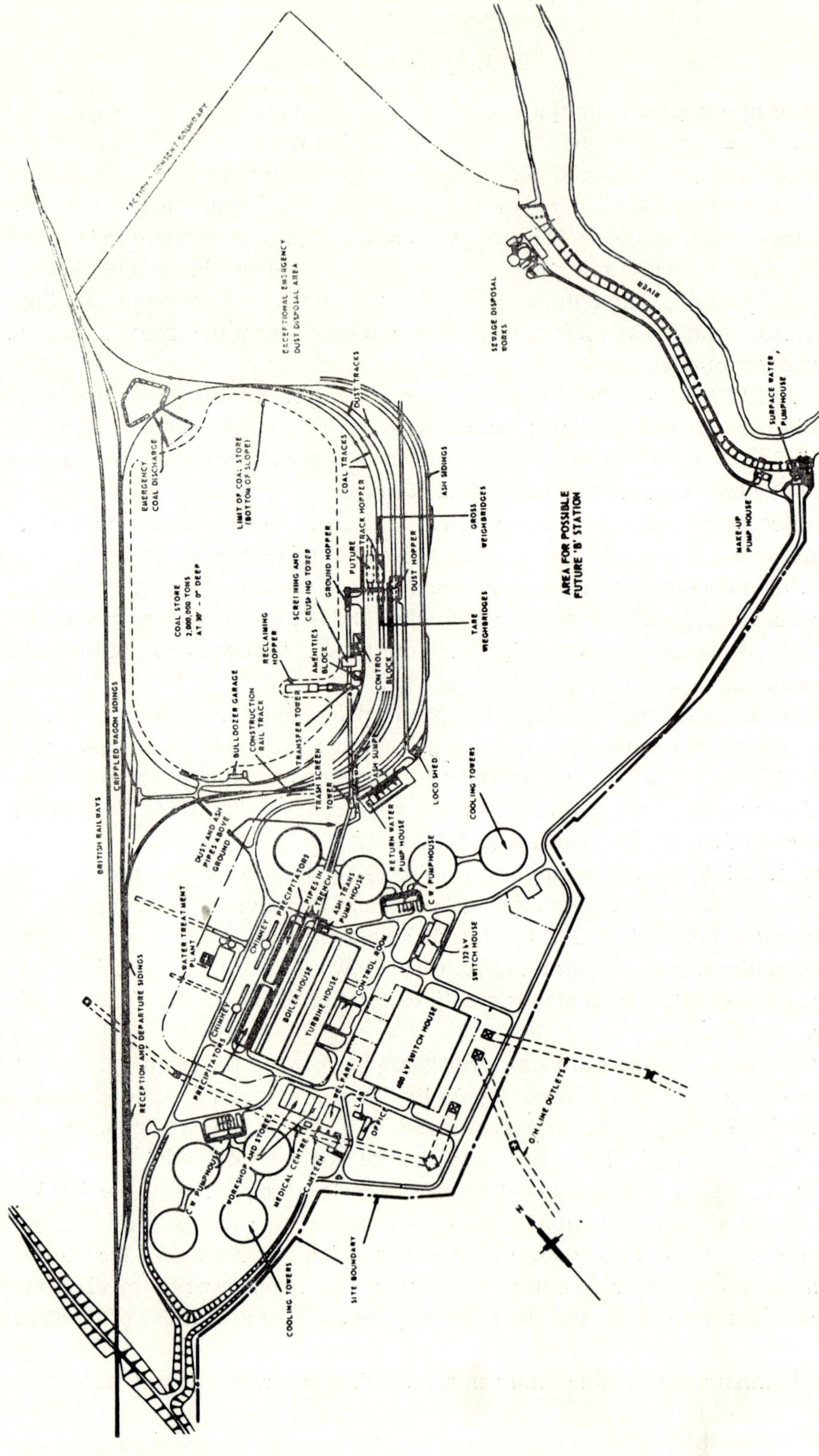

FIG. 1.12.13A. Site layout for 2000 MW coal-fired station—using cooling towers and with a railborne coal supply.

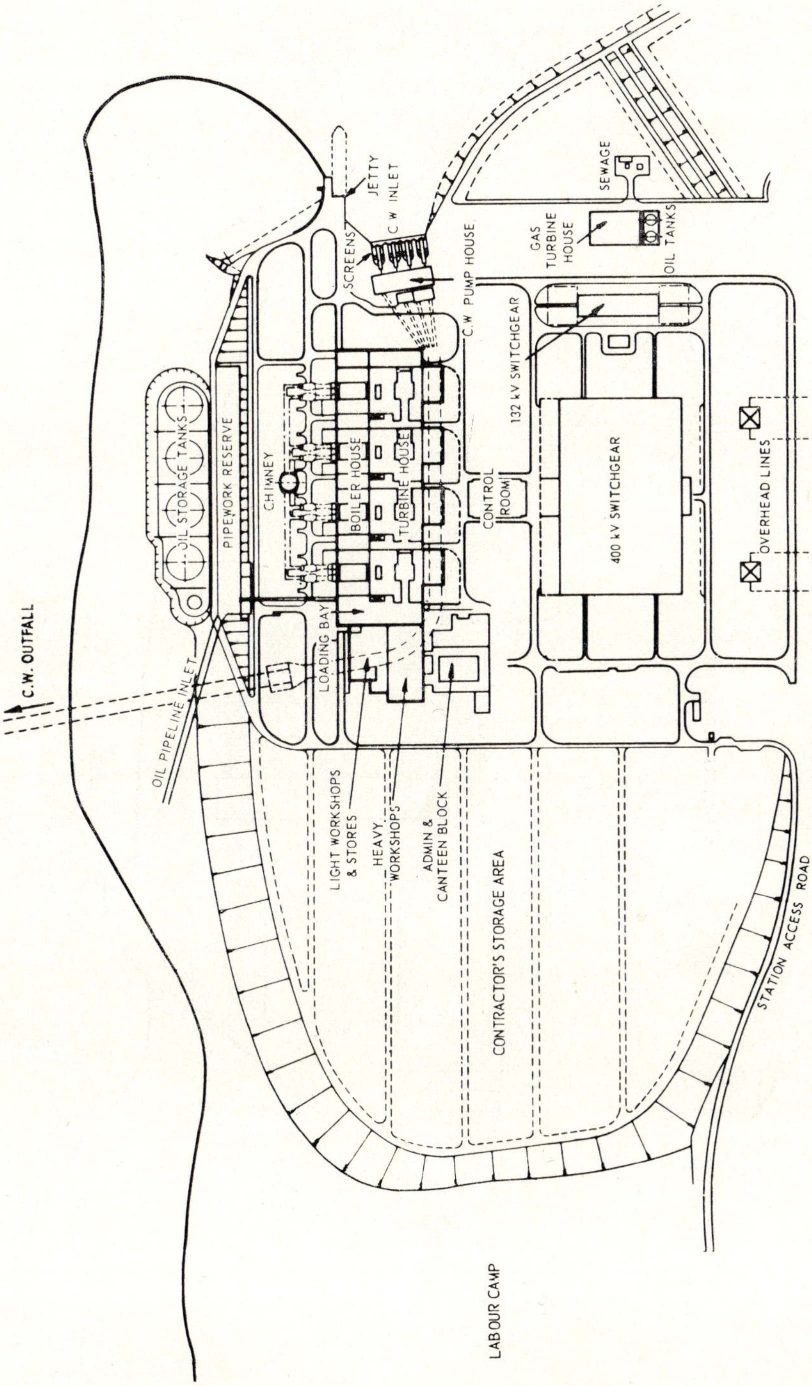

FIG. 1.12.13B. Site layout for 2000 MW oil-fired station—using direct cooling and with a pipeline oil supply.

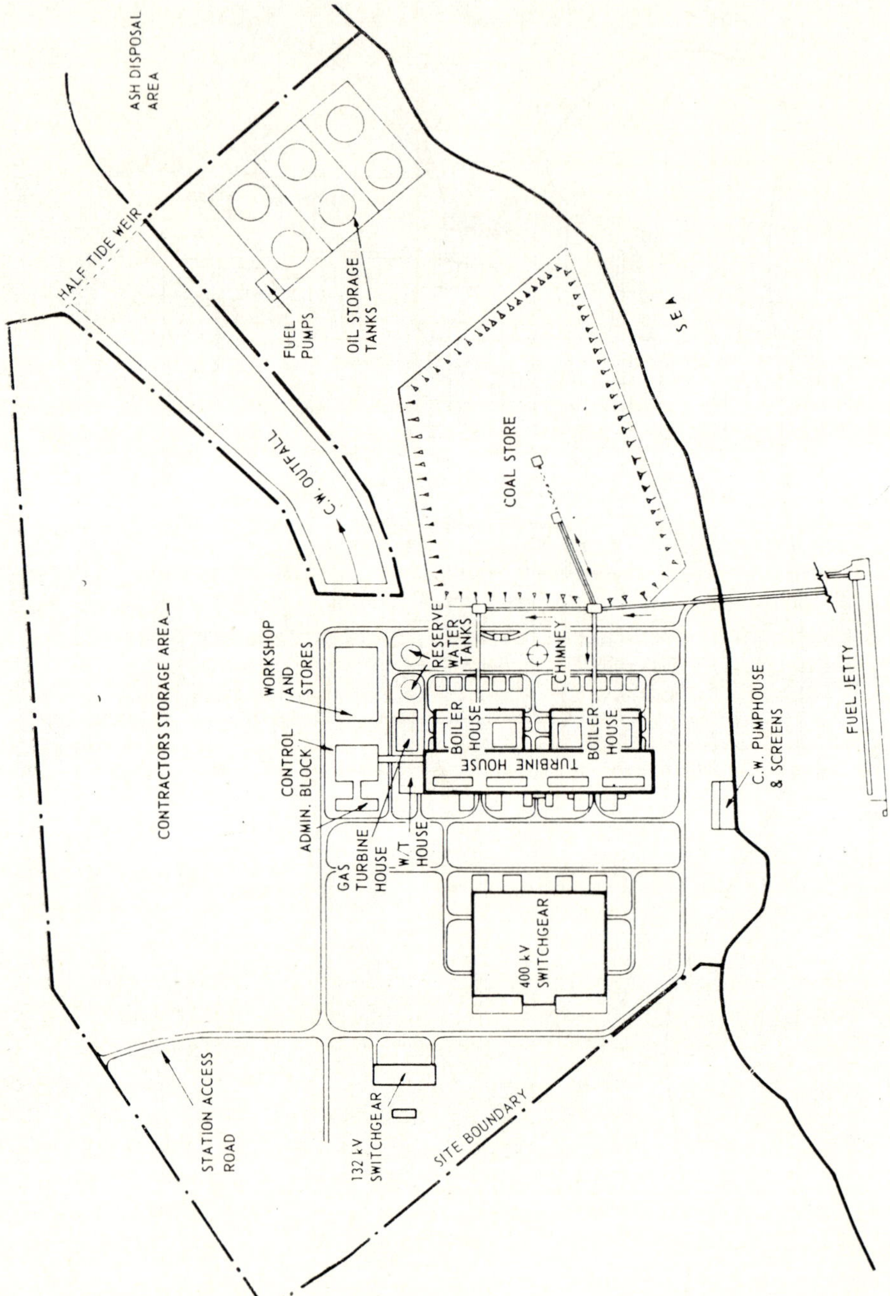

FIG. 1.12.13c. Site layout for 2000 MW dual-fired station—using direct cooling and waterborne coal and oil supply.

is connected by a bridge at operating floor level; it is situated on the main access road to the station at the "starting" end of the station.

Figure 1.12.13c is the layout adopted for a 2000 MW station comprising 4×500 MW units, using a direct-cooled circulating water system with a seaborne oil and coal supply from tanker and collier. A jetty sited in the deep water stream allows the berthing of the largest vessels used by the Board (up to 45,000 tons) and the coal conveyors and oil pipelines are in consequence rather long. The oil storage tanks and pumps are situated at the remote end of the site, risk of fire damage to other plant and buildings thereby being virtually eliminated. The coal store is at the rear and close to the boiler house thus giving minimum stocking out and reclaiming conveyor lengths. It will be noted that the boiler house is split into two, each served by its own conveyor.

Because of its close proximity to the effect of salt spray and atmospheric pollution the 400 kV switchgear is "indoors". The CW intakes are at a minimum length and the outfall is a dredged channel, along the course of an existing creek. This arrangement gives the greatest possible separation between the CW intake and outfall (over 4 miles) for this site and ensures that recirculation does not take place.

The administration and welfare block is adjacent to the starting end of the station just off the main access road and is connected to the detached control room by a bridge. A further bridge, at operating floor level, gives access from the control room to the turbine house. The workshops and stores are divorced from the main buildings.

Dust disposal is to another part of the site, which is mainly marshy; this will enable the land to be reclaimed for a possible future development.

1.13. STATION LAYOUT

1.13.1. General

Modern power stations are planned from the outset to be completed with a given number of units all of the same rating and it is usually assumed that all the units will be similar in general layout and overall dimensions. The advantages are that if each subsequent complete boiler/turbo-generator/transformer unit is in all major respects a repeat of the first, the detailed station design is to a great measure reduced to designing one unit only. There are also operational advantages in having a standard arrangement for each unit in the station and a reduction in the quantity of spare parts which have to be kept in stock. This concept of identical layout for each plant unit in a station has now become current practice for stations containing two or more units.

Over the years the layout of a station has developed into a general standard pattern. The boiler house and turbine house arranged side by side can be regarded as the central feature of the station layout. Very often between these two there is interposed a mechanical annexe accommodating auxiliary plant items. The building is usually completed by an auxiliary switch bay and generator transformers on the side of the turbine house and a bunker bay on the side of the boiler house. The other items required to complete the concept of this standard pattern are the grit arresting units, induced draught fans, main gas flue and chimneys arranged in that order from the boiler house.

1.13.2. Design Aims

As mentioned at the beginning of this section, a good design necessitates simultaneous consideration of all layout features, and included with this are the main connections between the various items of plant. These can be listed as follows:

(a) circulating water mains;
(b) main steam and feed piping;
(c) coal conveyors;
(d) flues;
(e) electrical connections.

It is the designers' aim on the grounds of economy to keep the building dimensions to a minimum, and the connections as short as possible. At the same time, adequate space has to be provided for plant maintenance and if steam and feed pipework are too short there may be insufficient flexibility and too great a thrust on the boiler or turbine. The best design, then, is one that strikes the correct balance between the lowest cost and the best arangement from both constructional and operational point of view.

1.13.3. Set Arrangement

In the design of the turbine house one of the most important considerations is the choice between longitudinal and transverse arrangement of sets.

The overall length of a turbo-generator of 500 MW and 660 MW capacity is usually in excess of the desired pitch of boilers, i.e. approximately 180 ft, and unless the boiler centres are increased the sets cannot be accommodated longitudinally in the turbine house. For this reason the majority of stations with sets of 500 MW have a transverse arrangement in the turbine house, with the steam end of the turbine opposite the centre line of the boiler. Where a longitudinal arrangement has been adopted the space between the boilers is utilised by accommodating plant that would otherwise have been placed at the rear of the boilers.

The first station with 660 MW units is now being designed. Because of the increase in the physical dimensions of the units it becomes apparent that neither the transverse nor the longitudinal arrangement is, from a station building dimension point of view, economic. A balance between the two is the obvious solution and the turbines have been arranged diagonally in the turbine house. This shortens the steam connections and turbine generator/transformer connections and also allows the set centres to be the same as the minimum practical boiler pitch. Another feature of the design of the new unit is that the turbine is now supported on a steel structure and not on the traditional concrete block, thus reducing the load to be supported by the main foundations. This will be explained in greater detail in another chapter.

For typical arrangements see Figs. 1.13.3A–E.

Until the introduction of the sub-critical 500 MW tandem turbines (i.e. all turbine cylinders being on a single shaft coupled to a generator), the largest tandem turbines in this country were 300 MW and 350 MW, both sub-critical units, followed by two super-critical

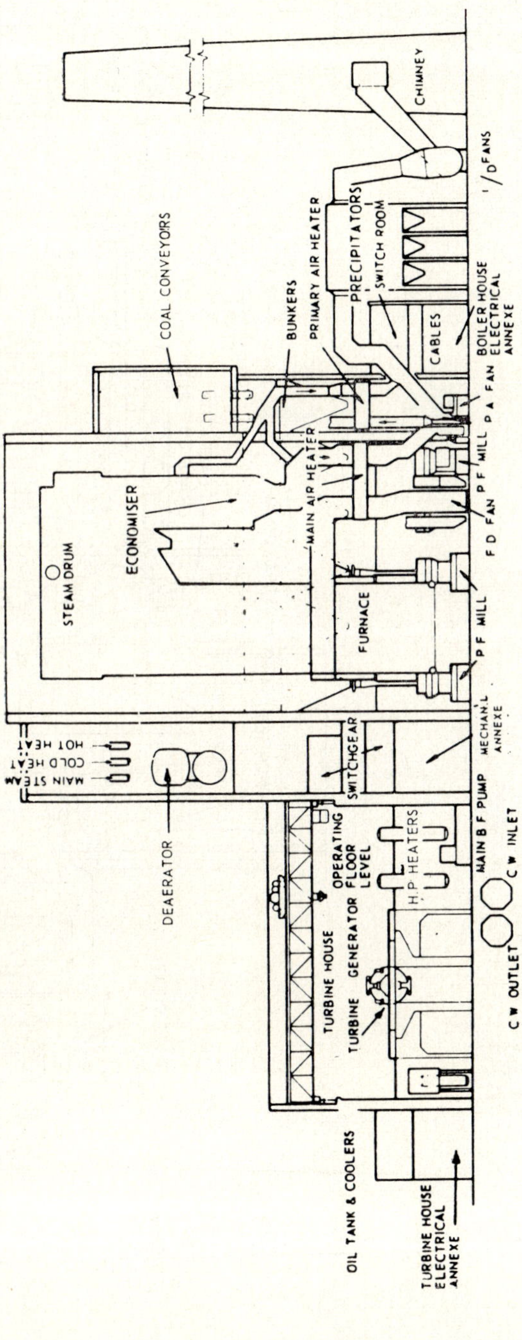

FIG. 1.13.3A. Cross-section of typical station containing 500 MW units with a longitudinal turbine arrangement.

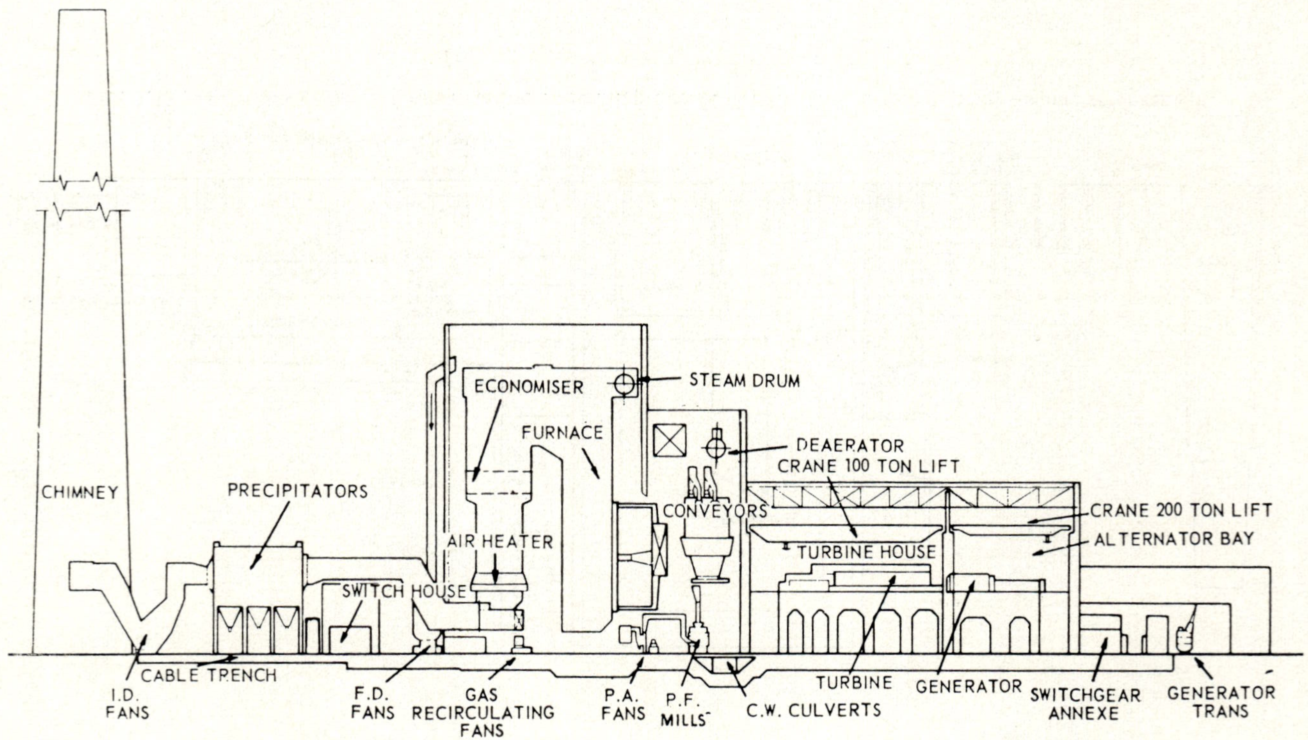

FIG. 1.13.3B. Cross-section of typical station containing 500 MW units with a transverse turbine arrangement.

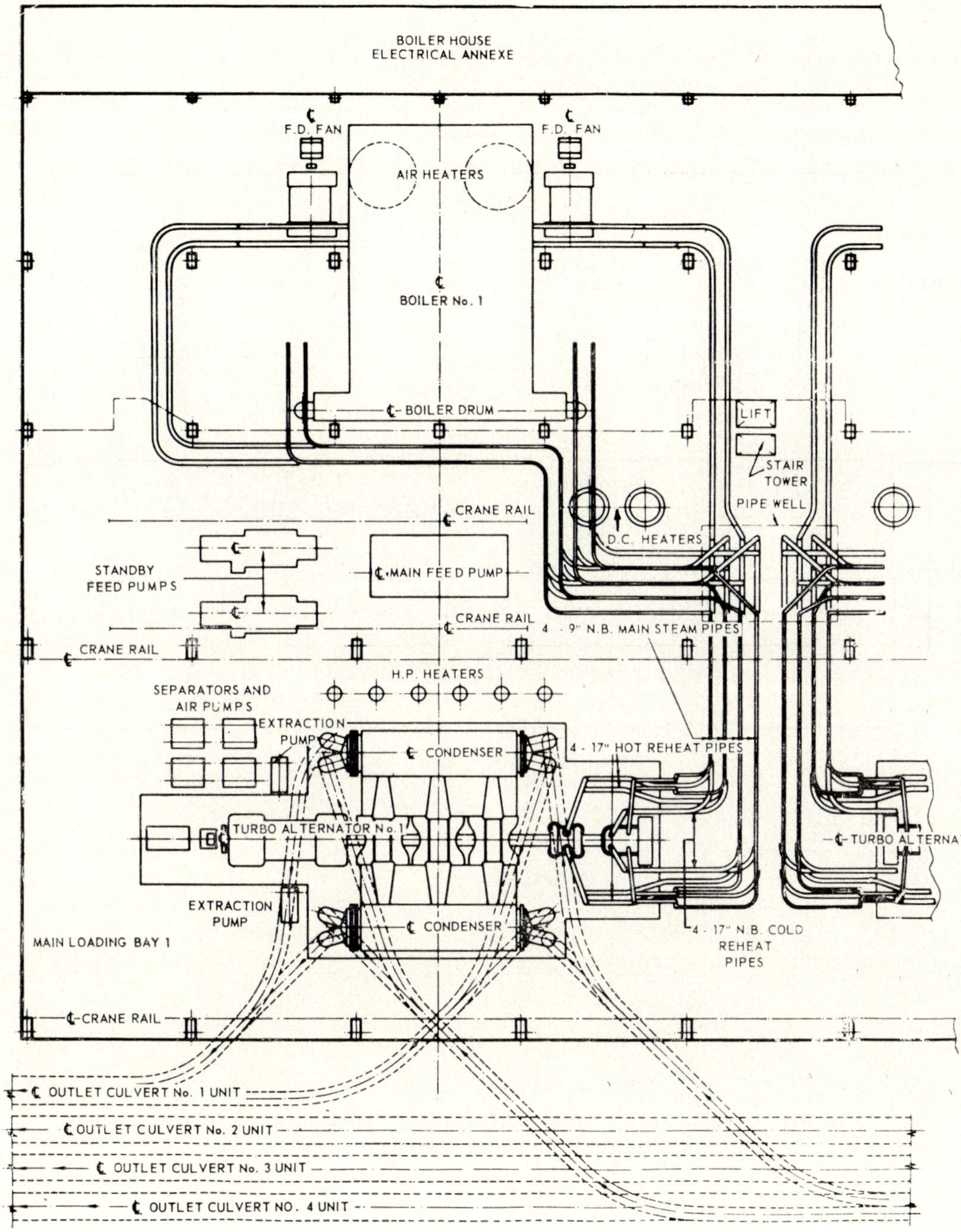

FIG. 1.13.3c. Plan of typical turbine house and boiler house containing 500 MW sets with a longitudinal turbine arrangement.

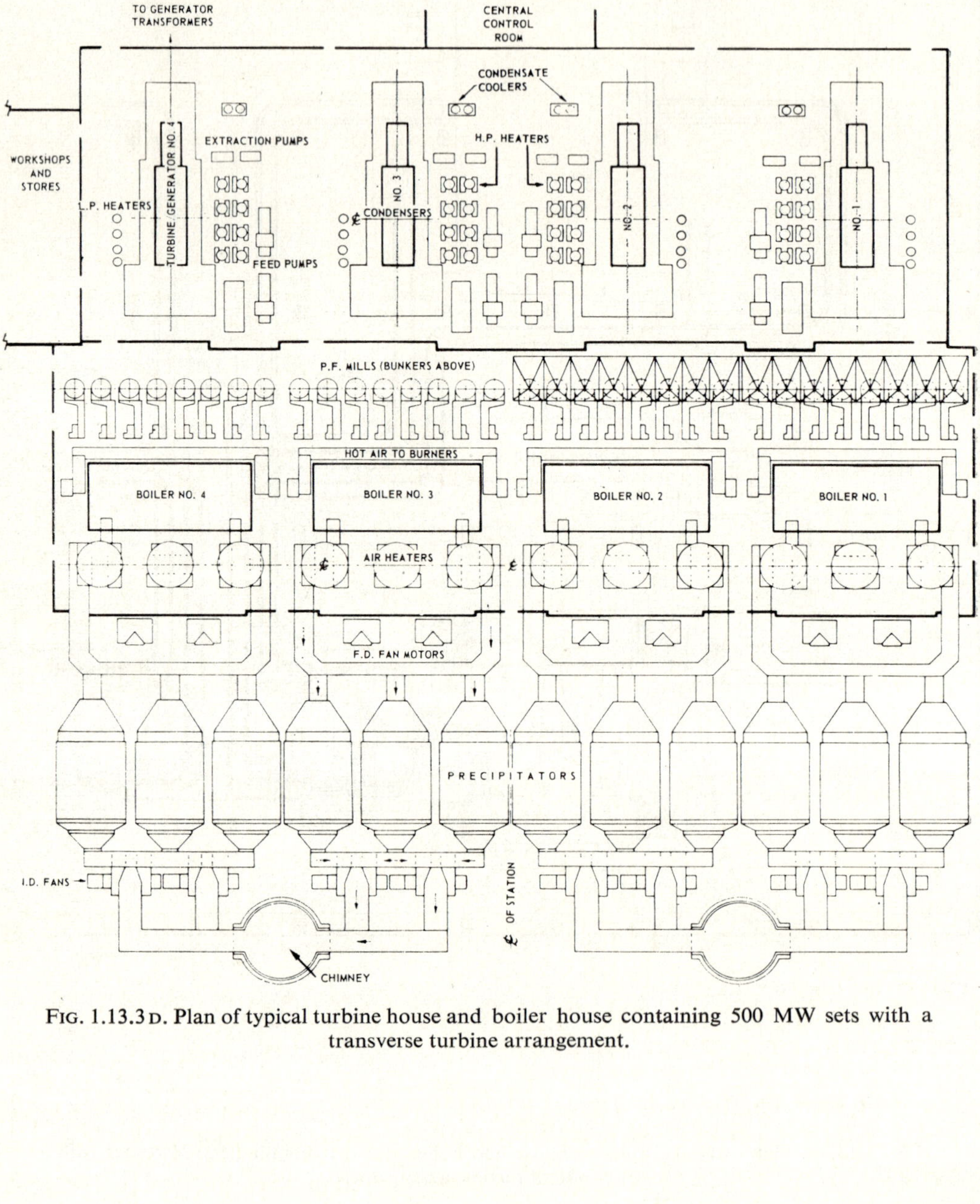

FIG. 1.13.3D. Plan of typical turbine house and boiler house containing 500 MW sets with a transverse turbine arrangement.

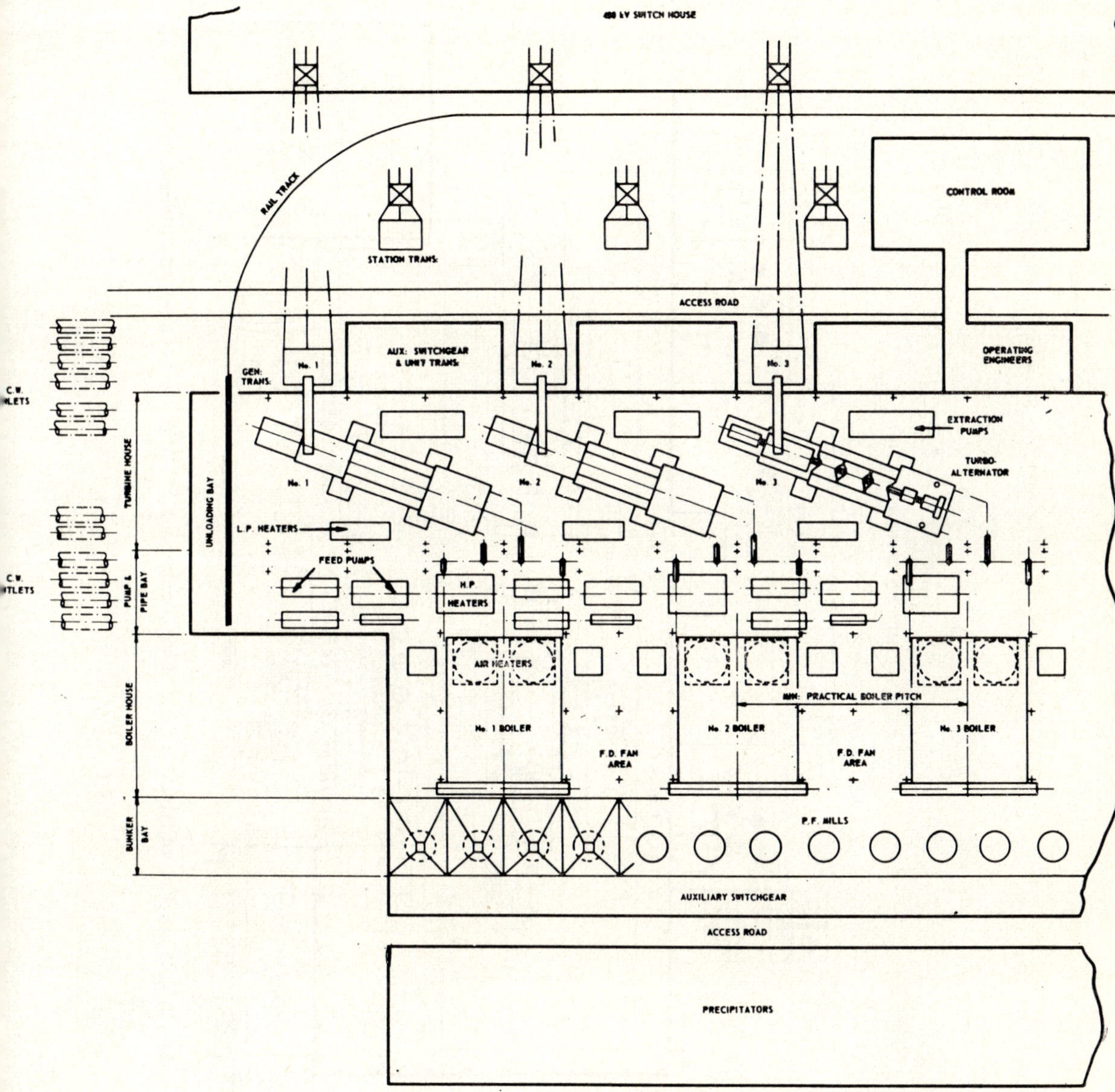

FIG. 1.13.3E. Plan of typical turbine house and boiler house for 2000–3000 MW station with a diagonal turbine arrangement.

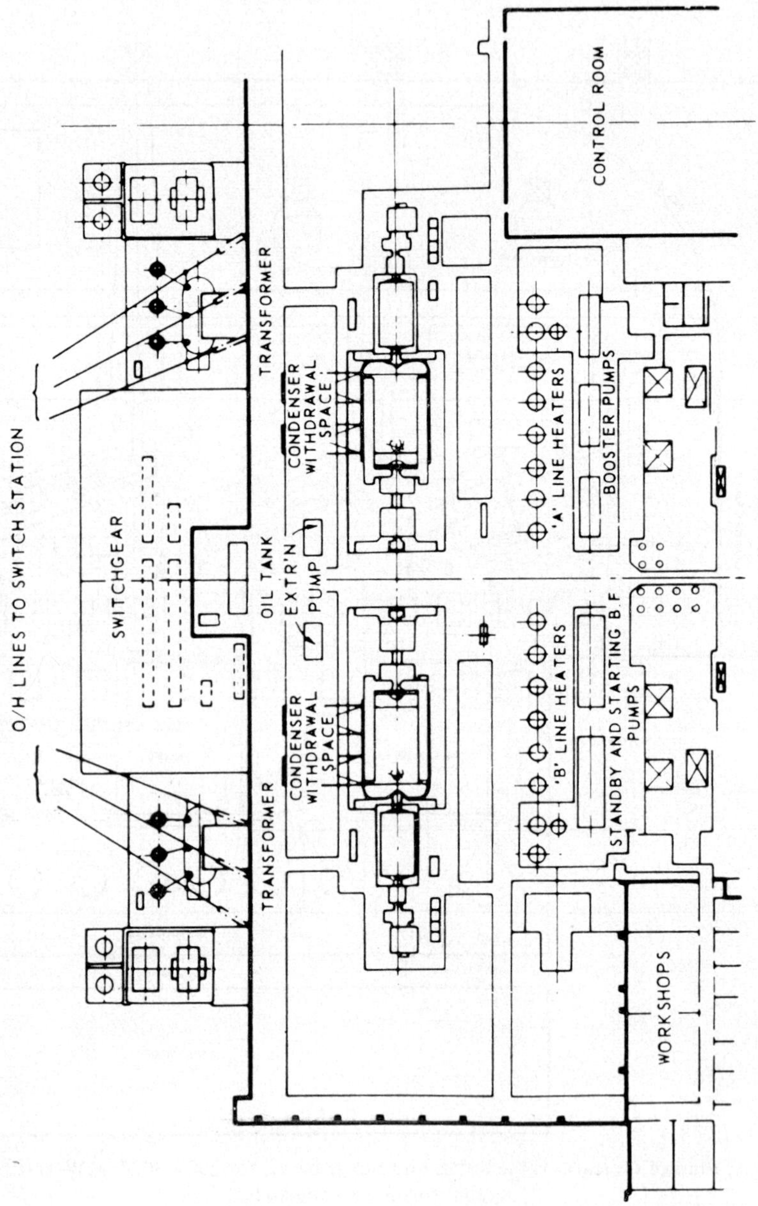

FIG. 1.13.3F (a). Plan of typical turbine house containing a 550 MW cross-compound turbo-generator.

375 MW units. One station was, however, equipped with 550 MW cross compound turbines (i.e. h.p. and i.p. cylinders on one shaft coupled to a 275 MW generator, with the l.p. cylinders on a separate shaft and coupled to another 275 MW generator). In America a 900 MW set has been designed of the cross compound type.

It is not known what the next step in unit size will be, or what arrangement will be adopted. The concept of a cross-compound turbine arrangement is given in a typical plan and section of a station containing a 550 MW set as shown in Fig. 1.13.3F. It will be seen that the set is arranged longitudinally but it can, of course, be placed transversely. In either case the rules applying to similar arrangements with tandem turbines will still apply.

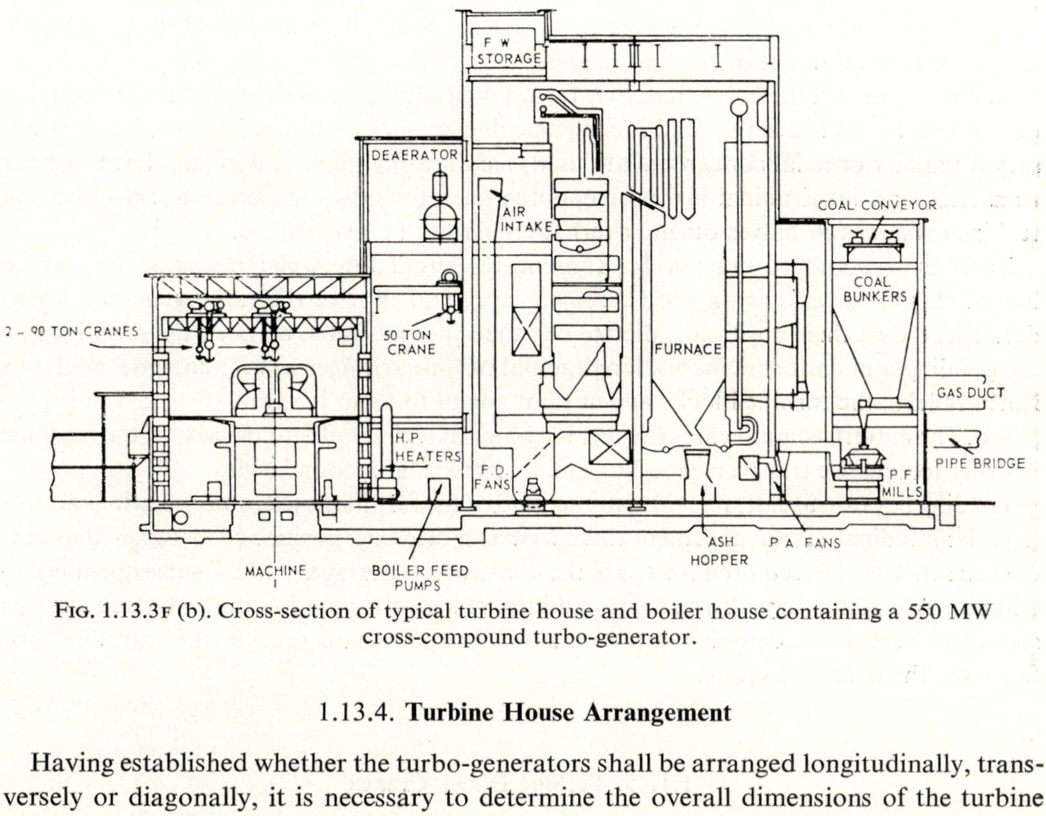

FIG. 1.13.3F (b). Cross-section of typical turbine house and boiler house containing a 550 MW cross-compound turbo-generator.

1.13.4. Turbine House Arrangement

Having established whether the turbo-generators shall be arranged longitudinally, transversely or diagonally, it is necessary to determine the overall dimensions of the turbine house. In the case of transverse arrangements the turbine house width is usually governed almost entirely by the machine length and the space required for generator rotor withdrawal. The need to use an auxiliary crane hook for lifting the bearing cap at the governor end of the turbine determines the location of the set relative to the boiler side crane rail and the minimum clearance required to remove the rotor determines the position of the outer wall. In many cases the building has been so designed that only skewed withdrawal is possible.

The feed heating plant is arranged between the sets giving the shortest and as near vertical as possible run of feed pipework. The booster feed pumps are usually located at basement level in the mechanical annexe.

It is also normal practice to locate the condensate extraction pumps adjacent to the condensers at basement level, usually at ground level (see Section 1.12.4.1). Operating floor level is then determined by the static head required on the extraction pump suction, the condenser height and the required depth of the turbo block beams spanning the condenser. Where pannier or axial condensers are used (where tubes are parallel to the shaft of the turbine and not at right angles) it is possible to reduce the height of the turbine house operating floor from about 40 ft using normal type condensers, to about 20 ft to 25 ft. Minimum crane hook height and hence minimum roof height with a transverse arrangement of set is usually governed by the need to lift a generator stator (for erection or repair purposes) over the top of an assembled machine from or into the loading bay. Having fixed the crane level in this manner the auxiliary crane hook usually provides sufficient clearance to lift the tube nests out of the feed heaters.

The width of turbine house required for a longitudinal arrangement of set is normally determined by the space required for the machine, condenser tube withdrawal (not necessary if pannier or axial condensers are used) feed heating plant and auxiliaries. For both transverse and longitudinal arrangements the need for adequate space to set down the turbine cylinder top halves during overhaul must not be overlooked.

There is no standard practice for running the circulating water mains in the turbine house. They are usually arranged at basement level, and clear of the turbo-generator foundations, so that their position is dictated by the space available between the machine and the building foundations. Where a longitudinal turbine arrangement is adopted the culverts run parallel to the turbine blocks and at right angles to them if a transverse layout is proposed. The culverts should enter the turbine block at right angles to the axis of the machine in order to equalise transverse loading and maintain stable foundations.

Loading bays are usually provided at each end of the turbine house, and on stations adopting a longitudinal set arrangement these bays can often be positioned between the ends of each set. They are required for use in the constructional stages and for subsequent maintenance. A loading bay also provides additional space for setting down turbine casings should two sets be undergoing overhaul simultaneously, and reduces the travelling distance for the overhead crane.

1.13.5. **Turbine House Cranes**

The majority of stations contain two electric overhead travelling cranes in the turbine house; one is a light duty crane instantly available for receiving materials at the loading bay; the other is for heavy duty work and fitted with a main hook of sufficient capacity to lift a generator stator, plus an auxiliary hook capable of lifting a generator rotor. Although stators have occasionally to be removed for repair, the usual main use of the heavy duty crane is for initial erection.

An attractive alternative arrangement is to provide three medium duty cranes any two of which can, by slinging a lifting beam from their heavy hooks, lift the heaviest load. This has the advantage of reducing the cost of the turbine house structure since the load is then spread over two bays and not one bay as in the case of a single heavy duty crane. The provision of three cranes in a turbine house comprising four or more units also speeds maintenance work on the machines.

In a transverse turbine arrangement the length of the set determines the crane span and in the case of the 500 MW sets this span is about 200 ft. To provide a crane for the maximum lift over this length would be an uneconomical proposition; therefore two cranes are provided, one of about one-third of the span over the generator the other about two-thirds the span over the turbine. The duty of these cranes could be halved if they were duplicated and loads, therefore, spread over two bays.

For a station with a diagonal turbine arrangement a special crane is used for rotor withdrawal. This crane has an additional travel motor which can be used in conjunction with the longitudinal travel motor to cause the crane hook to travel along the line of the centre line of the turbo-generator.

Where a station has a separate feed pump bay, i.e. mechanical annexe, it is usual to install cranes of sufficient capacity to lift pumps or motors during maintenance. The three-crane concept, described for the turbine house, can also be used in this bay with the same advantages.

1.13.6. **Plant Control Arrangements**

With the advent of unit boilers and turbines, the controls for these were brought together into one room known as the plant control room, sited between the boiler and turbine. The generator and transmission controls were all together in the station control room. However, modern practice now groups all these controls in one central control room.

Both the size and location of this central control room must be taken into account in the early stages of station design. A good position for the room would be between the turbine house and boiler house; but because of construction difficulties and the high ambient temperatures at this point the room is often situated adjacent to the turbine house, either as an integral part of the structure, or separately but with access via a connecting bridge at operating floor level. A separate control room is better from a construction programme point of view, since it allows completion of the room just prior to the commissioning of the first unit.

1.13.7. **High-pressure Pipework**

With the steady increase in set rating and steam conditions which have taken place over the past fifteen years, steam and feed piping have assumed an increasingly important role in basic station design, as shown in Table 5.

The comparatively modest size, thickness and working temperature of the h.p. piping for the 30 and 60 MW standard steam cycle units allowed the relative disposition of boilers and turbines to be settled with the minimum regard to piping requirements, since the pipework was sufficiently flexible to cause little difficulty in selecting suitable routes as a secondary design issue, that is after disposition of the main units has been fixed. With the advent of reheat units, greater attention had to be paid to pipework layout, not only because of the increase in pipe stiffness and total cost but also in order to reduce the reheater pipe lengths to a minimum to avoid incurring too great a pressure drop between the boiler and turbine. To obtain the necessary degree of flexibility in the connections for the larger

TABLE 5

	Main steam			Reheater inlet			Reheater outlet			Feed		
	No.	Bore, in	Approx. thickness, in	No.	Bore, in	Approx. thickness, in	No.	Bore, in	Approx. thickness, in	No.	Bore, in	Approx. thickness, in
30 MW	2	8	$\frac{3}{4}$	—	—	—	—	—	—	1	6	$\frac{17}{32}$
60 MW	2	9	$\frac{3}{4}$	—	—	—	—	—	—	1	8	$\frac{3}{4}$
	or 1	12	$\frac{15}{16}$									
120 MW	2	8	$\frac{17}{16}$	2	15	$\frac{9}{16}$	2	16	$\frac{11}{16}$	1	10	$1\frac{7}{8}$
200 MW	2	8	$2\frac{1}{2}$	2	18	$\frac{3}{4}$	2	18	$\frac{25}{32}$	2	$8\frac{1}{2}$	$2\frac{1}{8}$
550 MW	4	9	3	4	20	1	4	20	$1\frac{5}{8}$	4	10	$2\frac{1}{2}$
(2×275)						nominal			minimum			
						bend			bend	4	8	
500 MW	4	9	2·625	4	17	$\frac{11}{16}$	4	17	1·15	2	12	—

units, the pipe runs have to be carefully designed to ensure that the permissible thrusts on boiler and turbine are not exceeded. (*Note:* These limiting thrusts will be stated in another chapter.) This influences the layout considerably and the relative position of the boiler and turbine are determined very largely to meet the required pipe layout.

1.13.8. **Boiler House Arrangement**

The boiler house layout required to give the optimum arrangement of the interconnections depends on the boiler design and set rating and varies from one station to another. Wherever possible the boiler house basement level, usually at ground level (see Section 1.12.4.1) is arranged to coincide with the turbine house basement level, and the boiler operating floor level is adjusted where possible to coincide with the turbine room operating floor level. Open access gangways to boiler operating points, such as burners, etc., are required only at operating floor level.

The principal dimension to be fixed in the layout of the station, is the distance between the centre lines of two adjacent boilers. This is important because not only does it control the overall length of the boiler house, but it also influences the dimension between the sets in the turbine house since it is preferable to keep the centres the same to ensure repetition in design. This dimension is usually arrived at by placing two boiler units as close together as possible consistent with there being adequate space for construction, operation and maintenance and sufficient clearance for the withdrawal of the long stroke sootblowers. With the larger boiler units, the number of pulverised fuel mills and the space required for their maintenance has been the factor determining the boiler centres.

Another factor which must not be overlooked is the need to keep the main boiler flues to the precipitators as short and as straight as possible. In some instances the precipitator plant width is in excess of the minimum boiler pitch, therefore some opening out is necessary to accommodate them.

The depth of the boiler house, i.e. the dimension from front to back, is arranged to provide sufficient clearance for construction, operation and maintenance.

When considering clearances round a boiler, adequate space should be allowed for ventilation, particularly where the heat rising from a boiler surroundings is radiated and concentrates at the top of the boiler. Ventilation can be assisted by locating the forced draught fan suction in this area in order to collect the hot air.

The choice of position of the forced draught fans is influenced by the particular arrangement of boiler and air heaters and the associated air and flue gas ducting. Though it is sometimes a more economical arrangement of ducting to locate the fans at a higher level, they are arranged at basement level to avoid vibration troubles in the supporting steelwork.

The majority of recently designed power stations have a bunker capacity sufficient to steam the station at continuous maximum rating for 12 h. Closely associated with the bunkers are the pulverised fuel mills, which are fed by short travel scraper conveyors below the bunker outlet for feeding coal to each mill situated immediately below. The mills are, in turn, connected to the burners on the boiler by fuel pipes and it is desirable to keep these as short as possible.

Some typical boiler house arrangements are shown in Figs. 1.13.3C and D.

1.13.9. Auxiliary Switchgear

Where possible the auxiliary switchgear and transformers should be as near to the centre of gravity of the load as practicable. For this the obvious choice is the mechanical annexe; however, due largely to this space being occupied by feed heating plant and the need to keep cabling costs between switchgear and transformers to a minimum this is not always possible. A further consideration is the need to protect the switchgear from any other plant which may give trouble, or for such plant that can be endangered by any switchgear faults that may develop such as the fire risk associated with switchgear having oil circuit breakers instead of the more popular air breakers.

The practice has developed of locating the switchgear and transformers in an annexe that is an integral part of the main station buildings. Their location usually is immediately adjacent to the turbine house for controlling turbine auxiliaries and at the rear of the boiler house for controlling the boiler auxiliaries.

These positions are shown in Figs. 1.13.3A and B.

1.13.10. Operating Staff Offices and Amenities

When the administrative block and canteen is separate from the station it is necessary to provide office accommodation for the Operating Superintendent and Shift Charge Engineers together with messing, locker and toilet accommodation for the station operating personnel, within the confines of the station main buildings. This accommodation is often incorporated in the turbine house and boiler house electrical annexes or in other suitable situations (see also Section 1.12.10).

1.13.11. **Building Volume**

As mentioned in section 1.13.2, it is the designers' aim on the grounds of economy to keep the building dimensions to a minimum consistent with good facility for operation and maintenance. It does not, of course, follow that a smaller building is a cheaper one; it may be more expensive to have a number of different roof heights than one common level. With these limitations in mind, it has become common practice to compare building costs for different installed capacities, on a volumetric basis, expressed as cubic feet per kilowatt installed. Big reductions in building volume per kilowatt have been achieved recently but it must be emphasised that these reductions are due primarily to reductions in plant volume per kilowatt and only partly to refinement in station design. Table 6 shows typical figures of "specific building volume" for stations of different types.

TABLE 6

Unit size, MW	Station size		Boiler house, ft³/kW	Turbine house, ft³/kW	Station, ft³/kW	Remarks
	Sets MW×I	Boilers k×lb/h				
200	5×200	5×1400	16·7	11·1	27·8	
350	4×350	4×2350	14·9	8·8	23·7	
550	2×550	2×3750	12·8	8·8	21·6	Cross-compound 2×275
500	4×500	4×3450	11·87	7·42	19·29	Longitudinal turbine arrangement
500	4×500	4×3450	11·21	7·18	18·39	Transverse turbine arrangement

Although the absolute cost of a given building might be, say, 3/6*d*. per ft³ the incremental cost of extending the span of the building by 2 or 3 ft may be only in the order of 1/- per ft³. This yardstick, therefore, must be used with caution, although it cannot be too strongly emphasised that unnecessary space involves unnecessary steel and concrete and, therefore, unnecessary expenditure.

1.14. FUTURE POWER STATION LAYOUTS AND TRENDS

In the two preceding sections it has been broadly shown how a site is developed once it has been approved by the Board and the Minister of Power has given his consent to the construction of a power station.

The grouping of the various individual buildings and external plant items on site and the disposition of the major plant, or groups of plant, within the main station building, leading up to a determination of the shape of the building and its leading dimensions, has been based on the Board's experience so far on those stations designed and constructed by them.

That the present main buildings are nearly always rectangular in shape and associated cooling towers (if any) are hyperbolic concrete shells does not necessarily mean they will remain so.

Many attempts have been made to change the "inline" arrangement of plant, as described in sections 1.13.1 to 1.13.4 inclusive and 1.13.8 to improve good operational control, access, plant availability and clear lines of communication. As outlined earlier in the chapter, layout should be an optimisation of construction, operation, maintenance and administration requirements as well as the capital costs.

One way of satisfying these demands is a "radial" layout for a four unit station. Such an arrangement contains an operational focal point—a central complex containing the control room, auxiliary switchgear, workshops and water treatment plant. Grouped around this central complex are the four turbines and the four boilers.

Some disadvantages with the scheme appear to be the need for two chimneys and the difficulty in securing a satisfactory environment for control room personnel, i.e. freedom from dust, noise and vibration.

If unit sizes continue to increase, and there is no sign at the present time that they are not following an upward curve, then layout problems associated with such units will also increase. The search for economy in both site and plant layout, bearing in mind all the requirements to be optimised in attempting to achieve a satisfactory result, must continue and it will be appreciated that there are no simple solutions to the problems associated with layout.

Since no difficulty has been encountered in the manufacture and installation (in America) of tandem turbines of up to 660 MW, and no limitations in this respect are foreseen by the manufacturers it would seem possible that even larger tandem turbines will be built in the future to take advantage of the lower overall cost of this type of turbine against the cross-compound type (see Section 1.13.3). The cross-compound is reputedly easy to operate and offers a slightly higher overall efficiency. It is not possible, however, to predict with any certainty what the outcome of the tandem versus cross-compound turbine controversy will be for large units over 500 MW.

Similarly on boiler design the differences of opinion persist with the steam conditions being a major factor. The Board uses balanced draught furnaces, i.e. forced and induced draught fans, but it is known that in America there is now almost universal acceptance of the pressurised furnace which eliminates the I.D. fans and consequently affects layout.

In cooling tower design the aim has been to reduce their visual impact and to this end efforts are being directed to produce a design of comparable efficiency, but with a decrease in physical dimensions over the normal hyperbolic towers and, if possible, a reduction in the numbers required by increasing the individual tower capacity. So far attempts made in this direction have been on forced draught cooling as opposed to natural draught. This, however, reduces the "sent out" capacity of the station because of the increase in auxiliary power required for the fans. Another scheme is the construction of a large open top dome, possibly in steel with lightweight cladding, which is circular in plan and semi-elliptical in elevation. This tower, on which research is continuing, works on the principle of the "dry tower" explained in Section 1.12.5.3 and present thoughts are that the cooling surface contained within the shell would be sufficient for a 2000 MW capacity station; this structure is being known as an "ellipsoidal cooling tower".

Research work continues on the freon cycle, fuel cell, MHD (magnetohydrodynamic) and other methods of electrical generation. The possibility of utilising the natural gas supplies found off our coasts as an alternative fuel is also being investigated.

APPENDIX 1

RIVER BLACKWATER—COOLING CAPACITY

Object

The object of this study is to assess the cooling capacity of the River Blackwater when warmed by cooling water discharged at Bradwell, and from the results to determine the generating capacity that can be installed on this site without leading to excessive rises in the overall temperature of the river.

Local heating of the water in the vicinity of the intake which can affect the thermal efficiency of the station is not taken into account. This can be minimised by the careful location both horizontal and vertical of the relative positions of the cooling water intake and discharge works, the position of which can be checked on a hydraulic model.

Description of the River

Bradwell is situated on the south bank of the tidal estuary of the River Blackwater at a point slightly upstream of its entry into the North Sea.

The River Blackwater rises in the north-west corner of Essex. Near Maldon, about 32 miles from its source, it is joined by the River Chelmer and then enters a broad tidal estuary some 10 miles long. Seawards of the site there are wide and extensive mud flats enclosing a well-defined channel some 3000 ft wide at low water which is joined 5 miles below Bradwell, on the opposite bank, by the tidal estuary of the River Colne. For further details of the configuration of the estuary, reference should be made to Figs. A.1 and A.2 (C.E.A. Appendix 1 (a) and (b)).

The gathering grounds of the river are all in a flat and predominantly agricultural area which is one of the driest in the British Isles. Records of flow are available for the complete river system. Measurements being taken at Langford on the Blackwater, at Hatfield Peverel on the Ter and at Rushes Lock on the Chelmer.

Table A.1 gives (1) the mean, maximum and minimum monthly natural flows to the tidal estuary over the period 1937–53, and (2) similar figures for the river temperature at Paglesham which have been recorded by the Shell Fisheries Research Laboratory over a somewhat shorter period.

TABLE A.1. RIVER INFORMATION

1. *Natural flows* (source of information—*Surface Water Year Books*). *Figures in million gallons per hour*

Month	Average	Max.	Min.
Jan.	4·95	14·6	1·0
Feb.	5·53	14·1	0·9
Mar.	3·67	8·4	0·7
Apr.	2·36	6·8	0·8
May	1·30	3·9	0·6
June	0·89	3·5	0·4
July	0·63	1·6	0·3
Aug.	0·60	11·3	0·2
Sept.	0·54	1·0	0·3
Oct.	1·00	5·0	0·4
Nov.	2·78	10·0	0·6
Dec.	3·03	7·4	0·8
Annual	2·27		

2. *River temperatures* (recorded at Paglesham by Shell Fish Laboratories) *in °F*

Month	Average	Max.	Min.
Jan.	36	45	28
Feb.	37	47	27
Mar.	42	52	30
Apr.	49	59	42
May	57	69	48
June	63	73	54
July	63	76	57
Aug.	66	73	60
Sept.	60	70	51
Oct.	53	60	44
Nov.	46	54	35
Dec.	39	49	32

3. *Tidal range* (source—Admiralty Charts, etc.) in ft

	Springs	Neaps
Tollesbury	15·3	9·7
Osea Island	15·8	10·1
Heybridge	12·0	8·0

WATER ABSTRACTIONS (*Water Engineers Handbook*)

Chelmsford M.C.	1·6 million gallons per day (returned as sewage effluent) Storage 5·62 million gallons
Southend Waterworks Co.	3·0 million gallons per day (average) Storage 85 million gallons

The natural flow is, therefore, unlikely to be appreciably affected by extractions by the authorised water undertakers.

TIDAL LEVELS

Tidal data taken from Admiralty Charts are given (Figs. A.1, A.2) for Tollesbury (Mill Creek) and Osea Island. Above Osea Island there is little information available on tidal levels and the information has been obtained from yachting handbooks published by Imray, Laurie, Norrie & Wilson Ltd., and Reed's *Nautical Almanack*. Table A.1(3) includes details of the tidal rises assumed. At intermediate points a linear change in rise has been assumed.

From an examination of the Admiralty data and the initial hydraulic work supported by the views of the local fishermen, it was thought that the tidal ebb and flow past the station was the predominant water movement.

Later studies by oceanographers, biologists and engineers indicate that an exchange of water between the sea and estuary does take place; this is probably due to the channel configuration associated with water drainage from the shallow mud flats possibly added by the coriolis effect.

Method

If the tidal flow is predominant at Bradwell and little heat is carried away from the site downstream by fresh water flow in the river, the heat input must be dissipated mainly from the surface. To calculate this loss it is necessary to assess the surface area of both the upstream and downstream cooling ponds under Spring and Neap conditions from which this heat escapes.

For the calculation, it is assumed that plane vertical sections across the river remain vertical. This has been shown from previous calculations on the Thames, where temperature rise and heat input were known to be a reasonable method capable of giving realistic assessment. Float tests carried out on the Crouch have also produced evidence to support the validity of this assumption. In the initial study, except for the fresh water flow, no loss of water at the sea end of the plug is assumed. The second set of calculations allow a 10% exchange with the water from open sea.

The volumetric tidal flow past Bradwell has been calculated from Fig. A.1 in conjunction with the information on tidal ranges contained in Table A.1. The chart of the river has been divided into approximately equal sections of 5 cables (1000 yd approx.) in length and the surface area at LWS and HWS assessed by measurement with a planimeter. Readings of levels on the mud flats between high water and low water mark are very sparse, so it is not possible to plot the configuration of the banks at the section.

A straight bank of constant gradient has therefore been assumed and on balance this probably does not introduce a serious margin of error. The increase in volume for each section above the station can then easily be calculated for both springs and neaps and summated.

To calculate the surface area of the upstream cooling pond the volume below low water mark must be assessed, and cross-sections have been drawn for this purpose. If the volume between high water and low water and that below low water are summated, the progressive total volume at HW moving upstream from Bradwell can be calculated. The surface area and volume are then plotted as in Fig. A.3 from which the length and surface area of the cooling pond can be obtained by cross-reference of the curves. Similarly the figure for the downstream pond can be calculated from plotted cross-sections at low water as in Fig. A.4.

The four values calculated and the average surface area of the cooling pond in million square feet are as follows:

	Springs	Neaps
Upstream	222	165
Downstream	218	116
Average	170	

HEAT BALANCE

The calculations are based on the fact that:

Heat rejection = loss from surface + loss due to natural flow carried away by displacement at the seaward end of the tidal plug.

This is expressed mathematically by the formula:

$$H = Ak(I_m - I_{nr}) + 10Q(T_m - T_{nr})$$

where H = total heat rejected per hour,
k = constant depending on wind velocity and meteorological conditions,
A = surface area of cooling pond,
I_m = total heat per lb saturated air at elevated river temperature T_m,
I_{nr} = total heat per lb saturated air at natural river temperature T_{nr},
Q = flow in million gallons per hour.

In this equation, H, A, I_{nr} and T_{nr} are known, and the relationship between I_m and T_m can be obtained from hygrometric tables. The equation can therefore be solved for T_m and calculations made of the temperature rise under average, maximum and minimum flow conditions for each month of the year.

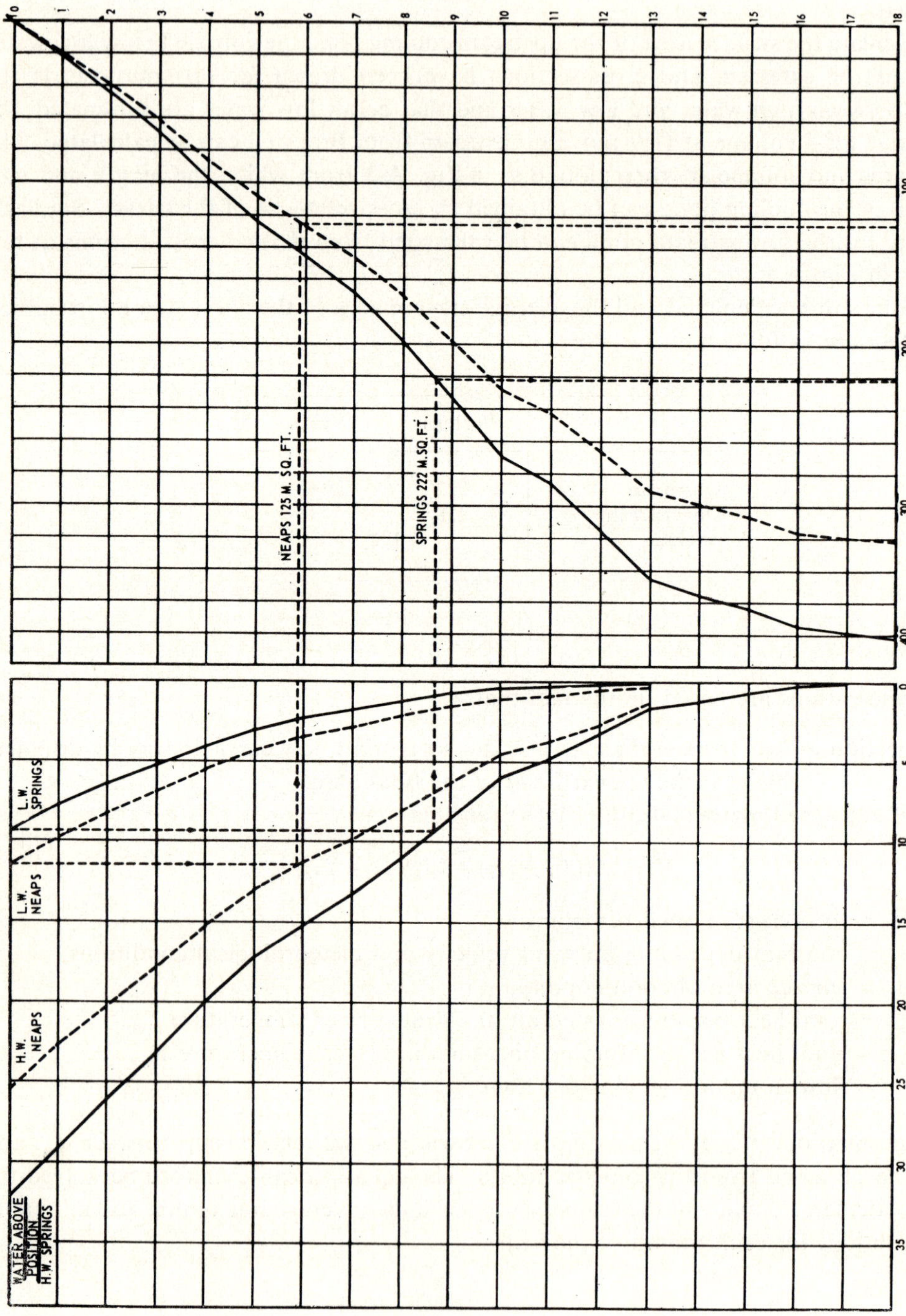

FIG. A.3. River Blackwater upstream cooling pond.

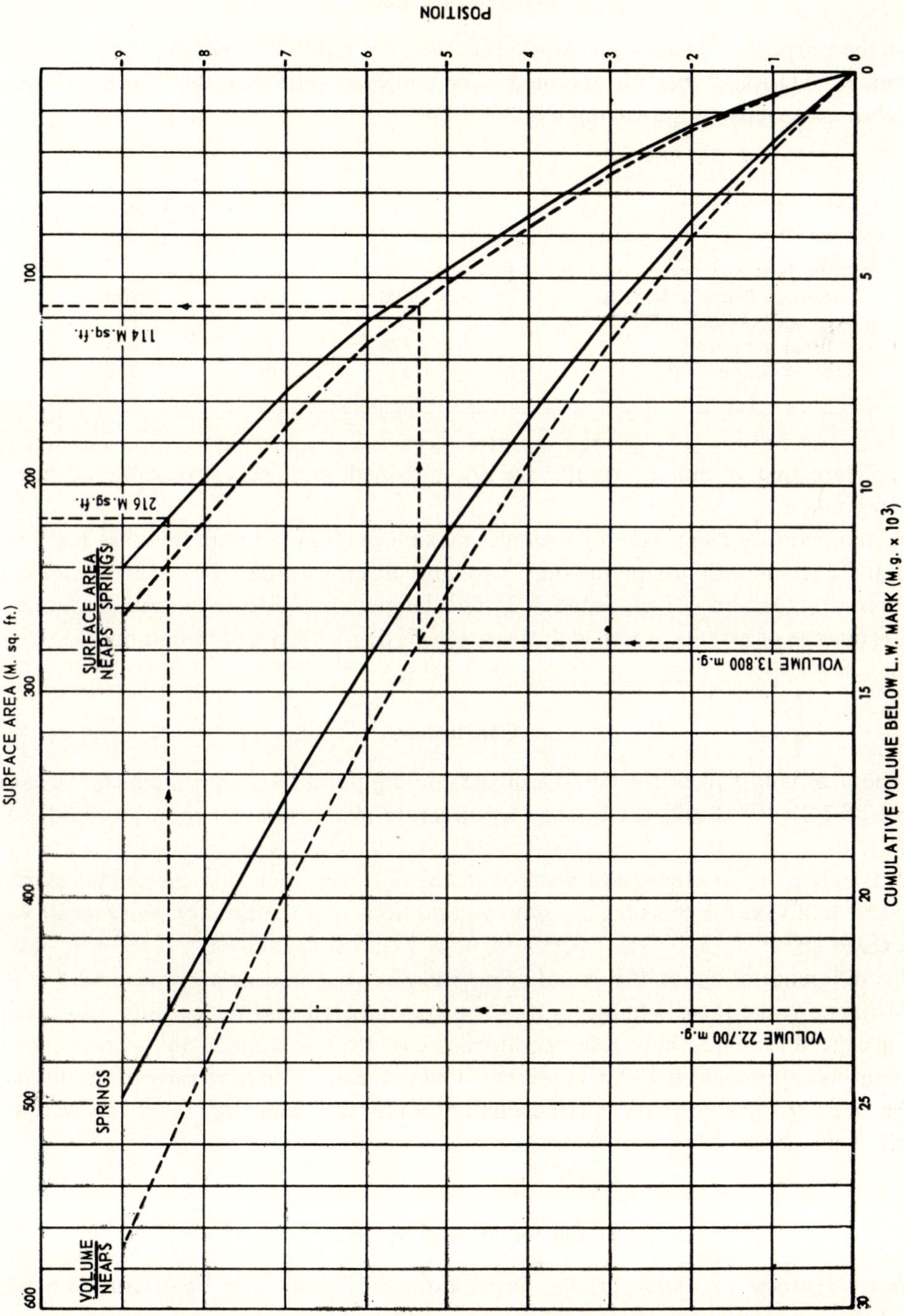

FIG. A.4. River Blackwater downstream cooling pond.

Temperature Rises

For the purpose of this simple study the overall temperature rise has been calculated in terms of "stations". Each is assumed to be a nuclear station of the "Magnox" reactor type with a capacity of approximately 350 MW.

	1 station	2 stations	3 stations
Total heat rejection, million Btu/h at C.M.R	2600	5200	7800
Average cooling surface area, million ft²	170	170	170
Heat rejection, ft²/h	15·3	30·6	45·9

To ensure that optimistic results are not obtained, a conservative value of k (= 10) has been taken.

The temperature rise assuming average, maximum and minimum flow has been calculated for each month assuming one, two and three stations. These are recorded in columns (2), (3) and (4) of Table A.2. The temperature rise with three stations and 10% exchange of estuary water with the sea each tide is recorded in column (5).

Conclusions

If the heat is lost from the surface of the cooling pond only with a constant heat discharge of $7{\cdot}8 \times 10^9$ Btu/h, the estuary temperature in the summer months could exceed 80° F.

With only a 10% exchange of water with the open sea each tide the temperature rise would be half that for the surface cooling conditions and the estuary temperature would not exceed 80° F.

This difference of about 4° F could be critical when negotiating a licence with the River Authority to make the discharge; it must be remembered that the cooling water passing through the station may have a temperature rise of 18° F and under the worst conditions temperatures approaching 100° F could be found in the immediate vicinity of the discharge.

The need to establish beyond reasonable doubt the tidal régime of any estuary is clearly demonstrated.

Table and Supplementary Information

Table of Hygrometric Data for Air, Institution of Heating and Ventilating Engineers.
C.E.R.L. Laboratory Report, No. 890.
The Cooling Power of a Lake, G. Spurr.

TABLE A.2

Temperature Rise of River Blackwater Figures in °F
1 station

(1) Flow	(2) Av.	(3) Max.	(4) Min.	(5) 10% tide exchange
Jan.	3·2	3·4	2·9	
Feb.	3·1	3·3	2·8	
Mar.	3·0	3·1	2·8	
Apr.	2·7	2·7	2·5	
May	2·3	2·3	2·2	
June	2·0	2·0	2·0	
July	2·0	2·0	2·0	
Aug.	1·9	1·9	1·9	
Sept.	2·2	2·2	2·2	
Oct.	2·5	2·5	2·4	
Nov.	2·8	2·9	2·6	
Dec.	3·1	3·2	3·0	

2 stations

Jan.	6·3	6·5	5·6	
Feb.	6·1	6·4	5·6	
Mar.	5·8	6·0	5·5	
Apr.	5·3	5·3	5·0	
May	4·5	4·5	4·4	
June	4·0	4·0	3·9	
July	4·0	4·0	3·9	
Aug.	3·8	3·8	3·7	
Sept.	4·2	4·2	4·2	
Oct.	4·8	4·9	4·7	
Nov.	5·4	5·6	5·1	
Dec.	6·1	6·3	5·8	

3 stations

Jan.	9·1	9·6	8·3	3·5
Feb.	9·0	9·4	8·2	3·5
Mar.	8·4	8·7	8·0	3·3
Apr.	7·6	7·7	7·1	3·2
May	6·6	6·6	6·4	3·0
June	5·8	5·8	5·7	2·8
July	5·8	5·8	5·7	2·8
Aug.	5·5	5·5	5·4	2·7
Sept.	6·2	6·2	6·2	2·9
Oct.	7·1	7·1	6·9	3·0
Nov.	8·0	8·1	7·4	3·2
Dec.	8·9	9·2	8·3	3·5

APPENDIX 2

TYPICAL DIRECT-COOLED CW SYSTEM

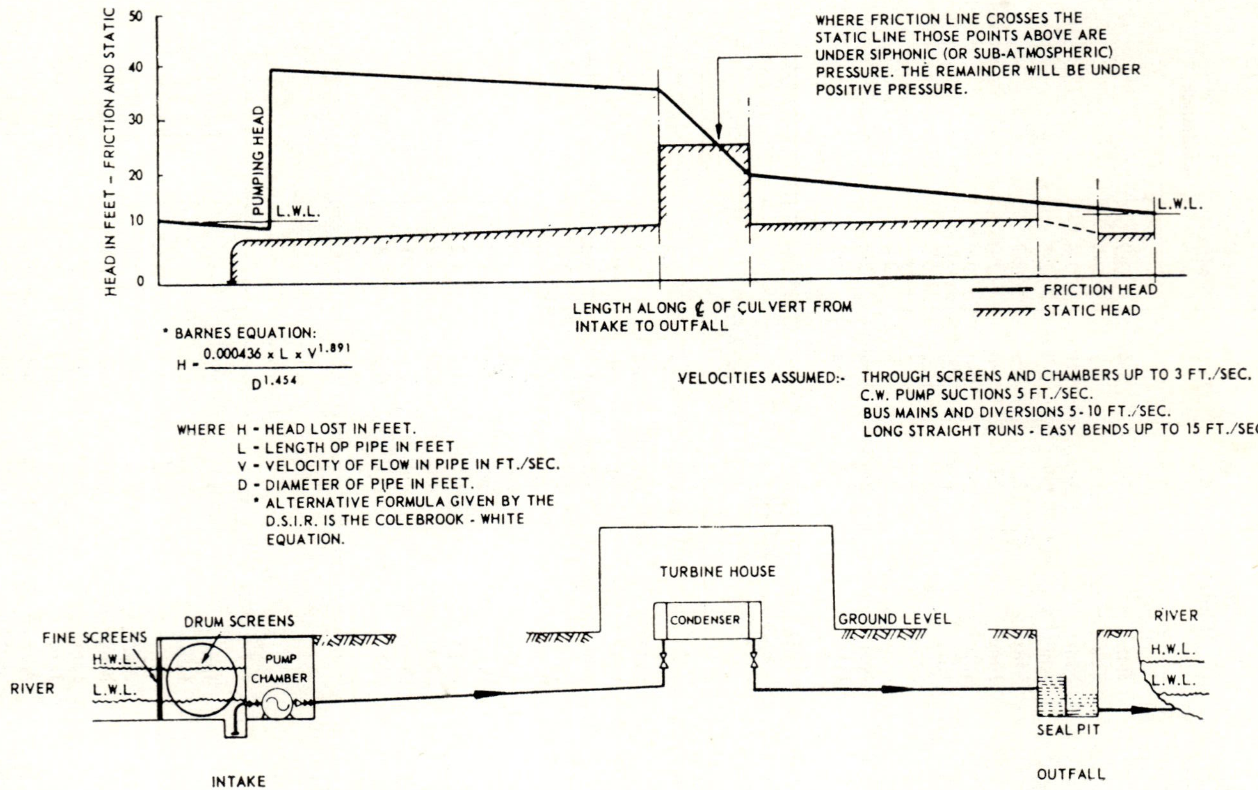

Fig. A.5. Hydraulic gradient for a typical direct-cooled CW system.

APPENDIX 3

TYPICAL CLOSED COOLING TOWER CW SYSTEM

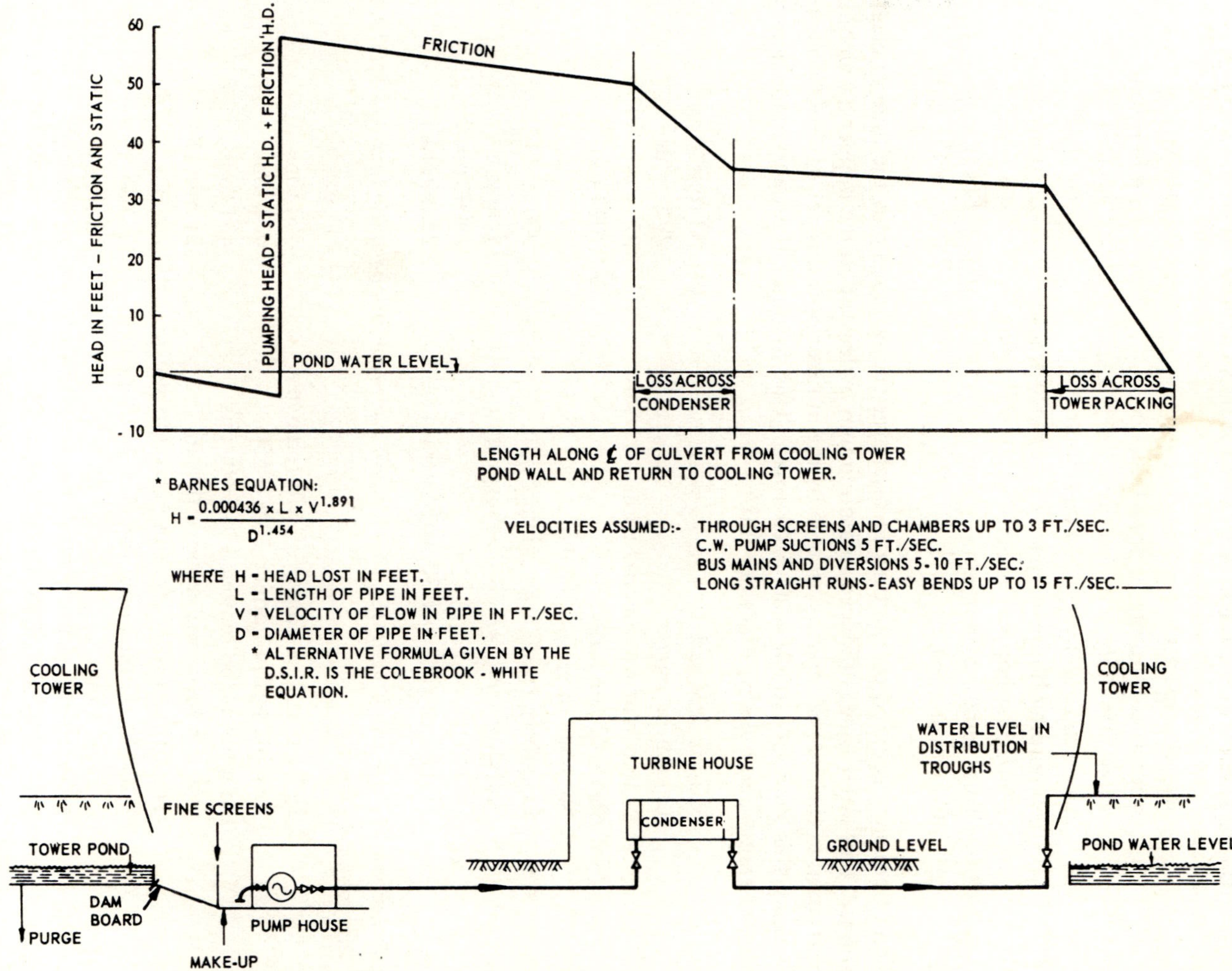

FIG. A.6. Hydraulic gradient for a typical closed cooling tower CW system.

APPENDIX 4

TIMING CALCULATION FOR A RAILBORNE COAL DISCHARGE CYCLE AT A TYPICAL 2000 MW STATION

Station capacity	2000 MW
Annual consumption, approximate	5·1 million tons
Maximum daily consumption, 100% L.F.	17,900 tons
Mean daily delivery, 250 days/year	20,400 tons
Unloading hopper capacity using two tracks	1220 tons
Unloading hopper capacity using one track	808 tons
Total length of hopper	178 ft
Conveyor capacity	2×1500 tons/h

Train frequency is based on the time to discharge the wagons, plus the time to clear the hopper. This latter time is reduced to a minimum by leaving in the hopper an amount of coal equal to the hopper capacity divided by the number of trains per day. With one track working this amounts to 808 ÷ 20 = 40 tons. (This assumes the day commences with hopper empty and ends with hopper full.)

Wagon payload	32 tons
Length of wagon	29 ft 8 in
Number of wagons per train	32
Payload per train	1024 tons
Number of trains per day	20
Length of train, approximate	950 ft
Speed of train whilst discharging	$\frac{1}{2}$ m.p.h.
Discharge rate	2800 tons/h
Discharge time per 32-wagon train	22 min (X)

Discharging on one track with one 1500 tons/h conveyor in operation and one 1500 tons/h conveyor as a standby

In 22 min, one conveyor will remove	550 tons
Residue left in hopper 1024 − 550	474 tons
Amount to be removed less 40 tons	434 tons
One conveyor will remove this in	21 min (Y)
Total time for a 1000-ton payload train to discharge plus time to clear the hopper 22 + 21 min	43 min
Minimum time for daily intake, i.e. 20 trains	14 h 20 min
Minimum interval between successive trains arriving at the track hopper, allowing 15 min running time for B.R. on site	58 min

NOTE: The theoretical time to discharge the train moving past a fixed point at $\frac{1}{2}$ m.p.h. is X and does not include the automatic weighing time Z, which is the time for covering the distance at $\frac{1}{2}$ m.p.h. from the leading end of the gross weighbridge to the trailing end of the tare weighbridge plus the length of the train.

Therefore the maximum total time to discharge and weigh the train is determined by $(X+Y)$ or Z, whichever is the greater.

BIBLIOGRAPHY

The following papers and articles contain the best information available:

Experience in planning and siting nuclear power stations for the C.E.G.B., by F. Faux and G. N. Stone. I.A.E.A. Symposium, Bombay, March 1963, pp. 435–455.

The duty and development of modern power, by F. H. S. Brown. I. Mech. E. paper, 12 December 1962, L3/63.

The evaluation of power reactor sites, by F. R. Farmer. U.K.A.E.A., DPR/INF/266.

400 kV grid system for England and Wales, by E. S. Booth, D. Clark, J. L. Egginton and J. S. Forrest. *Proceedings of the I.E.E.* Vol. 109, Part A, No. 48, December 1961.

Preserving amenities, by Sir C. Hinton and Sir W. Holford. Royal Society of Arts, 25 November 1959. C.E.G.B., G. 192.

Transport of fossil fuels to C.E.G.B. power stations, by J. Laing. *Electrical Review*, 18 June 1965.

Power station layout. *Electrical Times*, 10 February 1966, pp. 207–209.

Fuel Policy. H.M.S.O. Command Paper 2798.

The National Plan, Chapter II, *Energy*, pp. 118–125. H.M.S.O. Command Paper 2764.

Articles dealing with individual projects:

Dungeness "B". A.G.R. nuclear power station. C.E.G.B., July 1965, G. 384.

Dungeness "B". An appraisal of the technical and economic aspects. C.E.G.B., July 1965, G. 392.

The Ffestiniog pumped-storage development, *Water Power*, Vol. 15, pp. 185–191, May 1963, pp. 250–257, June 1963.

Ferrybridge "C". *Electrical Review*, Vol. 177, pp. 598–601.

West Burton. *Electrical Review*, Vol. 177, 26 November 1965.

132 kV and 400 kV totally enclosed switchhouses at West Burton. *Electrical Review*, 25 June 1965.

Annual Reports of organisations dealing with amenity, i.e. Nature Conservancy, National Parks Commission, Council for the Preservation of Rural England and the Royal Fine Art Commission.

EXERCISES

1. In preparing a "Development Plan" the first step is to estimate the demand in kilowatts and kilowatt hours each year covered by the Plan.

State clearly the purpose for which each estimated quantity is used; the factors taken into consideration when estimating the amount of generating plant to be installed on the Board's system and how you would estimate the capacity of plant to be installed in any one year.

2. To which Minister does the Board make its annual report and who gives consent to the construction of power stations?

In arriving at his decision, how does the Minister ensure other Ministries can make their interest known and by what process are the Planning Authority's views obtained? What decision can the Planning Authority force the Minister to make?

3. If you were planning an electricity system with a similar fuel economy to the C.E.G.B.'s what functions would gas turbines, nuclear and conventional power stations have? What are the facts which lead to this conclusion? Illustrate by figures given in the chapter.

4. What are the main technical features you would look for when selecting power station sites for conventional and nuclear power stations? State the reasons why you would look for these particular requirements and the one you would expect to have the most difficulty in providing inland. What means can be used to overcome this difficulty?

5. Over the last decade, amenity considerations have assumed an increasing importance in selecting power station sites and overhead transmission line routes. State (a) two basic reasons why this is so, (b) give two examples of the types of amenity which can be affected by the Board's activities, and (c) name one statutory and one private body who is active in the amenity field.

6. A site layout is being prepared for a four-unit, 2000 MW station adopting a cooling tower system. Each tower is 300 ft in diameter and rated at 250 MW, and the prevailing wind on the site is south-westerly. Illustrate by sketches:

(a) Where should the towers be positioned, relative to the switch compound, to avoid spray nuisance to the switchgear?
(b) What is the minimum pitch of adjacent towers and how close could they be sited to the main buildings?
(c) For minimum generator/switch station costs, where is the ideal position for this switchgear and what type of switch station, assuming it to be in an area of heavy atmospheric pollution, would be preferred?
(d) The annual coal burn of such a station is 5 million tons. What would the capacity of the coal stores have to be?
Assuming this store is stacked to a depth of 30 ft (the coal density is 40 ft^3/ton) with repose angle of 45° what would be the area of land required in acres?
(e) If each turbine requires $13{\cdot}5 \times 10^6$ gal/h of cooling water, what is the daily CW requirements for the station? If the purge/evaporation ratio is 2 : 1 and quantity of water required for make-up purposes is 3% of the total quantity circulated, what is the percentage and quantity in gallons per hour lost in evaporation?

7. Examine the power station layouts in this chapter.

(a) Give your views on the advantages and disadvantages of these layouts and possible improvements.
(b) Prepare an outline sketch plan at a scale of not less than 1 in 2500 of your proposals for a coal-fired, four-unit station of 2000 MW, having a single chimney, tower-cooled and connected to the 400 kV grid; with principal dimensions.

8. Describe briefly the method of determining the following dimensions for a turbine house containing a transverse arrangement of 4×500 MW turbine generators:

(a) Width.
(b) Length.
(c) Operating floor level.
(d) Overall height.

Illustrate by sketches two other possible arrangements of turbines and discuss their respective advantages and disadvantages. Give any other arrangement that you consider will be practicable.

CHAPTER 2

CIVIL WORKS: FOUNDATIONS FOR BUILDING AND PLANT

2.1. INTRODUCTION

This chapter covers mainly structures of reinforced concrete, which are constructed at, or below, ground level. Methods of soil investigation are reviewed first, followed by a general description of the basic types of foundations which may be used; typical examples of the reinforced concrete foundations of main and secondary structures are then discussed. The next section considers those general civil engineering site preparatory works which are not classified as "foundations" in the narrow sense, and, lastly, as far as land-sited works are concerned, the chief methods of construction are given.

The river and sea works normally associated with cooling water systems are then considered, followed by an outline of the works and methods of construction required for wharves where stations are fuelled by colliers and tankers. The chapter concludes with some notes on civil engineering contracts.

2.2. SOIL INVESTIGATION

2.2.1. General

Before final decisions can be made on the layout of the site and buildings and the detailed design of the foundations undertaken, it is necessary for the engineer to have adequate information about the subsurface strata. With this knowledge, and within the limits permitted by other major factors such as railways, roads, rivers and architectural and local amenity requirements, the layout of the site as a whole becomes a problem whose satisfactory solution is closely related to the best combination of the type, location, size and depth of the foundations for the various structures. The results of the site investigation may be sufficiently important to influence the design and choice of the type of superstructure. The site investigations may indicate difficulties which would normally make the site unsuitable for the construction of a large and complex building such as a power station, but these considerations may be overridden by the availability of cooling water, access for fuel and position relative to centres of power demand.

Information on the previous use of the site should be obtained to assist in the design, and also on the design and behaviour of any structures adjacent to the site.

In mining areas, the existence of old, new and prospective workings should be investigated and arrangements made if necessary to have pillars or islands of coal left under the site to avoid subsidence.

The geological survey will provide information about the main strata in the district, which may indicate any major difficulties. To provide the closely detailed information necessary for design, an investigation by trial pits and boreholes must be made of the soils beneath the site. The cost of a thorough investigation of the kind is not great, especially when compared with the total cost of foundation construction, and it is essential to enable an economic design to be prepared.

2.2.2. Trial Pits

These enable the soils or rocks to be sampled and examined *in situ*. They are preferable to boreholes in this respect, but are limited to shallow depths (10 ft or so). At depths over 20 ft, boreholes are more common, as the cost of supporting the sides and dealing with ground water in pits becomes excessive. Deep pits may be the only method for satisfactory examination of the surface of rock where it is badly disintegrated or where fissures exist.

2.2.3. Trial Bores

Initially a few boreholes are put down at a spacing of 1500 ft or so over the site. Should the site prove suitable for the construction of a station, further boreholes will be required, and these will be distributed generally around the areas to be occupied by the major structures. The number of boreholes will depend on the complexity of the site subsurface conditions, but should be sufficient to give a full picture of the probable variation of the strata under the site. The number has mounted to 60 on some sites.

The depth of boreholes should be sufficient to include all strata likely to be affected by foundation loads, particularly to prove the level of a bearing stratum for end bearing piles as distinct from friction piles, but again it is not possible to lay down a hard and fast rule. In theory the ground pressure under a foundation is reduced to approximately one-fifth at a depth $1\frac{1}{2}$ times the width or diameter of the foundation; the effect of loading from a piled foundation extends for some distance below the toes of the piles. The presence of a solid strata may limit the depth of exploration, but it is usual to take a few deep bores down to 120 ft.

Various types of boring equipment are available and a medium mobile rig in operation during a preliminary soil investigation for a proposed power station site is shown in Fig. 2.2.3A. The rig folds down and may be towed into relatively inaccessible areas of a site by a Land Rover or tractor.

The borehole is usually formed by a drop tool consisting of a steel tube about 3 ft long and 6 in in diameter, which is raised and lowered by a cable operated from a powered winch. The drop tools are a "claycutter" and a "shell"; the latter has a clack valve in the base and is used for penetrating sands and gravels. An auger may also be used for boring in clay and this is operated on boring rods and is rotated by tillers at ground level. Obstructions and boulders are broken up by chiselling and the fragments are removed

FIG. 2.2.3A. Medium mobile boring rig.

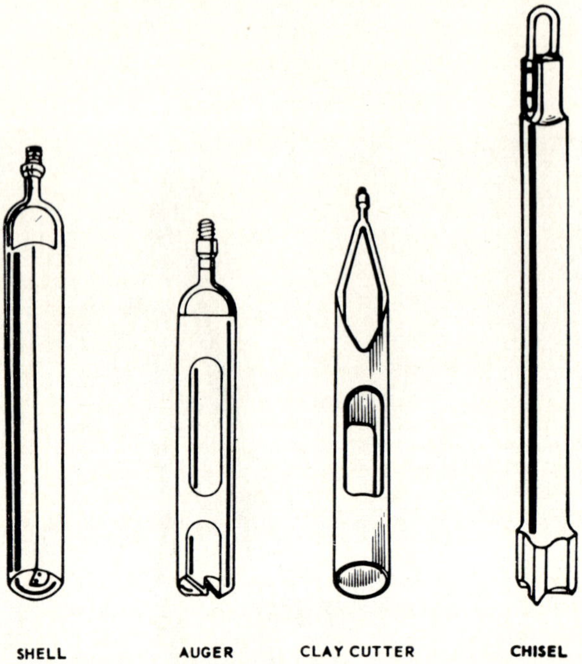

FIG. 2.2.3B. Site investigation boring tools.

by the shell. Boreholes in weak and cohesionless soils must be lined (cased) and tubes with screwed ends are used for this purpose. Boring tools are illustrated in Fig. 2.2.3B.

Note is taken of the depths of succeeding layers of soil, and of the presence of ground water. Samples—called "disturbed samples"—are taken periodically from all the various layers and sealed in air-tight containers for laboratory examination and test, and samples of ground water taken. "Undisturbed" samples of cohesive soils (i.e. clays, silts) are required for certain laboratory tests and these are obtained from the boreholes at intervals. A tube with a cutting edge and a closed top is pushed into the soil in advance of the boring tool. At the top of the tube is a non-return valve which opens when the sample enters and then closes when the tube is raised, thus retaining the soil. The cutting edge and the head of the sample tube are unscrewed and replaced by screw caps so that the moisture content of the sample is unaltered when the tube is received by the laboratory. Before screwing on the caps the ends of the sample are coated with wax as an added precaution to retain the moisture.

It is difficult to obtain undisturbed samples of non-cohesive soils (sand and gravel) from boreholes, although special sampling tools have recently been developed. Undisturbed samples can be obtained from pits by careful use of a sampling tube and hand trowel.

For the exploration of rock, boreholes are made using percussion or rotary drills, and water or drilling mud solutions circulated to clear the rock chippings and recover them for identification. Cores may be recovered using a core-cutter. Alternatively a hollow rotary drill, 2 to 4 in in diameter with a hardened cutting head, is used to cut a continuous core of rock. This method is not always successful in fissured and broken rock, and if necessary deep pits are sunk to the rock surface, or large diameter wells drilled into the rock for *in situ* examination to be made. This procedure, though costly, may be necessary where tunnels are to be driven through rock. An assessment of the probable leakage of ground water through the rock is also essential in this case.

Drilling of boreholes is not confined to dry land. In a direct cooling system with an intake and other river works to be constructed off-shore, boreholes are required to be drilled from temporary stagings or barges, into the river or sea bed.

2.2.4. *In Situ* Soil Tests

There are several soil tests which may be carried out on site. An estimate of the bearing strength of soils and settlement can be made by loading tests carried out in trial pits. A plate is loaded with weights, or by jacking from adjacent anchorages and the settlements measured with increasing load on a basis of time. It must be stressed that this test can only be used as an estimate of the subsequent behaviour of the sub-strata under the full-scale foundation load.

Penetration tests, the most common being the Standard Penetration Test, are carried out on the non-cohesive soils within the boreholes as boring proceeds. In this test the number of blows required to drive a 2 in diameter tube 12 in into undisturbed soil using a 140 lb weight falling freely through 30 in gives an indication of the classification and of the bearing strength of the layer of soil.

The Vane test is used to give a direct *in situ* reading of the shear strength of soft clays. A vane having four rectangular blades is attached to a rod and pushed into the undisturbed ground ahead of the boring tool at various depths. A spring-balance is used to measure the torque required to turn the blades, and thus give a measure of the shear strength.

The Deep Sounding test is used where piled foundations are being considered, to assess the end bearing capacity and skin friction of piles. A cone on a mandrel with an outer moveable tube is made to penetrate steadily under the application of load, or by jacks. Where a large scale piling contract is being considered in difficult ground, the driving and testing of various types of pile as part of the site investigation may become necessary. A pile test is described in Section 2.4.3.

The site investigation of ground water is carried out by measuring the depth of water in observation wells, and by measuring the pressure in the water bearing strata, using pore water pressure cells. These observations will indicate variations of water level with the tide, and the presence of water under artesian pressure. The permeability of the strata can be estimated by pumping tests from boreholes.

2.2.5. Site Reports

Samples taken on site are subjected to numerous laboratory tests, from the results of which, together with the borehole logs and results of site tests, a report is made by a civil engineer having a specialist knowledge of soil mechanics, often with the co-operation of a geologist.

Laboratory tests give information on classification and identification of soils from mechanical and sedimentation tests. Engineering properties of the soil are indicated by tests for shear strength, density, permeability, consolidation and compaction. Analysis of ground water samples will indicate the presence of harmful substances, such as sulphates, in the soil; the degree of impurity must be carefully considered in respect of chemical interaction with cement and aggregate.

The report will enable the type of foundation to be chosen and designed to exert safe bearing pressures on the various strata, and for an assessment of long-term settlement to be made. The stability of the excavations and earth slopes, and the pressures on retaining walls and basements can be calculated; and construction difficulties anticipated and allowed for. Consideration can be given to the methods of construction, i.e. dealing with ground water, the difficulties of excavation and disposal of excavated material, the choice and design of temporary works such as cofferdams, and the necessity for ground treatment. Pre-treatment of the ground by injection of cement, clay or various chemicals can increase the strength or reduce the permeability of various strata, either as part of the permanent works or to facilitate construction.

The report should indicate any major weaknesses of the site due to rock faults, steeply dipping strata, layers of very soft material, and unstable slopes both on site and due to the presence of adjacent excavations. It should enable consideration to be given to the long term affect of any alteration to the sub-strata drainage and water table.

The suitability of available materials for use in embankments and flood banks, for stabilisation in road bases, and for use as concrete aggregates will be indicated.

2.2.6. **Soil Classification**

Table 1 is based on British Standard Code of Practice 2001, Site Investigations.

The type or composite type of material is determined by a sieving analysis for sands and gravels. Finer particles cannot be sieved and the particle size is found by the rate of settlement in water, based on Stokes' Law.

The approximate shear strength applicable to the clays listed in the table are:

Stiff clay	above 2000 lb/ft^2
Firm clay	1000 to 2000 lb/ft^2
Soft clay	up to 1000 lb/ft^2
Very soft clay	200 to 300 lb/ft^2

Rocks cannot be classified in the same way as soils, but the more common are listed below:

Crystalline rocks: Granite, basalt.

Arenaceous rocks: Sandstone, quartzite

Argillaceous rocks: Shale, slate

Calcareous rocks: Chalk, limestone

TABLE 1. CLASSIFICATION OF SOILS

	Type of soil	Particle size and identification	Composite types	Strength	Structure
Non-cohesive soils	Gravels	Between 60 mm and 2 mm	Boulder gravels Sandy gravels	Loose. Can be excavated with spade Compact. Requires pick for excavation Slightly cemented Pick removes soil in lumps which can be abraded with thumb	Homogeneous or stratified
	Sands	Coarse sand. Between 2 mm and 0·6 mm Medium sand. Between 0·6 mm and 0·2 mm Fine sand. Between 0·2 mm and 0·06 mm Particles visible to naked eye. Possess no cohesion when clean and dry. May be uniform or well graded according to particle distribution	Silty sands Clayey sands		
Fine-grained cohesive soils	Silts	Between 0·06 mm and 0·002 mm. Particles invisible or barely visible to naked eye. Very slightly gritty touch. Exhibits dilatancy*	Clayey silts Sandy silts Organic silts	Soft. Easily moulded in the fingers Firm. Moulded by strong pressure in the fingers	Homogeneous or stratified
	Clays	Less than 0·002 mm. Dry lumps may be broken but not powdered. Smooth touch and plastic. Sticks to fingers. No dilatancy. Shrinks on drying, usually showing cracks	Boulder clays Sandy clays Silty clays Organic clays	Very soft. Exudes between fingers when squeezed Soft. Easily moulded in fingers Firm. Can be moulded by strong pressure in fingers Stiff. Cannot be moulded in fingers. Hard. Brittle	Fissured, intact, homogeneous, stratified or weathered
Organic	Peats	Fibrous organic material. Usually brown or black in colour	Sandy, silty or clayey peats	Firm. Fibres compressed together. Spongy. Very compressible and open structure	

* Dilatancy. If a pat of wet silt is shaken in the hand water is exuded. When the pat is pressed the water retreats leaving a matt surface.

2.3. TYPES OF FOUNDATION

2.3.1. Isolated Column Foundations

When the ground is of reasonably good bearing capacity, say about 2 tons/ft^2 and over, columns may be economically founded on isolated concrete bases. These are usually square or they may be rectangular or irregularly shaped because of interference of other bases or underground structures such as culverts. The area of the base is determined by the bearing strength of the ground and the load on the column.

The thickness of the base when made of plain concrete has to be sufficient to keep the stresses in the concrete low. If the base is reinforced, as shown in Fig. 2.3.1, the thickness may be reduced. For example, an 11 ft square base carrying a 2 ft square column at a ground pressure of 2 tons/ft^2 will be about 2 ft 6 in thick if reinforced. In the case of a plain concrete base the thickness should be 1·4 times the distance from the column face to the edge of the base and for the loading conditions above the thickness will be 1·4×4 ft 6 in, say 6 ft 3 in. It should be noted that in low bearing soils the high density of concrete forms an appreciable percentage of the total load and this will lead to a decision on the type of foundation.

In Fig. 2.3.1 "starter" bars are shown concreted into the foundation. These are the same diameter as the vertical reinforcing bars in the column and are lapped with them.

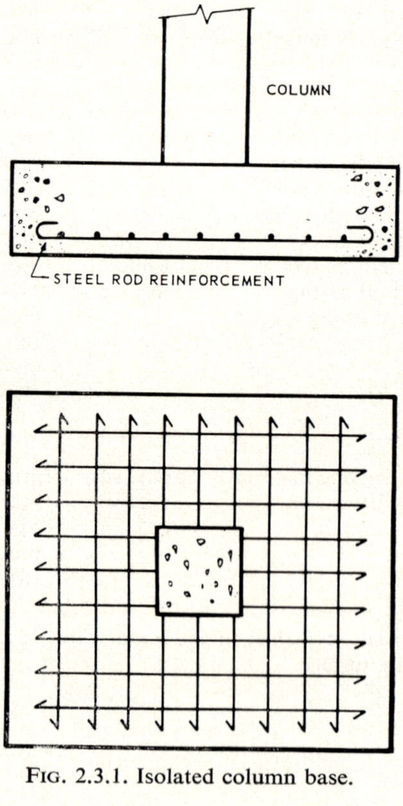

FIG. 2.3.1. Isolated column base.

The length of lap and embedment in the foundation is made sufficient to transfer the load in the column reinforcement to the starter bars and into the foundation as a result of the bond which exists between steel and concrete. For a round mild steel bar in compression this length requires to be about 30 times the bar diameter. This length is dependent on the grade of concrete used and is stipulated in British Standard Code of Practice, CP 114, Reinforced Concrete in Buildings.

The bottom of the foundation should be sufficiently deep to be beyond the influence of frost and the level at which there is an appreciable seasonal variation in the moisture content of clay soils; approximately three feet in each case. For heavily loaded columns and poor soil conditions the foundation will normally need to be deeper than for the above requirements, as bearing strength increases generally with depth.

2.3.2. **Steel Grillage**

This is a special form of isolated base which is used to spread the load from a steel column without considerable depth of construction. The load from the column is taken through the baseplate on to one or more tiers of steel beams; the tiers being at right angles to one another and completely encased in concrete.

A two tier grillage is illustrated in Fig. 2.3.2 and this grillage is capable of carrying 400 tons with a pressure on the soil of 2 tons/ft^2. The edge angles and the gas barrel separators hold the beams in each tier together whilst the grillage is being erected.

Usually steel column loads are transmitted by a thick steel "Bloom" baseplate to a reinforced concrete base. With the very high loads from boiler house columns now being dealt with (in the region of 4000 tons) the use of a grillage to spread the load to a reinforced concrete base becomes economical.

2.3.3. **Strip Foundations**

In its simplest form this is a very common foundation for a brick wall. In this case the loading is less and the foundation is therefore narrower. When the load of the wall is spread through footings as shown in Fig. 2.3.3 the base slab is not reinforced.

For heavy column loads or weaker soils isolated bases may almost touch. In this case it is structurally advantageous to carry a row of columns on a continuous strip foundation. A strip foundation is particularly suitable for excavation by mechanical means.

2.3.4. **Raft Foundations**

This is a common form of foundation for a light building on soft ground. For a larger structure, a raft foundation may be used where the ground bearing strength is below that which is economical for an isolated base or strip foundation, or where the weight of the loaded building approaches that which can be safely carried on the area of ground it is to occupy. The design of a large raft can be very complex requiring an assessment of the plant and column loads.

The "raft" is a layer of reinforced concrete covering the whole area of the building.

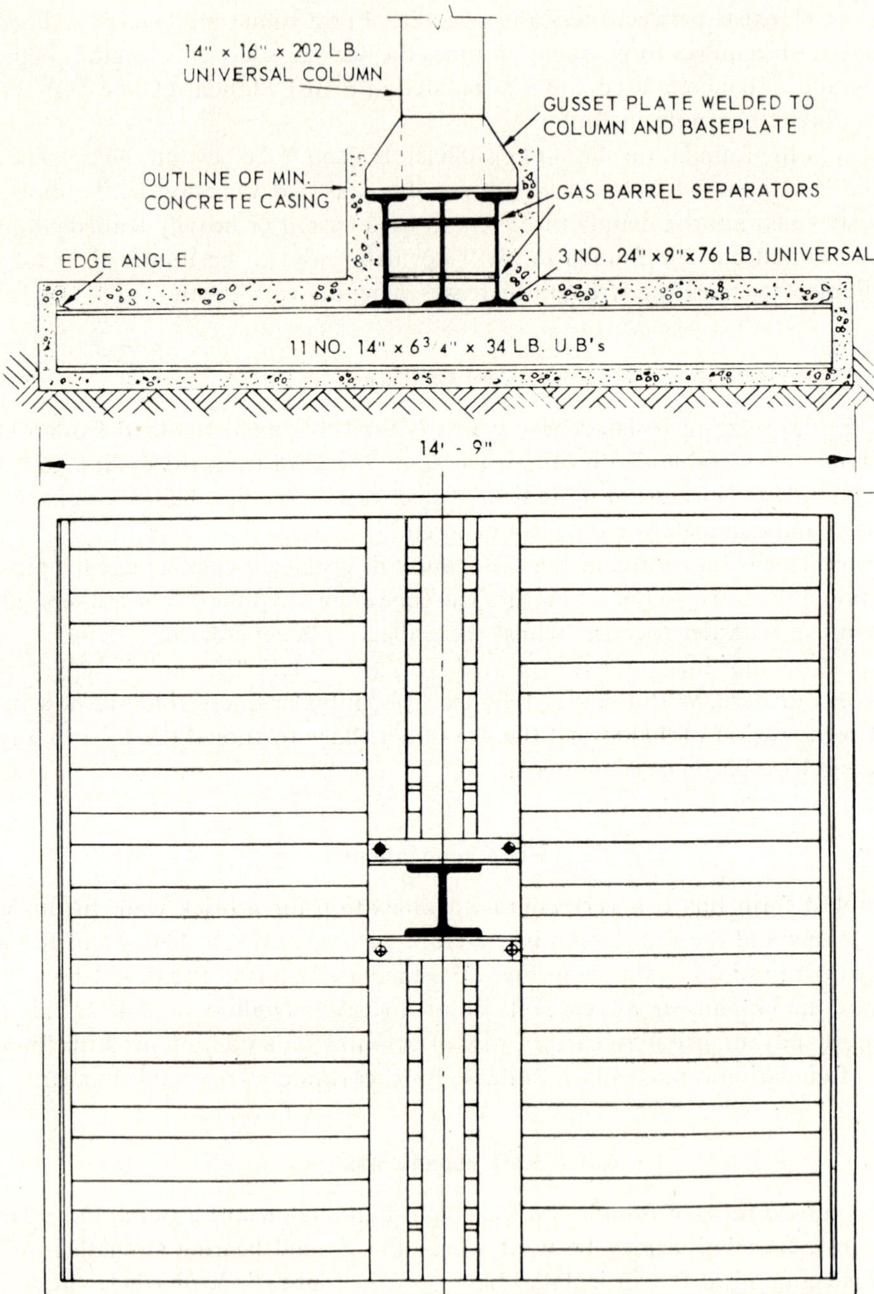

FIG. 2.3.2. Steel grillage foundation.

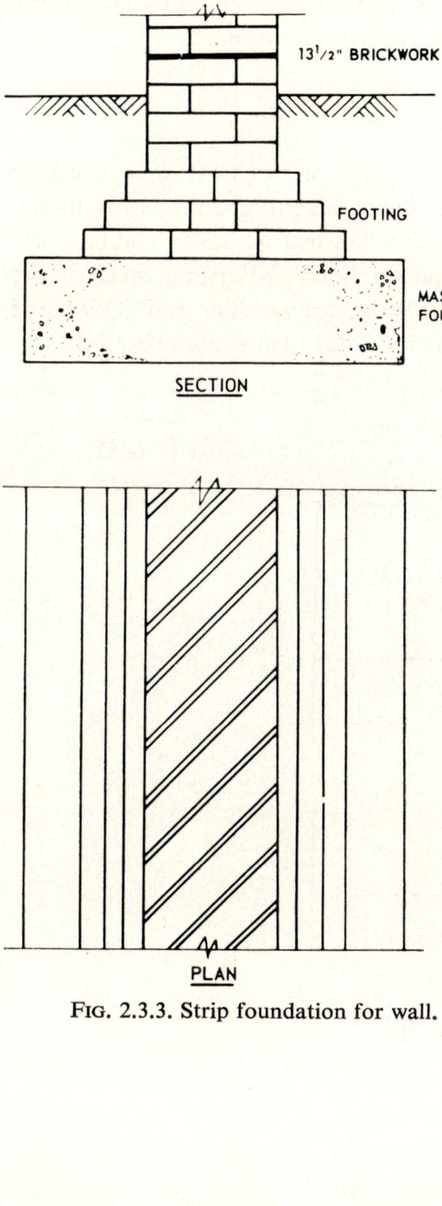

FIG. 2.3.3. Strip foundation for wall.

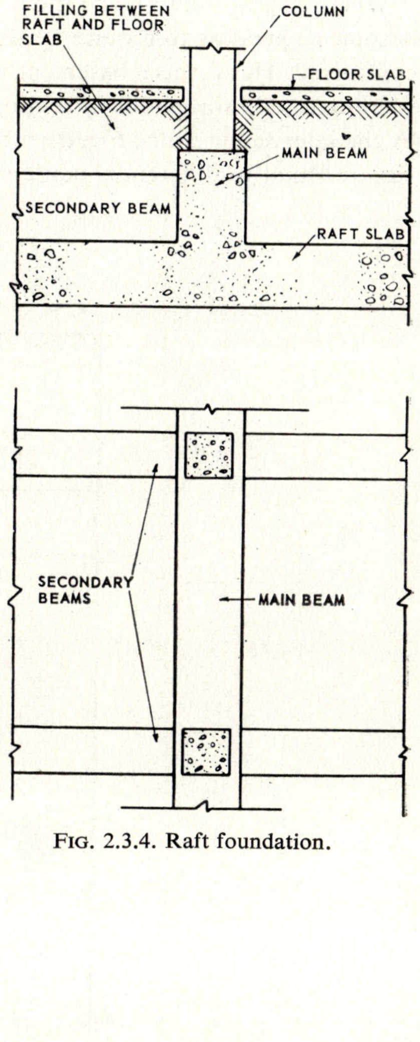

FIG. 2.3.4. Raft foundation.

It may be strengthened by increased thickness of concrete in areas of heavy loading, such as under columns, or by a system of main and secondary beams.

Raft foundations may be easily and quickly constructed, provided that a suitable stratum exists fairly close to the finished site level—say within 10 ft. In cases where the depth of excavation to firm ground exceeds the structural depth of the raft foundation, the excess depth over the raft can be filled with weak concrete or compacted broken stone. Figure 2.3.4 shows a raft foundation strengthened by main and secondary beams, and with filling placed over to carry the floor at the appropriate level.

2.3.5. **Cellular Foundations**

Where excavation to reach the foundation stratum is in excess of 10 ft, or where loads become so great as to require a raft of excessive thickness, a cellular foundation may be considered. This forms a basement to the building as shown in Fig. 2.3.5. The basement is constructed throughout in reinforced concrete and the floor and ceiling, spaced apart by the intervening walls, together form a structure of great strength and rigidity which is very resistant to differential settlement, and spreads the total load more effectively than

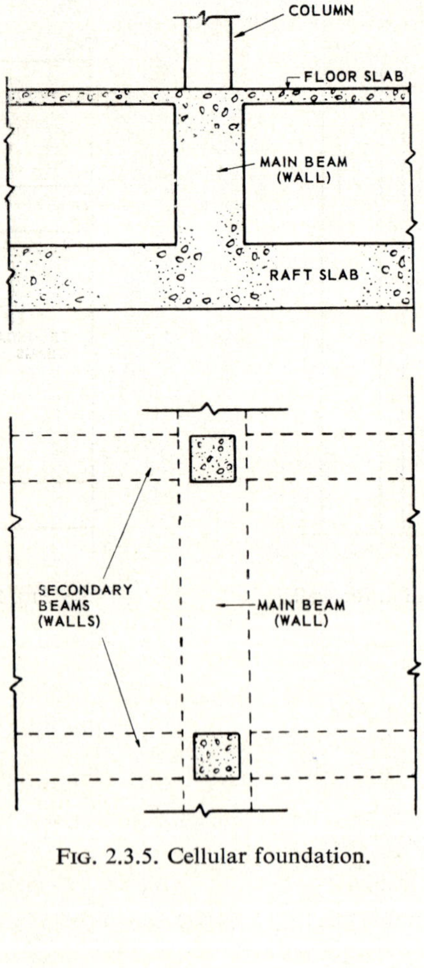

FIG. 2.3.5. Cellular foundation.

a raft foundation. The use of large quantities of weak concrete fill is avoided and the ground is relieved of the weight of soil which once occupied the void in the basement. For this reason cellular foundations generally give a lower intensity of loading on the ground than raft foundations, and so can be used on weaker soils. The designer must check that a cellular foundation constructed in water-bearing ground will not float due to the upthrust of water exceeding the weight of the foundation with or without its load. It is of interest that for one power station abroad foundations were required on ground so weak that the intensity of loading at any depth in the ground could not safely exceed that due to the weight of existing soil above. The solution lay in providing cellular foundations designed so that the weight of the foundation, with its superimposed load, was the same as the weight of soil removed for its construction.

This type of construction has the added advantage that drainage and service pipes and cables can be taken through the basement space.

2.3.6. **Piled Foundations**

When a sufficiently strong stratum does not exist fairly near the surface to support the imposed loading, piled foundations may be used. A pile is a column in the ground which is used to transfer the load to a stronger and deeper stratum.

Piles may be of timber, hollow steel section, prestressed or reinforced concrete; the last named being the most used for foundations of buildings. The load carried by a pile is normally from 30 to 80 tons depending on the size, type, and ground conditions. The load is transmitted from the pile to the ground either by end bearing or the pile toe on a hard strata underlaying soft ground, or by a combination of both.

There is no universal pile which is best in all conditions and all piles have their advantages according to circumstances. Concrete piles can be sub-divided into two main types:

(a) Driven piles.

(b) Bored piles.

(a) Driven piles are sub-divided into (i) "precast" piles, (ii) *in situ* piles.

(i) The precast piles in general use on power station sites are made with normal reinforced concrete and vary in size from 12 in square by 20 ft long weighing $1\frac{1}{3}$ tons, to 18 in square by 80 ft long and weighing 12 tons. Reinforcement consists of heavy longitudinal bars with square links or helical binders. The spacing of links and binders is decreased at the toe and the head to resist driving stresses; the longitudinal reinforcement provided is greatly in excess of that required to take the structural load, but is needed to resist the stresses of lifting, handling and driving. The toe is usually strengthened with a pointed cast-iron shoe, although this may not be necessary when driving in clay.

The casting, storing, transporting and driving of the precast piles for a major contract, which may call for some 20,000 piles, is a considerable undertaking. A large area of the site has to be set aside and levelled and accurately paved with concrete to provide a casting bed and further areas are required for storage. Travelling derricks are required for casting the piles and lifting and handling them, a batching plant is needed to produce the large quantity of concrete required, and a light railway system to transport the piles to the driving frames.

Piles precast in this manner have the advantage that they can be inspected before driving, and good control of the quality of concrete can be maintained. They are driven only after sufficient time has been allowed for the concrete to mature, and are thus resistant to attack from harmful substances in the soil.

Some wastage of concrete is necessary as the tops of the piles have to be cut away to expose the pile reinforcement for inclusion in the foundations. An accurate knowledge of the level of the bearing stratum is required to avoid further wastage through excessive cut-off. It is possible to drive piles below ground level using a wooden "Dolly" up to 10 ft long as a temporary extension to the pile. Permanent extension of the pile and subsequent re-driving is a tedious business as the head has to be stripped away, a new length of pile cast on *in situ* and left to mature before re-driving can be undertaken.

A mobile pile frame—a tall steel structure on rails fitted with diesel driven lifting gear—is employed to drive the piles. This hoists the pile into position, holds it and supports the pile hammer during driving, which continues until sufficient resistance is encountered. A variety of pile hammers are in use. Drop hammers varying in weight up to some four tons, dependent on the weight of pile, are still used, but diesel driven hammers, which are much quicker in operation, are largely replacing them. The pile head is protected during driving with a steel helmet lined with a wooden or hard plastic packer.

Prestressed concrete is also used to form precast piles and has the advantage that its very high quality concrete, in compression from the prestressing, gives a surface highly resistant to the attack of ground water and is thus particularly suited for use in river works. Piles formed by this method are less likely to be damaged during handling. They do not lend themselves to site fabrication on a large scale, however, because of the specialised stressing techniques required; and present special difficulties in driving, as horizontal cracks can be caused by shock waves rebounding up the piles.

Hollow steel piles, generally of hexagonal shape 18 in across and with pointed toes, are frequently used to support jetties, especially where long pile lengths are required. Although more expensive than concrete piles, they are much lighter and do not require such careful handling. Lengthening when required is a simple matter of trimming the head and welding on of further section. A protective external coating of bituminous compound, or protective paint is required before driving. Once driven, the piles are usually filled with concrete.

(ii) A number of systems for forming driven *in situ* piles are used, each specialist contractor having his own method. A steel tube fitted with a detachable cast-iron or concrete shoe, or a concrete tube formed of hollow precast sections and fitted with a concrete shoe is driven using a special piling frame generally mounted on a crawler-tracked excavator. A cage of reinforcement is lowered into the tube and the void concreted. The precast concrete shells are left in position and have the advantage that the outer skin of mature concrete is resistant to chemical attack. Where steel tubes are employed, these are withdrawn as concreting proceeds. Piles of this type need be only lightly reinforced to take the structural loads, and can be accurately finished to level, thus avoiding wastage of concrete. They are less certain in quality than the precast pile, particularly where a steel outer casing is used, as the ground pressure may cause the pile to close-up after the casing is removed.

The greatest part of the loading on power station foundations is vertical, but some horizontal forces due to wind and operation loads have to be resisted. These forces are resisted by driving some piles at a rake of up to about 1 in 3. Raking piles are commonly used on such structures as jetties, cooling towers, chimneys, coal-handling structures and transmission towers.

The very high column loadings now experienced in power station work cause the spacing of piles to be reduced to a point where the increase in ground pressure becomes a serious problem. 18 in square piles driven to a grid at 4 ft spacing represent an increase in volume of 14% of the subsurface material to a depth of perhaps 60 ft and the resulting ground heave in soft ground can be considerable, in the order of 6 to 8 ft. It is rarely that the piled areas can be left until conditions have become stabilised—construction programmes call for excavation to proceed soon after the piles are driven, and great care has to be taken if ground movement is not to occur, and piles be displaced. Pre-boring of a shaft in the ground at the location of each pile is sometimes undertaken to solve this problem. This procedure is also adopted to enable driven piles to penetrate thin hard layers of soil overlying softer material.

(b) Bored piles are constructed by specialist contractors, each with his own method. As with the driven *in situ* pile, a steel outer casing is driven into the ground, but without a shoe, and the soil from inside the tube removed with a rotary or shell auger. The reinforcing and concreting is then carried out in the same manner as for the driven *in situ* piles. Another method is to form the pile from precast units 3 ft long. The units are provided with holes to take the light longitudinal reinforcement, and cement grout is pumped into the pile as the casing is withdrawn.

These piles require only a comparatively light tripod rig and cause less vibration during construction than driven piles. They are particularly useful for smaller contracts and when piles are required adjacent to existing structures. Rigs needing only about 7 ft headroom are available and these are useful in the confined spaces often encountered during remedial works. Heavy reinforcement is not required and as in the case of driven *in situ* piles the concrete quality is inferior to that of precast piles.

Very large bored piles have been introduced in recent years; these have diameters from 2 to 7 ft and may be up to 100 ft in length. In good soils (stiff, unfissured clay, for example) an enlarged belled base may be formed by a rotating tool, thus permitting an increased load to be carried by the pile. Loads up to 1000 tons may be safely supported by one of these piles, depending on its size and soil conditions. A steel casing is still required when drilling through loose or weak soils.

It is rare for a single pile to be used on its own; piles are normally driven in groups at a spacing of $2\frac{1}{2}$ to 3 times their diameter and their heads incorporated into a thick capping of reinforced concrete designed to transmit the column load evenly to the piles. An arrangement of precast piles to support a column load of about 200 tons is shown in Fig. 2.3.6A and one to support about 2750 tons from a major column using large diameter bored piles is given in Fig. 2.3.6B.

If the piles are too close together there will be considerable overlap of stress from adja-

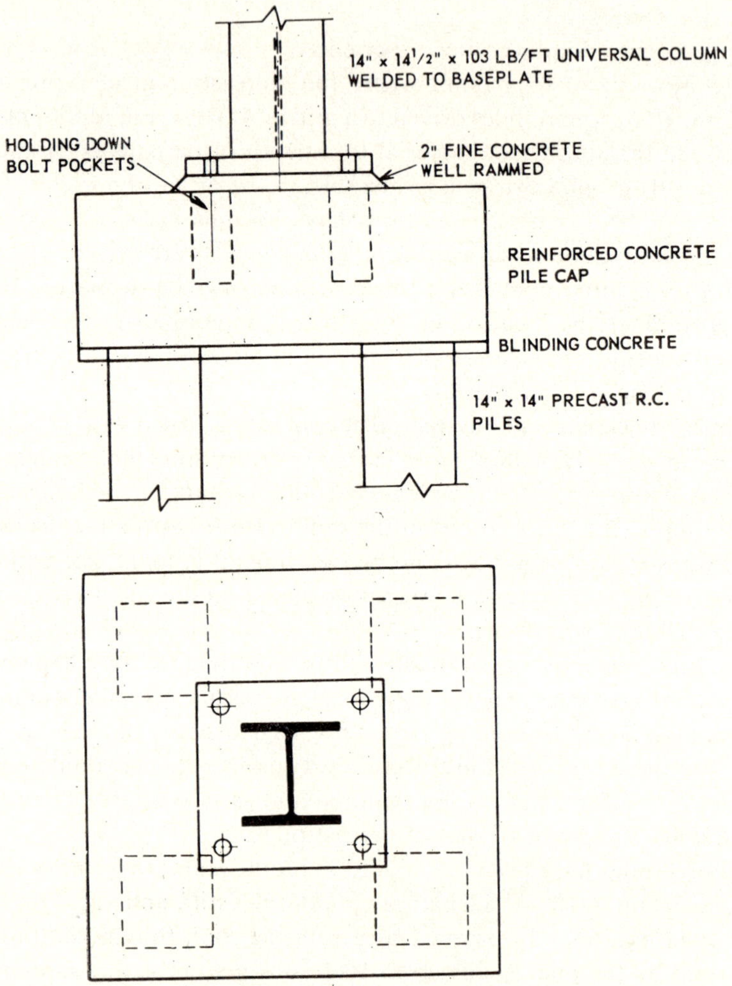

FIG. 2.3.6A. Piled foundation—precast concrete piles.

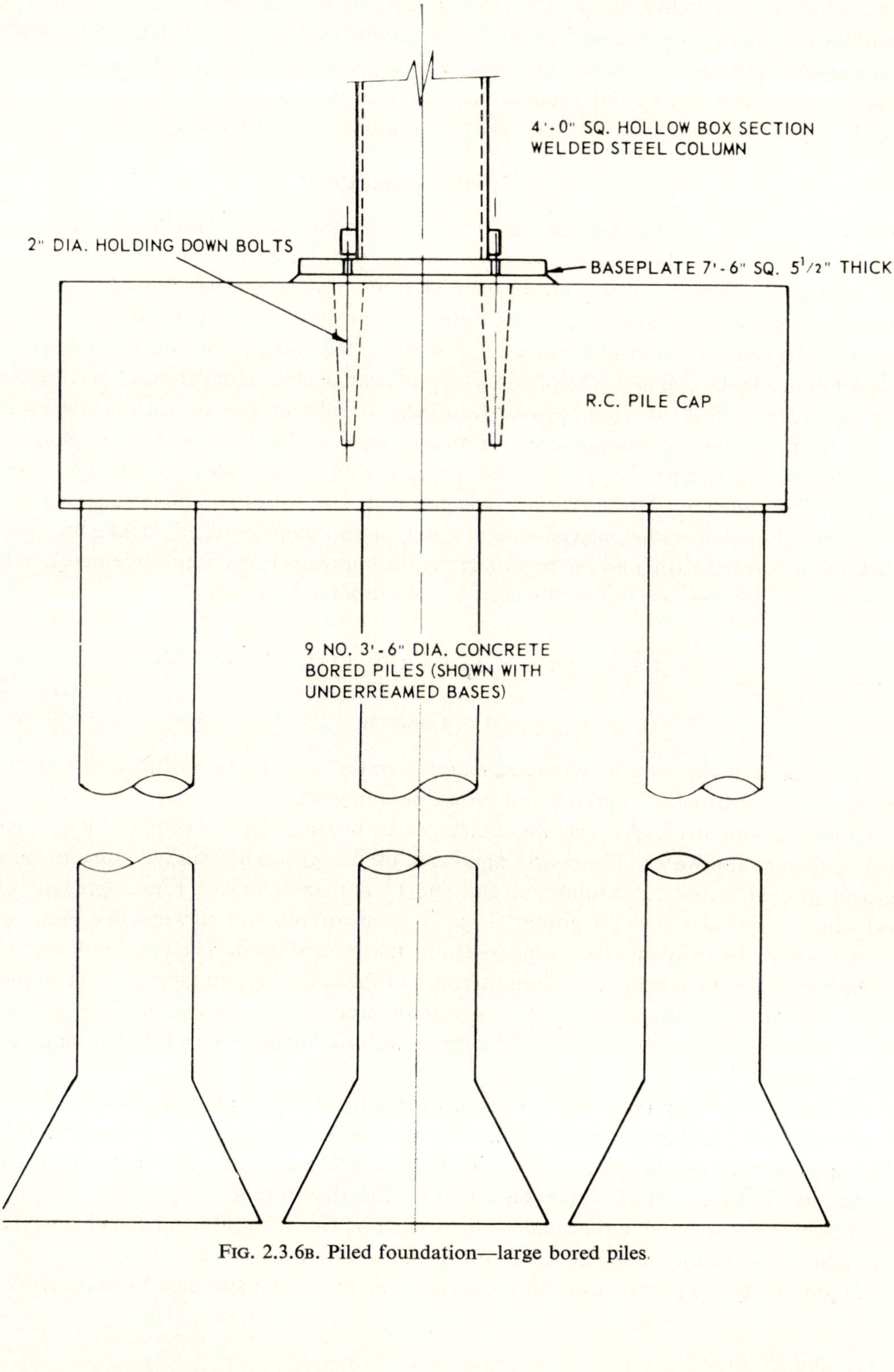

FIG. 2.3.6B. Piled foundation—large bored piles.

cent piles in the soil on which they rely for end bearing, and interaction will occur between piles which rely on skin friction. The bearing capacity of a group of piles is normally less than the capacity of one pile multiplied by the number of piles in the group. This reduction in capacity is greatest for piles in cohesive soils, and in large groups can be as much as one third. The estimated group capacity may only be determined with accuracy by a test, but this is expensive because of the difficulty in applying an adequate load.

2.3.7. **Cylinder Foundations**

For larger concentrated loads it may be necessary to use large diameter cylinders either singly or in groups. An example is given later in Section 2.6.6. The concrete shell of a cylinder is built from a steel cutting edge at ground level, and supported vertically by temporary guides. The walls are in the order of 6 in thick. Precast sections are used for cylinders between 6 ft and 10 ft diameter, but above that size, *in situ* concrete is used. The soil from inside the cylinder is removed with an "orange peel" circular grab, and the cylinder is forced to sink by loading with Kentledge. Grabbing can continue underwater if water bearing strata are encountered, or an air lock chamber fitted and excavation continued by hand in compressed air. Further precast sections are added at intervals, or the *in situ* walls extended until foundation level is reached. The bottom of the cylinder is plugged with mass concrete either placed under water or in compressed air, depending on the method of construction, and the remainder of the core concreted. Reinforcement is added if required. Cylinders are frequently used as the supports for jetties.

2.4. GENERAL NOTES ON FOUNDATIONS

2.4.1. **Concrete**

As concrete is the most widely used material in the construction of foundations, some notes on its constituents, making and properties are given.

Concrete is an artificial stone, manufactured by mixing together coarse and fine aggregates, cement and water. The coarse aggregate used is generally crushed stone or natural gravel, graded between $\frac{3}{16}$ in minimum size and $1\frac{1}{2}$ in maximum size. Fine aggregate, natural sand or crushed rock is graded from $\frac{3}{16}$ in maximum size downwards. Aggregates should always be clean and free from vegetable matter and earth. Their strength should be at least equal to the specified crushing strength of the concrete, and they should contain no substances likely to attack the cement or reinforcement steel. They should not be porous, or be liable to change volume with change in water content or be subject to damage by frost attack.

Cement is a factory made product conforming to British Standard Specifications, ordinary Portland cement being most widely used for foundation works. This is made by burning and then crushing a mixture of clay and chalk. Since it hardens after mixing with water, special care has to be taken on site to provide dry storage.

Water for concrete should be reasonably clean and free from silts, oil and chemical contamination. Normally water that is fit for drinking is fit for concrete.

Traditionally the proportions for a concrete mix have been specified by volume. A mix for an average quality concrete, giving a crushing strength at 28 days of 3000 lb/in^2, being one part of cement to two parts of fine aggregate to four parts of coarse aggregate.

The crushing strength of concrete is easily checked on site. Samples are taken from concrete about to be used in the foundations, and 6-in cubes made in steel moulds. These cubes, once set, are cured by immersion in water, the temperature being controlled by thermostat. The cubes are tested to destruction after 7 and 28 days, in a compression testing machine. On power station contracts the testing is carried out in the site concrete laboratory by a member of the resident civil engineer's staff. The crushing strength serves as a useful guide to the quality of the concrete. High compressive strength concrete is usually dense and impermeable, has good durability, and high tensile and flexural strengths.

Concrete for large civil engineering works today is specified both by the minimum strength required and by the relative weight of constituents. A maximum and minimum weight of total aggregate both fine and coarse—against a weight of cement is generally given. Concrete for foundation work is generally required to have a compressive strength of 3000—4000 lb/in^2 at 28 days. For special purposes this strength may need to be higher, in the region of 7500 lb/in^2 for prestressed work, for example.

The choice of proportions for a concrete mix, within the limits of the specification, is left to the contractor. The design of a concrete mix should aim at producing the specified strength in the most economic manner. The gradings of available aggregates are obtained by sieve analysis and the shape of the aggregate—whether rounded or angular—noted. With this information the proportions of cement, water, fine and coarse aggregate are determined. Other factors that have to be taken into consideration are: the workability of the mix, i.e. the ease with which it can be handled and compacted, and suitability for transporting by concrete pump, and compaction by mechanical vibrator, if these methods are to be used. It must also be suitable for placing around congested reinforcement, in constricted sections of the work, and for producing a satisfactory surface finish. Dense concrete with low shrinkage properties may be called for in water retaining structures, or for plant foundations such as turbogenerator blocks after cubes from trial mixes have been tested on site.

The contractor's design of mix is subject to approval by the Board's civil engineers, and final adjustments may be made after trial mixes have been carried out on site. These will assess the mix design under working conditions, which may give somewhat different results to those obtained in laboratory trials.

Slight variations in grading of aggregates can be expected when delivered in bulk. The mixing plant may retain fine material, the degree of workability may not be quite suited to the transporting and placing equipment, or to placing in the foundations, and some losses of fine material through the shuttering may be experienced despite care taken in assembling it. With a good design these affects should be marginal. Most contractors have the experience to design for the plant he intends to use, and to achieve strengths well above the minimum specified.

To achieve consistently high-quality concrete, good site control must be maintained throughout the duration of the contract. Proportions of materials must be measured by weight. Modern concrete plants are fitted with hydraulic weigh-batchers for solid materials, and pre-set dispensers for water, but these must be regularly checked for mechanical condition and calibrated over the range of weights and quantities used. Aggregate stock piles must be designed to drain properly, and to keep the aggregates clean and segregated from each other. Samples of aggregates must be taken regularly for analysis, samples of concrete taken for crushing tests, and simple site tests carried out on the concrete to assess workability. Particular regard must be paid to water content of the concrete. The aggregate should be left to drain as long as possible, tests taken of its water content and adjustment made to the water to be added at the mixer if necessary. The tendency of mixer crews to add water to make the mix more workable must be resisted, otherwise segregation of the concrete may result reducing its strength and density and increasing shrinkage. If greater workability is required in any part of the foundations, a properly judged change to the mix should be made.

Special grades and types of cement are available. Rapid hardening Portland cement gives high strength at an early date. Sulphate-resisting cement and supersulphated cement resist the attack of sulphates in the soil, supersulphated being used for high concentrations. High alumina cement produces concrete resistant to chemical attack, and gives high strength very rapidly—within 24 h. It gives off a considerable amount of heat during the setting process and in consequence needs to be used with great care. Only thin sections or lifts of concrete can be poured at a time—18 in maximum, and water has to be sprayed on for cooling until the setting process is complete. This cement is also highly resistant to heat, and is used with a refractory crushed firebrick aggregate to form the ash hoppers of power station boilers. These special cements must be used strictly in accordance with the manufacturers instructions, or the advice of specialists in this work.

There are also a large number of patent additives for concrete now being marketed. These perform a variety of functions—improving workability, accelerating or retarding the rate of setting, increasing surface hardness, and many others, and have useful application in certain circumstances.

Pulverised fuel ash has been successfully used as a replacement of 20% of cement in the making of concrete, particularly in very massive concrete works where a low heat of hydration is required. It has a rather slower rate of hardening than ordinary Portland cement concrete, but eventually reaches the same strength. It improves workability of the mix, and the concrete is ultimately more resistant to chemical attack.

Concrete is comparatively strong in compression and weak in tension. Thus the tension faces of members subjected to bending are reinforced, usually with mild steel round bars. Reinforcement is also used in compression members, e.g. columns, because the steel cross-sectional area will carry far more load than the concrete it displaces and will also resist any bending which may exist. Also, reinforcement is used to resist temperature stresses and prevent cracking of the concrete.

2.4.2. **Bearing Pressures and Settlement**

Foundations always settle under load and a limited amount of settlement is acceptable provided it is uniform. The allowable bearing pressure is the ultimate pressure divided by a factor of safety which is normally taken between 2 and 3 according to the importance of the structure and its sensitivity to differential settlement. For example, a structure having rigid beam to column connections will have high stresses set up in the framework if unequal settlement of the column bases occurs.

In the following cases it is assumed that the foundation is bearing on a uniform strata which is not overlying a weaker material:

For granular soils above the water table the allowable bearing pressure for a 10 ft wide footing and a factor of safety of 3 varies approximately linearly from about 5 tons/ft^2 when N is 50 in the Standard Penetration Test to 0·8 ton/ft^2 when N is 10. As the foundation width is increased there is a slight decrease in the allowable pressure.

For shallow foundations in purely cohesive soils above the water table, the theoretical *ultimate* bearing pressure for a long strip footing is 5·14 C, where C is the shear strength of the soil. For a square base the ultimate value is about 6·7 C. In these cases the bearing pressure may be increased by wD, where w is the density of the cohesive soil and D is the depth of the foundation. (*Note:* The bearing pressure is in lb/ft^2 when C is in lb/ft^2. Similarly w must be in lb/ft^2 and D in feet.)

The concept of allowable bearing pressure is a simplification of the problem, and may be employed only after careful assessment of the subsurface conditions. The load distribution under a foundation is not uniform; in the case of cohesive soils it is greatest at the edges of the foundations, and is greatest at the centre in non-cohesive soils. The loading effect from a foundation extends to a considerable depth below its underside. It is commonly assumed that the stressed zone extends to a depth of $1\frac{1}{2}$ times the breadth of the foundation, and a "pressure bulb" distribution is formed. The "pressure bulbs" from adjacent foundations, such as the columns of a boiler house, interact to form a single much deeper bulb. All the strata within this stressed zone must therefore be considered when arriving at a figure for allowable bearing pressure at foundation level. It is for this reason that loading tests on a small area of the soil at foundation level can give misleading results.

Assuming that a proper design has been carried out to avoid shear failure of the soil, settlement can occur in several ways:

(a) *Consolidation settlement.* This is caused by the expulsion of pore water from the soil due to the increased pressure from the foundation. On sands and gravels this occurs quickly, and may be complete by the time construction ends. With clays and silts it is a gradual process and can result in settlement over a number of years, depending on the thickness and compressibility of the layer.

(b) *Plastic settlement.* This is caused by underlying soft plastic clays and silts gradually being displaced from under the foundation. This may happen at loads considerably less than would cause shear failure of the soil. It may also occur where a thick layer of plastic material is sandwiched between layers of harder material and in consequence may be difficult to forecast.

(c) *Settlement due to changes in conditions.* Settlement may occur independent of the imposed loading. Lowering of the water level will cause drying and shrinkage of clays and silts, and raising of the water level in sands and gravels will cause a reduction in the safe bearing capacity, while long-term seepage may wash away the finer material. The drying out of clay soils beneath boiler houses due to the increased temperature may also cause shrinkage.

Differential settlement of foundations arises from three causes:

(a) The capacity of the ground to carry loads is unequal in different soils.
(b) The ground is unequally loaded under various parts of a structure—this should be avoided as far as possible in the design.
(c) Variation in thickness of a compressible stratum.

Mixed types of foundations are generally avoided in order to reduce the possibility of differential settlement.

2.4.3. Test Piling

An estimate of the load which a particular size of pile is capable of carrying may be made when the soil properties are known. The load which a pile driven in sands and gravels can carry may be determined fairly accurately from pile driving formulae when the "set" or penetration resulting from a known hammer blow is measured. (For further information see Appendix 3, Example).

However, the most accurate method of determining the capacity of a pile is to apply a test load to the head of the pile after it has been driven. This provides the most reliable source of information on which to base a design. Full information on test piles should be made available to companies tendering for piling contracts.

The test load may be applied to a pile by:

(i) Building a platform on the pile head and progressively loading it with kentledge.
(ii) Building an independent structure loaded with kentledge around the pile head and jacking from the structure down on to the pile.
(iii) Jacking down on to the test pile from a crossbeam; each end of the crossbeam is connected to a pile in order to carry the uplift reaction. The uplift piles must be sufficiently remote from the test pile to prevent interaction. This is usually the most economical method of applying a test load.

The load is applied to the pile in increments, each increment is made only when settlement from the previous load has ceased. Settlement is usually recorded by dial gauges. Although it is not necessary to be able to read settlement to a thousandth of an inch, the dial gauge is useful for showing when the settlement for each load increment is complete. Pile tests on slow draining soils, such as clays, may take several days to complete and the time is recorded as well as the settlement and corresponding load.

During test it is useful to remove the load at the design working load and measure the residual settlement. It must be realised that the pile is compressed during the test and the recovery is largely due to the release of elastic shortening caused by the load. Failure of the pile occurs when the applied load causes a continuing settlement. The factor of safety is the ratio of the failure (ultimate) load to the working load. This ratio depends on the permitted settlement of the structure, but is about 2 when pile testing is carried out. When pile tests are not done it is necessary to use a higher factor of safety, up to 3, on an estimated ultimate load. It can be seen that where a fairly large number of piles are required, pile testing is an economic proposition. The results of a pile test carried out by jacking down from a crossbeam are plotted in Fig. 2.4.3.

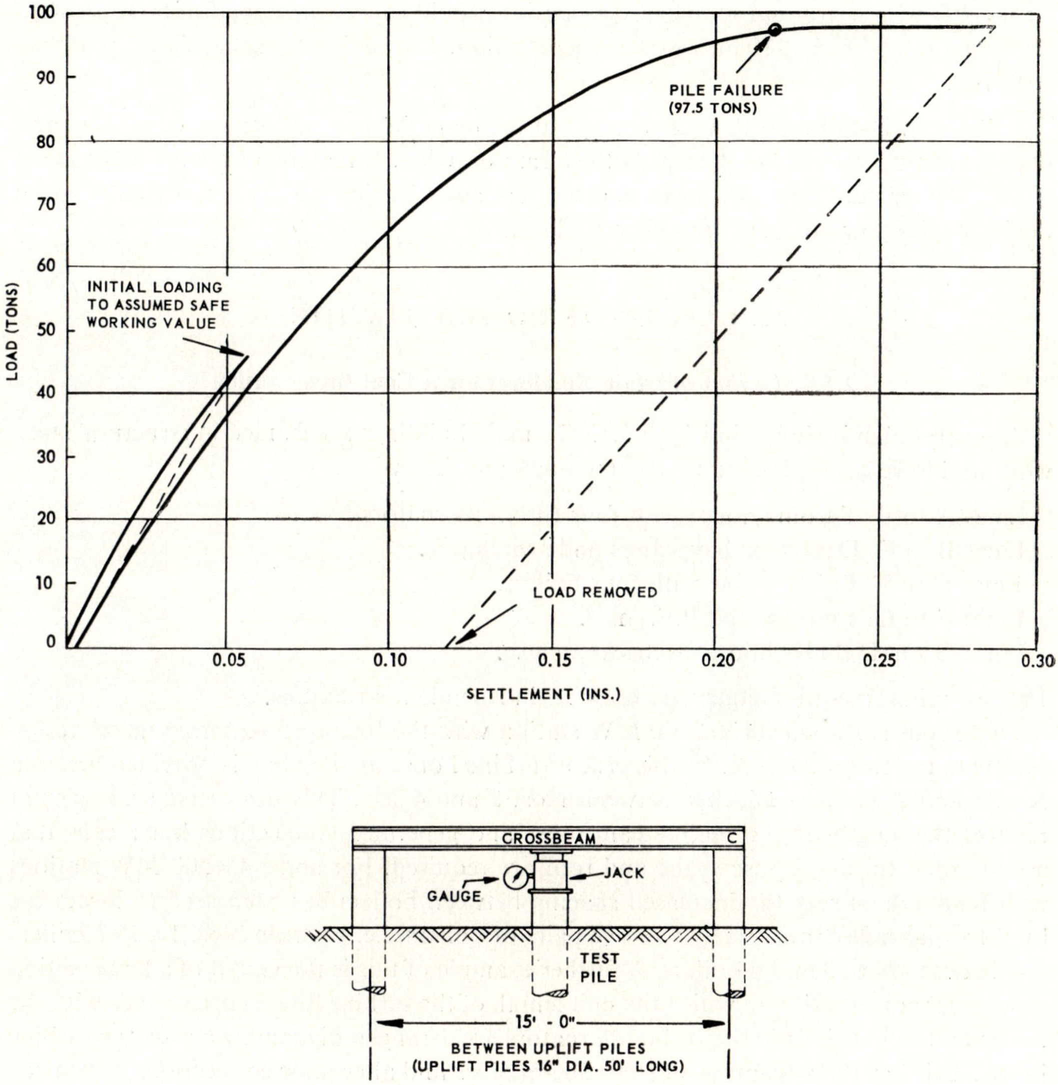

FIG. 2.4.3. Pile test on 16 in diameter bored pile 36 ft long in clay (test carried out continuously over a period of 72 h).

Over-driving. In the case of driven piles care must be taken to prevent damage to the pile by over-driving, and the safe load is related to the driving by what is known as the Set, i.e. the number of hammer blows necessary to produce 1 in of penetration. For example, a reinforced concrete pile 15 in × 15 in × 40 ft long driven with a hammer of 4 tons weight, stroke 4 ft with a gradual reducing penetration would carry a safe load of 50 tons when the Set had stabilised at 10 blows to 1 in penetration. The Set is established from the driven test piles.

Records. Detail records must be kept of the manufacture and driving of all piles and the penetration should be plotted on the strata drawing. It may be that when piles are in groups, the ground becomes so compacted that some boring will have to be done to get the later piles down to the required strata level.

2.5. LAYOUT OF POWER STATION

2.5.1. Layout of Main Buildings for a Coal-fired Station

A general plan is shown in Fig. 2.5.1. The main buildings are framed in structural steelwork and have been planned to accommodate the following:

Lines A to B. Turbine house with four turbo-alternators.
Lines B to C. Deaerator bay, tanks and switchgear.
Lines C to F. Boiler house with four boilers.
Lines F to G. Bunkers and P.F. mills.
Lines G to H. Boiler house electrical annexe.

The precipitators and chimney are external independent structures.

The layout is shown for a 2000 MW station with the four turbo-alternators set transversely in the turbine house. In this case a turbine house loading bay is provided between Nos. 1 and 2 sets and another between Nos. 3 and 4 sets. This transverse arrangement reduces the length of the main steam pipes and generator connections but means that a large span turbine house crane and roof are required. For some 4 × 500 MW stations with longitudinal sets the increased spacing between boilers has been used to locate the bunkers and mills transversely to serve pairs of boilers, i.e. between Nos. 1 and 2 boilers and between Nos. 3 and 4 boilers. A further example of this is the layout of a large station now under construction in which the unit length of the turbine house corresponds with the boiler house. In this case the turbo-alternators are arranged diagonally across the turbine house. This has the advantage of short steam pipes and alternator connections, a reduced crane span and lower civil work costs. This layout has been adopted for a station of 4000 MW with six 660 MW turbo-alternators. Other less important changes in layout are possible, but as indicated above it is necessary to correlate the turbo-alternator spacing with the boiler width.

Ideally the civil engineering contractor's plant should move progressively away from the completed work to leave unimpeded access to the plant contractors who follow.

In addition to the main buildings, precipitators, chimney, and cooling towers, the station layout will include other major works such as a switching compound, circulating water

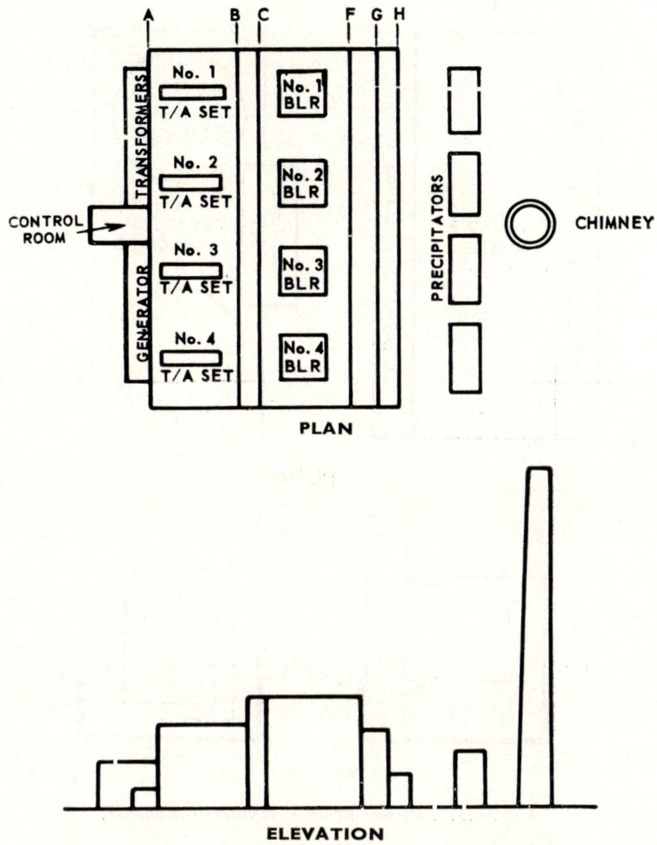

FIG. 2.5.1. Power station layout — coal-fired.

systems, roads, railway (and a quay or jetty if fuel is brought by boat), fuel storage and handling facilities, ash handling and disposal facilities for coal-fired stations, gas turbine houses, stores and workshops, water treatment plant, compressor houses and an administration block. The siting of rail tracks is largely controlled by the proximity of the main railway.

Great care is required at this stage in siting the ancillary buildings around the main buildings. They must be so arranged that it is practicable for the civil contractor to complete them in a sequence to suit the plant contractor's starting date, and they must be sited so as not to impede either civil and plant construction in the main buildings. These requirements frequently clash with the later convenience of operation and maintenance, but a properly considered compromise must be reached.

2.5.2. **Layout of Main Buildings for a Nuclear Station**

Apart from the elimination of all plant connected with the handling and storage of fuel, the major difference between a conventional and nuclear power station from the point of view of the foundation designer is the substitution of the reactors for the boiler.

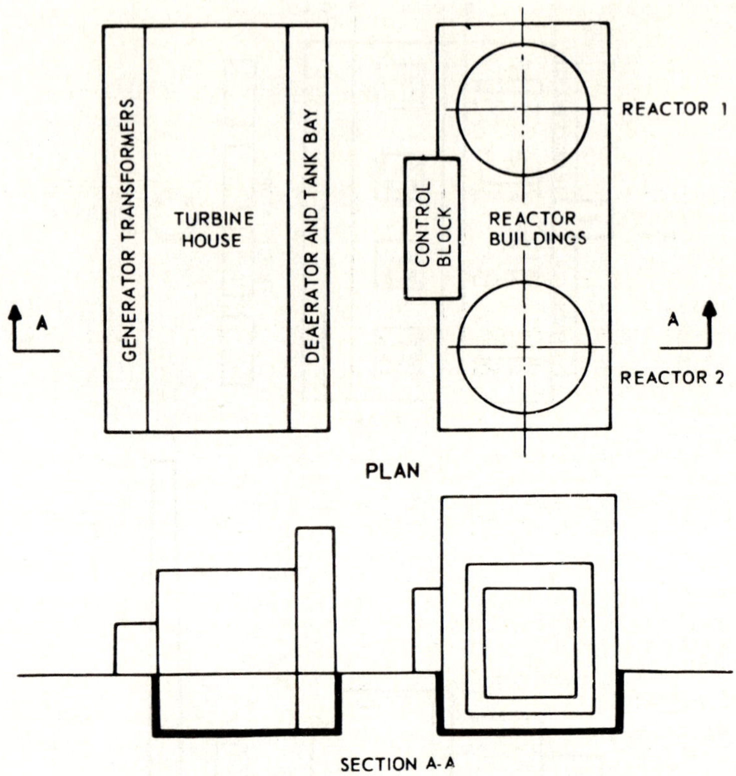

FIG. 2.5.2. Power station layout — nuclear.

Figure 2.5.2 shows a layout of the reactor building and turbine house.

Nuclear stations currently under construction have two reactors with the pressure vessel constructed in prestressed concrete.

A typical reactor foundation is described in Section 2.6.8.

2.6. FOUNDATIONS FOR MAIN AND SECONDARY STRUCTURES

2.6.1. Boiler House Foundations

Boilers are the largest items of plant, and in modern power stations they are suspended from a framework of heavy steel beams just below roof level. The boilers, with their ancillary parts and associated coal bunkers, comprise a very high proportion of the total load. The foundations provided are largely required to carry loads arising from the plant, the loads due to the buildings being only a relatively small proportion of the total.

Figure 2.6.1A is a cross-section appropriate to the station indicated in the key plan (Fig. 2.5.1). The columns in the boiler house are shown here as being carried on pre-cast concrete piles. It could be advantageous to use a smaller number of large diameter bored piles; the actual choice between these alternatives depending on the most economical type of pile for the specific ground conditions.

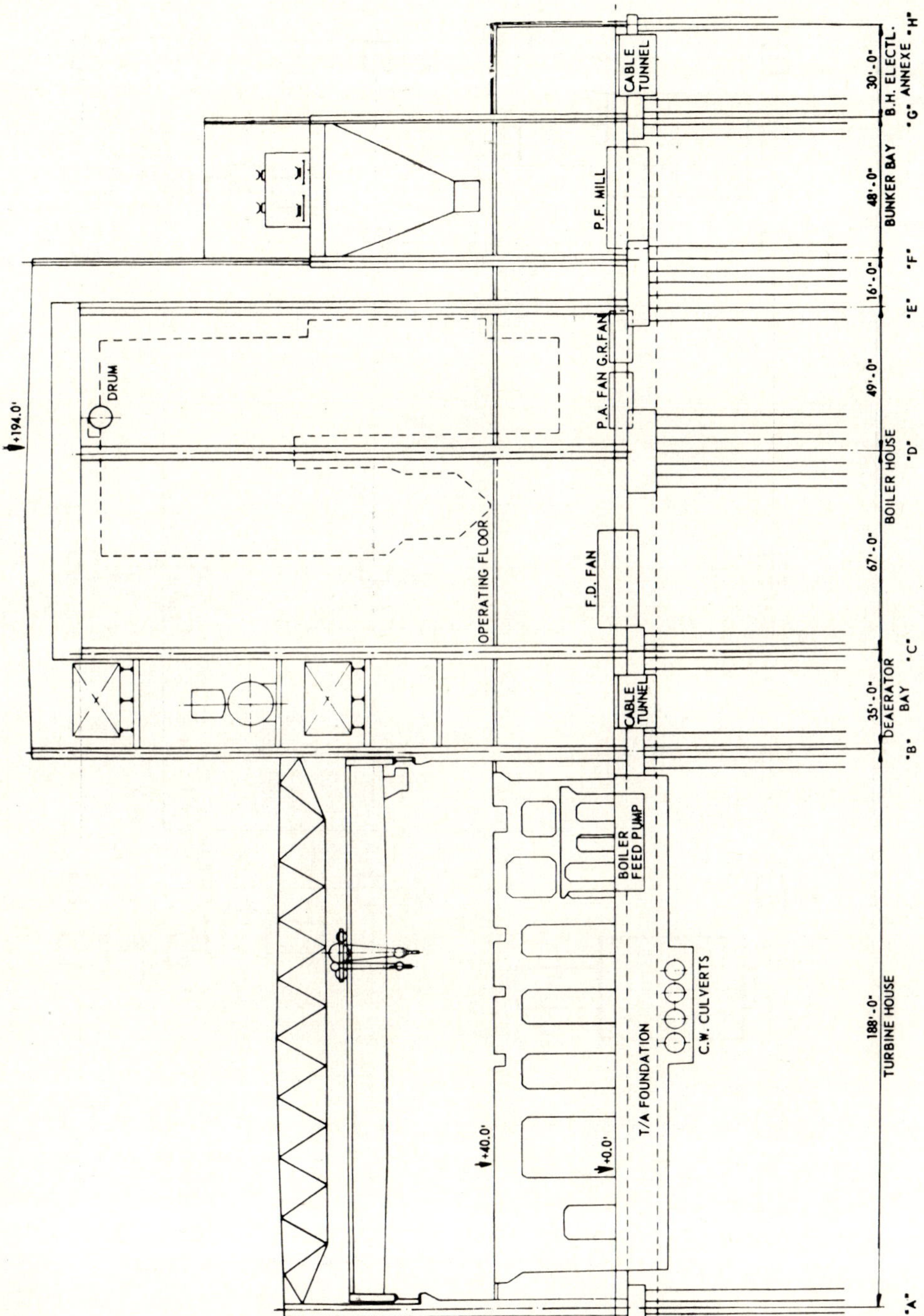

FIG. 2.6.1A. Cross-section through main buildings.

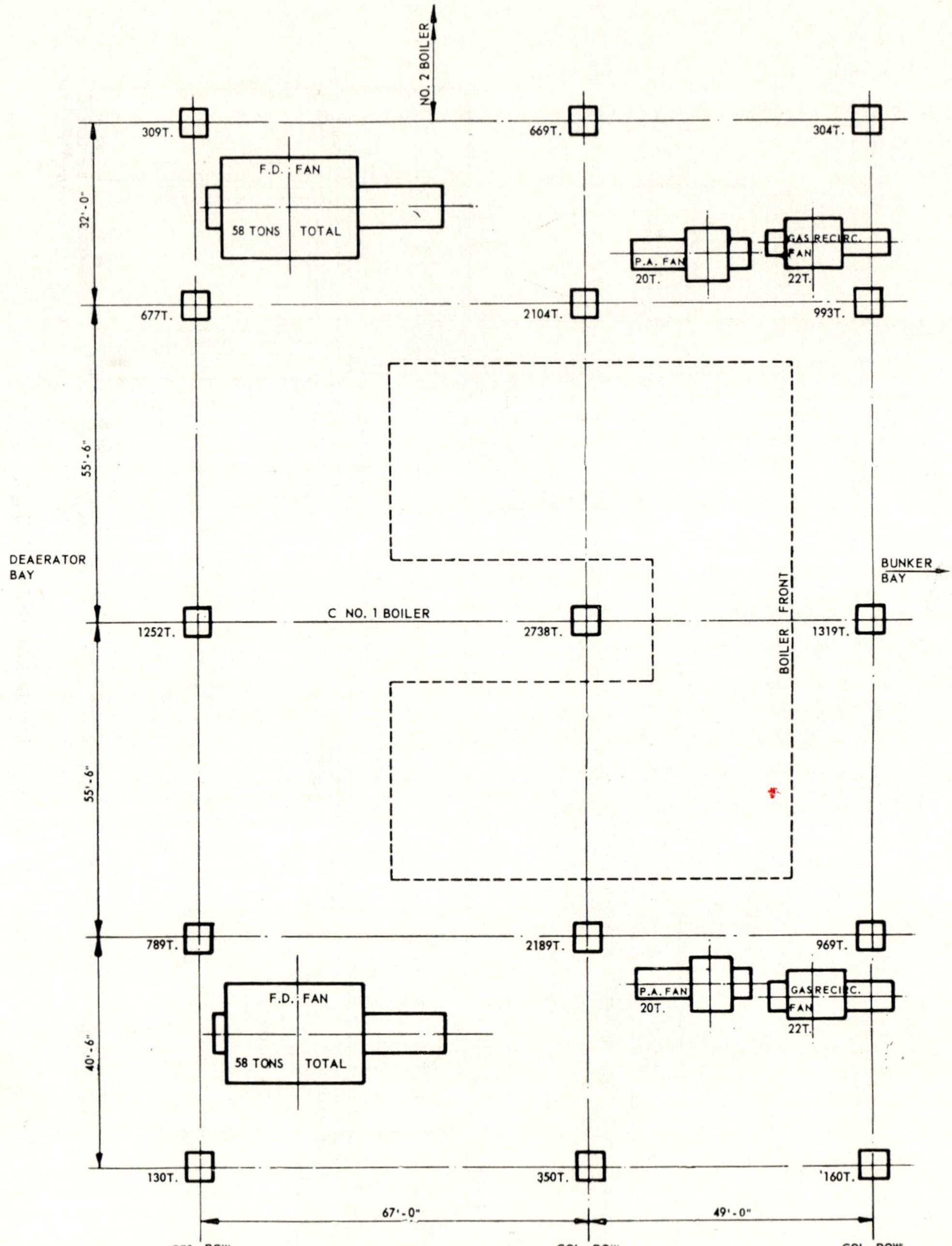

FIG. 2.6.1B. Boiler foundation loading plan.

The foundations for the P.F. mills and the fans are carried on isolated bases in this instance; the ground pressure intensity for these being very much lower than for column bases.

The high loads resulting from the boiler are shown in plan in Fig. 2.6.1B. These high loadings result in a condition where the individual pile caps or foundations have grown so large as to nearly touch. The interspaces are required to accommodate fan bases, mill-bases, cable tunnels, CW culverts and drainage culverts with the result that the boiler foundation assumes the form of a huge cellular raft into which the column pile caps intrude.

2.6.2. Turbine House Foundations

The turbine house main columns are on lines A and B. These too are heavily loaded, but not to the same extent as the main boiler columns. Thus, as shown in Fig. 2.6.1A, the turbine house columns require fewer piles than the boiler columns.

Each row of the turbine house columns carries a crane beam and rail of the overhead electric travelling cranes, which are installed for the erection and maintenance of the plant. The highest individual load on the column is due to the crane, consisting of the crane weight, dynamic loads and the maximum lift at the nearest point to the column. The main turbine house columns also carry the dead loads and superimposed loads of the roof, the weight of the wall cladding and the reactions from beams framing into them from adjoining floors.

2.6.3. Turbo-alternator Blocks

A turbo-alternator block is a rigid structure which extends from the basement up to the operating floor to carry the machine at that level. The block has openings within it to accommodate plant and pipework and is itself carried on a substantial foundation.

The turbo-alternator block has been for a long time made in reinforced concrete with the condensers placed within it on concrete plinths. This arrangement gives a height of about 40 ft from basement to operating floor for a 500 MW unit.

However, some 500 MW units are now being constructed with condensers not located under the machine. This allows the height of the block to be less, consequently reducing both the height and cost of the turbine house. One of the alternative arrangements is for the condensers to be placed on each side of the machine, these being known as side by side, or pannier condensers.

A recent development is to construct the block of structural steelwork. At present this is slightly more expensive, but may be expected to reduce the construction time. It reduces the foundation load, and being more slender permits better access and layout beneath the machine for circulating water pipes and other plant. In the case of a concrete block it is an advantage if it is completely built about nine months ahead of the machine erection in order to allow virtually all the concrete shrinkage to occur, whereas with a steel block this problem does not exist. The properties of the material are more accurately known, making the solution of differential settlement problems less difficult.

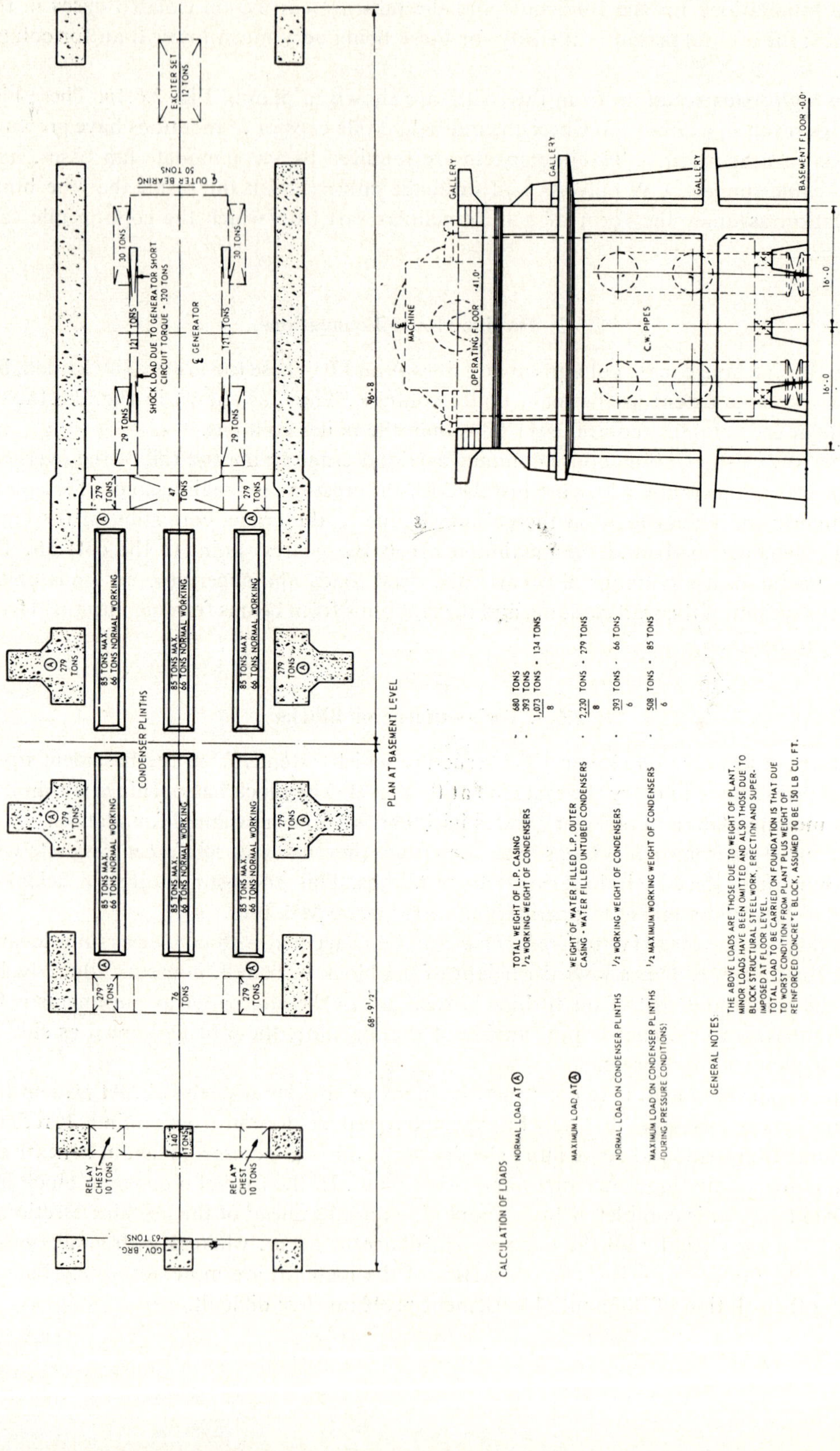

FIG. 2.6.3A. Simplified loading plan for 500 MW turbo-generator.

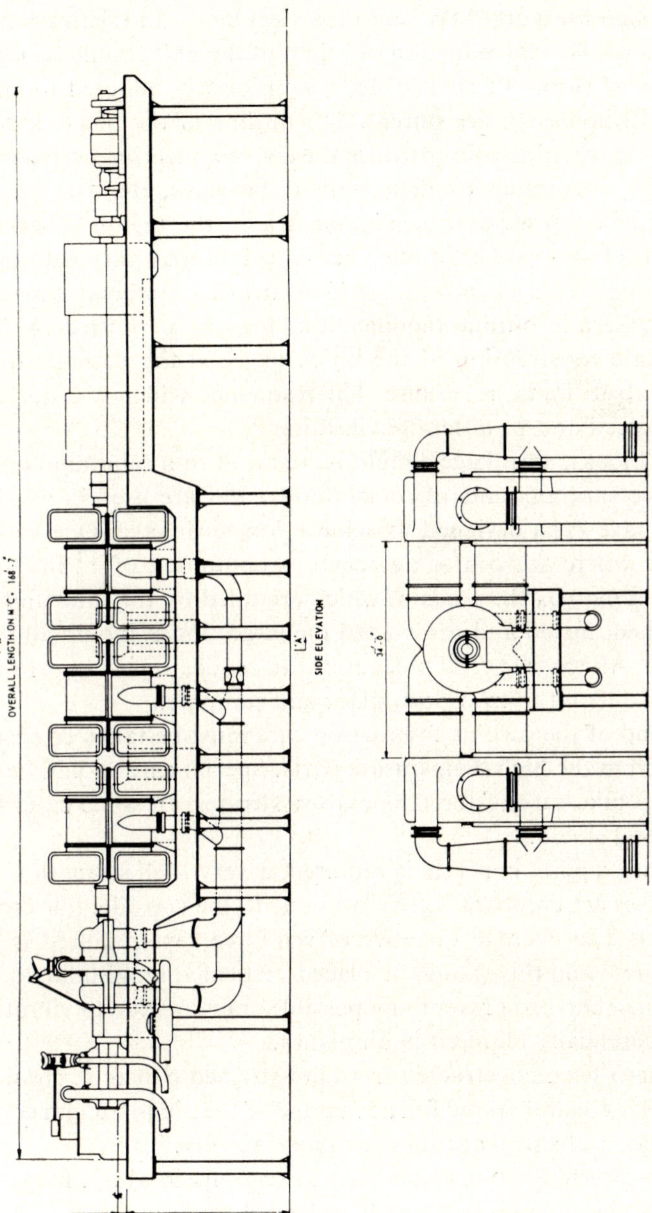

FIG. 2.6.3B. Arrangement of 500 MW turbo-generator on steel block with side-mounted condensers.

A loading diagram for a 500 MW unit is illustrated in Fig. 2.6.3A. This shows a concrete block with axial condensers beneath the machine.

A proposed design for a 500 MW unit on a steel block and with pannier condensers is illustrated in Fig. 2.6.3B. The reduction in height of the unit should be noted.

Whichever type of turbo-alternator block is ultimately adopted the requirements and basic principles of the design are similar. The outline of the block is determined by the manufacturer of the machine who provides the civil engineering designer with the loading conditions and the restrictions on deflections at bearings, etc. The maximum permitted differential movement subsequent to alignment is in the region of 0·004 in reducing to 0·002 in for points close together, though actual values may vary with different manufacturers. In this respect it is an advantage if all bearing supports can be mounted over piers or columns. Certain outline modifications may be permitted by the manufacturer in order to facilitate construction of the block to assist the designer to keep deflections within the prescribed limits, to ensure that resonance will not occur at the machine's normal running speed and to minimise vibration.

With concrete blocks, shrinkage should be reduced to a minimum and to achieve this the minimum necessary amounts of cement and water are used in making the concrete. Special mixes have been designed to achieve low shrinkage. One such is called "gap-graded" concrete where the coarse aggregate is composed of $1\frac{1}{2}$ in stones only, which compact to form a matrix, the voids of which are filled by the sand and cement mixture. If properly designed, mixed and compacted the large stones should all touch and shrinkage be eliminated. All these special precautions to reduce shrinkage of the concrete cause it to be extremely difficult to transport, place and compact.

At the steam end of the turbine foundation care must be taken to ensure that the concrete is shielded from the high temperature parts. Special care should be taken to provide reinforced steel to take care of the temperature stresses. In some cases it is necessary to provide a dura steel shield.

Although a good quality concrete is required, a very high strength is not necessary as the concrete stresses are comparatively low owing to the considerable cross-sectional area of most members. The average quantity of reinforcement required is in the region of 1 cwt/yd³ of concrete and this should be placed vertically, longitudinally and transversely in all structural members (to prevent any possible cracking due to vibration) even though it may not be theoretically required in all planes.

Some blocks have been constructed from pre-stressed concrete, considerably reducing the weight of steel used, and taking full advantage of the high-strength concrete. Shrinkage cracks are reduced, and shrinkage induced more quickly.

Concrete turbo-alternator blocks are cast in sections in about four vertical "lifts" in a sequence which gives an approximately balanced load on the foundation. If the programme time permits it is an advantage to allow a period of one month between concrete placing of "lifts".

The design of the block components has to be such that their natural frequency (or harmonic) is not the same, by at least $\pm 20\%$, as that of the machine at normal operating speed, i.e. 50 c/s for a 3000 r.p.m. machine, in order to avoid resonance. For a given material the frequency of vibration of any member may be raised by increasing the area of cross-section and vice-versa. Slender cantilevers and thin diaphragm walls are particularly liable to vibration and attention should be paid to their natural frequency and, if necessary, the section of these members should be increased.

In the case of the vertical columns, and other substantial members, there may be a choice between raising or lowering the natural frequency, and the latter choice may be the more suitable as it would represent a considerable saving in material, and still leave the structure sufficiently strong.

A massive monolithic foundation is essential in order to provide a rigid base for the turbo-alternator block and to overcome vibration. The thickness of the foundation should not be less than one-tenth of its length, and a foundation of this type is shown in Fig. 2.6.1A. It is not piled in this instance, as it is founded on a firm stratum at this extra depth. Discontinuity between the block foundation and the basement floor is attempted. The circulating water culverts passing through form a connection, but these can be made relatively flexible by using suitable joints.

Measurements are made to determine any change in the level of the block which could cause rough running of the machine as a result of movement of the foundation or of the block itself. Steel levelling plates are cast into the block around both the basement and operating floor level and a separate reference point provided in the turbine house. If necessary this is carried on an isolated pile driven through a hole in the basement floor and completely independent of the floor. Measurements from the reference point and around the levelling plates are carried out using an optical micrometer level or a micrometer water level. An Invar rod is installed to measure down from operating floor level.

In addition to these readings some blocks have been fitted with a large number of remote reading instruments buried at strategic points as construction proceeded. Strain gauges, thermocouples and moisture cells have been used in combination for this, and give some indication of the behaviour of the block under the influence of changing load and heat from the turbo-alternator.

2.6.4. **Basement or Ground Floor**

In a modern station a considerable part of the total plan area of the power house is occupied by the foundations for the main plant items and the structural columns which carry loads from both plant and buildings. When these foundations have been provided, and the superimposed framing erected, the remaining area is covered with a reinforced concrete slab which normally forms the basement floor. This floor is not designed to carry heavy concentrated loads, but it is strong enough to support minor fixed items of plant, and an overall superimposed load of not less than 500 lb/ft^2 is assumed for design purposes. This loading is adequate to allow items of plant to be set down on the floor during overhaul.

The foundations and floors have to be designed so that provision is made for other works associated with the installation of plant, such as the circulating water culverts, cable tunnels and the troughs for the ash disposal sluicing system. Other works, such as pipe trenches, drains and cable ducts, will also be incorporated. In some stations a 3 ft depth of infill concrete has been introduced above structural floor level to accommodate these services. Modifications during construction can then mostly be effected without alterations to the structure. Finally, when all installation work has been completed, trench covers are fitted and the entire basement floor covered with granolithic concrete, or quarry tiles, or other suitable wear-resistant finish. The floor is laid to falls to carry water from leaks, floor washing, etc., to drains. These should also be able to accommodate the drainage and blow-down from plant both during commissioning and subsequently.

2.6.5. Track Hoppers

Coal trains were formerly emptied by tipping each truck in turn over a ground hopper which discharged on to a conveyor. This method is inadequate for a 2000 MW station consuming 100,000 tons a week and requiring 20,000 tons delivered in one day.

In recently built stations the coal is discharged from bottom emptying wagons into reinforced concrete track hoppers as shown in Figs. 2.6.5A and B. The track hopper is nearly 200 ft long and has a capacity of over 1000 tons. The coal is removed from it on to conveyors by means of paddle feeders which can travel the length of the hopper.

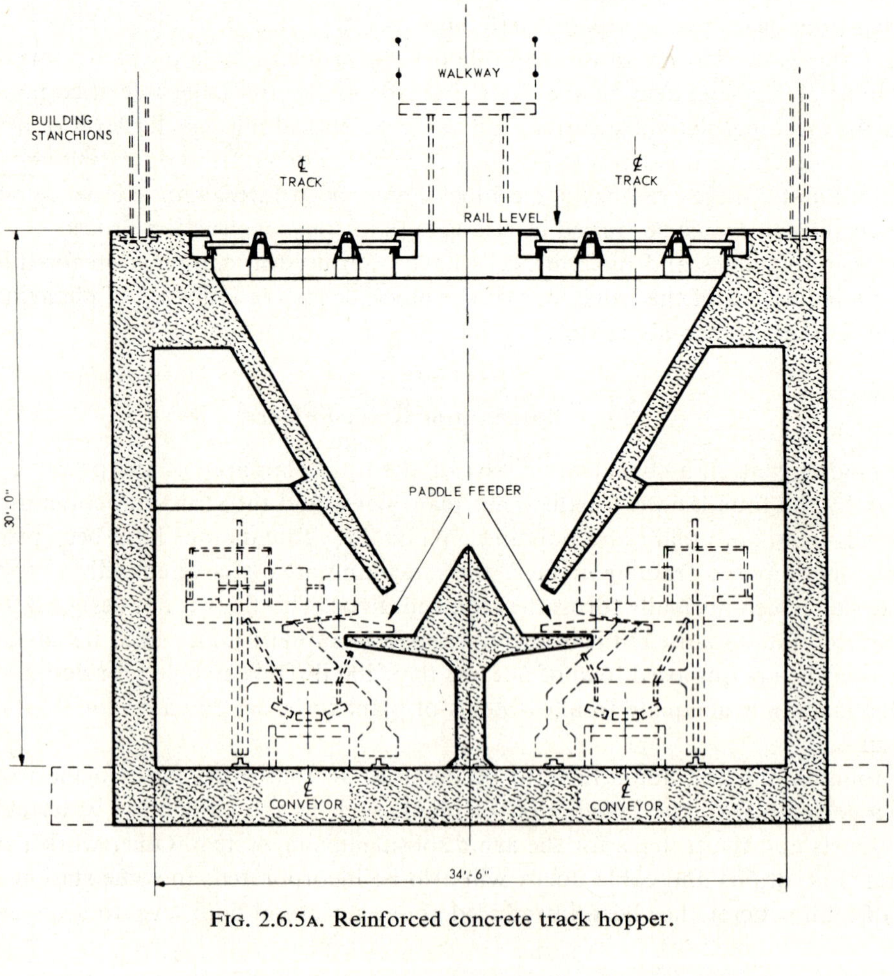

FIG. 2.6.5A. Reinforced concrete track hopper.

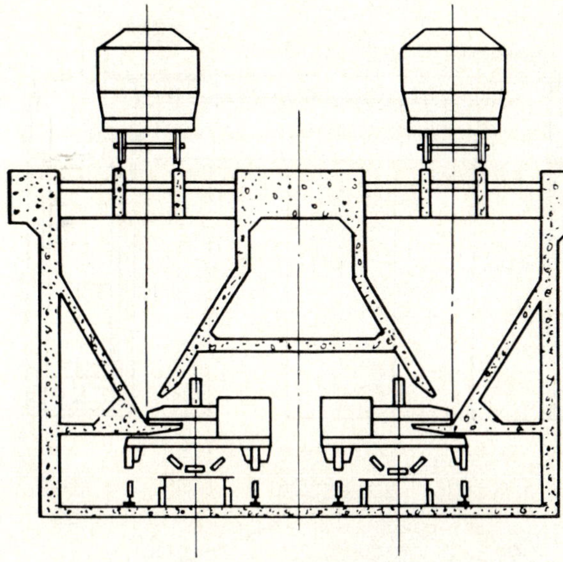

FIG. 2.6.5B. Alternative type of track hopper.

The track hopper extends rather more than 30 ft below ground level and has to withstand considerable earth pressure. Also, if the ground water level is high precautions have to be taken to prevent the empty hopper's tendency to float. This may be done by increasing the dead weight of the whole construction. Another way is to extend the width of the base slab, as shown dotted in the figure, and use the weight of soil above the projections to counteract the tendency.

The beams supporting the track over the hopper should be as narrow as practicable so that the discharge of coal from the railway wagons is not impeded. A smooth tile or $\frac{1}{2}$ inch thick cast glass lining to the hopper face assists coal flow.

An alternative type of track hopper is indicated in Fig. 2.6.5B and this should provide improved handling of coals which have poor flow qualities.

2.6.6. **Chimney Foundations**

A power station having four 500 MW units has either one or two chimneys, the height in each case being about 650 ft. When two chimneys are used each serves two boilers and consists of a single circular shaft. The more recent stations with 500 MW units have one chimney only and this has a separate internal flue for each boiler.

The weight of a 650 ft high chimney having four flues is about 20,000 tons. As the effect of wind on the chimney is to increase the downward load on the leeward side, a reliable subsoil will be necessary if the chimney is to be supported by a conventional foundation bearing on the ground. Chimneys are often for this reason supported on piles or cylinders as shown in Fig. 2.6.6A. In this drawing eight cylinders 15 ft in diameter carry the weight of the chimney down to rock about 35 ft below ground.

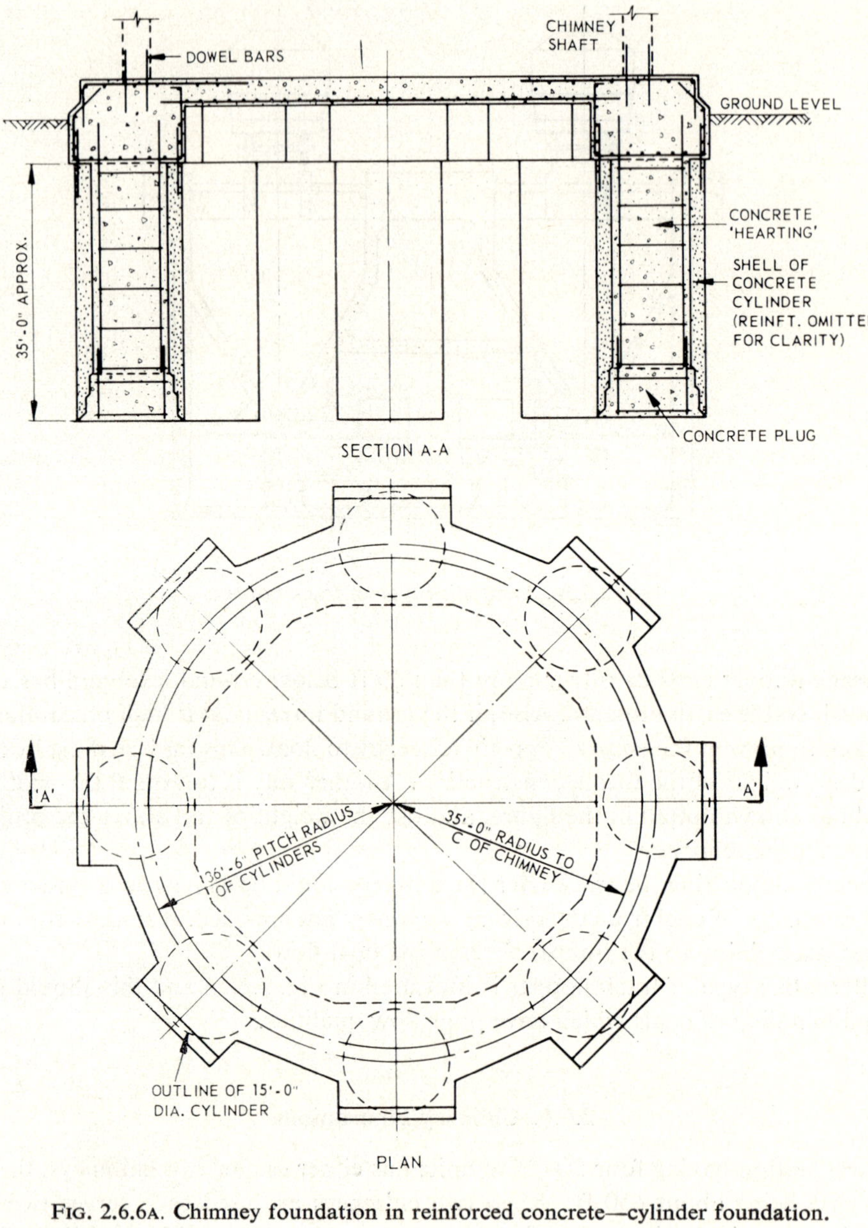

FIG. 2.6.6A. Chimney foundation in reinforced concrete—cylinder foundation.

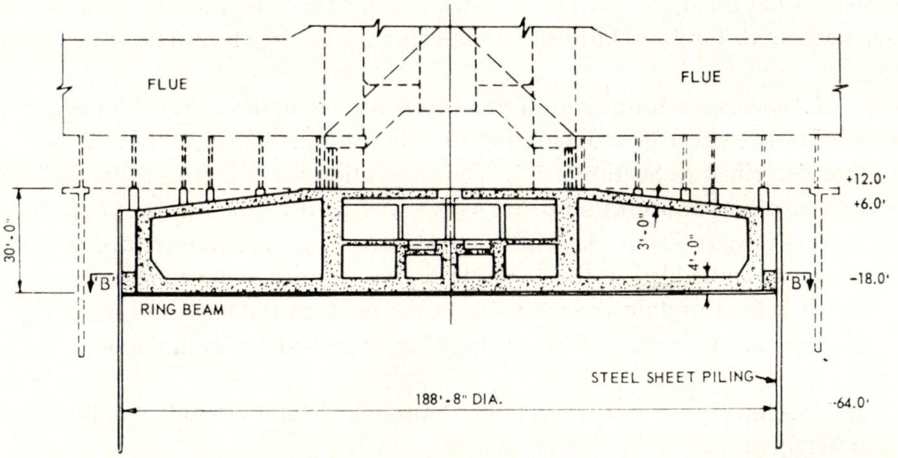

SECTION A-A

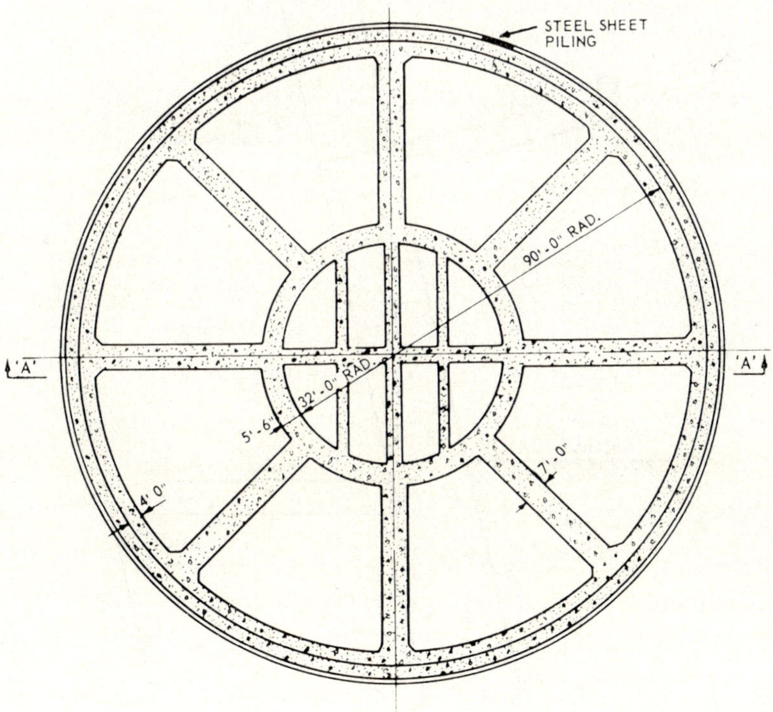

SECTION B-B

FIG. 2.6.6B. Chimney foundation in reinforced concrete—cellular foundation.

A conventional piled foundation requires a number of the piles to be driven as raker piles set radially to the foundation to counteract the horizontal force of the wind on the chimney.

Figure 2.6.6B shows a foundation for a 650 ft high chimney. This is a cellular foundation which has the effect of reducing the foundation weight below that of the excavated material (as described in Section 2.3.5). The excavation for this base is carried out within the steel sheet piling and ground water is lowered as described in Section 2.8.2. In this case the sheet piling is left in place after construction and because welded projections on its inner face are embedded in the concrete it gives some assistance in carrying the load. In addition it helps to retain the loaded subsoils beneath the base and resist their lateral movement. Circular foundations of this type have reached 180 ft in diameter and a depth of 30 ft.

For stations having an output greater than 2000 MW even taller chimneys may be built and heights up to 850 ft are envisaged.

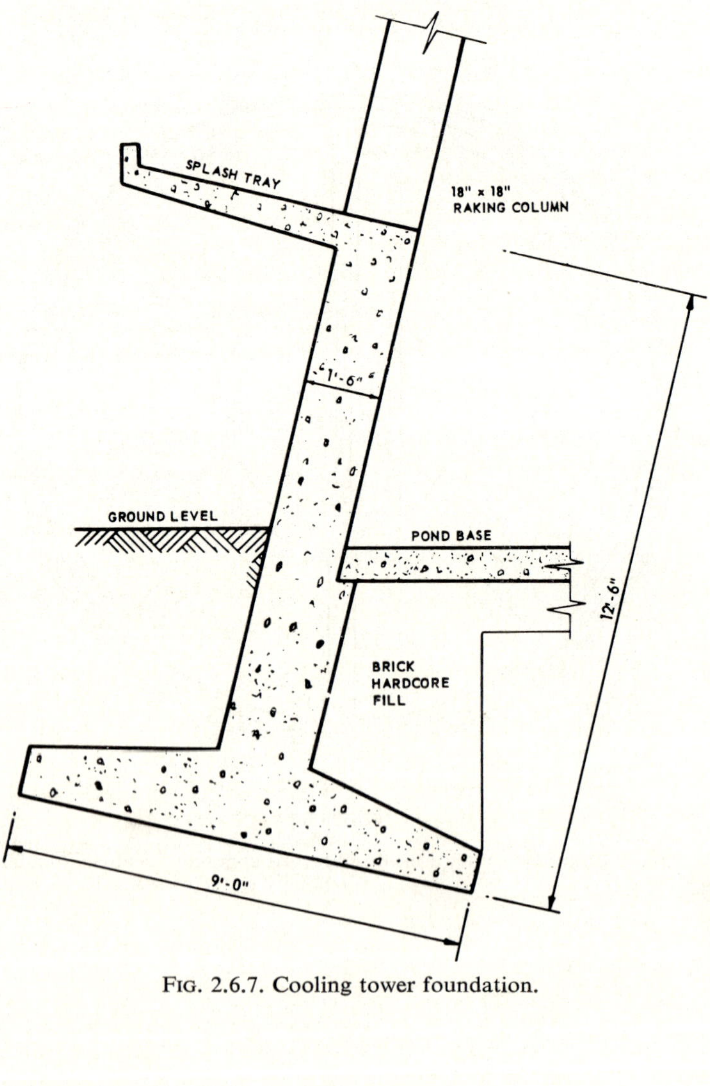

FIG. 2.6.7. Cooling tower foundation.

2.6.7. **Cooling Tower Foundations**

Modern cooling towers consist of a concrete shell supported on a triangulated lattice of concrete columns. (A full description of the superstructure is given in the next chapter.)

A foundation for a cooling tower is shown in Fig. 2.6.7 and some typical dimensions are indicated on the drawing. This is really a circular strip footing and it carries the shell of the tower and the triangulated system of columns. The water in the pond and the tower packing are carried by the base slab.

A piled foundation is less frequently required than for a chimney, as it is possible to distribute the weight of a cooling tower over a larger area, thus giving a lower ground pressure.

2.6.8. **Reactor Foundations**

Deep foundations have so far been used for concrete pressure vessels and these transmit a single very heavy load by means of direct bearing on to a large area of soil. This is in contrast to the loading conditions from boiler structures where a number of individual columns have to be supported over a wide area.

A foundation for a reactor having a prestressed concrete pressure vessel is shown in Fig. 2.6.8. Here the total weight of the loaded reactor vessel is transmitted on to the soil at a ground pressure of about 12 tons/ft^2 through a mass concrete base which is 85 ft in diameter. This high loading can be carried primarily because the foundation has been taken to a hard stratum, but some assistance is also given because the overburden pressure of nearly 2 tons/ft^2 has been removed. Stringent restrictions on the permissible

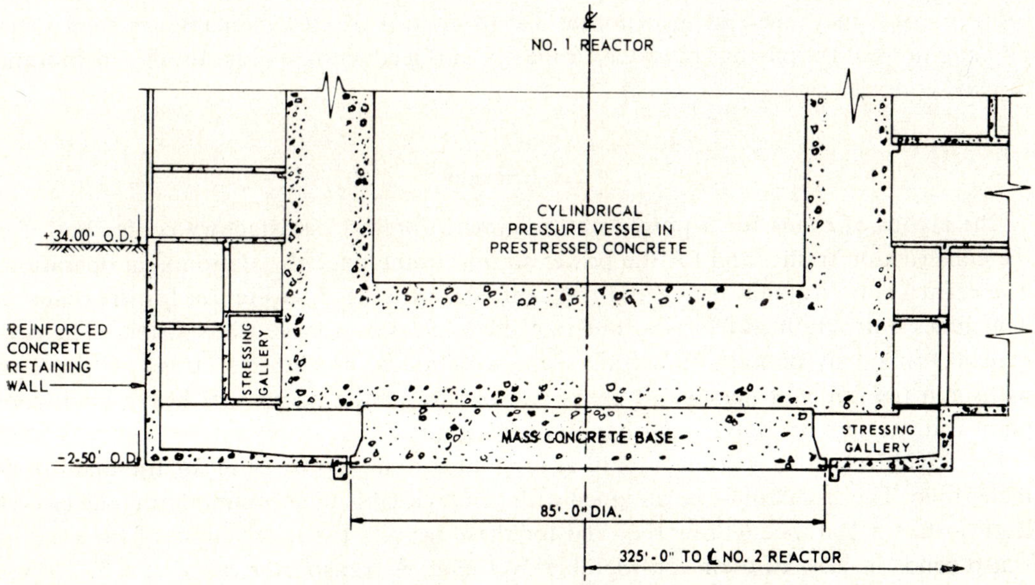

FIG. 2.6.8. Reactor foundation.

long-term settlement are imposed on reactor foundations, making the task of design very difficult.

The mass concrete base is 11 ft thick and was poured in three lifts of approximately equal thickness. Each lift was poured in bays and the vertical joints in each bay were arranged so that they did not coincide with those in an adjacent lift. A minimum period of 4 days was allowed between the pouring of adjacent bays so that the major part of shrinkage in the previous bay would have taken place.

Surrounding the circular base is a reinforced concrete retaining wall. This is 2 ft 6 in thick and supports the considerably lighter external portions of the reactor building. The retaining wall is a completely separate foundation and is not physically connected to the mass concrete base except by a continuous horizontal rubber water stop.

2.7. GENERAL SITE WORKS

2.7.1. Flood Embankments

In order to obtain an adequate supply of cooling water from natural sources many stations are located near to rivers on low-lying land which is subject to flooding. In such cases it may be possible to raise the whole area of the station above flood level by bringing in soil from outside sources. Pulverised fuel ash has been used successfully for this purpose. If the level has to be raised several feet, or there is a shortage of fill material in the district, only the area occupied by the buildings may be filled and the cooling towers protected by a flood embankment. Surplus soil arising from excavations for foundations may be suitable to form the embankment, which will vary from 3 ft wide at the top, with slopes on each side of 1 vertical to $1\frac{1}{2}$ to 2 horizontal, the top being 3 ft above the highest probable flood level. Impervious soils make good flood banks, but sands and gravels, being porous, may need to incorporate a central wall of clay, concrete or steel sheet piling to prevent the passage of water, or may be surfaced with an impermeable membrane of clay, concrete, etc.

2.7.2. Roads

The layout of roads for a power station aims to provide satisfactory routes both for the construction traffic, and for the power station traffic once the station is in operation. The construction traffic is heavy and frequent, and provides a severe test for the roads as it includes not only heavy lorries, but tracked vehicles and large transporters. The permanent traffic may be light on certain sections of roads, but other sections will have to withstand the frequent passage of heavy ash lorries, or the occasional heavy load from plant transporters or mobile coal handling equipment.

Road widths will vary from 9 to 10 ft for minor roads up to 24 ft for the main perimeter road. The layout must be designed with bends capable of accommodating the largest transporters. Clearance will be required for these beyond the inner curve of the road at sharp bends in view of their considerable overhang. A transporter carrying a 570 MVA transformer has a total weight of approaching 300 tons, is more than 100 ft long with six axles at each end with eight wheels to each axle. A road tractor at both front and rear is generally required.

The design of the road and the depth of construction will depend on the strength of the subsoil and its suitability for compaction, and the intensity of traffic loading. The road foundation must be kept well drained by subsoil drainage, or the roads built on shallow embankments if practicable. At this stage allowance should be made for cable crossings under the roads, for the roads to cross cable reserves, and the external surface water drainage to be constructed. This drainage will prove invaluable in helping to keep the site dry and clean during construction.

The nature of the subsoil will influence the choice of road design between two types—rigid or flexible. In both types the intention is to spread a high intensity of wheel loading at the road surface, through succeeding layers of progressively weaker material until the bearing load at foundation level is acceptable.

A typical example of flexible road construction would be a sub-base of rolled granular material, such as sandy ballast between 6 and 15 in thick, depending on the strength of the ground, underlying a base of 8 in of well-rolled brick hard core. The road surface 3 to 4 in thick, would consist of tarmacadam, bitumen macadam or asphalt, in ascending order of cost, laid in two layers of base course and a thinner wearing course of finer material. Asphalt provides the more impermeable surface and is more resistant to traffic wear. Materials such as cement-stabilised granular soils, pulverised fuel ash, and lean mix concrete are also used for bases and sub-bases, their choice being governed by local availability and economic considerations. Flexible roads are easy to repair where individual areas become damaged or subside, but may need extensive resurfacing after use by construction traffic.

Rigid roads are generally constructed of reinforced concrete 6 in to 10 in thick; the reinforcement being a welded mesh fabric. The base should consist of up to 6 in of rolled clinker ash, lean mix concrete or granular material. Rigid roads must be provided with suitable joints at intervals to allow for expansion and contraction. The surface of a good quality concrete road is easy to clean and is resistant to damage, but where uneven settlement occurs the slab will ultimately crack and effective repair is then expensive. This type of construction is often preferable for the heavily trafficked perimeter road.

2.7.3. **Drainage**

The water to be disposed of from a power station consists of:

(a) surface water which is the run-off of rain from roofs, roads and other areas;
(b) foul water, sometimes called "soil", which is discharged from lavatories, ablutions and kitchens;
(c) plant water which is used for cooling and other processes, and drained down or discharged to waste.

These effluents are disposed of through separate systems to avoid contamination.

(a) *Surface water.* This is normally uncontaminated and may be discharged without treatment to a river, estuary or the sea. (However, it should be noted that any discharge, and its quality, has to receive the consent of the River Authority or other statutory body.) The surface water drain for a power station has a diameter of about 30 to 36 in at its point of discharge.

(b) *Foul water*. The total discharge from a power station can be accommodated in a pipe not larger than 9 in in diameter. This is discharged to the local authority's sewer if it is not too distant and has sufficient capacity. When this is not the case a sewage treatment plant must be installed on the site.

(c) *Plant drainage*. The discharge from the various items of plant varies greatly in its purity and quantity. Where water is not contaminated this can be put into the surface water system. (The largest discharge of this nature is from condenser emptying and may be at the rate of 6000 gal/min for 10 min.) Water which may contain oil should be passed through an oil separator unit of sufficient size to allow separation to take place. Very acidic effluents from boiler washing and cleaning will either require to be neutralised on or off site or discharged to, say, the ash lagoon. Similarly discharges of acids and alkalis have to be dealt with from the water treatment plant and in these cases it may be necessary to use drain pipes of plastic or internally coated steel or cast-iron.

The "active" drainage at nuclear stations is carried in plastic pipes laid in larger diameter cast-iron pipes as a safety measure against leakage into the soils. If necessary the plastic pipe may be withdrawn and renewed.

Wherever possible drainage is by gravity, the slope at which drains are laid being such that the flow velocity is not less than $2\frac{1}{2}$ ft/sec with the effluent at a quarter depth in the pipe. The minimum gradient which will produce this varies from 1 in 70 for a 4 in diameter pipe to 1 in 900 for a 36 in pipe. However, pumping may be necessary on occasions for discharges from remote areas, and at the final discharge point from the station. These pressure portions of drainage systems are normally constructed in cast-iron or steel pipes, although asbestos cement pipes are suitable for smaller sizes, and concrete pipes suitably jacketed and reinforced have been used for larger sizes. Cast-iron pipes are also used for foul drains in buildings where extra strength is required and for blow down drains where the discharge may be nearly at boiling point. For the majority of discharges glazed vitrified clay or concrete pipes are adequate: the latter are available in longer lengths and are generally cheaper for diameters above 12 in. Both types are becoming increasingly used with flexible joints, especially in ground subject to settlement. This joint is made by a rubber ring placed round the spigot end of one pipe, which is pushed into the socket of the next. Although more expensive than the traditional cement mortar joint it is very quickly made, and the rigidity of the mortar joint has been a major cause of drain fracture resulting from ground movement. Drains are generally laid on a bed of concrete 6 in thick, and when they are located in buildings or subjected to load from plant, vehicles, etc., or high earth pressure as a result of being at a considerable depth they are surrounded by 6 in of concrete. The concrete surround must not continue over any flexible joint or the flexibility will be lost. The decision to provide protection to pipes in external areas has previously been by rule of thumb. However, the necessity for giving protection may now be determined from a building research station publication which tabulates the load the pipe will be subjected to under different conditions (see bibliography). Care must be taken to pro-

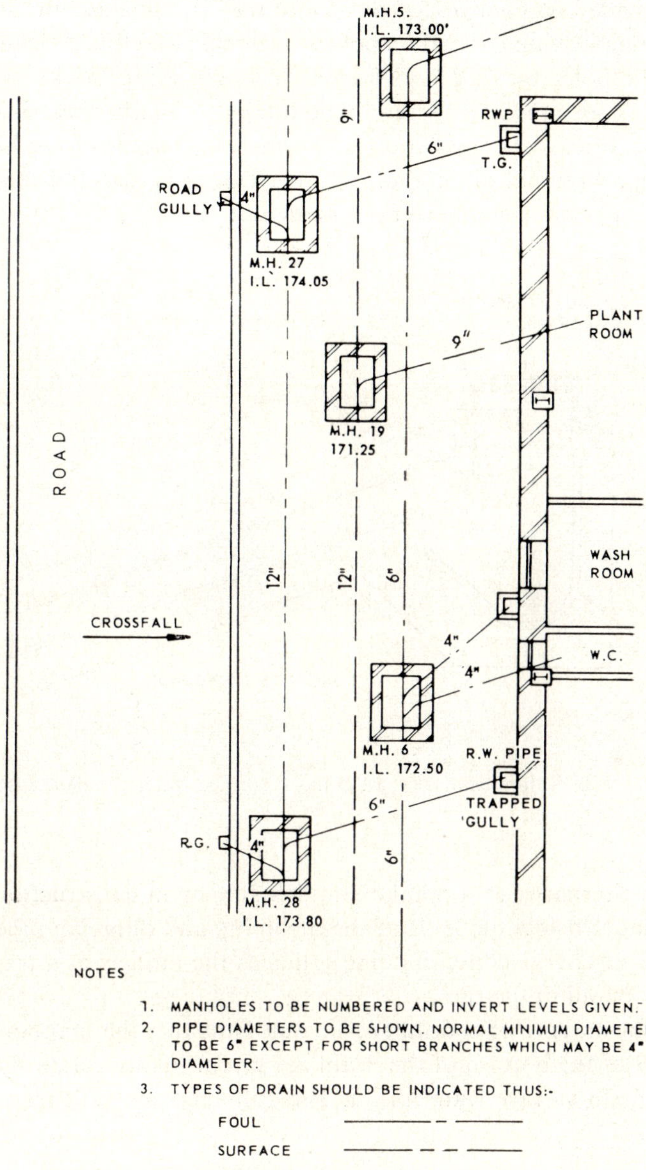

NOTES

1. MANHOLES TO BE NUMBERED AND INVERT LEVELS GIVEN.
2. PIPE DIAMETERS TO BE SHOWN. NORMAL MINIMUM DIAMETER TO BE 6" EXCEPT FOR SHORT BRANCHES WHICH MAY BE 4" DIAMETER.
3. TYPES OF DRAIN SHOULD BE INDICATED THUS:-

FOUL

SURFACE

COMBINED

PLANT

FIG. 2.7.3A. Typical drainage plan.

vide protective bitumous coatings to cast-iron and steel pipes laid underground, as they may be liable to attack by ground water.

Drainage pipes are laid in straight lines and at a constant gradient between manholes. Manholes are provided at changes of direction and at a maximum spacing of 250 ft, though a much closer spacing is usual on power stations, to facilitate cleaning of the drains when blocked. Manholes may be constructed in engineering bricks, with a minimum thickness of 9 in, or of precast concrete. The latter is usually cheaper, particularly for deep manholes, but brickwork is advantageous where it is necessary to break into a manhole to provide an additional discharge. In either case it is essential that the manhole is sufficiently large to permit cleaning to be carried out.

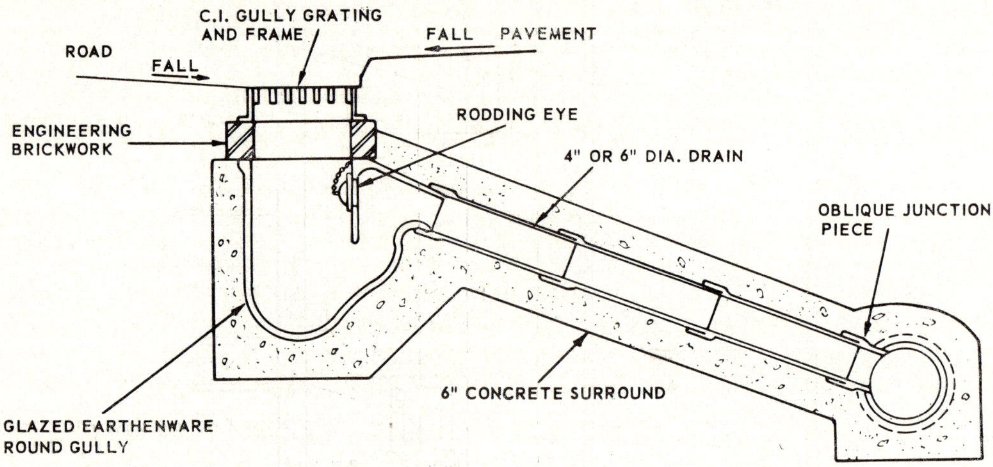

FIG. 2.7.3B. Connection of road gully to surface water drainage system.

A catchpit is an alternative to a manhole and is similar in construction except that the base is a foot or more below the level of the incoming and outgoing pipes. This is an advantage where the discharge contains some solids as the pit acts as a trap. It may not, of course, be used for foul discharges.

Figure 2.7.3A shows part of a drainage layout. Note that the manholes are numbered and that the levels of the bottom of the drain are given. Figure 2.7.3B shows a road gully discharging to a main surface water drain. The gully acts as a silt trap and will require periodic cleaning.

2.7.4. **Railways**

Until the introduction of close-coupled working, large sidings were necessary to accommodate coal trains.

Today, coal trains with bottom-opening trucks are rapidly unloaded without uncoupling and sidings are virtually eliminated except for crippled wagons, and construction traffic where materials are delivered by rail.

Trains are hauled by British Railways' locomotives and the track, loading gauge and minimum radius of curvature have to be suitable for the largest and heaviest diesel or steam locomotives.

Piped subsoil drainage of rail tracks is frequently required in order to prevent the foundation beneath the ballast becoming partially waterlogged and weakened. This is carried out by laying drains of open jointed porous clay, or other suitable pipes, and discharging these to a main drain run, having catchpits, located parallel to the track.

2.7.5. **Coal Storage Area**

The coal storage area is prepared by first levelling off the ground and then laying down a thick bed of rolled ashes or chalk, the level of which is above the highest water table. If the coal store is on an impermeable subsoil, some drainage should be provided, for example the centre of the coal store base may be sloped gently to the outside and drains installed at the perimeter. These measures are necessary to prevent coal retaining considerable quantities of water and thus making coal-handling extremely difficult.

The coal store is generally stocked out to a maximum height of about 30 ft and this gives a ground pressure of just under a ton a square foot. Sometimes the storage height has had to be limited to a lesser figure due to the presence of weak subsoils, and the consequent possibility of subsidence.

2.7.6. **Oil Tank Compounds**

Oil storage tanks, which may vary in size from 20 ft to 150 ft in diameter by 10 ft to 50 ft high, are of welded construction. The intensity of loading is due mainly to the weight of the contained oil and is thus moderate for the lower tanks which will generally be founded on a reinforced concrete raft, or on a bed of crushed stone hardcore and a thick layer of sand topped off by a sand/bitumen mix. For the larger and taller tanks piling is frequently required.

As a precaution against the risk of spread of fire, and against considerable leakage of oil from a damaged tank, the tanks are contained within earth retaining walls (called "bunds") or within reinforced concrete enclosures; these must be sufficiently high to contain the oil capacity of the tanks.

2.7.7. **Ash Disposal Areas**

A modern coal-fired station may have in the region of 4000 tons of ash and pulverised fuel dust a day to dispose of. These materials are in increasing demand by the construction industry, but the demand may fluctuate or the quality vary and the disposal of the surplus material become a problem. Worked-out quarries or pits near the station may be available as a reserve tipping ground, but facilities on, or adjacent to the site may be needed. These can vary from the stripping of top soil from an area and the provision of rudimentary drainage and access roads for lorries, to the construction of ash lagoons for the disposal of water-borne ash and dust. This may require an area of 100 acres or more and involve considerable excavation and building of banks and the provision of drainage to dispose of the decanted water.

2.8. METHODS OF CONSTRUCTION

2.8.1. Site Clearance, Access Roads and Construction Offices

The construction of the power station foundations is carried out in accordance with a detailed programme drawn up to provide the various foundations and general site works in the sequence necessary to enable the building work and plant installation to proceed in accordance with the overall construction programme.

The first steps in the civil engineering work are the clearance of old buildings, hedges and other obstacles from the site, followed by any general site levelling which is necessary. This may involve only excavating soil from high areas and filling low areas with it, or excavating selected material to form flood banks, or importing fill to raise the whole or part of the site level. To provide material for landscaping which is almost invariably necessary, the top soil should be carefully stripped off and stock-piled for re-use when construction is complete. Preparatory work for the installation of an electric power supply with ring mains follows.

The permanent main drainage of the future power station is then installed, and arrangements made to dispose of surface water from the system if the permanent outlet or drainage pumphouse cannot be constructed until later. The soil drainage is connected to the local authority sewers, or if none are available, a small sewage treatment plant will need to be built. Site lavatories for construction workers will be built at various points around the site at this stage.

A new access road from the public highway to the station will be built, and an access bridge if required. The permanent roads of the station are laid down for use during the construction period, and any temporary roads necessary for the construction of the station are built. These may be removed following completion of the works. Hard storage areas for the plant contractor's material and huts are provided by installing subsoil drainage and then surfacing large areas with rolled hardcore and ash and making suitable temporary access roads.

A site boundary fence is built to improve security, and a car park built for the construction workers, preferably outside the boundary fence.

Site offices are built for the Board's site staff and other buildings erected such as stores, garages, concrete laboratory and a site canteen for all the construction personnel. Water and electricity supplies for domestic and constructional purposes are installed. Welfare facilities, such as a site first aid centre, are provided, and in some areas where accommodation is difficult, a hostel or camp may be made available for men who are working away from home.

Completion of the preliminary works will permit later the efficient deployment of the plant and labour required to construct both civil works and plant installation. A separate contract is frequently let for these preliminary works, so that they may be put in hand while the detailed design of the foundations is undertaken and contracts let for their construction.

2.8.2. Construction in Cofferdam and Caisson

All ground is saturated with water below a certain depth depending on natural conditions in the locality. The top surface of the ground water is called the "water-table", and if a hole is dug below that level water will flow into it until its surface reaches the "water-table", the rate of flow being largely controlled by the permeability of the ground itself. Before any deep excavations can be carried out it has to be decided how ground water will be dealt with—clay strongly resists the flow of water through it; silt is more porous than clay, and ordinary coarse sand is even more porous. Gravels are very porous, and water flows easily through them.

From the results of the soil investigation the level of the ground water and the soils which lie below will be known, and the methods to be adopted to keep the excavations dry enough to work in can be decided. For excavations above the water-table and for conditions of low flow through clays or silts it is sufficient to introduce ordinary suction pumps or lift-and-force pumps as the excavation proceeds. This equipment will also remove such surface rain-water as finds its way into the excavation. This method is often sufficient for the main foundations excavation for a power station to proceed in open cut, i.e. with the sides sloped back to maintain stability. For deeper excavations this is seldom practicable and the stability of the sides has to be maintained with sheet piles.

Cofferdam. Figure 2.8.2A shows a diagrammatic arrangement for the construction of a typical closed sheet pile cofferdam used for a construction in water; the same arrangement would hold for a construction on land. Sheet piles are obtainable in various sections, and up to 60 ft in length and have interlocking clutches along their edges, so that when driven they form a water-tight wall. Diesel, steam, and compressed air hammers are used for driving the sheet piles, and electrically driven vibrators have been recently developed for this purpose. The sheet piles are driven down into an impervious layer, or into a firm foundation such as chalk or hard ballast.

Excavation, or dewatering in the case of an offshore cofferdam, proceeds in stages so that the various levels of bracing can be assembled. In large cofferdams the long struts may need supporting at mid-span by temporary piles called "King" piles. The bottom bracings are spaced more closely together to take the increased load with increased depth. Excavation and bracing is continued until the foundation level is reached, when concreting of the permanent structure follows in the normal way. Precast piles may be driven below the excavated level if needed. If the sheet piles are to form part of the permanent structure, concrete is poured against them, and the temporary bracings removed as work progresses. This practice is not recommended, however, for should there be any ground pile movement or any flexing of the sheet piles when removing struts serious damage can be caused to the recently poured concrete structure. The sheet piles are otherwise driven wide of the permanent structure, so that as it rises within the cofferdam, load from the sheet piles can be transferred to the concrete walls by short struts and wedges. The space between the sheet piles and the concrete is finally backfilled or flooded and the piles extracted.

This form of construction is widely used for such structures as pumphouses, intakes and outfalls, ash plants, and the deeper parts of large open cut excavations. Circulating water culverts are frequently constructed within continuous sheet pile cofferdams. Intake and outfall structures lend themselves to construction within a circular cofferdam. In this type cross-bracing can be eliminated by using reinforced concrete circular walings in compression to support the sheet piles, leaving the centre free of obstructions.

The design of a cofferdam is a difficult exercise of soil mechanics not lending itself to precise analysis, and failures have occurred. Failure may be due to the mechanical failure of the lower struts, or in a plastic soil the unbalanced earth pressures may cause the sheet piles below the bottom strut to buckle and the bottom of the excavation to heave up. Unbalanced hydrostatic head may cause water and fine material to flow under the toes of the sheet piles, and the bottom of the excavation to "boil up".

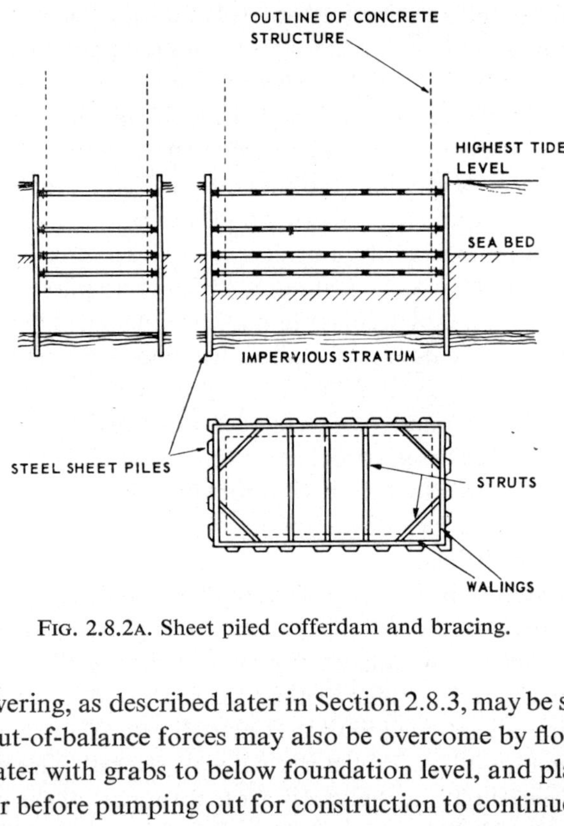

FIG. 2.8.2A. Sheet piled cofferdam and bracing.

Ground water lowering, as described later in Section 2.8.3, may be sufficient to deal with this problem. The out-of-balance forces may also be overcome by flooding the cofferdam, excavating under water with grabs to below foundation level, and placing a layer of mass concrete under water before pumping out for construction to continue. This mass concrete must be thick enough to balance the forces likely to cause failure, and to form in effect a permanent bottom strut to the cofferdam.

Injection grouting techniques may be employed to increase the strength of the soils and reduce their permeability. Depending on the nature of the soil, the materials used vary; examples are cement grout, special clays (sometimes mixed with cement), P.F. ash, bitumen compounds, and a variety of chemicals. These have been so effective that excavation within a grout curtain has been carried out without recourse to sheet piling in some power station work on the Continent.

In very large cofferdams, which may be required to enclose the entire area of bulk excavation for a power station, it is impracticable to support struts right across such a large area. In this case the sheet piles are tied back at the top with steel rods to anchor blocks in firm ground, or an earth bank is left immediately at the edge of the foundations to support the sheet piles. When the permanent foundation has been built sufficiently close to the sheet piles, the earth bank is removed and replaced by raking struts braced off the foundations.

An interesting recent development in cofferdam construction has been the use of reinforced concrete diaphragm walls below ground level, in place of sheet piling. A special grab is used to excavate a trench 1 ft 6 in to 2 ft 6 in wide which is filled with a solution of Bentonite as excavation proceeds. Bentonite is a mud having gel-like properties. The Bentonite solution holds the sides of the trench apart and forms a thin impervious membrane at the surface by interaction with the soil. Excavation is continued through the Bentonite solution until sufficient depth is reached. A reinforcement cage is lowered into the solution and concrete poured from the bottom upwards using a trunking as in normal underwater concreting, the Bentonite solution being expelled from the top. A water-tight wall is thus formed within which, with suitable bracing, excavation may proceed. Walls made in this manner have been incorporated in permanent works with little or no surface treatment.

The depth to which construction can proceed in cofferdam is limited according to soil conditions. Two further methods for forming deep foundations are given below. Both these methods have been used in construction of the intake structures and pumphouse structures for power stations, and are equally suitable for construction on land or through water.

Compressed air caisson. A steel or concrete working chamber is constructed at ground level, its shape corresponding to the shape of the completed structure. The working chamber rests on a cutting edge, and the construction of the permanent reinforced concrete walls is commenced. One or more working shafts are bolted to the roof of the working chamber, air locks provided and excavation commenced by hand under compressed air much as described later in Section 2.9.4. Air pressure is reduced from time to time to allow the structure to settle under its own weight. Construction of the walls is continuous with the excavation and its weight is balanced by skin friction between the walls and the soil, and the uplift from the compressed air. When foundation level is reached, the bearing surface is cleaned up, the working chamber filled with mass concrete and then pressure grouted to ensure good contact. The practical depth to which this type of construction can be continued is 120 ft due to the increased pressures of air required. Two or more working chambers can be used in conjunction, as demanded by the size of structure. For work over water, the working chamber is constructed in a steel cylinder and launched and floated into position, or assembled on a temporary staging which is later removed.

Open grab monolith. A steel cutting edge is assembled at ground level on temporary foundations and the permanent concrete walls built up for a height of 15 to 20 ft. The soil is grabbed out from below the cutting edges and the monolith descends under its own weight. Excavation is halted from time to time while extra sections of wall are built up. The process is continued until foundation level is reached, excavation being carried out under water when water bearing strata are encountered. The monolith is finally plugged with mass concrete placed under water. Divers using high pressure water jets and air lift pumps are used to clear the slurry from the foundation level. The monolith is finally dewatered and the reinforced concrete work completed.

The permanent structure has to be made strong enough to withstand the additional stresses involved in this type of moving construction. In a large structure temporary bracing walls may have to be built and later removed. Pumphouses have been successfully constructed using this method, their "egg crate" internal structure being particularly suitable from the point of view of strength. In one instance the moving structure was some 130 ft×170 ft in plan and 60 ft deep and weighed in the order of 60,000 tons.

Both these types of construction are illustrated diagrammatically in Fig. 2.8.2B.

2.8.3. **Ground Water Lowering**

When major excavations are required where a high rate of inflow through the coarser sands and gravels may be anticipated, the ground is generally dewatered. This is achieved by encircling the area to be excavated with a series of wells from which water is pumped continuously until the whole of the work within the excavations is completed. When water can be pumped from the wells faster than it can flow through the ground, the water-table in the vicinity is lowered, and the layout of the wells and pumps is so designed that the water-table will be lowered beneath the bottom of the excavations, leaving the whole area within the ring of wells dry for working.

Suction pumps installed at ground level may only be relied upon to lower the water-table to about 15 ft below the pump level; the vacuum in the pumps is unreliable to raise the water from a much greater depth. One suction method of dewatering is known as "well-pointing". Here the wells are 2 in diameter tubes with filters at the lowest points, sunk into the ground at about 6 ft centres around the area of excavation, and connected to a common main at ground level. The object of the filters is to prevent fine soil particles being taken out of the ground.

If the pumps have to raise water much more than 15 ft, a two-stage well-point system may be used, with the sides of the excavation being taken down in two steps. The outer row of well-points is sunk from the original ground level and the inner row from the intermediate level.

For greater depths submersible electric pumps are installed at the bottom of larger wells. This is the "deep-well" system, which in suitable soils will be adequate for most of the requirements for power station construction. A "deep-well" is made by driving a perforated steel tube to the required depth in the ground, removing the encased soil and inserting an inner tube, also perforated. The intervening space between inner and outer tubes is filled with a properly graded sand to act as a filter to the inflowing water. The submersible electric pump, with its delivery pipe and electricity supply cable attached, is lowered to the bottom of the inner casing and set to work. The diameter of the well tubes varies according to the size of the pump which has to pass down inside, but the outer tube is generally about 2 ft in diameter. Deep-wells are installed much further apart around the excavation than well-points, the exact spacing depending on the rate of pumping required.

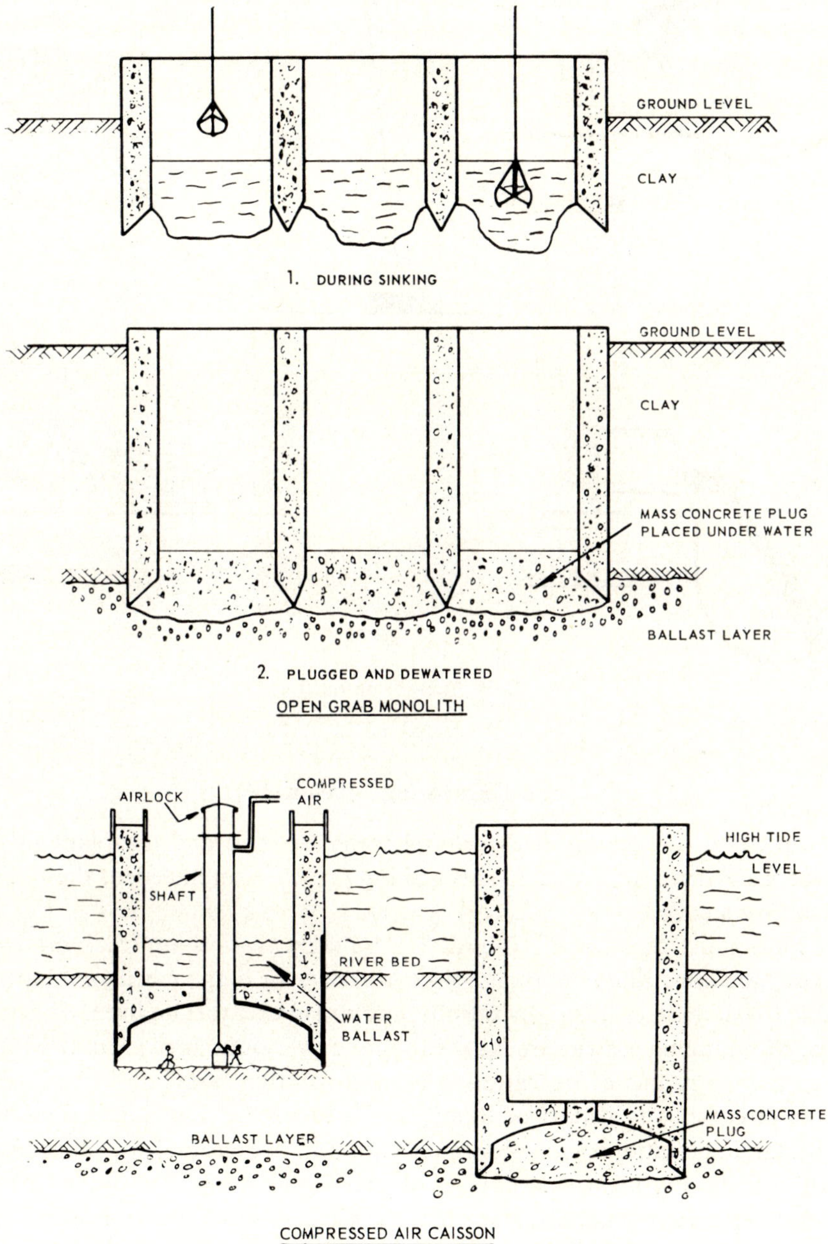

FIG. 2.8.2B. Caisson and monolith construction.

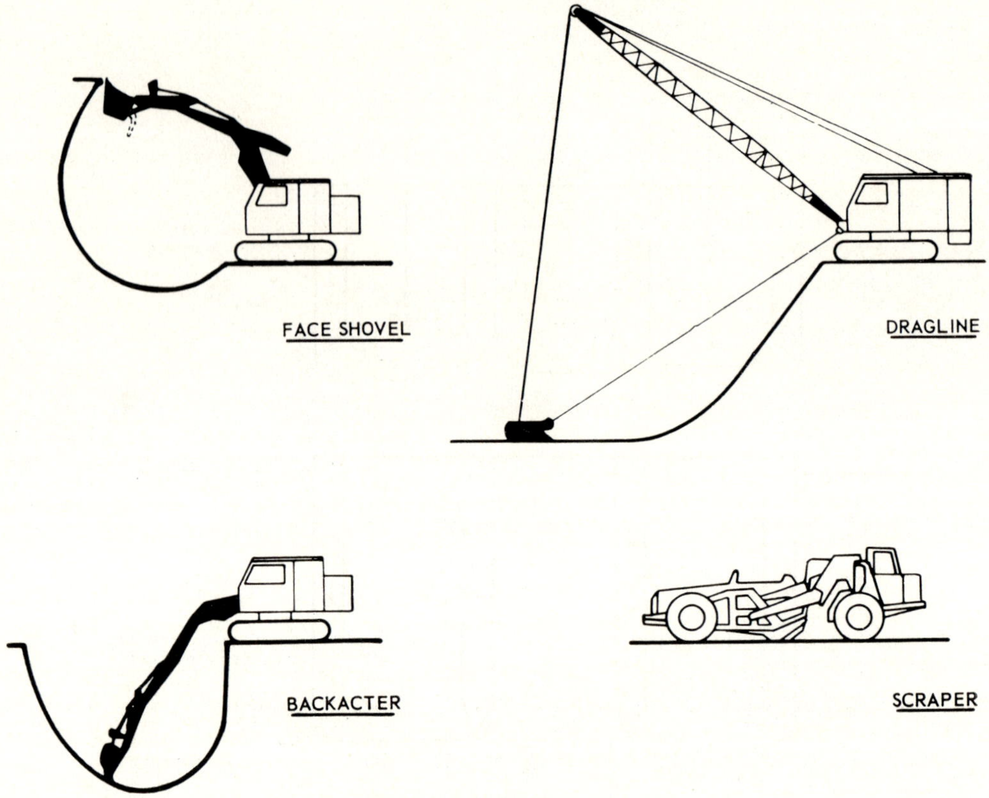

FIG. 2.8.4. Excavating machinery.

2.8.4. Excavating Machinery

With the method for dealing with ground water decided and the plant installed, or available for use when required, and sheet piling driven where needed, the bulk excavations can be commenced.

The machines most used for the foundation bulk excavation are track mounted excavators. These are illustrated in Fig. 2.8.4. A variety of digging arms and attachments are available to tackle the various operations involved in excavation, but the basic machine is common. Mounted on crawler tracks it can move over quite soft ground without bogging. It has a wide radius of operation, having a rotating body, and can dig and then turn to load direct into transport or dump onto a stockpile. Digging is done whilst the machine is standing stationary, avoiding churning up the ground by continuous movement of the tracks. Frequent small shifts of the machine must be made to enable the cutting tool to dig at its best position into the excavation. These machines have until recently been built as a crane body, with winch drums for operating the various attachments by wire ropes and pulleys. All the attachments are thus capable of fitting to the one machine. A recent development has been the introduction of all-hydraulic operation for the digging arms, which limits the choice of equipment to backacter, face shovel or a grab having a limited reach. The hydraulic type backacter and face shovel are illustrated.

The various front attachments are as follow.

Dragline. The dragline bucket is attached to the crane rope by means of a yoke, to which a hauling rope or "dragline" is shackled. By using the two ropes an operator can cast the bucket some distance beyond the reach of the machine jib. Digging is achieved by winding in on the dragline. Some skill is required to load direct into transport; the material can also be side cast to form banks or spoil heaps. This machine can be used for bulk excavation well below track level in soft to medium ground or well-blasted rock. With a special perforated bucket it can carry out below water excavation. It is also used for ditch digging and bank building, and may be used for loading concreting aggregate into the batching plant from stockpiles. Although the digging arrangement may seem haphazard, in skilled hands the machine is capable of very accurate work and high output.

Grab. A hinged two-part bucket attached to the crane rope is dropped vertically onto the soil, and is closed on lifting, taking a "bite" of material. This is the machine most frequently used for very deep excavation, and is well suited to underwater work. It is also used for excavation within cofferdams or sheet piled trenches, where the bracing is too close to allow for other forms of excavation, and between closely spaced piles for the same reason. Loading aggregate into the batcher plants is another task for which the grab is often employed.

Backacter. The backacter bucket is mounted on hinged arms, and as its name implies, digs straight back towards the machine, and below track level. It can dig soft to fairly hard material, and can work to very fine limits, leaving a minimum of hand trimming. It is used for foundations of all kinds, but is especially useful for trenches up to a medium depth, and excavation between closely spaced piles.

Face shovel. The face shovel is mounted on hinged and sliding arms and digs away from the machine. It can operate to slightly below track level, but is most productive when digging into a vertical face above track level. The face shovel is capable of digging in materials far firmer and harder than the other types of equipment. The bucket has a hinged back, operated by a trip wire for dumping. This machine is frequently used in quarries and pits to excavate and load fill material, aggregate or blasted rock into transport.

The above machines, even when fitted with hardened steel teeth, are capable of operating only in soft to medium ground, with the exception of the face shovel, and this is not capable of cutting into rock. A considerable amount of drilling and blasting is required where large buildings are founded on rock, the loosened material being subsequently excavated and loaded into transport.

The excavator. Rigged as a crane, this is of great use on site for such work as handling skips of hand excavated spoil, removing cut-off pile heads, lifting reinforcement steel and shuttering material, and placing concrete from skips. It is also extensively used for handling, pitching and driving sheet piles, and for boring and driving piles when fitted with suitable rigs.

A range of machine capacities are available, the sizes most often seen on power station sites varying from $\frac{1}{2}$ yd^3 to 2 yd^3 capacity. The suitable size and number of machines needed varies from site to site and depends on such things as the nature of the ground, the size of the excavation and the programmed time for completion. Quantities of excavation in large power station work can now be expected in the region of 50,000 yd^3 or so.

Production is also governed by the rate at which the excavated material can be transported to the dumping grounds, which may be situated some miles off site. Off-site dumping is generally done by road lorries of 5 to 8 yd capacity: on-site dumping may enable dumpers and tractor-drawn dump-trucks to be employed. The sizes of vehicles will depend on the size of the excavators, but the number used should be sufficient to keep the excavators working with no uneconomic standing time.

Tractor shovels or loaders. These are used frequently for attendance on the main excavation work. They are tracked or rubber-tyred tractors fitted, not with a bulldozer blade, but with a hydraulically operated bucket. The bucket is sometimes hinged in two parts to form a grab. These machines are invaluable in trimming bulk excavations, clearing and loading spoil and slurry, laying temporary access roads for the muck lorries, carrying out minor excavations, as well as being used as bulldozers for spreading and levelling. They range in size from $\frac{1}{2}$ yd^3 capacity upwards, the smaller loaders frequently being fitted at the rear with a hydraulically operated small backacter arm. These machines are suited to minor trenching for drains, and then backfilling the excavation and clearing away surplus material. Tractor shovels suffer from the disadvantage that they dig by driving forward, and must then be backed and manoeuvered to a stationary lorry to off-load so churning up the ground.

Scraper. Bulk excavation is also carried out using scraper equipment. A scraper is a large bucket or bowl mounted on rubber tyres, and either tractor drawn or self-propelled It digs by lowering a wide cutting edge into the ground while travelling forward; the depth of cut being a few inches depending on the nature of the ground and the work. These machines are employed for topsoil stripping, site levelling, shallow excavations over a wide area and building of earth embankments. Tractor-drawn scrapers have a limited economic distance over which they can travel to off-load, as they are slow and have high powered engines for loading. Self-propelled scrapers run entirely on rubber tyres and have a high travelling speed, and in consequence have a longer economic haul distance. Any deficiencies in self-loading ability can be made up by having a high-powered crawler tractor at the digging area to assist the loading by pushing the scraper. Scraper sizes range up to 32 yd^3 capacity. They are limited by ground conditions in that they will not operate successfully in ground that is either too soft or waterlogged or too hard. In suitable ground conditions the whole, or the substantial part of the bulk excavation for some power station foundations has been dug using scraper equipment.

2.8.5. **Formwork and Reinforcement**

As soon as an area of excavation is completed and trimmed to the correct profiles, a layer of concrete usually from 3 in to 12 in thick but occasionally thicker, dependant on soil conditions, is laid on the exposed surface. This layer, known as "blinding concrete", serves several important functions. It protects the exposed subsoil from the effects of weather—drying out by wind and sun, or softening by rain—and helps to control seepage of ground water. The subsoil is protected by it from the later activities of construction, such as pile stripping, and handling and fixing of reinforcement, timber, pipework, etc., and can allow light construction traffic to cross. The shapes of the underside of the foundation can be accurately formed, and the working area can be kept clean and dry for all the activities of foundation concreting to proceed.

Before concrete can be poured, the reinforcement steel has to be cut, bent and fixed and the formwork or shuttering constructed. These are major operations in the foundation construction and employ a great deal of hand labour. The amount of reinforcement used throughout the whole of the foundation construction may be in the order of 10,000 tons, most commonly in mild steel rod ranging in size from $\frac{3}{8}$ in to 2 in diameter, from a few feet to 60 ft long, and either straight or bent into complicated shapes. The cutting and cold bending is generally carried out on site in a reinforcement yard equipped with power cutters and bending machines, and having a derrick crane to handle the steel.

The prepared reinforcement is transported and off-loaded near the place of work, or handled by crane, but the final lifting into place of the rods and binding together of their intersections with iron wire is done by hand. There is limited scope for pre-fabrication, although mesh reinforcement is used in roads and for lightly loaded flat areas. Spacers, in the form of small concrete blocks, are used to maintain the correct clearance, or "cover" between the steel and the outside of the concrete to prevent corrosion of the bars. Short rods are used to space one layer of steel from another, and the top layers of steel in a foundation are held up by bars bent into "stools". When completed the reinforcement forms a rigid cage; care must be taken to ensure that space is left to allow concrete to be poured effectively around the steel, and for poker vibrators to be inserted.

The concrete may be poured against permanent sheet piling, the open sides of excavation in some circumstances, or rough blockwork walls and rough shuttering below ground level, but where the finished surface is exposed to view it has to be contained by accurately made and closely fitting moulds called formwork or shuttering. The internal faces of the formwork are made from planed timber, plywood panels or steel panels fitted closely together to prevent the escape of fine materials, and framed and braced with walings and props to withstand the considerable pressure exerted by the wet concrete. The faces of the formwork are coated with a mould oil before the concrete is placed to prevent adhesion at the surface. The formwork materials are intended for re-use as many times as they remain serviceable, but have to be left in place long enough for the concrete strength to be developed. The shuttering assists in preventing premature drying out which would otherwise take place due to the heat generated in the setting concrete.

FIG. 2.8.5. Main foundations showing reinforcement prior to concreting.

Besides forming the permanent outline of the concrete, a considerable amount of rough shuttering is used to divide the foundations up into convenient sections or "pours". This dividing has to take due account of the points of maximum stress and shear in the foundations when permanently loaded. The sizes of pours in the main foundation areas vary up to 300 yd^3 in size, depending on the capacity of the concrete mixing and transporting plant. It should be the aim of the contractor to have sufficient pours, both large and small, ready to keep his plant producing a steady average amount of concrete each day, and to have other areas of the foundations progressing steadily at various stages of preparation to maintain this average concrete production over a long period.

Such things as drains, cable ducts, starter bars for reinforced concrete plinths and columns, holding-down bolts for plant and steel column pockets for bolts and steelwork, have all to be firmly fixed accurately in position before concreting can take place. It is essential that the bars are free of loose rust, oil or earth (or a satisfactory bond will not take place with the concrete) and that the inside of the formwork is clean. Before concreting is carried out these points are inspected and the whole arrangement is checked to see that it complies with working drawings.

Figure 2.8.5 is a photograph of the main foundations for a nuclear power station. The reinforcing bars are clearly shown in position prior to concreting and the formwork may be seen in the foreground.

2.8.6. Mixing and Placing of Concrete

The amount of concrete now required in power station foundations totals over 250,000 yd^3, calling for very high average weekly productions of up to 5000 yd^3. In one case recently 10,000 yd^3 were poured in one week.

The choice of size and number of mixing plants depends on the rate at which concrete is required to achieve the programmed construction for the foundations, and the size of pours which can be prepared by carpenters and steel-fixers as excavation proceeds. The choice of plant to transport the concrete economically is dependent on the same factors, and on the type of foundation being constructed.

The problem of regularly producing amounts of concrete of this nature economically is generally tackled by installing a central mixing plant of high capacity to meet the average demand throughout the contract. This central mixing plant is intended to operate throughout most of the life of the contract. The extra capacity required to meet periods of maximum demand is obtained by setting up other mixing plants around the periphery of the work using more mobile equipment. This arrangement has the advantage that these individual smaller plants can be moved from time to time to follow the construction without disrupting production, and similarly a breakdown of plant will not cause so serious a stoppage. The travelling distance for the concrete to individual pours can be arranged more economically, and special requirements of mix and aggregates can be more readily accommodated.

The central mixing plant is composed of one or more 1 yd^3 mixers of the pan type or non-tilting drum type set high enough above ground level to discharge into transport, or the hopper of a concrete pump. The mixers are built into a substantial structure containing aggregate hoppers, pre-set weigh-batching machines, bulk cement silos and water tanks. Large stock piles of aggregate are built near the plant and from these the aggregate is loaded into the hoppers by an excavator or chain elevator. The materials are pre-weighed and fed into the mixer pan or drum.

A pan mixer consists of a large diameter open top stationary cylinder in which paddles rotate. This type produces a satisfactory concrete mix in as little as 30 sec, compared with the customary $1\frac{1}{2}$–2 min for the rotating vertical drum mixer. In theory a one yard mixer may be able to produce 60 yd^3 of concrete per hour, but in practice this figure is likely to be more like 40 yd^3. The smaller mixing plants around the site will consist of one or more non-tilting drum mixers of $\frac{1}{2}$ yd^3 to $\frac{3}{4}$ yd^3 size, each with an elevating hopper incorporating a hydraulic weighing device, fed with aggregate from stockpiles by a winch-operated drag shovel, and with weighed amounts of cement from a silo.

The main building foundations on power station sites cover a large area and are shallow in comparison to their plan dimensions. This type of construction favours distribution by pump, as it is not easily accessible to wheeled traffic. The concrete pipelines vary from 4 in to 7 in in diameter; the individual lengths of pipe are designed to be manhandled across scaffold stagings, and the joints are quickly made and broken. Bends are available so that the pipeline can reach any part of the pour, and the breaking and re-laying of pipelines is carried out frequently as work proceeds in order to deposit the concrete exactly where required.

Mechanical pumps are most frequently used to push the concrete along the pipelines. Concrete is fed into a hopper from the mixing plant, and gravitates through an automatic valve in the bottom of the hopper into the horizontal cylinder of the pump. A steel piston operated by an electric or diesel engine forces the concrete past a second valve and along the pipeline. The pumps may be single or double cylinder. Pneumatic pumps of various types are also available. The largest pumps have a capacity of about 45 yd^3/h. Horizontal distances of 1200 ft can be pumped, or heights of 120 ft or so.

The types of aggregate and the concrete mix design have to be suited to being handled by these types of equipment; the concrete must be neither too wet nor too dry, nor too stony, and the quality must be controlled to within fairly fine limits to achieve success. A blockage in the pipeline due to bad concrete may result in every joint having to be broken and the concrete tipped from each pipe before it sets! The pipeline is cleared at the end of concreting by forcing a rubber ball through by compressed air. Careful calculation has to be made of the volume of concrete required, and allowance made for the concrete remaining in the pipeline if wastage is to be avoided.

Concrete is also transported about the site in front tipping dumpers, in crane skips mounted on specially adapted dumpers, or in lorries fitted with large hydraulically tipping skips holding up to 5 yd^3. Temporary roads and ramps are required to reach the places where concrete is required. It is not good practice to tip concrete from a great height into the work, as segregation of the materials occurs and the reinforcement and shuttering may be damaged. Hoppers with trunking or chutes are arranged to take the discharge of the concrete and distribute it within the pour area. With deep foundations cranes are used to handle skips of concrete into the correct points of discharge; the skips are either brought to the crane full, or are filled at the crane by tipping from the lorries or dumpers. For small foundations mobile cranes would be used, but for larger structures one or more derrick cranes would be built to carry out all the work, from excavation to concreting.

Concrete for power station foundations is now invariably vibrated to compact it into the formwork and around the reinforcement. This produces dense concrete free of voids and air pockets, with a good surface appearance and with maximum strength. Vibrators may be of the internal poker type, or external type working on the outer face of the formwork. The poker type is mostly used in large foundation pours. They vary in size from one to four inches diameter and are some 20 in in length. They are operated by electricity, by compressed air, or by flexible drive from small petrol engines.

Although an apparently simple operation, some care is required on the part of the operators to ensure that the concrete is properly compacted. The depth of concrete layers should be kept to not more than 2 ft to allow air bubbles to escape. When using vibrators care should be taken to ensure against over vibration causing segregation of the concrete.

The contractor often chooses to install a stationary compressor plant on site, with a galvanised steel main fitted with tapping points laid over the foundation area. A great deal of compressed air plant is now used for vibrators, for cleaning up the face of concrete ready for the next pour, for cutting away unwanted spillage, and for blowing the formwork clean of dirt and rubbish before concreting is carried out.

2.9. RIVER AND SEA CIRCULATING WATER SYSTEMS

2.9.1. Civil Engineering Structures in Direct Cooling Systems

In Chapter 1 reference is made to the various circulating water systems. In a direct-cooled system the water is taken from a river or the sea and then passes in sequence through four structures:

(a) An intake built at such depth in water as to ensure that adequate water can be drawn under all conditions of flow or tide; and of sufficient width that the entrance velocity of the water is not sufficient to cause erosion of the bed nor bring in an excessive amount of floating debris or silt. A coarse screen is installed across the front of the intake to exclude large debris.
(b) A screen chamber to house the fine screens which remove from the water the debris which has passed the coarse screen at the intake. The screenings are automatically jetted off into a trough from which they are sluiced into a metal cage for disposal.
(c) A pumphouse to contain the pumps which take water from its natural level in the screen chamber and distribute via concrete culverts or metal pipes to the condensers.
(d) An outfall to discharge the water heated by use in the condensers back into the river or sea at a low velocity which will not cause erosion of the bed or banks nor currents dangerous to craft.

The cooling water passes through these structures in the order given. It flows in "intake" culverts of reinforced concrete or steel between the intake and the condensers, and after use in the condensers it flows in "discharge" culverts to the outfall.

In a tidal river care must be taken in positioning the intake and outfall to ensure that recirculation of the heated outfall water into the intake, and its consequent bad effect on operating efficiency, does not take place, or is minimised. This is sometimes easy if the station is sited so that water can be drawn from a river and discharged into a creek or into the sea. Where no such solution is possible a model test may be necessary to ensure that circulation is as small as possible at all states of the tide. Several arrangements and designs of intake, outfall and jetty can be tried until the best one is found. Such tests will also help to indicate whether dangerous currents or excessive scour will be caused by any particular design.

2.9.2. Culverts

The overall design of the culverts is governed by the layout of the station, but has to take into account the hydraulic design of the CW system and condensers and the costs of pumping. The cross section size is based on the allowable water velocity and depends on the total flow to each set; sizes of 8 ft to 10 ft square can be expected. Care has to be taken in the design to avoid sharp corners which would set up turbulence in water flow, to avoid areas of dead water which would encourage marine growth and siltation and to allow for the escape of trapped air from release valves or surge pipes. Access points are required for inspection and cleaning. The culverts have to be designed to resist the maximum discharge head of the circulating water pumps on a closed condenser inlet valve.

Culverts are generally constructed under the main building foundations, or below ground level in shallow excavations, either open cut with sloping sides, or within a continuous sheet pile cofferdam. Separate culverts are allowed for each set, and this results in up to four being built side by side. A check should be made to ensure that the culverts when empty are not subject to flotation caused by the hydrostatic head of water in the ground.

Steel culverts up to 10 ft diameter and with walls up to $\frac{3}{4}$ in thick are used, and where buried in the open ground require the external protection of bituminous compounds reinforced with hessian or glass fibre. Difficulties have been experienced with internal protective linings as these deteriorate with age and use, and tend to break up and clog the condensers.

Reinforced concrete culverts are constructed with walls up to 2 ft thick and rectangular in section, with splayed corners to distribute stress. In the case of inlet culverts which operate under pressure, an octagonal section may be adapted with reinforcement steel, in the form of hoops, to resist bursting pressure. Where multiple culverts are laid side by side the structure is made monolithic, with division walls between the culverts. The culverts are concreted in lengths up to 30 ft to allow shrinkage to take place as the concrete sets. These construction joints are sealed by casting across them in the concrete a continuous water stop. A good quality dense concrete is required to minimise leakage and to obtain a smooth surface, and special care is required to maintain a good surface on the shuttering and good alignment between sections to cut down hydraulic losses due to surface friction, and inhibit marine growth. Large rigid travelling shutters are used to form the internal shape of the culverts. These are often designed to be dismantled and erected using jacks and wedges and are re-used throughout the length of the culvert, requiring occasional re-surfacing. Special formwork is required at corners. The internal surface of the culverts is sometimes formed using thin steel sheets, welded to form continuous pipes which are braced internally to hold them rigid until the concrete is set.

2.9.3. **Pumphouse and Screen Chamber: Intake**

In order to draw adequate supplies of water at all states of tide or river flow, an intake at some considerable depth below general site level may be necessary, and in consequence the pumps are installed in a deep basement. This results in very deep excavations for massive underground structures, and causes great difficulties in construction. Techniques for building these structures have been discussed in Section 2.8.2.

The location of the pumphouse, screen chamber and intake varies from site to site. Where there is sufficient depth of water, the structures can be combined into one and situated at the river bank. Such an arrangement is shown in Fig. 2.10.2c where the cross-section shows the pumphouse screen chamber and intake built immediately behind the coal wharf. The coal wharf is kept as a separate structure to avoid transferring ship berthing loads to the pump chambers. Pump and screen chambers are equipped with overhead cranes for constructing the plant, and subsequent maintenance.

Where sufficient depth of water exists only at some distance from the shore, a separate intake is required. This is sometimes combined with an off-shore jetty, but may be an independent structure. Tunnels are required to carry the circulating water to the inshore screen chamber, which may be combined with the pumphouse, or the pumps may be situated close to each individual set. This complicates the turbine house foundations, but reduces the length of expensive pressure culverts; in this case water flows under gravity from the intake to the pumps.

2.9.4. **Circulating Water Tunnels**

Circulating water tunnels are required to carry the water from offshore intakes to the screen and pump chambers on shore.

These tunnels are frequently constructed 100 ft or more below ground and river level, in order to find suitable strata for tunnelling in or through where the design specifies a permanent intake shaft. The construction programme often calls for work on pumphouse, screen chambers, intake and tunnels to be carried out simultaneously to meet commissioning dates. Where the design specifies a permanent shaft below the screen chambers, there is insufficient time for this to be built and used as a starting point for the tunnel. A separate temporary shaft has to be sunk for this purpose.

A variety of techniques are used for tunnelling, depending on the nature of the ground. In hard rock a number of holes are drilled forward, explosives placed and fired and the rock removed. The tunnel is afterwards lined with mass or reinforced concrete pumped and vibrated behind formwork in the normal manner. Tunnelling in rock is fraught with difficulties. Water enters through fissures and faults and considerable pumping is required; where the roof of the tunnel is near river bed level, compressed air working may have to be resorted to. Weak rock and steeply dipping rock may result in roof falls and extensive propping with steel arches may be required, or even continuous lining with cast-iron segments bolted together.

Where soft ground is to be tunnelled through, and beds of pervious materials will be encountered, it may be considered expedient to drill down and consolidate these layers with cement or chemical grouts before work starts. Compressed air working will nevertheless be required to prevent water entering the tunnel, although the rates of entry of water and escape of air will be reduced by the grouting, and the danger of a "blow", or rapid escape of air through weak ground, will also be reduced. Pressures of up to 40–45 lb/in^2 maximum are permissible and these govern the maximum depth of working; variations of pressure must be made to suit the tide. A large compressor plant has to be built on site to supply sufficient air to keep the tunnel dry. Ventilation is seldom a problem as losses of air through the face are usually sufficient. The plant must be in multiple units, with sufficient stand-by units available to cope with breakdowns. Diesel driven stand-by generators are required in case of failure of the main electricity supply. Compressors are also required to supply high pressure air for pneumatic picks and spades, and for grouting the tunnel linings.

Passage of men and materials into and out of the tunnel is through separate air locks — chambers with air tight doors in which the air pressure can be raised or lowered without

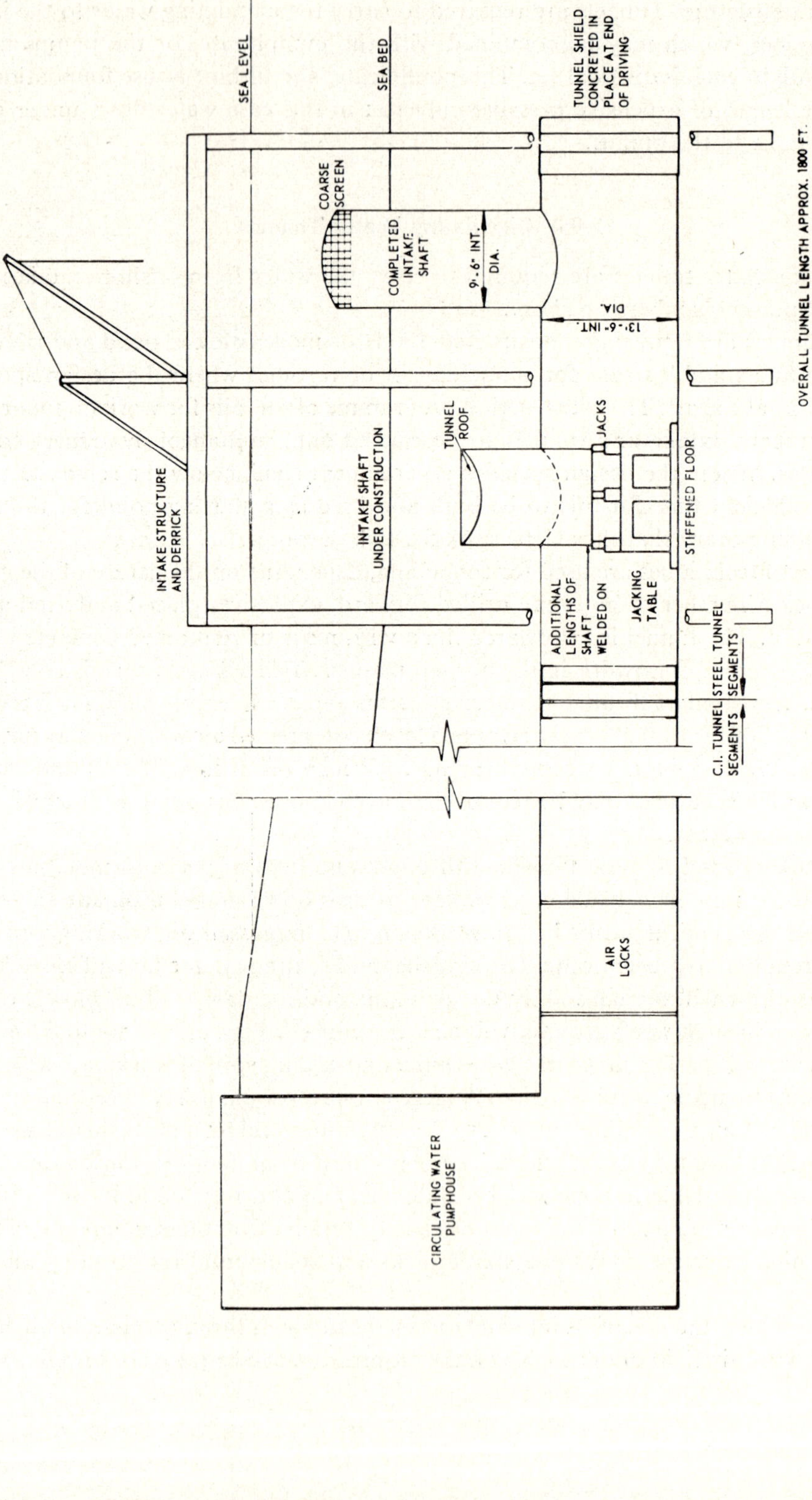

FIG. 2.9.4A. Diagrammatic arrangement of intake shafts constructed from inside the tunnel.

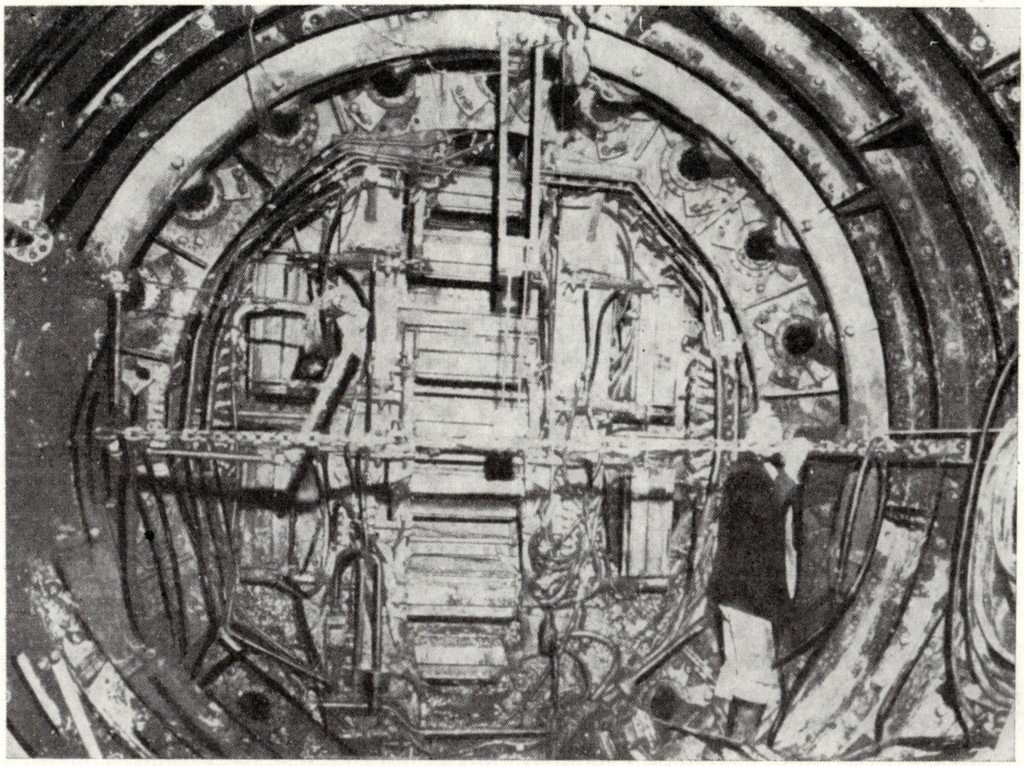

FIG. 2.9.4B. Shield driving circulating water tunnel.

affecting the pressure inside the tunnel. The air locks are either bolted to the top of the working shaft, or are formed with double bulk heads in the tunnel.

Stringent regulations of the Factories Act govern the length of time a man can work, and the period of his subsequent decompression. The time of decompression rises rapidly as the working pressure increases, e.g. 15 min for 20 lb/in^2 and over 90 min for 35 lb/in^2. In spite of these precautions, and in spite of stringent medical checks for fitness, decompression sickness may occasionally occur. This can be painful and dangerous and is treated by decompressing the men in a medical lock on the site under medical supervision.

Tunnels in soft ground are normally driven using a circular steel shield. This is fitted with a cutting edge, and is driven forward by jacks acting on the tunnel lining. A short heading is first dug by hand or compressed air tools in front of the shield, the shield is moved forward on the jacks, the excavated material is removed, and a further ring of lining erected when the jacks are withdrawn.

The structural tunnel lining may be formed with ribbed cast iron or concrete segments bolted together. The meeting faces are a close fit, but the joints are caulked with lead wool to make the lining water-tight. The segments have screwed plug holes in the bosoms, and grout is pumped through these to ensure full contact between the iron and the soil outside. When the lining is complete, caulked and grouted, air pressure is reduced to atmospheric pressure and the tunnel lined with concrete.

As with the culverts, the circulating water tunnels and shafts require a smooth surface to the lining, and bends have to be properly designed to eliminate hydraulic losses. The size of tunnel has to take into account the velocity of flow, but in some cases may depend on the size of tunnelling shield and cast-iron lining available.

The tunnel is completed when the shaft sunk below the intake structure is reached. This shaft must be at the same pressure as the tunnel. The shaft is designed with a section of its cast-iron lining removable to take the tunnel lining. This operation presents a considerable problem of setting out. The intake can be sighted from the tunnel shaft, but this line of sight must be transferred vertically down to the tunnel using plumb bobs or vertical optical plumbing instruments and the tunnel then kept to line and level to meet with the intake shaft to within an inch or so. An alternative sometimes adopted is for the intake shaft to be sunk down on to the culvert which has been designed with a removable section of roof for this purpose.

A further ingenious alternative which has been used on two nuclear sites is for the intake shafts to be constructed inside the tunnel in sections and jacked up into the bottom of the intake structure. The top of the shaft is removed by divers and replaced with coarse screens after the tunnel is flooded. A diagram of this arrangement is shown in Fig. 2.9.4A.

A shield in operation during the driving of a circulating water tunnel is shown in Fig. 2.9.4B. Note the hydraulic jacks around the perimeter of the shield and also the tunnel segments.

Rotary shields and new types of cast-iron and precast concrete linings have been recently developed to extend the London underground railway system. In suitable soils these methods could be employed on power station sites in future.

2.10. WHARVES AND JETTIES

2.10.1. General

Estuary or riverside stations designed to be fed with fuel by ship must be provided with a wharf or jetty to berth tankers or colliers, and this must be equipped for the speedy unloading of the fuel to tanks, or to the coal store and bunkers.

The dimensions of the wharf or jetty are governed by the number and size of the ships to be accommodated, the size and number of unloading cranes and conveyors, and by the buildings such as offices, stores, lavatories, etc., which are to be built on the deck. The deck has to be a safe height above the highest tide or flood level, at least 3 ft, to avoid interruption of working due to the deck being awash. The draught of the ships will govern the location of the wharf or jetty. The berthing area can be dredged to allow the deeper vessels to berth close to the shore, but there is a limit to this dredging or the stability of the banks will be put in jeopardy. This may lead to the adoption of offshore unloading facilities, with a jetty connected to the shore by an approach bridge. A conveyor bridge will also be required for a coal fired station.

The heaviest loads on coaling jetties come from the large travelling cranes used to grab the coal from the colliers and feed the conveyors. This load is not present on oil jetties.

Lighter equipment is used to handle the oil pipe connections, and the structure of the jetty can be made less massive. In siting oil fired stations close to oil refineries, the need or a jetty may be eliminated, oil being transported to the station through pipelines.

2.10.2. **Types of Wharves and Jetties**

The simplest type of wharf is illustrated in Fig. 2.10.2A. Here the river bank is supported by a line of steel sheet piles with their toes driven well below the ultimate level of the river bed, and with their heads supported by walings which are held back by steel tie-rods to concrete anchor blocks set well back in stable ground. This type of wharf is most

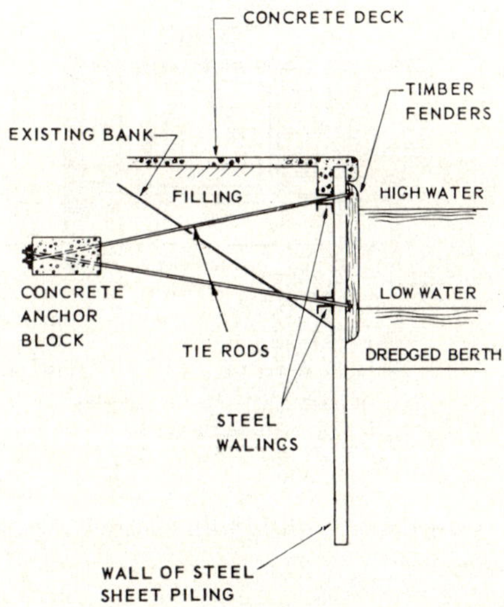

FIG. 2.10.2A. Cross-section of wharf with steel sheet piling retaining wall.

suitable for low ranges of tide and moderate depth of water, due to the limited earth pressures which can be supported.

Fgiure 2.10.2B illustrates a type of wharf used where deeper berthing is required. The stability of the bank is again maintained by a sheet pile wall and by a reinforced concrete sub-deck and retaining wall supported on piles. The coaling cranes are carried on a reinforced concrete beam and slab deck supported partly on the reinforced concrete piles, and on its outside edge by a row of concrete cylinders. These are spaced at intervals of about 20 ft, and besides supporting the front of the wharf, also resist the impact loads from ships berthing. In this instance piles are driven down through the base of the cylinders into the chalk, and the cylinders are braced from the sub-deck and concrete retaining wall by the reinforced concrete trusses. Replaceable timber fenders and rubbing timbers are fixed to the face of the wharf and cylinders to minimise the damage from impact loadings. The berth has been dredged out, but the river bed is left at its natural profile behind the row of cylinders.

Figure 2.10.2C shows a cross-section of a coaling wharf taken in a position where the sub-deck has been replaced by coarse screen chambers. The cylinders, fendering and deck are indicated, with one of the travelling cranes. The river bed at the wharf's frontage has been dredged to give sufficient depth of water at all stages of the tide for laden colliers to remain afloat, and consequently enter or leave the berths at any time.

Deeper berthing facilities may call for the construction of jetties far enough off shore to be on the edge of the deep water channel. A reinforced concrete deck of beam and slab construction is supported on rows of raking and vertical piles, or on two or three rows of cylinders. The cylinders are connected across the jetty by deep beams. The junctions between beams and cylinders are heavily reinforced, as are the cylinders, so that a rigid structure is produced. This removes the need for deep cross braces between beams

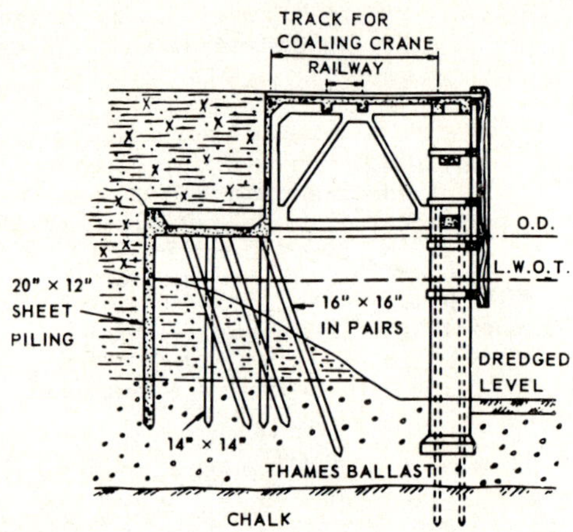

FIG. 2.10.2B. Cross-section of wharf with concrete piles and cylinders.

and cylinders, and the consequent difficulty of working between tides. The cylinders may be taken down to final foundation level or may have steel piles driven down through their bases to take the vertical load.

A typical fendering system may consist of hollow steel piles driven between the front row of cylinders, and supported off the deck by strong rubber buffers. Timber protection is also provided, fixed to the steel piles, and spanning between them. This type of jetty may also be built with intake, or outfall chambers built integral with it.

A reinforced concrete road bridge, supported on piles, is provided to connect the jetty to the shore, and this will carry electricity supplies, water sewage pipes, telephones and fresh water. Coal conveyors will be supported on isolated bases of piles and reinforced concrete pile caps. The jetty will have to accommodate offices, switch house, amenities, and stores for the operating staff.

2.10.3. **Construction of Wharves and Jetties**

Construction of both jetties and wharfs involves working over water, with the problem of working between tides. A floating pontoon is often employed to drive piles, particularly temporary piles, but the bulk of the construction is carried out using derrick

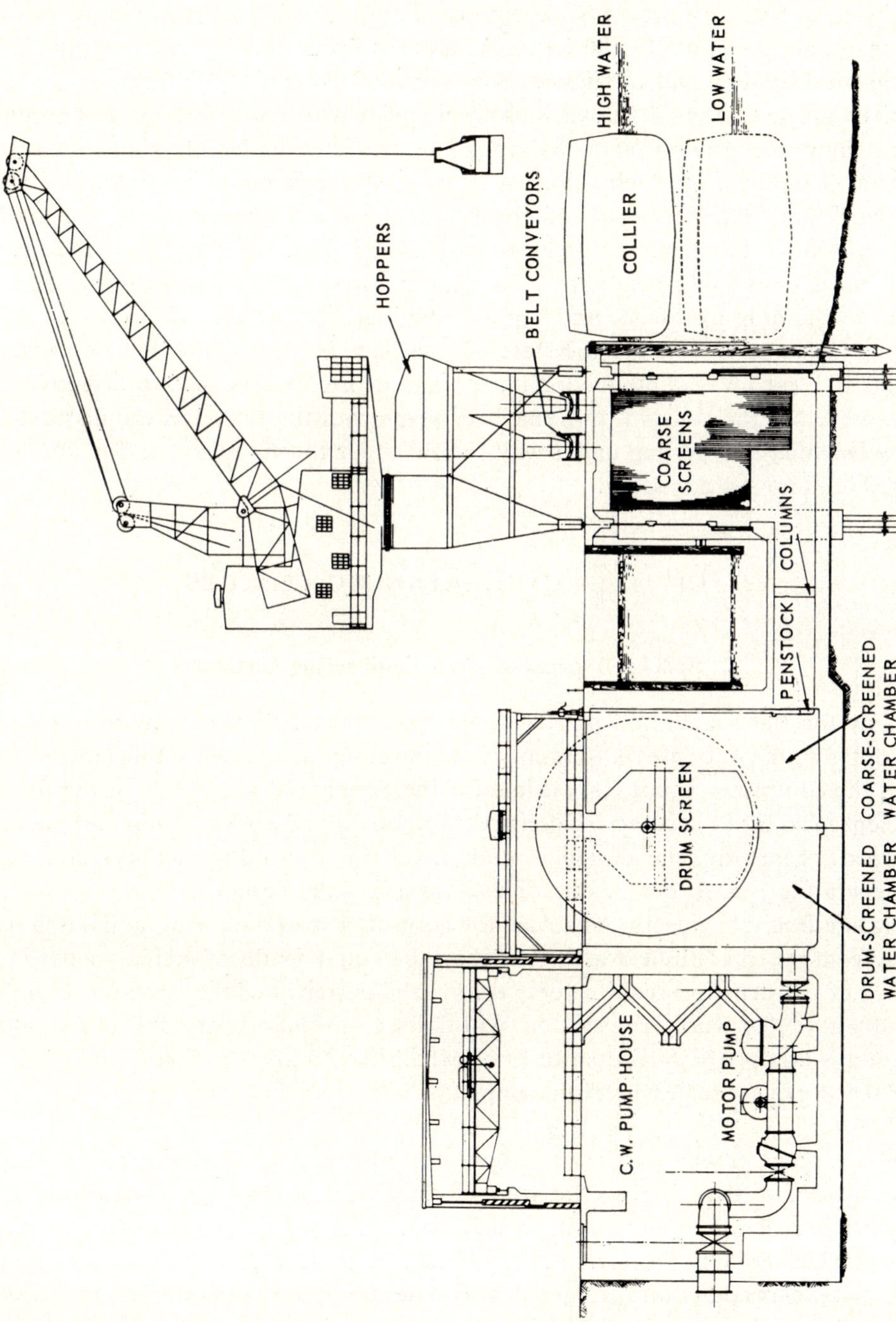

FIG. 2.10.2C. Cross-section of wharf showing intake, screen chambers and pumphouse.

cranes capable of lifting up to 20 tons, and with a maximum reach of 120 ft. Temporary piers are built, using timber or steel piles, to support these derricks which are moved to fresh positions as the work proceeds. A temporary approach jetty is required for offshore work to transport materials to the jetty. Temporary frames are erected to support the cylinders during sinking, and to support piles during driving if a pile frame is not used. The derricks are used for all the various works of construction: extending temporary works—pitching and driving piles—handling the precast cylinder units and excavating within them—handling and placing precast units, reinforcement and concrete. The beams for jetty work are frequently made of precast concrete, and precast panels are used to form the bottom of the deck slab. An *in situ* concrete deck is formed over the precast unit to form a monolithic beam-and-slab deck construction. This method avoids the difficulties of building formwork and support stagings over water.

Dredging is carried out before the start of construction, and again after construction is complete to clear away subsequent siltation and debris. A dredged berth tends to silt up if not used, but the constant movement of ships with the tide, and the disturbances caused by berthing and getting under way should scour the silt away, and maintenance dredging should not be necessary.

2.11. CIVIL ENGINEERING CONTRACTS

2.11.1. Special Aspects of Civil Engineering Contracts

Contracts for the foundation works which have been referred to in the preceding sections differ from contracts for items of plant and equipment in two important respects.

Firstly, the contractor is not responsible for the design and sufficiency of the foundations (except in so far as accuracy of construction, quality of workmanship and materials is concerned); generally, the design is undertaken by consulting engineers on behalf of the Generating Board, but in one or two instances the design has been carried out by Generating Board engineers. As a rule, the contract is based on provisional quantities measured from general outline drawings of the works, and details of design are developed in advance of construction requirements during the currency of the contract. It follows that the quantities of work, as executed, will differ from those in the Bill of Quantities, which contains the provisional quantities on which the contract is based; therefore, the whole of the works are remeasured on completion and are paid for at the unit rates of the accepted tender.

Secondly, a much greater part of the labour cost of the civil engineering works is spent on site construction than is the case with mechanical and electrical equipment, where the site work is generally limited to its installation. It follows that the careful supervision and testing, which the latter receives in the workshops, must be given to the civil engineering work as it is carried out on site, and an adequate supervisory organisation is necessary. On the one hand, the contractor provides his own engineers and foremen to supervise the quality of work produced, and to organise his labour and equipment to the best advantage, all under the general direction of the contract manager, or an agent, who is the contractor's representative on site. On the other hand, the site representative of the civil engineers, known as the Resident Civil Engineer, with the assistance of other engineers and experienced inspectors, exercises quality control of the work as it is carried out by the contractor's workmen, and maintains a systematic check on the quality of all materials incorporated in the works.

2.11.2. **Monthly Payments**

During the contract period, the amount of work completed by the contractor is measured each month by the staff of the Resident Civil Engineer and its value is calculated according to the rates contained in the contractor's tender. Payment is then made to the extent of 90% of the value of the work; 5% is retained until completion of the contract, and 5% is retained for twelve months thereafter as security for carrying out such remedial work as may become necessary to put right any defects of workmanship or materials which become apparent within the maintenance period (usually twelve months from the completion of the contract).

2.11.3. **Workmen's Camp and Transport**

In executing the civil engineering works for a new power station, the contractors may jointly employ up to 2000 men at the peak. The availability of labour varies with the locality, but in rural areas perhaps one-half of the total are drawn from towns within daily travelling distance of the site, and the remaining half are drawn from farther afield. In order to accommodate the latter it is normal practice for the Generating Board to erect a workmen's temporary camp adjacent to the power station site, to be operated by the contractor under the supervision of the Resident Civil Engineer with regard to the standard of accommodation and charges. Accommodation in the camp is also open to the workmen of the mechanical and electrical contractors.

For those men who can travel daily from nearby towns to the site, it is customary for the contractor to arrange special services of transport to fit in with their normal working hours.

APPENDIXES

Worked examples for the design of two types of foundation are shown and these are included to show the general nature of the necessary calculations. The various steps are explained but it is not possible to cover in this chapter the basic design theory, a knowledge of which is essential in order to prepare an adequate design.

Up-to-date textbooks on the design of structural steelwork and reinforced concrete should be obtained by any reader who is new to these subjects and who may wish to learn the underlying principles.

The strength of members in reinforced concrete may be determined by means of the elastic theory or by the load factor method. In structural steelwork either the elastic or plastic theory may be used. Elastic design is used in each of the following examples.

APPENDIX 1

DESIGN OF ISOLATED COLUMN BASE

EXAMPLE

A reinforced concrete base is required as a foundation for a 12 in×12 in universal column having a baseplate 24 in square in plan. The column carries an axial load of 200 tons and the maximum permissible ground pressure is 2 tons/ft² at a depth of 3 ft.

The appropriate stresses and design requirements are covered by British Standard Code of Practice, CP 114, *The Structural Use of Reinforced Concrete in Buildings.*

Permissible stress for compression in bending of concrete grade to be used	1000 lb/in²
Permissible shear stress for this concrete	100 lb/in²
Permissible tensile stress for mild steel bars	20,000 lb/in²

The effect of the column load is to cause tension in the lower face of the foundation and compression in the upper face. In reinforced concrete design steel reinforcement is used to carry all the tension as concrete is assumed to be capable of resisting compression only.

Column load plus own weight	200 tons
Foundation weight (say)	20 tons
Total load	220 tons
Area of base required at 2 tons/ft²	110 ft²
Size of square base	10 ft 6 in×10 ft 6 in

The maximum bending moment in the base occurs under the centre of the column and this value is determined by taking moments about the centre line, i.e. the point at which it is desired to find the bending moment. The base is symmetrical and, therefore, forces on either side may be considered. There is an upward force of $W/2$ acting midway between the column centre line and the edge of the foundation and a force of $W/2$ acting down at the face of the column.

$$\begin{aligned} \text{Bending moment } (M) &= (100\times 31{\cdot}5)-(100\times 6) \text{ tons in} \\ &= 2550 \text{ tons in} \\ &= 5{,}712{,}000 \text{ lb in} \end{aligned}$$

The effective depth (d) is that measured from the top of the foundation of the centre of the tension reinforcement. The minimum value of d (in) so that the compressive stress of the concrete in bending at the top of the concrete base is not greater than 1000 lb/in²

$$= \sqrt{\frac{M}{Qb}}$$

where b = foundation breadth (in),

Q = 184 (a factor which is dependant on the permitted steel and concrete stresses)

$$d = \sqrt{\frac{5{,}712{,}000}{184 \times 126}} \text{ in}$$
$$= 16 \text{ in approx.}$$

Allowing for concrete cover to the steel reinforcement the minimum overall depth (D) would be 20 in.

The depth of the foundation is also dependant on the "punching shear". This is assumed to occur in a vertical plane along the perimeter of the column baseplate and in this case the shear stress permissible in the concrete is taken as twice the normal value. (Punching shear is more justified by practice than theory.)

$$\text{Punching shear} = \left(\frac{\text{Area of foundation} - \text{Area of baseplate}}{\text{Area of foundation}}\right) \times \text{Total load}$$
$$= \left(\frac{10{\cdot}5^2 - 2^2}{10{\cdot}5^2}\right) \times 220 \times 2240 \text{ lb}$$
$$= 475{,}000 \text{ lb}$$

$$D = \frac{\text{Punching shear}}{\text{Baseplate perimeter} \times \text{Twice permissible shear stress}}$$
$$= \frac{475{,}000}{4 \times 24 \times 2 \times 100} \text{ in} = 25 \text{ in approx.}$$

(25 in is an unusual foundation depth, so increase D to 2 ft 3 in).

The value of D for punching shear is greater than that required for bending and the shear governs the required depth.

$$\text{Area of tensile steel } (A_t) = \frac{\text{Bending moment}}{\text{Safe steel stress} \times \text{Lever arm}}$$

(The lever arm is the distance from the steel reinforcement to the centre of compression in the concrete. It is approximately 0·86d, though it varies slightly depending on the ratio of the steel to actual concrete stress. As D is 27 in, the value of d, allowing for concrete cover, is about 23 in.)

$$A_t = \frac{5{,}712{,}000}{20{,}000 \times 0{\cdot}86 \times 23} \text{ in}^2 = 14{\cdot}7 \text{ in}^2$$

Adopt 19 No. 1 in diameter bars giving 14·92 in². These bars are required in each layer as the moment is the same in each direction.

Note. An increase in the thickness of the concrete slab may be made where it is required to take the foundation to a greater depth. This will result in less reinforcing steel being required and a reduction in the compressive stress in the concrete, e.g. if the overall depth is to be 4 ft then the effective depth will be about 3 ft 8 in.

Thus $$A_t = \frac{5{,}712{,}000}{20{,}000 \times 0{\cdot}86 \times 44} \text{ in}^2 = 7{\cdot}6 \text{ in}^2$$

This is just over half the steel required previously. In this case to prevent the volume of concrete becoming excessive the top of the concrete base may be sloped downwards from the column baseplate, though a minimum depth of 12 in should be maintained at the edge of the concrete base.

APPENDIX 2

DESIGN OF STEEL GRILLAGE FOUNDATION

EXAMPLE

A steel grillage is required as a foundation for a 14 in×16 in universal column carrying an axial load of 400 tons. The permissible ground pressure is 2 tons/ft² at a depth of 4 ft 6 in.

The appropriate stresses and design requirements are covered by British Standard 449, *The Use of Structural Steel in Building.*

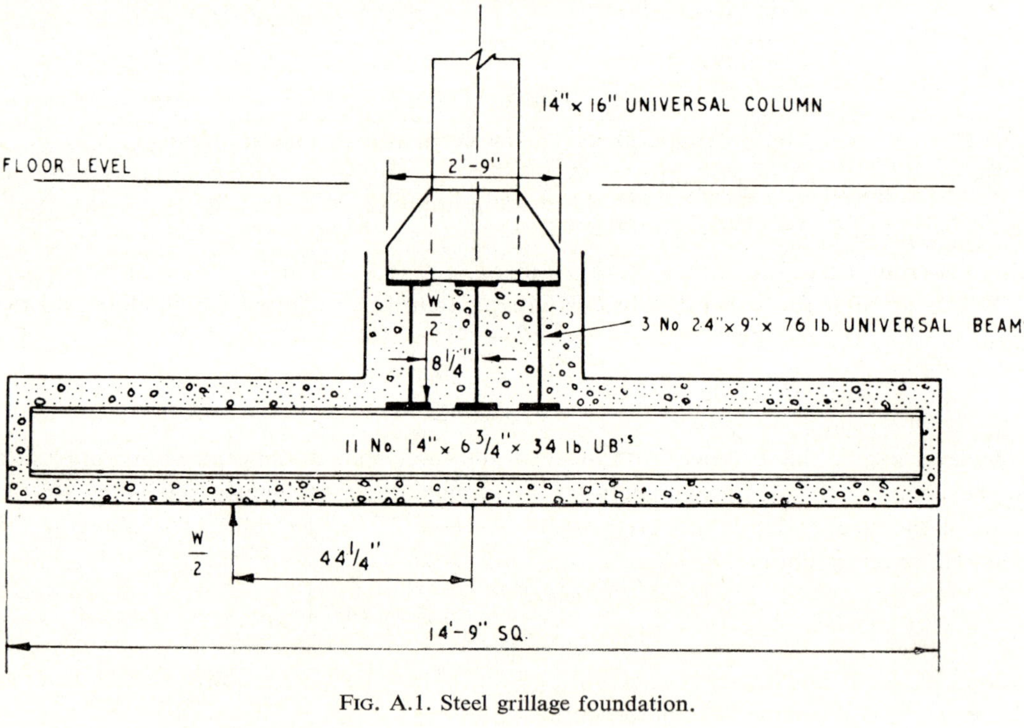

FIG. A.1. Steel grillage foundation.

Column load (W)	400 tons
Foundation weight (say)	35 tons
Total load	435 tons
Area of base required at 2 tons/ft²	217·5 ft²
Size of square base	14 ft 9 in×14 ft 9 in

Consider bottom tier of beams.

The maximum bending moment occurs under the centre of the column and this value is determined by taking moments about the centre line. The base is symmetrical and forces on either side may, therefore, be considered. There is an upward force of $W/2$ acting midway between the column centre line and the edge of the foundation and a force of $W/2$ acting down midway between the centre line and the edge of the top tier of beams.

Bending moment (M) = $(200 \times 44{\cdot}25) - (200 \times 8{\cdot}25)$ tons in = 7200 tons in

Also $$M = fZ$$

where f = bending stress (tons/in^2),

Z = section modulus of steel beam (in^3).

B.S. 449 permits the maximum value of f (under certain conditions) as 10·5 tons/in^2 for steel to B.S. 15. However, in the case of steel beams in grillages adequately surrounded in concrete the stress may be increased by one-third, i.e. to 14 tons/in^2.

$$Z \text{ required} = \frac{7200}{14} \text{ in}^3 = 514{\cdot}3 \text{ in}^3$$

Refer to Safe Load Tables published by British Constructional Steelwork Association or structural engineering handbook for value of section modulus.

Z of 14 in$\times 6\frac{3}{4}$ in$\times$34 lb U.B. = 48·5 in^3

Try 11 No. 14 in$\times 6\frac{3}{4}$ in$\times$34 lb U.B.s (Z = 533·5 in^3)

Check this section for shear.

From Safe Load Tables permissible shear for 14 in$\times 6\frac{3}{4}$ in$\times$34 lb. U.B. is 24·11 tons. This value may be increased by one-third for grillage steelwork.

Permissible shear for one beam = 32·14 tons

Permissible shear for 11 beams = 353·5 tons

Shear force = $W/2$ = 200 tons

Thus, 11 No. 14 in$\times 6\frac{3}{4}$ in$\times$34 lb. U.B.s are satisfactory.

Consider top tier of beams.

The bending moment is caused by $W/2$ acting upwards 44·25 in from the column centre and $W/2$ downwards and assumed to act on the face of the column flange, 8 in from the column centre.

Bending moment (M) = $(200 \times 44{\cdot}25) - (200 \times 8)$ tons in = 7250 tons in

$$Z \text{ required} = \frac{7250}{14} \text{ in}^3 = 517{\cdot}9 \text{ in}^3$$

From Safe Load Tables,

Z of 24 in×9 in×76 lb U.B. $= 175{\cdot}4$ in^3

Try 3 No. 24 in×9 in×76 lb U.B.s ($Z = 526{\cdot}2$ in^3)

Check this section for shear.

Station construction is such that grillages by virtue of later ground works are often left exposed, i.e. without their concrete casing. Care should be taken to ensure that the grillage beams are encased with concrete or alternatively a lower stress worked to, i.e. as uncased beams.

Safe Load Tables give shear value as 63·12 tons.

Adding one-third for grillage steelwork permissible shear = 84·16 tons
Permissible shear for 3 beams = 252·5 tons

Shear force $= W/2 = 200$ tons

Thus, 3 No. 24 in×9 in×76 lb U.B.s are satisfactory

(The webs of the beams sometimes require stiffening. This should be checked by reference to Safe Load Tables. However, stiffeners are not necessary in this case.)

APPENDIX 3

ESTIMATION OF CARRYING CAPACITY OF PILES

(a) Driven Pile in Non-cohesive Soil

One of the best known of the dynamic formulae is the Hiley formula. This is based on the impact of elastic bodies and equates the energy of the hammer blow to the resistance of the ground to the penetration of the pile. Allowances are made for loss of energy due to elastic contractions of the pile, dolly and subsoil as well as the losses due to the inertia of the pile.

$$R = \frac{Wh\eta}{S + \dfrac{c}{2}}$$

where R = ultimate driving resistance in tons,
W = weight of hammer in tons,
h = height of free fall of hammer in inches,
S = final set or penetration per blow in inches,
c = sum of temporary elastic compressions in inches of pile, dolly and ground; these values are also dependent on whether driving is easy, medium or hard, and are obtained from tables,*
η = efficiency of blow; this value is dependent on the ratio of the weight of the pile to the hammer and the dolly type. This value may be obtained from tables.*

* Civil Engineering Code of Practice, No. 4, *Foundations*.

The factor of safety to be applied to the ultimate driving resistance obtained by the formula should not be less than 2. However, if resistance is reduced on redriving this should not be less than $2\frac{1}{2}$. An increase in these values should be made for structures sensitive to settlement.

(b) Driven and Bored Piles in Cohesive Soil

The carrying capacity may be estimated when the cohesive value of the soil is known. The first term in each of the following formulae is that due to skin friction and the second term is due to end bearing support.

For a driven pile the estimated safe load is:

$$\pi dL \times \frac{0{\cdot}6C}{F} + \frac{\pi d^2}{4} \times \frac{9C}{F} \text{ tons}$$

and for a bored pile is:

$$\pi dL \times \frac{0{\cdot}45C}{F} + \frac{\pi d^2}{4} \times \frac{9C}{F} \text{ tons}$$

where d = diameter of pile in feet,

L = effective length of pile in feet,

C = cohesive value of soil in tons/ft^2,

F = factor of safety, about 2 on skin friction and 2·5 for end bearing.

EXAMPLE

Determine the probable safe bearing capacity of a 16 in diameter pile driven to a depth of 45 ft in a clay having an average shear strength of 1800 lb/ft^2 (say 0·8 ton/ft^2)

Approximate safe load =

$$\left(\frac{\pi \times 1{\cdot}33 \times 45 \times 0{\cdot}6 \times 0{\cdot}8}{2}\right) + \left(\frac{\pi \times 1{\cdot}33^2 \times 9 \times 0{\cdot}8}{4 \times 2{\cdot}5}\right) \text{ tons}$$

$$= 45 + 4 \text{ tons}$$

$$= 49 \text{ tons}$$

APPENDIX 4

DESIGN OF PILE CAP

EXAMPLE

Using the 16 in diameter piles from the last example design a pile arrangement and pile cap if the total load on the piles is 240 tons. The pile cap carries a 21 in square concrete column reinforced with 8 No. $1\frac{1}{8}$ in diameter mild steel bars. Adopt same permissible stresses as in Appendix 1, Example.

Allowing for the group effect of the piles six piles will be required, each carrying 40 tons.

Pile spacing should not be less than $2\frac{1}{2}$ times pile diameter. However, it is preferable to allow not less than 3 diameters. In this example pile centres are 4 ft.

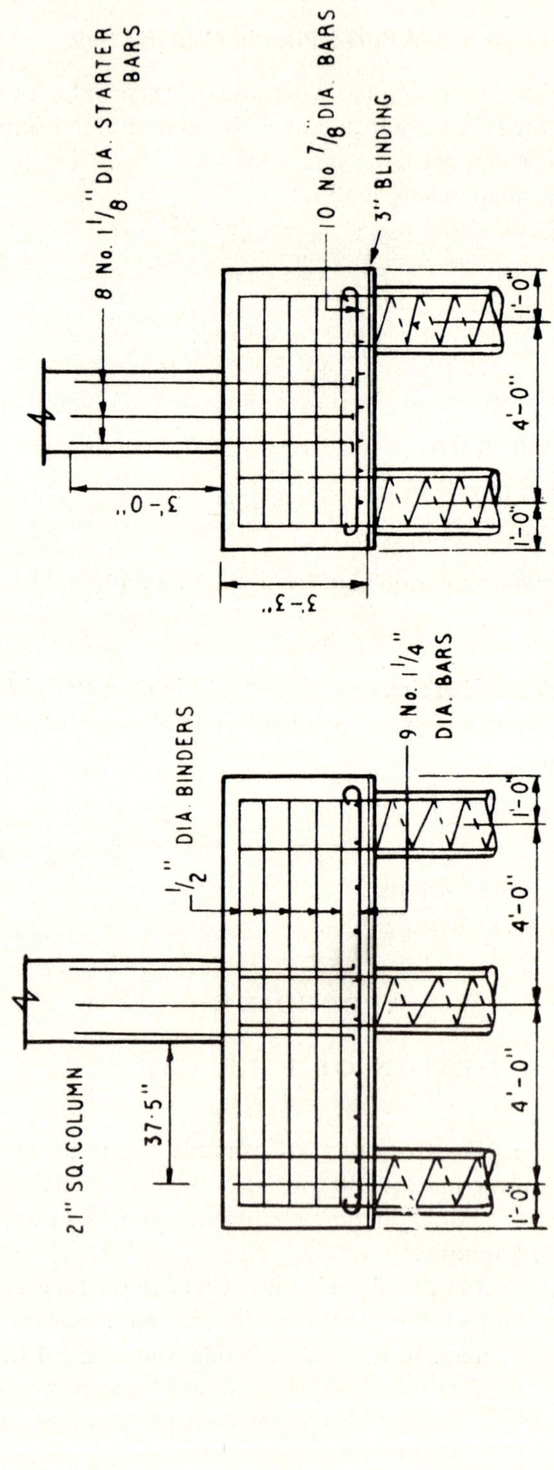

FIG. A.2. Piles 16 in diameter with 5 No. $\frac{1}{2}$ in diameter longitudinal bars and $\frac{5}{16}$ in spiral binding at 6 in pitch.

Depth of pile cap must be adequate to permit the starter bars to be embedded a minimum of 30 bar diameters in order to permit transfer of compression in bar. Similarly the bars must project the same distance above the top of the pile cap. The starter bars require to be the same diameter as the column bars, i.e. $1\frac{1}{8}$ in.

Depth of pile cap must also be adequate to resist punching shear—see Appe dix 1, Example.

Adopt depth of pile cap of 3 ft 3 in.

The longitudinal bending moment is the greatest and for a reinforced concrete column connected by starter bars to the pile cap it may be assumed that the maximum moment occurs at the column face.

$$\text{Approx. maximum moment} = 2\times 40\times 2240\times 37{\cdot}5 \text{ lb in}$$
$$= 6{,}720{,}000 \text{ lb in}$$

Check that depth of pile cap is sufficient to withstand the above bending moment without exceeding permissible compressive stress of this grade of concrete.

$$\text{Required minimum depth of cap } (d) = \sqrt{\frac{\text{Bending moment}}{184\times\text{Breadth of pile cap}}}$$
$$= \sqrt{\frac{6{,}720{,}000}{184\times 72}} \text{ in}$$
$$= 22{\cdot}5 \text{ in}$$

Allowing for concrete cover, minimum permissible depth of pile cap to resist bending—26 in approx. Overall depth provided is 3 ft 3 in and, therefore, acceptable.

$$\text{Area of tensile reinforcement required} = \frac{\text{Bending moment}}{\text{Safe steel stress}\times\text{Level arm}}$$
$$= \frac{6{,}720{,}000}{20{,}000\times 0{\cdot}86\times 36} \text{ in}^2$$
$$= 10{\cdot}8 \text{ in}^2$$

Use 9 No. $1\frac{1}{4}$ in diameter bars giving 11·04 in^2 to resist longitudinal bending moment.

The transverse bending moment in the pile cap is very low as the depth of the pile cap is greater than the horizontal dimension of the centre piles from the column. The steel area cannot be determined from normal bending considerations. Allow 10 No. $\frac{7}{8}$ in diameter bars for transverse bending. These provide 6·01 in^2 which is over 50% of the steel area to resist longitudinal bending.

The size of the horizontal binders in the pile cap is not calculated and this amount of reinforcement is purely nominal.

BIBLIOGRAPHY

1. British Standard Codes of Practice:
   - CP 2001. *Site Investigation.*
   - CP 2003. *Earthworks.*
   - CP 114. *The Structural Use of Reinforced Concrete in Building.*
   - CP 301. *Building Drainage.*
   - CP 101. *Foundations.*
   - CP 2005. *Sewerage.*

2. British Standard No. 449. *The Use of Structural Steel in Building*.

3. *The Mechanics of Engineering Soils*, by P. L. Capper and W. F. Cassie. Published by E. and F. N. Spon.

4. Design and construction of a reinforced concrete foundation block for a 200 megawatt turbo-generator, by F. A. Rae. *Proceedings Institution of Civil Engineers*, May 1962.

5. *Simplified Tables of External Loads on Buried Pipelines*, by N. W. B. Clarke. Building Research Station. Published by H.M.S.O.

6. *Modern Civil Engineering Practice*. Edited by Rolt Hammond. Published by Newnes.

7. *P.F.A. (An introductory booklet on the uses of pulverised fuel ash)*. Published by C.E.G.B. Further information may be obtained from Regional P.F.A. Marketing Officers.

EXERCISES

1. If you had to design a foundation for a boiler house square column 30 ft long loaded with a load of 500 tons, describe how you would investigate the soil conditions, and put forward three proposals for the design of a foundation for varying soil conditions. Assume a water table of minus 3 ft. For the allowable compressive stress in a steel column take 6 tons/in^2, for a reinforced concrete column take 600 lb/in^2.

2. Describe what action you would take to ensure:
(a) A concrete suitable for a turbo-alternator foundation block.
(b) The construction of such a foundation block.
(c) The measurement of shrinkage or settlement of the foundation block.

3. Describe the principal items to be considered in the construction of a cofferdam. Prepare a sketch to illustrate your proposals for a cofferdam required to construct circulating water. Riverside screen chambers 120 ft long to river, 30 ft wide, 30 ft deep. The highest water level being minus 2 ft, the lowest minus 12 ft and the river bed minus 20 ft.

4. A coal-fired 2000 MW power station has a 1000 ton coal train discharging at $\frac{1}{2}$ mile per hour. Prepare a design illustrating your views on the shape and capacity of the receiving hopper together with your recommendation for the extraction of the coal from the hopper. Assume a station load factor of 80% and two-shift working of the coal plant under severe winter conditions.

5. The coal for a 2000 MW P.F. station has 25% ash content. Consider ash removal to fill low-lying land 10 miles away. Describe your views on the handling of this material and what civil works and plant you would require. Assume the storage area would have a 20 years life. Take the furnace ash as 15% of the ash content.

6. Describe and sketch your proposals for the layout of:

(a) a construction office;
(b) roads;
(c) electric power supplies;
(d) erection areas;
(e) sidings;
(f) stores;
(g) hostels;
(h) car parks;

for a 2000 MW station under construction.

7. In a change over from coal to oil for an existing stoker-fired power station, of 2000 MW with a 70% load factor, one week's oil storage is required. Describe with a layout sketch your proposals for the civil works and storage tanks.

CHAPTER 3

CIVIL WORKS: SUPERSTRUCTURE

3.1. INTRODUCTION

The function of power station buildings is to house plant and protect it from the weather and ensure that operatives work in a reasonable environment. Structures such as chimneys, precipitator casings and cooling towers which function as items of plant are also included in civil engineering contracts.

This chapter, therefore, covers buildings and plant items (excluding their foundations) which are constructed as building and civil engineering works.

The first section defines the various parts of buildings and explains the factors which must be considered before a particular form of construction is adopted.

The next sections describe various methods of construction and although each trade has not been dealt with individually, a pattern relating to groups of trades is established.

These are followed by the sections covering the power station buildings. The factors influencing the development of the layouts and elements of construction adopted are given at the commencement of the section dealing with a particular building.

Finally a section dealing with the basic landscaping requirements and a further one relating to Factory Acts and other Statutory Requirements are included.

3.2. DESIGN CONSIDERATIONS

3.2.1. General

For convenience and ease of reference the superstructure of most buildings can be classified in the following way:

(a) *Frame or structure*. Usually steel or reinforced concrete.

(b) *Cladding of fabric*. Consists of roofs, floors and walls.

(c) *Finishes*. Includes floor, wall and ceiling finishes, doors, partitions and similar items.

(d) *Services*. These include the plumbing systems, heating, air conditioning, lifts, cooking equipment, dining facilities and similar items. Although other services such as lighting, telephones and similar items do not form part of building or civil engineering contracts on power stations, provision has to be made for them during construction.

3.2.2. Considerations

Before a decision on the type of frame and cladding to be used is made or the finishes or services selected the following basic requirements must be considered:

(a) *Use of building*. This decides overall dimensions and influences the methods of construction adopted.

(b) *Architecture*. Internal and external requirements.

(c) *Economics*. The capital outlay and maintenance costs. A financial appraisal based on the capital cost together with anticipated life span and consequent maintenance charges of the building is necessary to determine the most economic method of construction.

(d) *Heat insulation*. Although high insulation values for cladding are normal to reduce heating costs, in boiler houses where large amounts of heat are given off by plant items the reverse applies.

(e) *Sound insulation*. The noise from gas turbines, transformers and similar plant can be a nuisance to operating personnel and in certain cases the general public outside the confines of the station. Where this occurs the plant must be installed in sound insulating buildings or located in positions away from the main buildings.

(f) *Fire risk*. This is very largely influenced by the use of the building and also adjacent equipment and buildings.

(g) *Durability*. In view of the limited life span of power station buildings, materials are often used which are less durable than those from which other more permanent structures are constructed. Plant buildings are usually designed on a 30-year life span whereas control rooms may be built to last 50 years.

(h) *Maintenance*. The disruptive effect of building maintenance on plant availability must be considered as well as the problems associated with the building itself.

(j) *Plant considerations*. The degree to which the superstructure is required to support items of plant and services.

3.3. LOADINGS

3.3.1. Definitions

When designing any structural element in a building it is necessary to ascertain the imposed loads in addition to the dead load of the frame, cladding and other components making up the building. The imposed loads are the result of static plant, dismantled plant being placed on floors, impact operation of plant, vibration and dynamic forces, foot and wheeled traffic and also wind and snow loads.

3.3.2. Imposed Loads Due to Plant

Loads due to plant items may vary from less than one ton on small pipe supports and similar items to those up to 13,000 tons for slung boilers, turbine and generator components.

If maintenance of the heavier plant is not carried out by station personnel no provision is made for accommodating plant components on the boiler and turbine house operating floors, and the crane is used to move equipment directly to transporters in the loading bay. The loading bay floor in the turbine house is designed to carry transporters and loads up to 250 tons.

3.3.3. Distributed Imposed Loads

Although it is necessary to limit the positions on which heavy loads can be set down on floors, it is not practical to do this with smaller items and it is therefore usual to decide on a floor loading per square foot which will cater for this form of loading. Foot traffic and the movement of materials are also taken into account when assessing the imposed distributed load. At the time the design of the frame is prepared it is not possible to ascertain all the smaller individual imposed plant loading requirements, especially loading on floors due to underslung pipework and similar items and it is common practice to increase the distributed imposed load per square foot by an amount which will cater for these requirements.

3.3.4. Cranes

The largest cranes provided on power stations are those in the turbine houses. Although single cranes are used to lift the heaviest loads, it is sometimes advantageous to use two smaller cranes individually for the smaller loads and linked together with a lifting beam on the two main hooks for the heavier loads. The latter arrangement gives a greater spread of loading on the crane girders with considerable reductions in the cost of foundations and structural frame. The use of two cranes for the smaller loads improves the erection programme and a typical arrangement of this type of loading is shown in Fig. 3.3.7.

Examples of typical imposed distributed loads are as follows:

Floor location	Loading in lb/ft²
B.H. Operating level	250
B.H. Basement	500
T.H. Operating level	250
T.H. Basement	500
T.H. Loading bay	1000
Conveyor house	100/150
Deaerator bay	100/150
Galleries	100/150
Switchgear room	250
Stores and heavy workshop	300
Light engineering and electrical workshop	150
Control room	80

The recommendations given in B.S. Code of Practice 3, Chapter 5, *Loadings*, are normally used to establish the design imposed loads for offices and similar buildings.

Typical imposed distributed loads for office from C.P. 3. are 50 lb/ft^2 for normal offices and 70 lb/ft^2 for offices accommodating data processing and similar equipment.

The imposed loads referred to above are used in the design of floors and secondary members. For main beams and columns it is usual to allow a reduced loading as discussed in Section 3.3.6.

3.3.5. **Wind and Snow Loads**

The allowable pressure and suction on the vertical faces of walls and also sloping and horizontal surfaces of roofs due to wind are also covered by recommendations in C.P. 3, Chapter 5. Pressure depends upon the wind velocity, the conformation of individual buildings and also the proximity of other buildings. Wind velocities are related in C.P. 3, Chapter 5, to sites with varying degrees of exposure, i.e. *A* being the most sheltered areas with *D* being the more exposed areas. Exposure *D* relates to a wind velocity of 72 m.p.h. and a corresponding pressure of 31 lb/ft^2 where the building is over 200 ft high. These conditions are not severe enough for many of the Board's sites and in many cases structures are designed for wind in excess of 100 m.p.h. It is necessary to consult local meteorological records to ascertain the exposure classification.

Due to the inertia of large buildings when considered as units, gusts have little effect on the main frames. Gust pressures however do affect cladding, sheeting rails and purlins and may in certain cases warrant design pressures up to 50 lb/ft^2.

Flat roofs are usually designed for an imposed load of 30 lb/ft^2 where they are subject to foot traffic although this may be reduced to 15 lb/ft^2 for buildings such as air blast switch houses where minimum access for maintenance is required.

Suction due to wind forces on the roof and also the lee side of structures and internal pressure set up when the wind enters buildings partly erected must be considered.

3.3.6. **Reduced Loadings in Main Beams and Columns**

Where floors or roofs are designed on a load per square foot basis to accommodate random loading it is usual to allow a reduction in the imposed load only on the main beams and columns, for although individual areas of floor slabs or roof decks may be fully stressed by the maximum loading it is unlikely that the maximum design load will be applied to the whole area of the roof or floor on one occasion. One example of these recommendations is in roof construction where the roof decking and purlins may be designed to 30 lb/ft^2 whereas the main beams and columns are designed for an imposed load of 15 lb/ft^2. If a turbine house has main beams which span 200 feet and are at 30 feet centres, the total imposed load on the basis of 30 lb/ft^2 would be:

$$200 \times 30 \times \frac{30}{2240} = \underline{80 \text{ tons}}$$

This is a very high load and by reducing the imposed load to 15 lb/ft^2 for the main beams and columns the more realistic figure of 40 tons is obtained for the total imposed load on one beam.

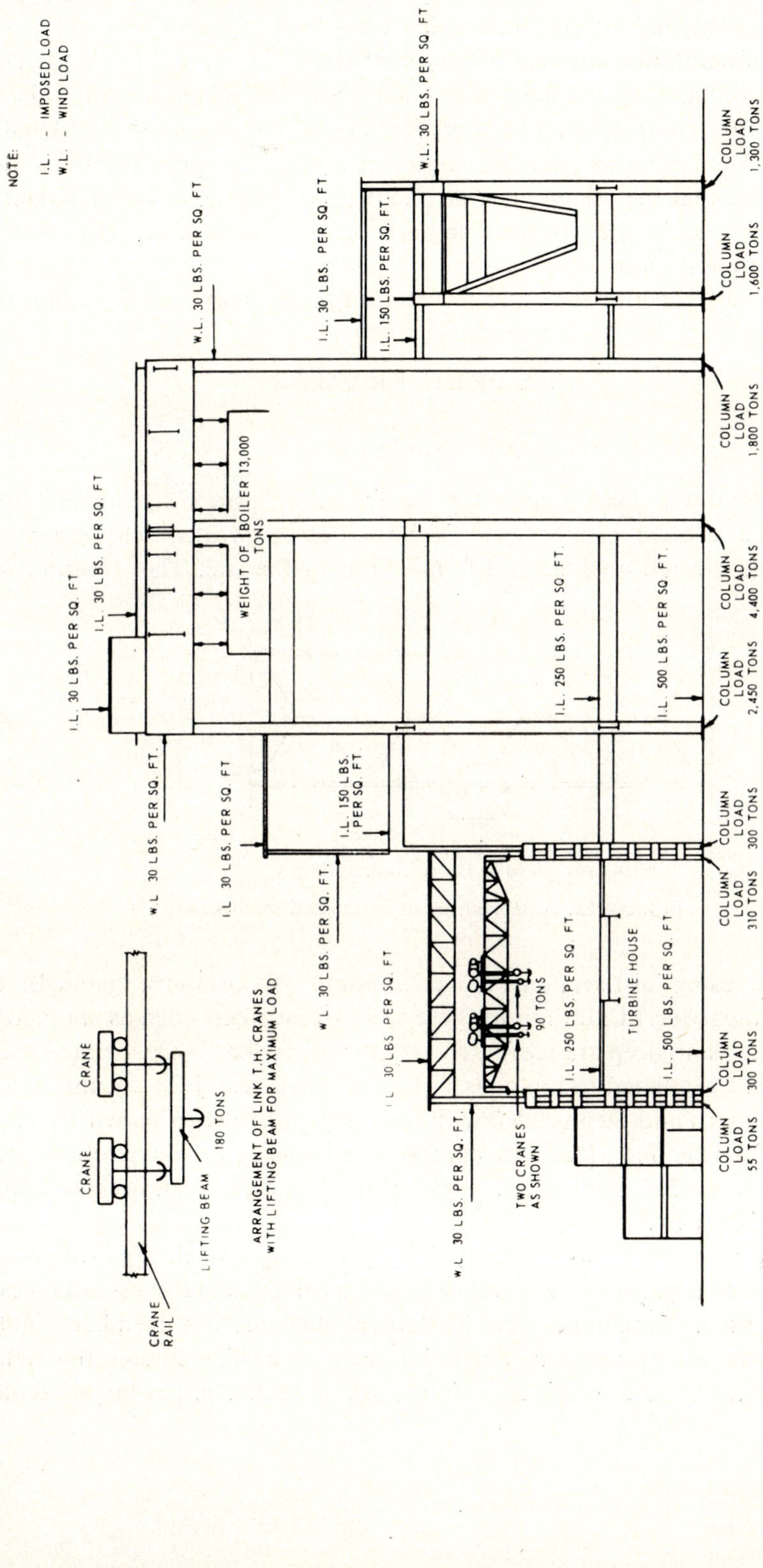

FIG. 3.3.7. Cross-section of typical station with loadings.

3.3.7. **Design Loadings**

The frame for a building is subject to a continually changing load system even after completion due to thermal expansion, varying wind loading, snow loading, dismantling and re-erection of plant, machinery, traffic and other similar varying loadings. The frame must be capable of carrying the worst combination of any of the loadings at all stages of erection and after completion. The complications of the distribution of bending moments in rigid joints structures and continuous members, limiting deflections and shear forces, and the fact that the design must be on a three dimensional basis, make the computer an invaluable aid for the structural engineer when he designs the frame for the main buildings of a power station.

Figure 3.3.7 shows some of the more relevant imposed loads on a frame for the main buildings.

3.4. STEEL FRAMES

3.4.1. **Steelwork**

Structural steelwork is largely used for the frames to power station buildings. The beams, columns and other members are fabricated at the works, delivered by road or rail transport and erected on prepared foundations at the site. The structural engineer

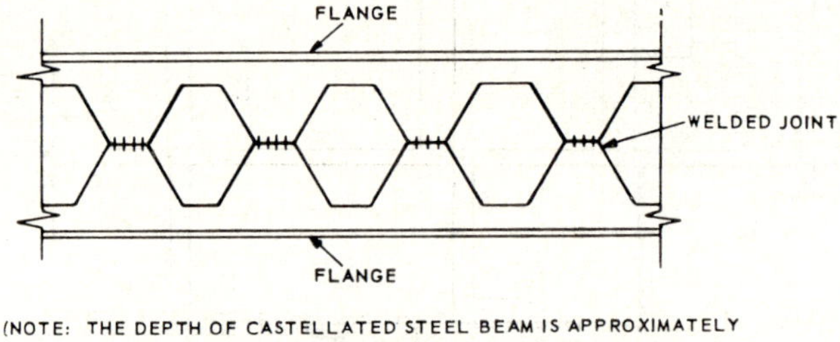

FIG. 3.4.1A. Arrangement of castellated steel beam.

designs the frame using joist sections, universal beams and columns, channels, ties, angles, flat plates and tubes. In addition to these sections members such as standard welded plate girders up to 78 in deep are useful to carry the heavy loading associated with power station buildings. Castellated beams consisting of a rolled steel joist cut through the web to a saw tooth pattern and then welded with the points together as shown Fig. 3.4.1A are particularly useful when light loads are carried on a long span and deflection rather than bending stresses is the design criterion. They are used in many roof designs as purlins to support the roof decking.

Although mild steel is used extensively, high yield strength steels are available and the higher working stresses permitted with this material means that in certain instances smaller sections can be used. Compound sections, lattice members, box members light trusses are also used. For the columns in boiler houses and turbine houses the welded box column using plates up to 5 in thick, as shown in Figure 3.4.1B, has largely replaced the

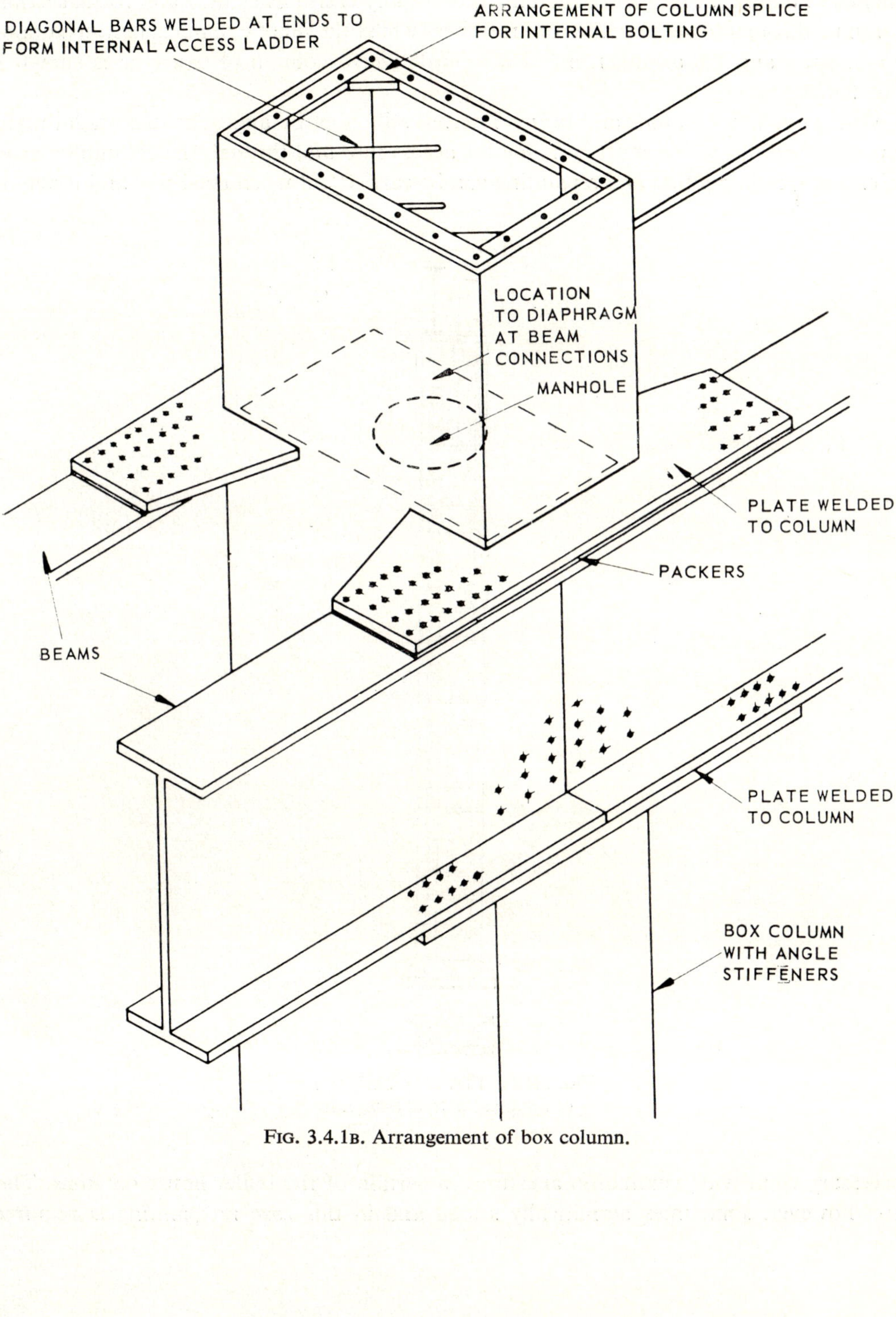

FIG. 3.4.1B. Arrangement of box column.

lattice column, although battened columns consisting of two rolled sections held together by flat plates welded or riveted at intervals along the length of the column are still used in the lighter frames for gas turbine houses, stores, workshops, pumphouses and other ancillary buildings. This form of column is especially useful when the outer rolled sections continue through to support the roof members whilst the inner column is used to support a crane rail and the arrangement of a typical riveted column of this type is shown in Fig. 3.4.1c.

With large box columns and beams internal stiffening by angles or diaphragm plates must be used. Provision must be made for access into and through the column or beam to ensure that inspection and repainting can be carried out when necessary and it may be

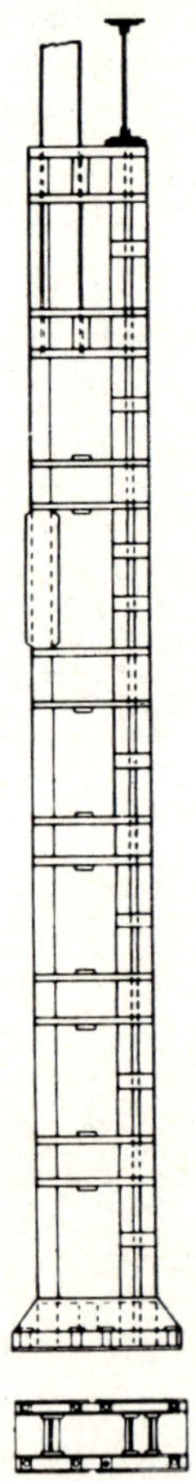

FIG. 3.4.1c. Elevation and section of a typical battened column.

necessary to provide ventilation apertures in certain of the boiler house columns. They are, however, sometimes hermitically sealed and in this case no painting is required.

3.4.2. **Design of Members**

The design of a building frame can be based on simple, semi-rigid or rigid design methods. If the structural engineer uses a simple design basis all beams are assumed to be freely supported at the ends. In most cases considerable savings in the weight of steel can be effected by the application of rigid or semi-rigid design methods. These allow for the distribution of bending moments introduced into a beam or column under given loading conditions throughout the structure.

The total loading on a frame is the result of the imposed loads and the dead loads of the structure. As stated in an earlier part of the chapter it is possible at an early stage in the design calculations to ascertain the imposed loads on the frame. Although experienced structural engineers can make close approximations for the dead loads of the structure it is only after the member has been designed that his first assumptions can be checked to ensure that the dead load has not been under-estimated. A modern technique in steel beam design where the beams are intended to support reinforced concrete floors is to use a large number of studs $\frac{1}{2}$ in diameter 2 to 3 in long welded to the top flange. These allow horizontal shear to be transmitted to the reinforced concrete floor which can be considered to take some compression.

The basis for the design of steel frames to buildings are given in B.S. 449/58, *The Use of Structural Steelwork in Buildings.*

The British Constructional Steelwork Association give typical examples of design in accordance with this B.S. in their publications. For information regarding the properties of steel sections and other useful design information reference should be made to the *Handbook of Structural Steelwork* using tables prepared and issued jointly by the British Constructional Steelwork Association and the British Steel Makers.

3.4.3. **Design Example—Steel Beam**

The following example covers the design for a steel beam spanning 30 ft. The beams are at 12 ft 6 in centres and the floor which itself weighs 60 lb/ft^2 carries an imposed distributed load of 100 lb/ft^2. The beam supports in the centre of its span a runway beam which will give a maximum point load of 20 tons. No reduction is allowable on the imposed load when designing the beam.

Load per foot run of beam due to imposed load and slab = (60+100) 12·5 =	2000 lb/ft^2
Weight of beam (assumed)	200 lb/ft^2
Total load per foot	2200 lb/ft^2
	Say 1 ton/ft

$$\text{Maximum bending moment} = \frac{wl^2}{8} + \frac{Wl}{4}$$

$$= \frac{1 \times 30 \times 30}{8} + \frac{20 \times 30}{4}$$

$$= 112{\cdot}5 + 150$$

$$= \underline{262{\cdot}5 \text{ tons/ft}}$$

B.S. 449 states that the buckling effect of the top flange must be considered. Lateral restraint can be given by concrete floors, intermediate members and similar methods and all these factors are considered when establishing the effective length of the top flange.

In this case it is assumed that although the compression flange is free to rotate in plan at the bearings there are no intermediate supports and the coefficient which must be applied to the actual length to establish the effective length is 1·0.

$$\text{Effective length} = l = 30\times 12$$
$$= \underline{360 \text{ in}}$$

Try a 36 in×12 in×170 lb/ft universal beam, $Z_x = 579{\cdot}1$ in³, $r_y = 2{\cdot}45$ in and $\frac{D}{T} = 32{\cdot}9$.

In this case there are no holes in the tension flange for bolts. Should the arrangement of the steelwork necessitate bolt holes in the tension flange then allowance must be made to the above factors.

$$\text{Slenderness ratio} = \frac{l}{r_y} = \frac{360}{2{\cdot}45}$$
$$= \underline{147}$$

Therefore from tables in B.S. 449, where $\frac{l}{r} = 147$ and $\frac{D}{T} = 32{\cdot}9$,

$$\text{Permissible bending stress} = p_{bc} = 6{\cdot}48 \text{ tons/in}^2$$

$$\text{Actual bending stress} = f_{bc} = \frac{\text{Bending moment}}{\text{Elastic modulus}}$$
$$= \frac{262{\cdot}5\times 12}{579{\cdot}1}$$
$$= \underline{5{\cdot}44 \text{ tons/in}^2}$$

As the actual stress f_{bc} is less than the permissible stress p_{bc} the beam is suitable.

Where the web of a beam is unstiffened the permissible shear stress in the web of the beam must not exceed 6 tons/in².

In this case the maximum shear force on the beam

$$= \frac{wl}{2}+\frac{W}{2} = \frac{1\times 30}{2}+\frac{20}{2} = \underline{25 \text{ tons}}$$

Shear stress is taken to act on an area of the beam equivalent to the depth of the beam multiplied by the web thickness.

Resistance to shear of beam $= p_q\cdot D\cdot t$

where p_q = permissible shear stress (in this case 6 tons/in²),
D = depth of beam,
t = web thickness.

$$\text{Therefore resistance to shear} = 6\times 36{\cdot}16\times 0{\cdot}680$$
$$= \underline{147 \text{ tons}}$$

This is ample to resist the applied shear force.

The assumed weight for the beam is in excess of that for the section selected which means that the beam is suitable.

Although the section to be used has been selected and is adequate to resist applied bending and shearing forces, the resistance of the beam to web buckling and bearing under concentrated loads and at the reactions, require investigation. A check should also be made that the deflection of the beam is not sufficient to be unsightly or damage finishings and deflection due to the super load is not more than $\frac{1}{360}$ of the span.

3.4.4. Design Example—Steel Column

This example is based on the design of a column which is required to carry given loads. Columns may be axially loaded or they may be subject to axial loads and also bending stresses, the latter being due to wind loading, bending moments transferred from beams

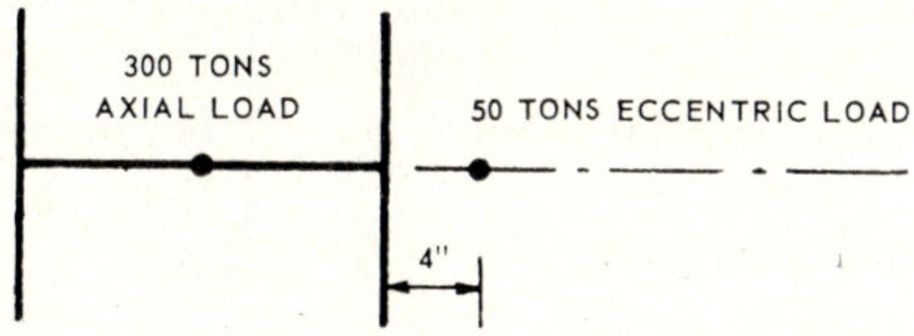

FIG. 3.4.4. Arrangement of loads on column.

by rigid or semi-rigid connections or by eccentric loads. The column being considered has an axial load of 300 tons and an eccentric load of 50 tons from a cleated connection on the flange of the column.

The column is 24 ft high and from B.S. 449 it will be seen that the end conditions will probably result in a factor of 0·85 being applied to the actual length to determine the effective length of the column.

The minimum eccentricity permitted by B.S. 449 for a cleated connection of the type used in this example is 4 in.

It is necessary to assume a column section to ascertain the dead load of the column and also other information relevant to the calculations. Safe load tables for columns axially loaded for various effective lengths are available in handbooks on structural steelwork and although it is not possible for these tables to include for all combinations of eccentric loads, experience and the tables normally enable the designer to assume a section which is near to the one proved by calculation, thus obviating repeated calculations.

In this case try a 16 in $\times$ 14 in$\times$193 lb/ft universal column, $A = 56{\cdot}73$ in^2, $r_y = 4{\cdot}05$ in and $Z_x = 310$ in^3 and $\frac{D}{T} = 10{\cdot}8$.

$$\begin{aligned}
&\text{Axial load} && = 300 \text{ tons}\\
&\text{Eccentric load} && = 50 \text{ tons}\\
&\text{Weight of column} = \frac{24\times 193}{2240} && = 2{\cdot}1 \text{ tons}\\
&\text{Total load} && = 352{\cdot}1 \text{ tons}
\end{aligned}$$

$$\text{Effective length} = 0{\cdot}85L = 0{\cdot}85 \times 24 \times 12 = \underline{245 \text{ in}}$$

The permissible stress in columns is obviously related to its effective length and stiffness. The slenderness ratio is defined as the effective length divided by the least radius of gyration and having established this ratio it is possible to obtain from tables in B.S. 449 the allowable stresses on the gross section of the column for axial compression.

$$\text{Slenderness ratio} = \frac{\text{Effective length}}{\text{Least radius of gyration}} = \frac{245}{4{\cdot}05} = \underline{60{\cdot}5}$$

From tables in B.S. 449:

$$\text{Permissible axial stress} = p_c = 7{\cdot}98 \text{ tons/in}^2$$

$$\text{Actual axial stress} = f_c = \frac{\text{Load}}{\text{Area}} = \frac{352{\cdot}1}{56{\cdot}73} = \underline{6{\cdot}21 \text{ tons/in}^2}$$

There is no bending moment about the y axis.

$$\text{Bending moment about } x \text{ axis} = \text{Load} \times \text{Eccentricity} = 50\left(\frac{15{\cdot}5}{2}+4\right) = 50 \times 11{\cdot}75 = \underline{587{\cdot}5 \text{ tons/in}}$$

$$\text{Actual bending stress} = f_{bc} = \frac{\text{Bending moment}}{\text{Elastic modulus}} = \frac{587{\cdot}5}{310{\cdot}0} = \underline{1{\cdot}9 \text{ tons/in}^2}$$

Now $\frac{l}{r_y}$ as previously found is 60·5 and from tables of section properties (B.S. 4) $\frac{D}{T} = 10{\cdot}8$

Therefore from tables in B.S. 449

Permissible bending stress $= p_{bc} = \underline{10{\cdot}5 \text{ tons/in}^2}$

B.S. 449 states that for column loadings of this type the ratio $\frac{f_c}{p_c}+\frac{f_{bc}}{p_{bc}}$ must be less than unity.

$$\text{In the case } \frac{f_c}{p_c}+\frac{f_{bc}}{p_{bc}} = \frac{6{\cdot}21}{7{\cdot}98}+\frac{1{\cdot}9}{10{\cdot}5}$$

$$= 0{\cdot}778+0{\cdot}181$$

$$= \underline{0{\cdot}959}$$

As this is less than unity, the column selected is suitable.

3.4.5. **Connections**

When using rigid or semi-rigid design methods of construction the connections of the beams to the columns must be capable of transferring the bending moments from the beam to the column, which in turn must have sufficient stiffness to withstand the bending moments induced by the beams. Also illustrated in Fig. 3.4.1B is a typical rigid joint between a beam and box column using welded cleats and friction grip bolts. Riveting was until recently the best method of obtaining rigid connections as the comparatively large clearances used for black bolts did not allow this type of bolt to be used for other than location purposes and to withstand direct shearing and bearing stresses. Welding and turned or friction grip bolts, however, have replaced riveting, largely because riveting methods demanded good access platforms to the joints, compressed air equipment and forges with consequent high labour costs. Equal or better results can be obtained using a combination of welded cleats and friction grip bolts, the latter being bolts which are tightened sufficiently to ensure that friction between the plates transfers all loads and the bolts are not therefore subject to shear or bearing. The correct load on bolts required to induce the necessary friction is measured by means of torque wrenches on the nuts of the bolts or alternatively by the use of load indicating washers or bolts. Load indicating washers are crimped and the heads of load indicating bolts are shaped in such a manner that deformation of the washer or bolt head takes place under load. The amount of deformation is related to the force applied to the bolt which means that the tension in the bolt can easily be checked by observation or using feeler gauges. Another type of friction grip bolt is designed in such a way that the top section of the bolt shears when the necessary tension is obtained leaving the bolt fully stressed. One advantage of load indicating washers and bolts and also shear bolts is that visual inspection can check the tightness of bolts and torque measurement is not necessary.

Turned or close tolerance bolts are also used for connections and the use of these bolts permits higher stresses in bearing and shearing than is permissible with black bolts. Although they are an improvement on black bolts in certain instances they are being superseded by friction grip bolts.

Welding must be carried out at the works where ultrasonic and X-ray equipment is available for testing welds and site work should be restricted to bolting.

3.4.6. Tubes

Steel tubes can also be fabricated to form structural members and although the circular faces do not lend themselves to bolting, they can be welded at the joints and to flat plates with holes for site bolted connections. This type of construction is very suitable for long span roof girders of the type used for turbine house roofs.

3.4.7. Protection of Steelwork

A big disadvantage with steel is its lack of resistance to corrosion. Although non-ferrous metals, mainly of the aluminium alloy groups, have been used successfully in certain instances for structural frames, their high cost does not make them economic for power station frames. When comparing the economics of a steel structure with a reinforced concrete structure the cost of the original protection and subsequent treatments must be considered. As will be seen from other parts of the chapter, other factors must also be considered before the type of frame is decided.

In the past steelwork, after fabrication at the works, was painted one coat of red oxide or red lead paint before being despatched to the site. It was found, however, that in many cases the paint was being applied over loose mill scale and although the steelwork was erected and painted in the normal way, repainting was very soon required because the mill scale and paint film became detached. It has been the practice, therefore, to deliver steelwork to site unpainted and allow it to weather, thus removing loose mill scale before cleaning and painting commences. The paint treatment usually consists of a coat of rust inhibiting primer followed by one or two undercoats and a finishing coat of gloss paint. The provision of large steel frames to indoor sub-stations has revealed the need to reduce the frequency of maintenance painting where there is live electrical equipment. A better finish in the first instance is one way of reducing the frequency of maintenance painting, and grit blasting followed by a good paint sequence is one method of obtaining this finish. One of the best specifications is for a normal paint treatment using a special primer on a galvanised or other form of zinc or aluminium dipped or spray coating, and up to 20 years' protection before repainting is necessary can be anticipated with this treatment. These treatments are expensive and it is unlikely that with high interest charges on the capital expenditure that shot blasting will be economic for other than external steelwork and other small sections of the internal surfaces of the steelwork in the main building.

In certain areas the steelwork is entirely encased in brickwork or concrete reinforced to prevent cracking; this is mostly in offices and amenity areas and provides fire protection to the steel frame and makes it easier to build filler walls and apply decorations.

3.5. REINFORCED CONCRETE

3.5.1. General

The principles including mixing, transporting, placing and vibration of concrete in foundations are equally applicable to concrete used in superstructures. A larger proportion of high quality concrete is used on superstructures as most of the work involves columns, beams, slabs, walls and similar structural members, which in general are more highly stressed than foundation concrete.

3.5.2. **Formwork**

In addition to timber and steel panels as used in foundations, shutters lined with hardboard, plastic and similar materials are used to obtain better finishes to the concrete.

3.5.3. **Reinforcement**

Although mild steel reinforcement is used extensively, high tensile steel reinforcing bars and mesh are also used. Because of the higher stresses which may be used, the saving of steel in a structure using high tensile reinforcement is quite large and the reduced diameter or smaller number of bars of this type helps to simplify complicated arrangements of reinforcement and consequently the placing of concrete. The design of reinforced concrete allows for the transfer of stress by bond from steel to the concrete and vice-versa throughout the length of the bars. Shrinkage of the concrete whilst setting develops this local bond between the circular outer face of mild steel bars and the concrete. High tensile steel bars have deformed outer faces for which higher bond stresses are permitted. In addition to local bond the ends of bars may require hooks or bends to ensure sufficient anchorage. A typical arrangement of reinforcing bars in a slab beam and column is shown in Fig. 3.5.3.

3.5.4. **Design of Reinforced Concrete**

Many excellent books have been published on the theory and practi of reinforced concrete designs. Reference can be made to the *Reinforced Concrete Designer's Handbook* by Chas. E. Reynolds for examples on the design of structure in reinforced concrete. In reinforced concrete design it is possible to ascertain the actual imposed loads at an early stage, but assumptions must be made regarding the dead load of the structure. The problem is further complicated when designing rigid jointed structures in steelwork or reinforced concrete because the transfer of bending moments between members having different second moments of areas is related to the ratio of these values. As for steelwork, the sections assumed at the commencement of the design must be checked on completion.

The basis for the design of reinforced concrete in buildings is given in British Standard Code of Practice 114, *The Structural Use of Concrete in Buildings*. C.P. 115, *The Structural Use of Prestressed Concrete in Buildings*, gives recommendations on the use of prestressed concrete and C.P. 116, *The Structural Use of Precast Concrete*, gives recommendations regarding the design, manufacture and use of precast concrete.

The following design examples are based on the information given in C.P. 114.

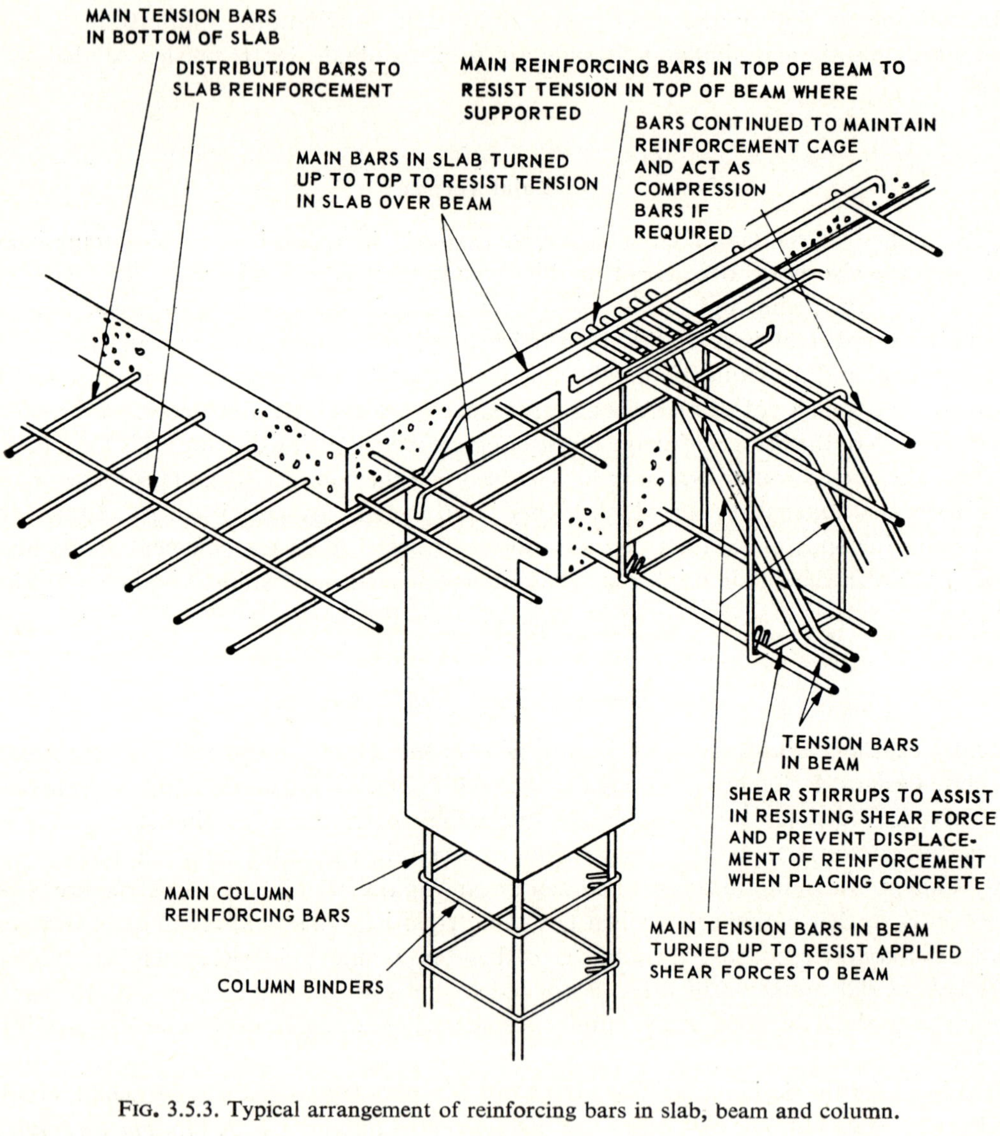

FIG. 3.5.3. Typical arrangement of reinforcing bars in slab, beam and column.

3.5.5. **Design Example—Concrete Beam**

Design a reinforced concrete beam to carry the floor and loading shown in Fig. 3.5.5A.

Weight of floor and finishings	=	75 lb/ft^2
Superimposed load on floor	=	50 lb/ft^2
Total load	=	125 lb/ft^2

Area of floor carried by one beam = 20 ft×15 ft = 300 ft^2

Load on beam due to floor	= 300×125 =	37,500 lb
Weight of beam (assumed)	= 20×288 =	5760 lb
Total load on one beam	=	43,260 lb

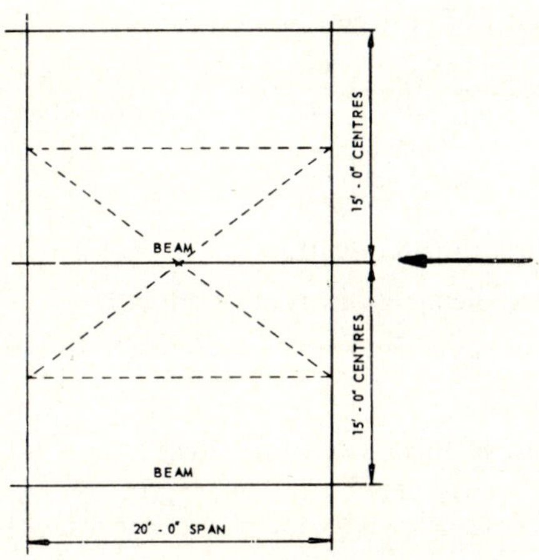

FIG. 3.5.5A. Arrangement of floor and beams.

Concrete beams may be designed on the elastic theory based on modular ratio or on the load factor method. The load factor method does not involve a knowledge or use of modular ratio and does not assume a linear relationship between the stress and strain in the concrete. It assumes that as failure is approached the compressive stresses will adjust themselves to give a total compression greater than that deduced from the elastic theory and the extent of this adjustment has been proved by experiment. The load factor method has been used for the following example.

Certain assumptions must be made at this stage, i.e. assume a beam 12 in wide by 24 in deep and the tension reinforcement to be 6 No. $\frac{7}{8}$ in diameter reinforcing bars.

The permissible stresses in concrete and steel depend upon the grade of concrete and type of reinforcement used, and in this example p_{cb} = permissible compressive stress in concrete in bending = 1000 lb/in^2, p^{st} = permissible tensile stress in the reinforcement = 20,000 lb/in^2.

C.P. 114 states that where the slenderness ratio for beams (i.e. span divided by breadth of beam) is less than 30 no reduction is to be made to the permissible concrete stresses.

In this case slenderness ratio $= \frac{20 \times 12}{12} = \underline{20}$

Therefore no reduction is required.

C.P. 114 also states that for simply supported beams the ratio of span ÷ depth of beam shall not be greater than 20.

In this case $\frac{L}{d} = \frac{20 \times 12}{24} = \underline{10}$

Therefore the beam is suitable.

The following simplified formulae are used to determine the moment of resistance of reinforced concrete beams having rectangular cross-sections and are given in C.P. 114.

(i) Based on the tensile reinforcement:

Moment of resistance $= M_r = A_{st} \cdot p_{st} \cdot l_a$

(ii) Based on the strength of concrete in compression:

Moment of resistance $= M_r = \frac{p_{cb}}{4} \cdot b \cdot d_1^2$.

where l_a is the lever arm $= d_1 - \frac{3A_{st} \cdot p_{st}}{4b \cdot p_{cb}}$

and A_{st} = area of tensile reinforcement,

p_{st} = permissible tensile stress in reinforcement,

p_{cb} = permissible compressive stress in concrete in bending,

b = breadth of section,

d_1 = effective depth of tensile reinforcement,

M_{max} at centre of beam $= \frac{Wl}{8} = \frac{43{,}260 \times 20 \times 12}{8} = \underline{1{,}297{,}800 \text{ lb/in}}$

$A_{st} = 6 \times 0{\cdot}601 = \underline{3{\cdot}61 \text{ in}^2}$

$d_1 = \text{depth of beam} - \left(\frac{\text{diameter of bars}}{2} + \text{cover}\right)$

$= 24 - \left(\frac{0{\cdot}875}{2} + 1\right)$ say $\underline{22{\cdot}5 \text{ in}}$

$l_a = 22{\cdot}5 - \frac{3 \times 3{\cdot}61 \times 20{,}000}{4 \times 12 \times 1000}$

$= 22{\cdot}5 - 4{\cdot}5$

$= \underline{18 \text{ in}}$

Based on tensile reinforcement:

$M_r = 3{\cdot}61 \times 20{,}000 \times 18{\cdot}00 = \underline{1{,}299{,}600 \text{ lb/in}}$

Based on strength of concrete in compression:

$M_r = \frac{1000}{4} \times 12 \times 22{\cdot}5^2 = \underline{1{,}518{,}750 \text{ lb/in}}$

From the above results it will be seen that although on the assumptions made the reinforcement is fully stressed, the concrete is understressed. The depth of beam selected, however, affects the lever arm and it follows that a reduction in depth whilst reducing the concrete stress also necessitates an increase in the amount of reinforcement.

Where a concrete beam is used in conjunction with an *in situ* floor slab it is permissible to design the beam on the basis that a portion of the slab width can be used to assist in resisting compression forces in the top of the beam. This results in an L or T beam section, and the assumptions for the design of beams on this basis is indicated in Fig. 3.5.5B.

The arrangement of the tension reinforcement in the bottom of the beam must now be considered. C.P. 114 states that the distance between bars shall be equal to the diameter of the largest bar with a minimum space equal to the size of the aggregate plus $\frac{1}{4}$ in, and that

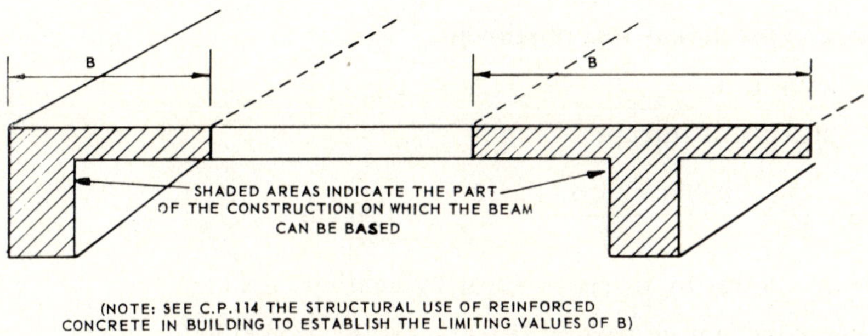

FIG. 3.5.5B. L and T beam construction.

adequate cover must be provided between the outside of the bars and the face of the concrete. This cover assists in protecting the bars against corrosion and the minimum recom mendations are given in C.P. 114 for varying conditions. In this case:

$$\text{6 bars} + \text{5 spaces} = 11\times\tfrac{7}{8}\text{ in} = 9\tfrac{5}{8}\text{ in}$$

$$\text{1 in cover both sides} = 2\times 1\text{ in} = 2\text{ in}$$

$$\underline{\overline{11\tfrac{5}{8}\text{ in}}}$$

The 12 in beam selected therefore allows all bars to be placed in one layer.

In addition to designing the tension reinforcement in this beam, a check must be made on the shear.

$$\text{Maximum S.F.} = \frac{W}{2} = \frac{43{,}260}{2} = \underline{21{,}630\text{ lb}}$$

Based on the concrete alone, the resistance to shear of the concrete section is given by the following equation from C.P. 114.

$$\text{Shear stress } q = \frac{Q}{bl_a}$$

where Q = total shear force across section,

b = breadth of beam,

l_a = lever arm.

In this case $q = \dfrac{21{,}630}{12 \times 18} = \underline{100 \text{ lb/in}^2}$

For the grade of concrete selected the permissible value of q is 100 lb/in² and no shear reinforcement therefore is necessary. On most applications, however, it is preferable to assume that no shear will be taken by the concrete, and that reinforcement must be provided to withstand all shearing forces.

Assume $\frac{3}{8}$ in diameter stirrups at 6 in centres. From C.P. 114 resistance to shear by stirrups is given by the equation

$$Q = \frac{p_{st} \cdot A_w \cdot l_a}{s}$$

where p_{st} = permissible tensile stress in shear reinforcement,

A_w = cross-sectional area of stirrups,

l_a = lever arm,

s = spacing of stirrups.

In this case $Q = \dfrac{20{,}000 \times 2 \times 0{\cdot}11 \times 18}{6} = \underline{13{,}200 \text{ lb}}$

Total shear = Shear by stirrups + Shear by bent up bars

Therefore shear by bent up bars = 21,630 − 13,200 = $\underline{8430 \text{ lb}}$

Shear by two $\frac{7}{8}$ in diameter bars bent up at 45°.

$$Q = A \times p_{ct} \times \sin \theta$$

where Q = shear by bent up bars,

A = cross-sectional area of bent up bars,

p_{st} = permissible tensile stress in reinforcement,

θ = angle between bent up bars and horizontal.

In this case $Q = 2 \times 0{\cdot}601 \times 20{,}000 \times 0{\cdot}7 = \underline{16{,}828 \text{ lb}}$

This is sufficient and the arrangement of reinforcement is shown in Fig. 3.5.5c.

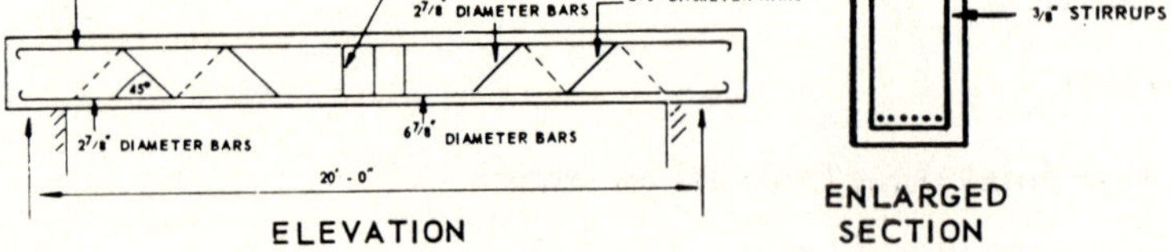

FIG. 3.5.5c. Arrangement of reinforcement to beam.

3.5.6. **Design Example—Concrete Column**

Design a reinforced concrete column 20 ft high to carry an axial load of 180 tons if the column is properly restrained at the top and bottom in position and direction.

Columns may be subject to stresses due to axially applied loads and/or bending. Bending in a column increases the compressive forces due to vertical loading in one face of the column with a corresponding decrease in the stresses on the opposite face, and although excessive bending in conjunction with comparatively small vertical loading can induce tension in a column, reinforcement is usually included to assist in resisting compressive forces or is of a nominal nature for lightly loaded columns where the concrete carries the whole of the load.

Although theoretically the stresses in the concrete and reinforcement are directly related to the modular ratio of concrete and steel, experiments have shown that when a reinforced concrete column is loaded, the reinforcement and concrete are stressed to their respective permissible stresses and the following formula is used to ascertain the cross-sectional area of longitudinal reinforcement and concrete.

$$P_0 = p_{cc} \cdot A_c + p_{sc} \cdot A_{sc}$$

where P_0 = permissible load,

p_{cc} = permissible compressive stress in concrete,

A_c = cross-sectional area of concrete less area of steel,

p_{sc} = permissible compressive stress in steel,

A_{sc} = cross-sectional area of longitudinal reinforcement.

For "short" columns (i.e. where the ratio of the effective length of the column to the least lateral dimension does not exceed 15) the maximum permissible stresses given in C.P. 114 may be used. For "long" columns (where the ratio exceeds 15) reductions in stresses used must be made in accordance with the coefficients included in C.P. 114.

The permissible stresses in concrete and steel depend upon the grade of concrete and type of reinforcement used, and in this example:

$$p_{cc} = 760 \text{ lb/in}^2$$

$$p_{sc} = 18{,}000 \text{ lb/in}^2$$

The effective length is the actual length multiplied by a factor which depends upon the end conditions of the column and in this case (see question for conditions) it is 0·75.

Therefore effective length $= 0{\cdot}75\,L = 0{\cdot}75 \times 20 = \underline{15 \text{ ft}}$

Certain assumptions must be made at this stage and these may be based on physical features of the structure.

Assume 18 in × 18 in square column with 8 No. $1\frac{1}{4}$ in diameter longitudinal bars.

The ratio of effective length to least lateral dimension

$$= \frac{15 \times 12}{18} = 10$$

Therefore no reduction is to be made in permissible stresses.

Load on column	$= 180 \times 2240$	$= 403{,}200$ lb
Weight of column	$= 1{\cdot}5^2 \times 20 \times 144 =$	6480 lb
	Total load	409,680 lb

$$P_0 = p_{cc} \cdot A_c + p_{sc} + A_{sc}$$

$$= 760[18^2 - (8 \times 1{\cdot}227)] + (18{,}000 \times 8 \times 1{\cdot}227)$$

$$= 760(324 - 9{\cdot}82) + (18{,}000 \times 9{\cdot}82)$$

$$= 760 \times 314{\cdot}18 + 18{,}000 \times 9{\cdot}82$$

$$= 238{,}800 + 176{,}800 = 415{,}600 \text{ lb}$$

This is in excess of the applied load and the column is therefore suitable.

C.P. 114 states that transverse reinforcement shall be provided in columns to prevent buckling of the main reinforcing bars and to retain the cage in position whilst concrete is being placed, the diameter of this reinforcement being not less than a quarter of the diameter of the main bars are in no case less than $\frac{3}{16}$ in. For this column links are being used as transverse reinforcement and the diameter of the links is $\frac{5}{16}$ in.

C.P. 114 also states that to avoid excessive spacing between binders the pitch of the transverse reinforcement shall be not more than:

(1) The least lateral dimension of the column (i.e. 18 in).
(2) 12 times the diameter of the smallest longitudinal reinforcement (i.e. 15 in).
(3) 12 in.

In this column 12 in is the pitch of the reinforcement.

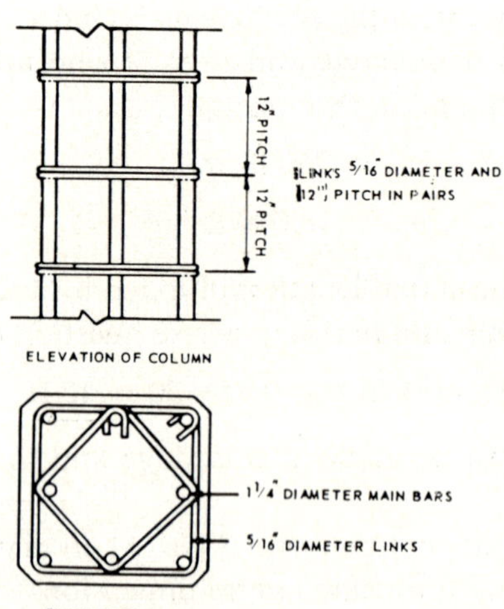

FIG. 3.5.6. Arrangement of reinforcement to column.
(Reproduced by kind permission of The Cement and Concrete Association)

To retain the main bars in position and prevent side movement during concreting, all transverse reinforcement must change direction at longitudinal bars and it is necessary therefore to introduce additional links to meet this requirement on the centre bars.

The details of the column and reinforcement are shown in Fig. 3.5.6.

3.5.7. **Prestressing**

When a beam is loaded it deflects and compressive stresses are introduced into the top fibres and tensile stresses into the lower ones. With prestressed concrete, high tensile steel wire or cables are inserted into tubes cast in the tension section of the beam and held in permanent wedge type anchorages; after tensioning by means of hydraulic jacks the tubes are grouted up. This has the effect of compressing the concrete in the lower fibres and introducing tension into the upper fibres. Loading of the beam then has to remove the induced reverse stresses before normal compression of the top section and tension in the lower section commences. Figure 3.5.7 shows the stresses which are introduced into the concrete by loading and prestressing in a concrete beam. All the stress diagrams are taken on section *AA* at the centre of the beam and the first diagram shows the method of arriving at the stresses resulting from the dead load together with the prestress and the second diagram shows the final resultant stresses after the imposed load is applied. It will be seen from the last diagram that even when the beam is subjected to all stresses resulting from the dead loads, the imposed loads and the prestressing forces, no tension stresses remain in the concrete. Using this method members of a section much lighter than that required for normal reinforced concrete may in certain instances be used and much larger spans for beams can be obtained.

Beams are sometimes tensioned through the supporting columns of a concrete structure to form rigid joints. Prestressing is used for valley beams of barrel roofs, bridges, tanks, pressure vessels.

For precast items such as cooling tower stack members or floor and roof beams, the wires are stretched on a permanent bed, possibly several hundred feet long and a number of units are then cast around each section of wires. After the concrete has set the jacks are released and the wires between each unit are cut, the wires remaining tensioned inside each unit.

Where tensioning of wires is carried out after the concrete has set as in the case of concrete reactors, the method is termed "post tensioning" and where the wires are stressed before the concrete is placed it is called "pretensioning".

3.5.8. **Thermal Movement**

The coefficient of linear expansion for steel is 0·000012 per °C and for concrete it varies between 0·000007 per °C where limestone and similar aggregate are used to 0·000012 per °C for most gravel aggregates. The latter aggregate is normally used in power station structures and stresses set up by differential temperature movement are therefore negligible. Temperature changes do, however, cause expansion and contraction of the structure and unless the design allows it to expand and contract uniformly expansion

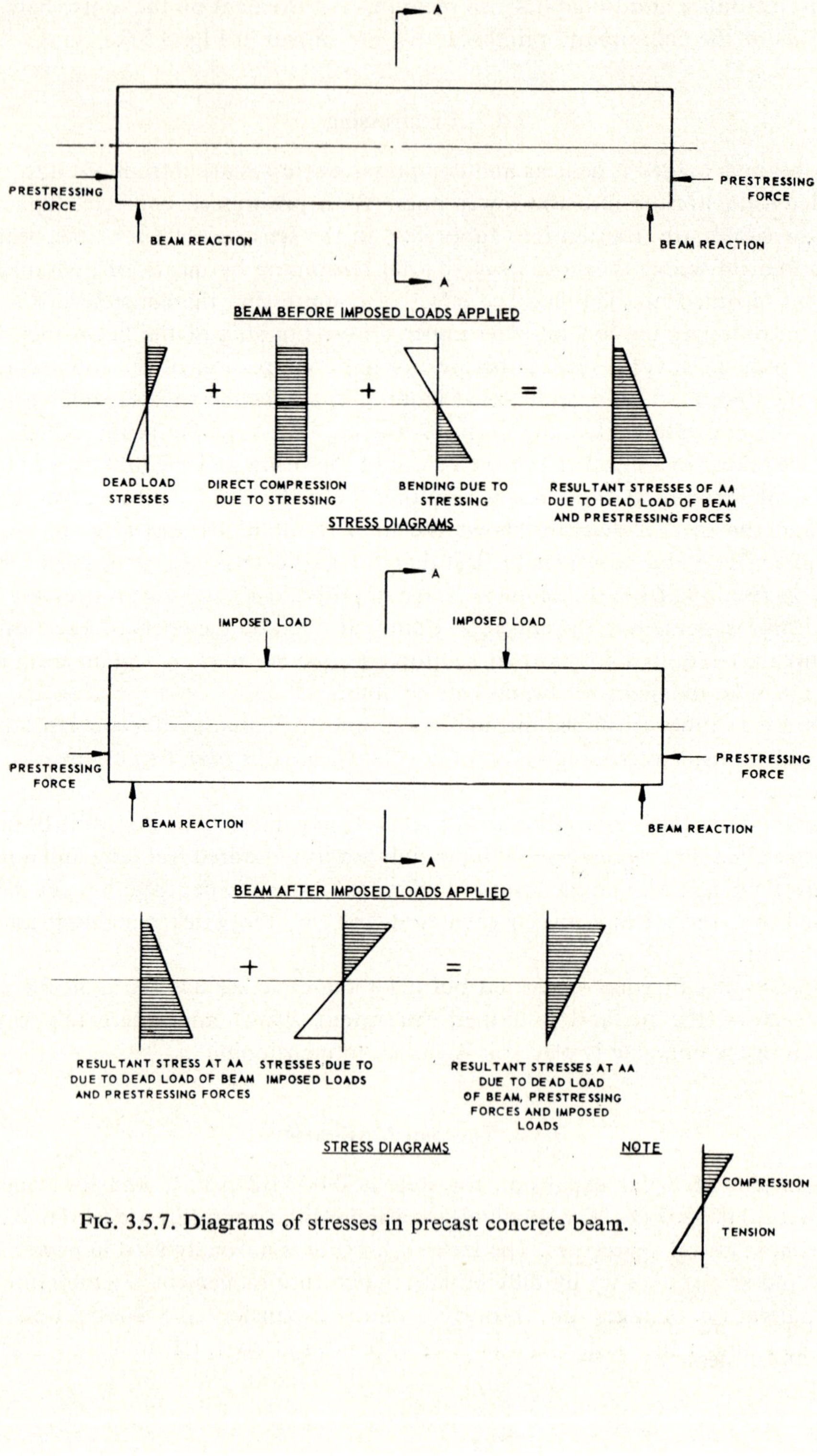

FIG. 3.5.7. Diagrams of stresses in precast concrete beam.

joints must be provided. These joints vary from a simple gap filled with rot-proof compressible boarding as used in concrete retaining walls, slabs and similar structures to the complicated roller, rocker or sliding joints provided at the end of bridge or similar beams. Expansion joints in buildings must be carefully located and made weatherproof by providing cappings at joints in roofs and the use of copper water bars or similar methods of sealing in vertical walls. Care must be taken with the treatment of internal finishes at the positions where expansion joints are located, plates being used on the floors and cover strips or other features being used on walls or ceilings to permit sliding and prevent cracking of plaster or other finishings.

3.5.9. **Contraction Joints**

The amount of shrinkage in setting concrete is usually about 1/10 in over 20 ft in the first year. Correct curing can reduce the amount of drying shrinkage but cannot completely prevent it. Good design permits this contraction to take place by providing planes where the concrete may crack and at the same time allow for sealing of the joint after the shrinkage has occurred.

3.5.10. **Curing**

Cement sets by hydration, i.e. it combines with water and sufficient water must be kept in the concrete to ensure complete hydration of the cement. The amount of water necessary to make the concrete sufficiently plastic to place and compact is in excess of that required for hydration of the cement and the removal of a small proportion of the mixing water by evaporation, after the concrete is compacted, is of no consequence. Drying out of the concrete or curing as it is usually called is carried out by retaining formwork in position, keeping the surface wet by spraying or puddling with water, use of plastic sheets, covering with wet sand, spraying with solution to form impervious plastic membrane or similar methods. Curing is usually carried out for 7 days after which period the concrete has hardened sufficiently to retain enough water to complete the hydration process under normal weather conditions in this country.

3.5.11. **Precast Concrete**

Concrete poured into moulds which retain the concrete in its final form and position is termed *in situ* concrete. If the concrete is cast in locations which are different from the final positions it is referred to as "precast" concrete. Although on large contracts precasting yards may be established on the site, factory made units are most popular. Factory manufacture and handling processes can usually more than offset the additional expense necessitated by transporting units from the factory to the site. The principal advantages of *in situ* concrete are that it can be cast monolithic, i.e. there are no joints to make good, and the size of members is not limited by the capacity of the handling equipment. Some of the advantages of precast concrete over *in situ* are as follows:

(a) Good concrete of uniform quality is easily obtained under factory conditions.
(b) Curing is more easily carried out on precast units.

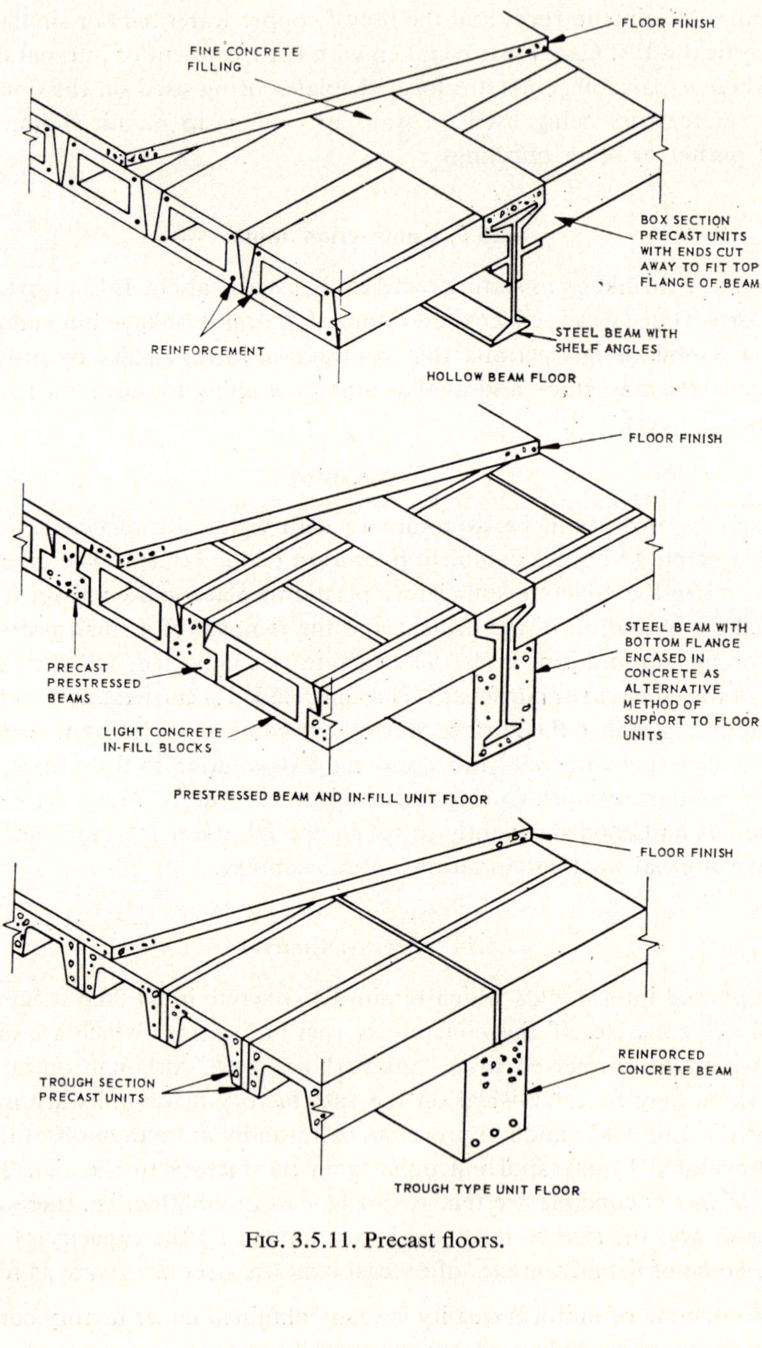

FIG. 3.5.11. Precast floors.

(c) Units can be protected against frost more easily during the early stages.

(d) It is not necessary to delay construction of a further stage of the structure to allow completed sections to obtain a strength sufficient to withstand additional load.

(e) Congestion on the structure is reduced by the manufacture of units in more suitable areas and because less support formwork and scaffolding is required.

(f) There is no delay due to striking and erection of framework or fixing reinforcement.

Advance planning is necessary to enable the units to be manufactured and delivered to site in accordance with a programme and adequate lifting equipment is essential.

Precast concrete floors, one type of which is shown on Fig. 3.5.11, have been used for many years but are unsuitable where they may be subject to the moving of heavy plant loads. Precast wall panels have been used successfully on several newer power stations. Effective architectural effects have been obtained on the face of wall panels by using pebbles or partially set, by water jets or other method, thus exposing the aggregate on the face. Lintels, coping stones and similar items are often made from "artificial stone" or "reconstructed stone" which is concrete manufactured using stone aggregates and coloured cement where the colour of the stone necessitates its use. Very often the finished block has only a face of this type of concrete, the backing being concrete mixed from ordinary gravel and sand.

Large panels of reinforced lightweight aerated concrete are used for roof units and infill wall panels.

3.6. BRICKWORK AND BLOCKWORK

3.6.1. General

Although on power stations brickwork is used mainly as a fabric material, some types of bricks do have high compressive strengths and are used in load bearing wall or pier construction.

3.6.2. Bricks

Bricks are normally divided into one of the following groups:

(a) *Common bricks.* These are usually manufactured by wire cutting extruded lengths of clay, the bricks being dried and then burnt in a kiln normally fired by coal. Forming bricks by pressing instead of extrusion and wire cutting is also carried out at some brick yards. For internal walls bricks are now manufactured containing up to 90% P.F.A.

(b) *Engineering bricks.* These are usually manufactured from shale by a pressing process and are hard, dense bricks of good strength, low permeability with good acid and frost resistance, and excellent weathering characteristics. As the name implies they are suitable for load bearing walls, piers, mass retaining walls and other similar locations. They are also used for the internal face of walls where a high resistance to mechanical damage is required.

(c) *Facing bricks.* These are cut or pressed bricks finished on the exposed faces by sand blasting or other method of roughening to give the required surface effect. They can be coloured before burning and typical descriptions include the words "sand faced", "multi-coloured" and similar descriptive terms. They are used for external walls as they are resistant to weather and have a good appearance.

(d) *Sand-lime bricks.* These are made to normal brick sizes from a mixture of sand, lime and water which is pressed and then subjected to high temperature steam in an autoclave which accelerates and ensures complete hydration of the lime. They are uniform in size, have a pleasant surface texture and colour and are used extensively unpainted for the surface of fair faced internal walls.

(e) *Concrete bricks.* Made from concrete these bricks can be obtained in qualities similar to sand-lime bricks and are used in similar locations.

(f) *Lightweight bricks and blocks.* These are obtainable in various sizes and are manufactured from pulverised fuel ash, aerated concrete, coke breeze and similar materials. Their light weight together with high thermal and sound insulation properties make them very suitable for internal partition walls. They are not suitable for external walls except where protected on the outer face.

3.6.3. **Mortar**

The main constituents of mortar are sand, lime, cement, mixing water and other substances such as plasticisers, the proportions being varied to give a mortar suitable for the work in hand. In general it should not exceed the bricks in strength. For engineering brickwork a hard mortar composed of one part cement to three parts sand would be used whereas for normal external brickwork one part cement, one part lime and four parts of sand would be suitable. For internal block walls one part cement, two parts lime and eight parts of sand is used. Proprietory cements especially for mixing mortars are now available which can be used without additives.

3.6.4. **Brickwork**

Wire cut bricks have flat surfaces and pressed bricks have flat surfaces or frogs on one or both sides. Stronger joints are obtained in walls using bricks with frogs and engineering bricks normally have frogs on both faces.

The bond used in the brickwork is most important for the appearance and strength of the finished brickwork. There are many variations on the bonds used in brick walls, but the two most common are English Bond and Flemish Bond (see Fig. 3.6.4).

For external wall cavity construction using two leaves of brickwork with a space usually 2 in wide between the inner and outer skin, gives a wall with the best damp resisting properties. These walls have better thermal insulation properties than a solid brick wall of equivalent thickness.

A great disadvantage with brickwork is its own dead-load. High walls without restraint at intermediate levels are subject to buckling and these effects can be reduced by forming piers on the face of the wall at suitable intervals along its length or by tying back to the structural columns and floors using ties which allow vertical movement.

The strength of brick walls can be increased by the use of steel reinforcement laid in the bed joints and small diameter steel bars in the vertical joints.

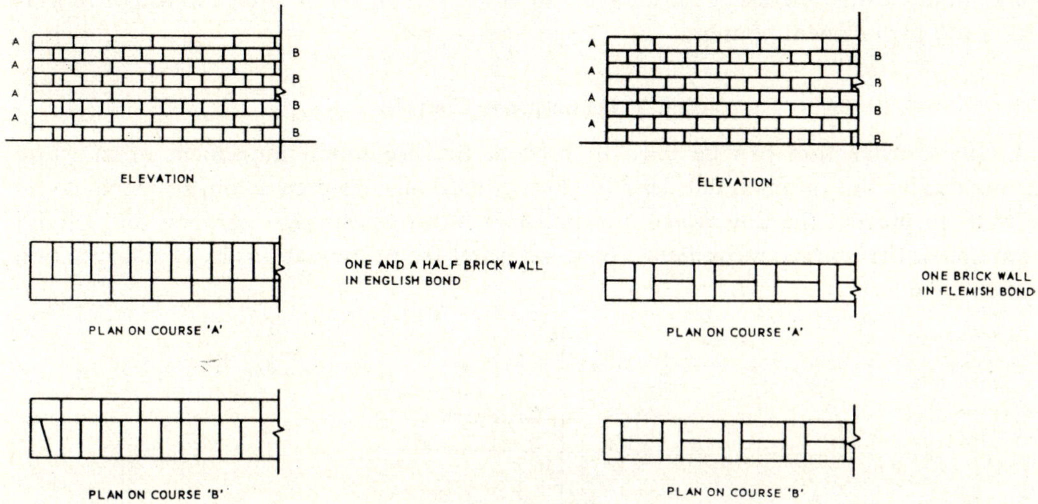

FIG. 3.6.4. Brickwork bond.

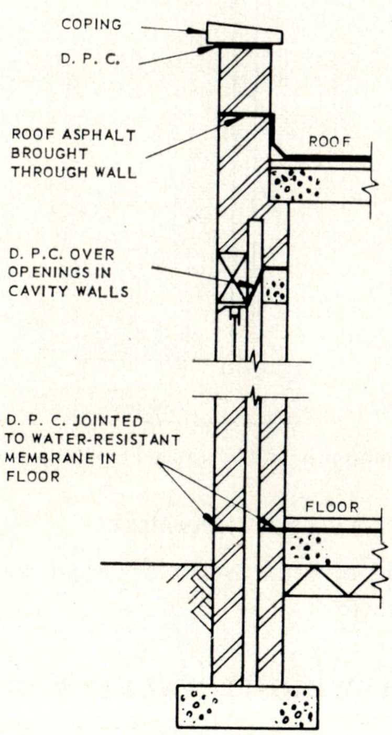

FIG. 3.6.5A. Damp-proof courses in walls.

The correct mortar mix for the type of bricks or blocks and also the location of the work should always be used. Shrinkage and cracking can result from the use of mortar with a too high cement content.

3.6.5. Damp-proof Courses

Cavity construction can be used to prevent the horizontal movement of moisture through walls, but bitumen felt, lead or similar materials are used as impermeable layers in walls to prevent the downward movement of moisture through parapets and similar locations or the upward movement of ground water from foundations by capillary action (see Fig. 3.6.5A).

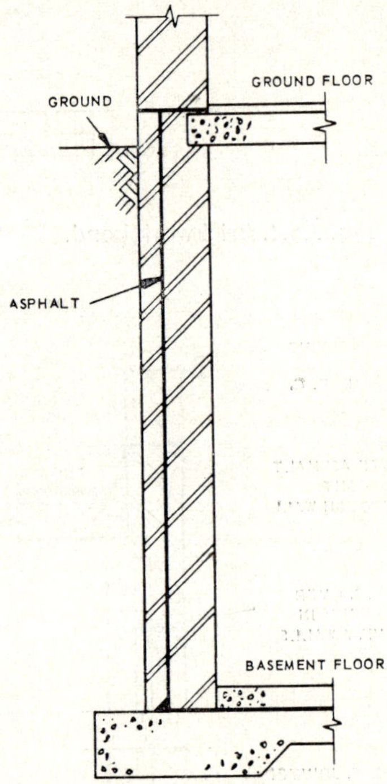

FIG. 3.6.5B. Damp-proofing basements by tanking with asphalt.

3.6.6. Openings

Concrete lintels, steel beams or brick arches are used to carry the brickwork above openings for doors, windows, etc.

3.7. LIGHTWEIGHT WALLING SYSTEMS

3.7.1. Sheeting

Vertical cladding usually consists of aluminium or steel sheets, the latter being plain or zinc coated by galvanising or other approved process. The sheets are normally corrugated or profiled and although galvanised sheets are sometimes used with no further treatment, wall cladding including aluminium is usually coated with a film of P.V.C., alkyd amino-based paint spray, bitumen felt or other approved material. Asbestos sheets are sometimes used especially in locations where damage is of little consequence. All sheets can be obtained in various colours.

Steel sheets are normally 22 gauge and aluminium sheets 18 gauge. The zinc coating of the steel sheets can be a 2 oz/ft^2 treatment for self-finished galvanised sheets although a much thinner zinc coating is often used in conjunction with a plastic coating. Plastic films can be applied to sheets by dipping, brushing, roller, spray treatments or by application in a laminate form, the plastic sheet being secured by a suitable adhesive. Drying may be carried out using stove techniques or air drying processes depending on the type of coating used. The film thickness may be from approximately 0·001 in for the coated finishes to films in excess of 0·01 in for the laminated finishes. When comparing different materials, however, film thickness is not a measure of the relative durability characteristics of different materials.

Sheets can be cut by the manufacturers to the client's requirements and in many cases it is transport and erection difficulties which limit the length of sheet. Nevertheless, sheets up to 40 ft long are commonly used for large walls to reduce the effects of horizontal joints. The cladding is also more rigid where sheets are continuous over a number of sheeting rails.

3.7.2. **Insulation**

Where exclusion of weather is the only consideration then single sheets without a lining are sufficient. In locations where human occupancy and the necessity to preserve temperatures above those at which condensation forms (i.e. dew point temperatures) require some form of heating, then to prevent excessive heat losses a lining of fibre board, porous polystyrene sheet, plaster board, asbestos board or similar materials is used.

3.7.3. **Fixings**

The external sheeting is normally fixed by self-tapping screws, hook bolts, clips or other suitable means direct to the angle sections forming the sheeting rails. The insulation layer may be fixed to the back of the outer sheeting or it may be supported on the inner face of the sheeting rails, thus forming a cavity which increases the thermal insulation value. Although cladding of this type can be used on a structure with a reinforced concrete frame, the spacing of the sheeting rails and other characteristics make it most suitable for use with a structural steel framework. Figure 3.7.3 shows a typical arrangement of insulated cladding used in conjunction with a steel frame.

To obtain natural light inside a building with this type of fabric, windows of the normal pattern or alternatively patent glazing consisting of wired cast glass in aluminium or lead covered tee shaped steel glazing bars, may be used. Alternatively panels of extruded PVC sheeting or transparent sheets made from polyester resin reinforced with glass fibre may be incorporated in the sheeting. These sheets can be obtained with similar profiles to the main cladding material and may be incorporated directly into the cladded area or alternatively sheets of a different profile may be used with flashings at the top and bottom of the line of sheets.

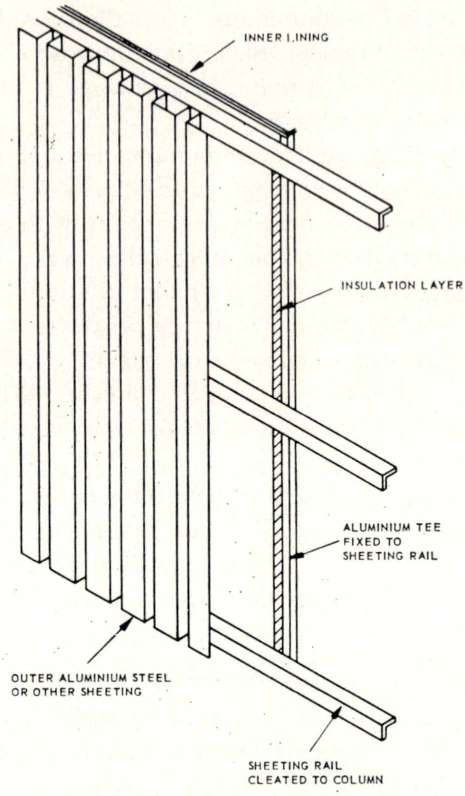

FIG. 3.7.3. Lightweight wall cladding.

3.7.4. **Application to Power Stations**

These methods of construction have been developed to meet the need for a lightweight cladding to structures. Their main advantages over traditional materials such as brickwork, stonework or similar materials is:

(a) The lightness of this form of cladding reduces considerably the dead load on the structure.
(b) Speed of erection to enable plant erection to proceed in sheltered conditions.

3.7.5. **Durability**

Some of these forms of cladding, especially the plastic faced metal sheets, are comparatively new and this together with the development of techniques involving fixing from one side only have given rise to problems which have resulted in some failures. The development of these materials has been very rapid and although accelerated weathering tests as developed for paint testing can be carried out using closed carbon arc lamps to simulate sunlight, exposure to ultraviolet rays to determine colour fading and washing with artificial solutions of sea water and water containing dissolved sulphur dioxide to determine resistance to corrosion, the final proof of a material's durability is found only after use in a particular location. Care must be taken when using different metals in close proximity if they are different in the electrochemical series for corrosion cells will be formed if adequate insulation and sealing of fixings and joints is not used. The use of bitumen felt, PVC tape and plastic washers together with layers of insulating board between purlins and sheets helps reduce the risk of corrosion electrolytic action.

3.8. LIGHTWEIGHT ROOFING SYSTEMS

3.8.1. Structural Elements

These usually consist of profiled sheets (of aluminium or steel protected in a similar manner to wall cladding) which are laid across steel or concrete beams to form the roof covering. The sheets are fixed together and to the roof beams by self tapping screws, cartridge fixings or other approved fixing clips.

Metal decks are designed in the same way as any other structural unit, the panels being designed as beams spanning between purlins. Panels with different elastic moduli can be obtained, and it is only after considering the spans of purlins and decking that the most economic arrangement of roofing can be obtained. Excessive deflection can occur due to loading which will not overstress the metal decking and it is usual to limit deflection to $\frac{1}{240}$ of the span.

3.8.2. Insulation and Weatherproofing Layers

An insulating layer of fibre board about $\frac{1}{2}$ in thick is bedded in bitumen on top of the metal decking and on this is laid a weatherproofing layer consisting of three layers of asbestos or glass fibre based bitumastic felt. The joints and individual felt layers are stuck together using hot bitumen and a final surface of small white pebbles set in bitumen is provided to reflect heat and increase the resistance to spread of flame in case of fire. Insulation board is kept clear of steam pipes passing through the boiler house roof and steel plates with collars and flashings are used to exclude the weather. Asbestos tiles set in hot bitumen are sometimes used on the principal walkways across the roof and to the areas of felt around steam pipes.

A vapour barrier consisting of a sheet of bitumen felt is used between the metal decking and the insulation board to prevent moisture being absorbed into the insulation board.

The purlin levels are arranged to give a minimum gradient of $1\frac{1}{2}$ in in 10 ft.

3.8.3. Application to Power Stations

Provided the felt is laid properly on this type of roof no trouble should be encountered for many years. These roofs, however, are easily damaged and care must be taken to restrict traffic to the areas which are specifically designed for it.

The advantages applicable to wall cladding regarding lightness and speed of erection apply equally to this form of roof construction. Figure 3.8.3 shows a typical form of roof construction. There are limitations in its use, however, for example it should not be used

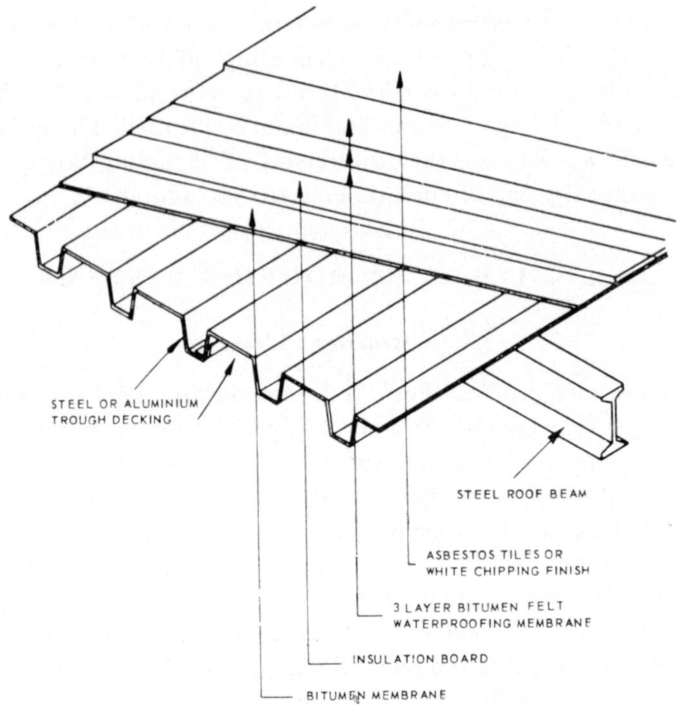

FIG. 3.8.3. Lightweight deck roof.

within 5 ft of boiler escape pipes. One necessary precaution to be undertaken by the structural engineer is to provide with one of the high steel columns a 6 in bore M.S. dry vertical fire main with standard fire brigade fittings at the ground floor and turbine house and boiler house roof level. There should be one main to each unit. The same main can be used for boiler and tank fillings.

3.8.4. **Durability**

As previously stated roofs of this type if constructed properly require little maintenance when only subjected to light traffic conditions. The insulation layer, however, deteriorates rapidly if the waterproofing layer is damaged, and efforts are being made to develop an insulating board which as well as being harder, will also be more moisture resistant.

3.8.5. **Rainwater Disposal**

In view of the large areas of roofs to be drained and the consequent large amount of water to be dealt with, roof drainage and rainwater pipes require careful consideration at the design stage. If rainwater pipes are spaced at large intervals water can be running quite deep in the vicinity of the rainwater outlet during heavy storms unless large box gutters are formed to take rainwater discharged by cross falls, the gutters being sloped to carry the water comparatively large distances to rainwater pipes. On tall power station buildings large quantities of rainwater are deposited on the face of the building.

On two stations water from the roof is allowed to cascade down the glazing forming the elevations of the main buildings and is collected in large gutters at operating floor level. This has the effect of washing the glazing and reduces the cost of rainwater pipes.

Rainwater pipes are located inside or outside the buildings. Ducts are provided in welfare, offices and similar buildings to conceal internal pipes and the large pipes used for turbine and boiler houses are often fixed in the box columns of the structure. This arrangement conceals the pipe and reduces the chance of damage by water due to a stopped pipe bursting at a joint. Cast iron, aluminium, and rigid plastic pipes are used for rainwater dispersal.

3.9. DOORS, WINDOWS AND OTHER FITTINGS

3.9.1. Doors

There are many types of doors used on power station buildings, some of the more usual types being as follows:

(a) *Sliding folding doors and roller shutters:* These doors with hand or electric motor operation are used on the boiler houses, turbine houses, garages and plant rooms to give access for plant and vehicles. Roller shutters have been used, but damage due to high winds has shown on a number of power stations that sliding folding leaf doors of galvanised or sherradised steel give the best service.

(b) *Timber doors:* Flush doors with hollow or solid cores are used extensively on power stations. The latter construction is preferred for doors subject to damage by heavy use. Although normally supplied with a plywood finish this type of door can be provided with a metal, plastic, asbestos board or other surfaces to suit special requirements. The modern resin adhesives used in plywood manufacture and door construction give good water resisting properties and doors of this type can be obtained suitable for external use. Timber doors consisting of a framing filled in with boards or solid or plywood panels are also used. Hardwood doors of this type are often used in entrance halls and similar locations and if made from teak they are very durable and suitable for external doors in switching stations. There are many types of hinges, sliding gear and fasteners used for timber doors and it is necessary to decide upon the duty required from the hinges or opening gear and the degree of security required before selecting the ironmongery required for a particular door.

(c) *Fireproof doors.* Sheet steel is often used for fireproof doors although the insulation value of doors consisting of a core of asbestos, timber treated with approved fire proofing compound or similar material, faced both sides and edges with sheet steel, may be useful in locations where there is a possibility of long exposure to fires. Fireproof doors can be hinged or fixed on sliding tracks and they are usually equipped with weight-operated mechanism or sloping tracks to ensure that the doors will automatically close after the separation of a fusible link by the heat from the fire.

The frames of all such doors must be equally fire resistant.

(d) *Other types of door.* There are many other types of doors including solid glass or aluminium framed entrance doors, rubber doors to allow for the passage of trolleys and

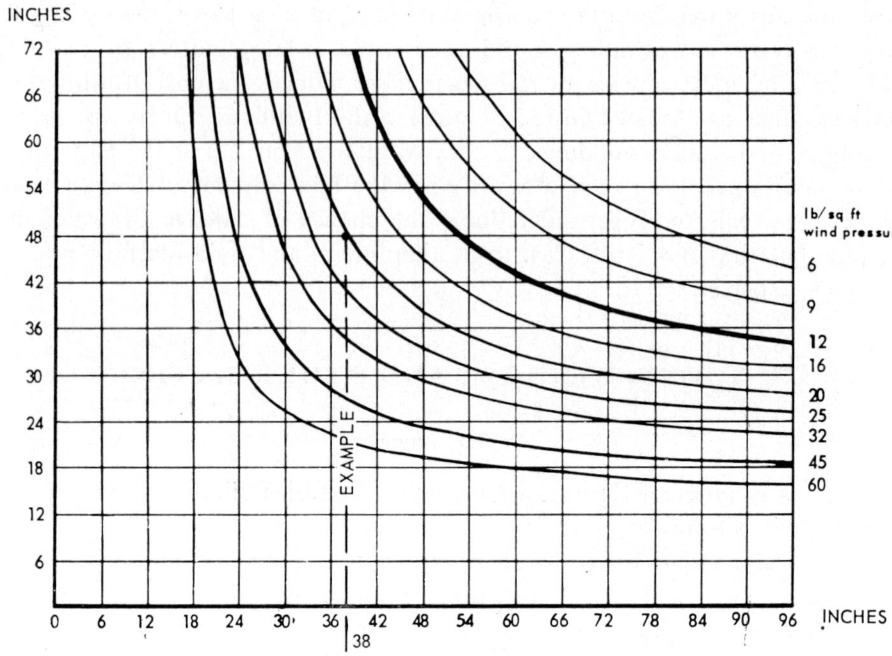

FIG. 3.9.2A. Graded exposure chart for 32 oz sheet glass.
(Reproduced by kind permission of Pilkington Bros. Ltd.)

other doors to suit similar special requirements. It is also possible to introduce automatic opening and closing of doors by use of photo-electric cells, remote switches and similar automatic equipment.

3.9.2. **Windows**

The type of frame used for windows on power station buildings is often determined by architectural requirements. In brick walls and similar locations they are usually of soft-wood, hardwood, aluminium or zinc coated steel construction. The glass used may be sheet, cast plate or float glass. Plate and float glass is obtainable in heavier grades than sheet glass, and toughened or laminated glass is available for special circumstances. Clear or rough cast plate glass reinforced with wire is particularly suitable for use in windows and doors where falling glass due to a breakage is likely to prove dangerous.

Typical charts from which the type of glass suitable for a particular location can be selected are shown in Figs. 3.9.2A and B. Having established the overall dimension of a rectangular glazed aperture it is possible after assuming the type of glass to be used to refer to the appropriate chart and establish what wind pressure can be safely withstood. If the permitted pressure obtained from the chart is too low then the provision of additional glazing bars to reduce the size of the clear openings can be considered or alternatively the use of a heavier glass can be investigated by reference to the appropriate chart. Although only two charts are illustrated they can be obtained for all types of glass.

It follows that the weight of glass is a function of wind pressure and dimensions and this

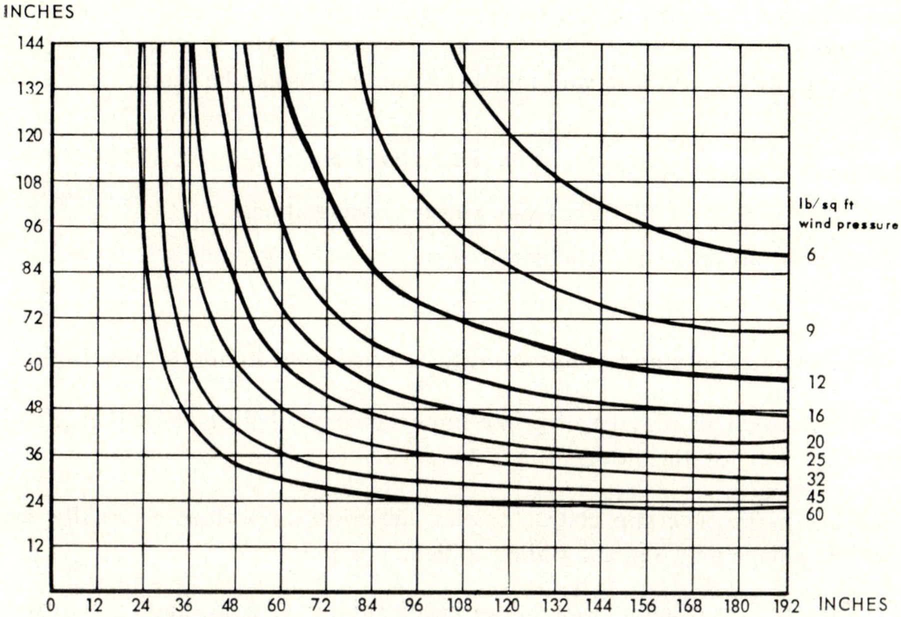

FIG. 3.9.2B. Graded exposure chart for $\frac{1}{4}$ in clear plate glass.
(Reproduced by kind permission of Pilkington Bros. Ltd.)

example is based upon the assumption that a line of glazing 4 ft high is built into the wall of a building.

What is the maximum spacing of the vertical glazing bars if the wall at glazing level is subject to a maximum wind pressure of 20 lb/ft^2 and 32 oz clear sheet glass is used? From Fig. 3.9.2A it will be seen that the maximum spacing of glazing bars is 38 in.

Opening windows are usually of the side hung, top hung, pivoted or hopper type, the latter being most suitable if ventilation in adverse weather conditions is required. To operate inaccessible opening lights, remote operated opening gear based on wire or rod linked system can be used and to prevent excessive long linkages or to operate a number of windows motor operated gear can be used. In some buildings, especially those with air conditioning where opening windows are not necessary or in other suitable locations such as offices, double glazing can help in reducing heat losses, improving sound insulation and by avoiding cold internal glass surfaces they prevent condensation on windows.

3.9.3. Roof Lights

Patent glazing can be used to give natural lighting in sloping roofs, a common method being to introduce a continuous bay of glazing midway up the slope of the roof. For large areas of glazing in flat roofs, lantern lights are often used. These consist of a framework of metal windows covered on the top by patent glazing. For smaller circular, square or rectangular openings in the roof, dome roof lights of plastic or glass in the shape of an inverted saucer are bedded by means of a suitable gasket onto curbs formed round the opening and secured by metal clips. For concrete roofs, lights consisting of glass blocks set in concrete with or without wall panels of glass bricks can be used.

3.10. FINISHES

3.10.1. Floor Finish Considerations

Before a floor finish for a particular area is selected the following basic factors should be considered:

(a) *Cost*. Maintenance and cleaning must be considered in addition to capital cost.

(b) *Appearance*. This may be of prime importance or of little consequence depending upon the location of the floor.

(c) *Surface finish*. The slip characteristics are most important especially where the presence of spillage may aggravate the problem.

(d) *Durability*. It is essential that a floor has the ability to withstand the wear and traffic imposed by a particular location.

(e) *Acoustic properties*. Sound absorbtion properties are essential in certain locations, i.e. control rooms, etc.

(f) *Fire resistance*. In areas of high fire risk it is essential to avoid a floor finish with high flame spread characteristics.

(g) *Special requirements*. Floor finishes used in conjunction with under floor heating or those subject to contamination by oils, acids, alkalis and similar substances require special materials. Expansion and contraction joints should be introduced to allow for movement of the structure and where floors are laid adhered to or separate from the structural slab.

3.10.2. Types of Floor Finish

Having assessed the requirements of a floor finish in a particular location a selection can be made and some of the more common types available are listed below.

(a) *Trowelled concrete*. This is a method of obtaining a reasonable finish to a concrete floor. The concrete is trowelled smooth by hand or mechanical means and the surface can be treated with silicate of soda or a proprietary solution to prevent dusting.

(b) *Granolithic finish*. This finish is applied to concrete floors and consists of a layer of concrete made from fine granite chippings and cement. The granite aggregate is graded in size from chippings down to dust; it gives a superior finish to concrete being harder and less absorbent. Although the finish can be applied to set concrete, it is best laid on fresh concrete to ensure a good bond and to avoid differential shrinkage of the floor and finish. The thickness laid varies from 1 in to 3 in and depends upon the location on the floor. The thickness can be reduced if it is laid monolithic with the concrete. One disadvantage with monolithic laying is the difficulty of protecting the surface from damage during plant erection.

(c) *Terrazzo*. This is an expensive finish consisting of marble chippings set in cement, the surface being ground and polished to a flat surface after the cement has hardened. It can be laid *in situ* or as tiles. This finish is very impermeable and hygienic but is so brittle that cracking occurs very easily. Expansion and contraction joints must be very carefully located.

(d) *Quarry and other vitreous tiles*. These are made from burnt clay and have good acid resisting properties. Tiles can be obtained in a variety of colours with rough or smooth surfaces. Purpose made tiles can be obtained to form coves, skirtings and similar features. They can be set and jointed in cement or alternatively in a proprietary jointing compound which remains slightly plastic after setting and thus prevents failure by differential movement of the finish and slab. The fully vitrified tiles are impermeable and are suitable for areas subject to contamination in reactor buildings.

(e) *Wood*. Hardwood boards or blocks are the timbers usually used for floor finishes in power stations, although softwood boards are sometimes used to provide a base for a further covering. Timbers vary and although good results can be obtained from the best hardwoods such as maple boards or oak blocks, other types of wood floors are not suitable for heavy traffic. Wood blocks laid with their grain vertical make an excellent floor finish for workshops and similar locations. The end grain gives good wearing characteristics and the nature of the surfaces means that there is a smaller chance of damage to castings and similar items placed on or moved across the floor. These blocks are often set in pitch and should not be used where the risk of fire is high.

(f) *Wood fibre blocks*. These are made from reconstructed timber and are very similar to wood blocks.

(g) *Sheet floorings*. These include linoleum, rubber and PVC. They are usually secured to a timber or cement scored base by means of a suitable adhesive. They give a good impervious hygienic finish which is not difficult to keep clean but are not usually suitable for heavily trafficked areas.

(h) *Thermoplastic tiles*. They are similar in many respects to sheet floors except that in tile form they are more easily laid and their resistance to wear is usually greater.

(j) *In situ composition floors*. These include asphalt, magnesium oxychloride, epoxide resins and many other types of finish. The quality and type of finish can be varied to resist acids, alkalis and other conditions.

3.10.3. **Finishes to Walls and Ceilings**

For many locations in Power Station buildings of brick and concrete construction a fair finish and flush pointing to the brickwork together with a good finish to the concrete is adequate.

Plastering, rendering and similar treatments to walls and ceilings are called "wet processes" and although satisfactory results are obtained, the time taken by application of these finishes together with the drying out periods which must be allowed before decoration can commence have led to the development of "dry finishes". These include ceilings constructed from plastic sheet, insulation board, hardboard, metal trays and similar materials which can be supported by hangers from the roof to provide a space above the ceiling for ducts and services. Insulation boards on the inside of sheet cladding and unit partitions in office blocks are typical examples of this type of finish. The panels can be supplied ready decorated if required.

Special finishes such as asbestos or vermiculite spray can be applied in thicknesses usually between one and two inches to concrete, steelwork or other surfaces. Although occasionally used for insulation and other purposes the main function of these finishes is the protection of steelwork against fire.

3.10.4. **Wall Tiling and Other Special Finishes**

In kitchens, toilets, medical centres or similar locations where walls are subject to contamination by grease, dirt or corrosive substances and a high degree of hygiene is necessary, the conditions and frequent washings require the use of a better surface than is obtainable with a paint finish. In these circumstances glazed or unglazed wall tiles, *in situ* terrazzo or terrazzo tiles, plastics or similar materials are used for the finish to wall surfaces.

3.10.5. **Internal Painting**

Apart from the steelwork, internal painting is mainly for decorative reasons and is usually carried out in accordance with schemes prepared by architects. For painting on timber, concrete, brickwork, plaster, metal or similar surfaces, the basic system is usually as follows

(a) *Cleaning down and surface preparation.* Before painting all loose material such as unused nuts and bolts should be removed and all surfaces thoroughly cleaned. A vacuum plant is useful for the removal of dust.

(b) *Priming coat.* This seals and prepares the surface and also provides a key for subsequent coats of paint. It can also be used to protect a material during the constructional stages before the final coatings are applied. White lead primers are normally used for timber and alkali resisting primers for the surfaces of new plaster, concrete, brickwork and similar surfaces. For ferrous metals a red lead or other rust inhibiting primer is suitable, calcium plumbate primers being used if the surface is galvanised.

(c) *Undercoats.* One or two coats of paint is the usual specification for undercoats, and a paint which is compatible with the primer and finishing coat should be used.

(d) *Finishing coats.* These are normally alkyd based resin paints or similar materials which give a hard durable finish. They can be supplied to give a matt, semigloss or high gloss finish. Cellulose spray decorative finishes give a high degree of resistance to new dirt and grease.

(e) *Other paints.* A cheaper finish can be obtained on brickwork, plaster, concrete and similar surfaces with oil bound water or emulsion paints.

3.10.6. External Painting

The amount of external painting on a power station is usually small when compared with internal painting. Although the decorative aspect must receive some consideration especially on the main buildings, office and welfare blocks and similar buildings, there are many locations such as precipitator steelwork, coal gantry steelwork, external plant and other similar items where protection is of prime importance and appearance of no great consequence. Most external painting is carried out using preparations and materials similar to those used for internal work.

3.11. SERVICES

3.11.1. Plumbing Systems

These comprise the cold and hot water supplies to the various sanitary and other fittings, the sanitary fittings and waste and soil disposal systems from fittings to the connections with the station drainage system.

3.11.1.1. Cold Water Supplies

With the exception of drinking water, cold water supplies and the cold feed to the hot water systems are generally taken from header tanks located in the buildings or blocks of buildings they serve. The header tanks are supplied by rising mains from the town's water main and the amount of storage capacity for each tank is influenced by the reliability of supply to the tank and the relative importance of the locations fed by the tank.

Taps, fountains, coolers and other drinking water fittings are normally supplied by a separate system connected direct to the town's water main and are often pump assisted.

3.11.1.2. Hot Water Supplies

If water or steam space heating installations are installed, the hot water supplies can be obtained from steam, or water heated calorifiers. For comparatively small requirements in isolated locations or if steam or hot water supplies are not available it may be more economic

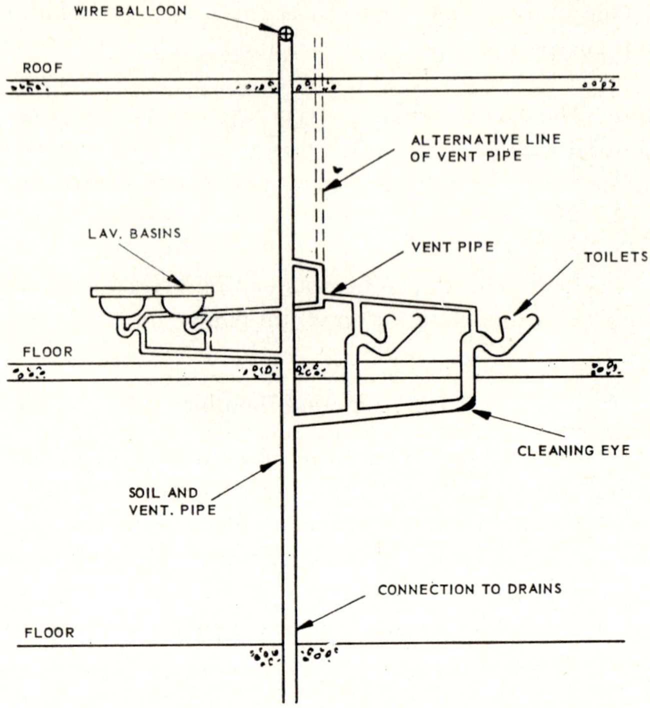

FIG. 3.11.1.4A. Soil system using anti-siphon pipes.

or necessary to use independent electric water storage heaters. Circulation systems are usually installed in conjunction with centrally located calorifiers. The rating and storage capacity of electric heaters and calorifiers must be carefully calculated to meet peak loads in canteens, amenities and similar locations.

3.11.1.3. Sanitary Fittings

Most of the usual fittings such as wash basins, toilets, showers, etc., are used in the amenities provided in power stations and the allocation of fittings is based on the number of personnel in a particular location. Showers are generally of the wall mounted type and are fed by thermostatically controlled supplies. Low level W.C. suites or alternatively concealed flushing cisterns are normally installed. Bowl type urinals are used in preference to the stall type.

3.11.1.4. Waste and Soil System

As stated in Chapter 2, separate or combined systems are provided for the disposal of soil and surface water. The discharge from sinks, lavatory basins and similar fittings on the ground floor of a building is carried through the outer wall by means of waste pipes and discharged into open gullies, which in the case of separate systems form part of the soil drains. If a combined system is used gullies may also receive the discharge from

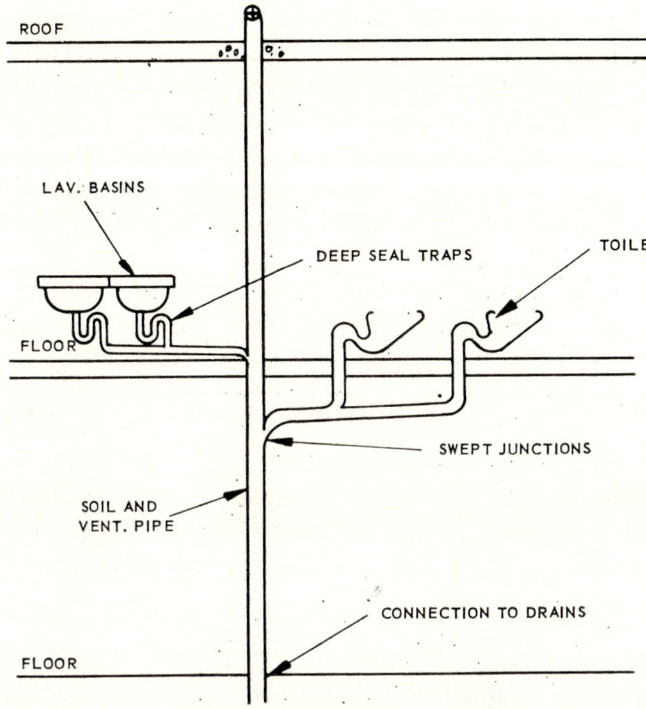

FIG. 3.11.1.4B. One-pipe soil system.

rainwater pipes (and where a combined drainage system is used, sinks, lavatory basins and similar fittings may be discharged into heads fixed at intermediate positions on rainwater pipes). W.C.s on the ground floor are connected direct to the soil or combined drainage system. W.C.s above the ground floor must be connected by means of soil pipes to the drains. Sometimes on a combined system and always on a separate system all the fittings on the first floor and above discharge to the drains by means of a common soil pipe (see Fig. 3.11.1.4A). When a number of trapped appliances are connected by branches to a single pipe, the flow of water from one appliance would tend to siphon the contents of the other traps and leave drains ventilating into rooms. This is prevented by fixing anti-siphon pipes in the top of all traps, these pipes being open to atmosphere. The alternative to the separate soil and anti-siphon system is the one pipe system which is shown in Fig. 3.11.1.4B which incorporates deep seal traps, no anti-siphon pipes and carefully designed bends, connections and branches to ensure that turbulence of the water in the pipes and consequent siphonic action due to air movement inside the pipe is reduced sufficiently to prevent the deep seal traps being siphoned.

3.11.1.5. PIPEWORK

Pipes are usually made from lead, copper, iron, steel, asbestos or plastic.

Copper, because of its neat appearance and strength, is commonly used for water services and is equally suitable for use with hot or cold water. It is also used for waste pipes.

Cast iron and asbestos are used for external water services. They are not normally made in sizes suitable for internal water pipes apart from rising mains. These pipes, however, with spigot and socket joints are used extensively for soil and anti-siphon pipes.

Polythene and PVC pipes with solvent or heat welded joints or joints made with compression fittings are used for cold water mains and services. PVC pipes are more rigid than polythene and are more suitable for waste and soil pipes. Because of their flexibility they are unlikely to be damaged by frost and are less affected than other materials by certain waste discharges.

Pitch fibre pipes made from bitumen impregnated paper are used occasionally for soil pipes, although they are more suitable for use in buried locations as drain pipes.

3.11.2. **Heating Systems**

The system of heating adopted depends upon economics, the arrangement and location of the plant and buildings, the frequency of occupation and the standard of control required. The main heat sources are steam from the main boilers, oil fired independent heating boilers or electricity. The utilisation of the heat is normally by one of the following methods.

(a) *Steam.* Steam as low-grade flash steam or bled steam is used in unit heaters, radiators or similar equipment. The pipes, radiators and other equipment used in a steam heating system are because of the high temperatures most useful for heating stores, workshops and other buildings with high ceilings where the pipes and heaters can be fixed at high level.

Steam can also be used to heat water by means of calorifiers, the water being circulated in low or high pressure systems of pipes and radiators, unit heaters, or underfloor heating coils. The high temperatures associated with high-pressure hot water systems make them dangerous in offices and similar buildings unless adequate grilles and guards are provided to pipes and radiators. Low-pressure systems are most suitable for normal heating systems.

(b) *Oil-fired and electrode boilers.* These are mostly used with high- and low-pressure hot water systems and they have the advantage that the heating system is independent of steam supplies from the main boilers. They are sometimes installed as standby equipment when the primary source of heating is low-grade flash or bled steam.

(c) *Electric heating.* In addition to electrode boilers, used in conjunction with hot water circulation systems, electric convection and radiant heaters are used in isolated buildings or in rooms where heating is required for comparatively short periods.

3.11.3. **Ventilating Systems**

Air movement to remove concentrations of carbon dioxide, moisture vapour, fumes, odours and heated air is necessary if comfort conditions are to be maintained. Although special consideration must be given in special locations, a general rule under normal conditions is to design a ventilating system to provide approximately 1000 ft^3 of fresh air per person per hour. The ventilation systems employed are as follows.

(a) *Natural ventilation.* These systems employ convection currents and wind pressure to circulate the air, opening windows, roof ventilators, inlet grilles, cowl roof ventilators and similar fittings being used to control the flow.

(b) *Forced ventilation.* Paint booths, canteens and other buildings where high concentrations of warm air, dust, moisture vapour, fumes or odours are likely to form usually require some form of forced ventilation system. A system of this type employing fans to extract air is known as an Induced Plenum system, the inlet air being supplied through grilles, windows and other apertures.

A Forced Plenum system has the fans in the inlet ducting and the air in the building is maintained at a pressure slightly in excess of atmospheric. The outward flow of air from the room can be used to create dust-free conditions inside the room especially if filtered air is supp ied through the inlet ducting. If heaters are fixed in the inlet ducts, warm air can be supplied to heat the rooms.

If the ventilating system is used for heating it may be advantageous to recirculate a proportion of the exhausted air to avoid excessive heat losses and a layout of this type is usually designed as a Balanced Plenum system, inlet and outlet air supplies being controlled to obtain balanced pressures.

(c) *Air conditioning*. Full air conditioning is provided by a system of ventilation similar to a Balanced Plenum system except that in addition to facilities for filtering and heating the air, humidity control and refrigeration equipment are included. It is used for special areas, e.g. control rooms, control equipment rooms and other special rooms requiring dust-free atmosphere.

3.11.4. **Design Example—Heat Losses from Building**

The following example is used to show how the total heat losses and consequently the heating requirements of a building are calculated.

The building on which this example is based is a typical workshop or stores building 100 ft long by 50 ft wide by 50 ft high. It has a metal deck roof and lightweight metal cladding to the walls above a brick plinth wall 10 ft high. There are five windows each 10 ft wide by 12 ft high located in the metal wall cladding in each long wall of the building. A large sliding door 12 ft wide by 15 ft high is fixed in one end wall and there are two personnel access doors, each 3 ft wide by 7 ft high.

The design thermal transmission (design U values) established for the various forms of construction and exposure conditions are given in Table 1 which also gives the heat loss in Btu/°F for the building.

Heat loss by ventilation = Volume × Number of air changes per hour × Specific heat of air

$$= 100 \times 50 \times 50 \times 0{\cdot}5 \times 0{\cdot}02 = \underline{2500\ \text{Btu/h/°F}}$$

(Note it is assumed that the ventilation provided would give one air change every 2 h under the coldest conditions.)

$$\text{Therefore total heat loss per hour per °F} = 5179+2500 = \underline{7679 \text{ Btu/h/°F}}$$

If the conditions in the workshop required a temperature of 60°F and the external temperature is assumed at 35°F then total heat requirements

$$= \text{Heat losses per hour} \times \text{Temperature difference}$$
$$= 7679 \times (60-35)$$
$$= \underline{191{,}975 \text{ Btu/h}}$$

and if unit heaters are provided for heating purposes they must be capable of delivering this amount of heat.

TABLE 1. HEAT LOSSES BY CONDUCTION

	Design *U* valves, Btu/ft²/h/°F	Areas of different forms of construction, ft²	Heat loss, Btu/h/°F Σ area×*U*
Main sliding door	0·8	15×12 = 180	180×0·8 = 144
Personnel access door	0·5	2×3×7 = 42	42×0·5 = 21
Windows	1·0	10×10×12 = 1200	1200×1·0 = 1200
Roof	0·25	100×50 = 5000	5000×0·25 = 1250
Plinth wall	0·3	2(100+50)10−12×10−42 = 2838	2838×0·3 = 851
Wall cladding	0·22	2(100+50)15−12×5−1200 = 3240	3240×0·22 = 713
Floor	0·2	100×50 = 5000	5000×0·2 = 1000

Total = 5179 Btu/hr/°F

3.12. DEVELOPMENTS IN DESIGN OF BUILDINGS

The walls in the older power stations were usually of brickwork, built with a solid plinth course and with brick piers to carry the loads from floors, roofs, etc. Roofs were pitched and consisted of wrought iron and, later, steel roof trusses carrying timber or angle iron purlins with slated roofs. Floors were supported on cast iron joists the ends of which were supported on padstones built into the brickwork or supported on circular cast iron columns. Floors were of timber joist and board construction except for those required to carry heavy loads where cast iron plates or continuous brick arches spanning between cast iron beams were used. The introduction of rolled steel members and later reinforced concrete enabled the main buildings to be constructed using a structural frame in which the walls ceased to form structural elements but served mainly to insulate and exclude the weather. Flat roofs of reinforced concrete or precast concrete units covered with asphalt became common and floors were of reinforced concrete, or where heavy loads were to be carried, concrete reinforced by small section rolled steel joists. The latter is termed "filler joist" construction and is still used for floors required to carry very heavy loads.

Compound sections using one or two rolled sections with additional plates riveted to the flanges and riveted plate girders made up from flat plates jointed by angles were introduced to carry the increased loads on columns and beams. The cladding became a plinth of brickwork to withstand mechanical damage with bitumen coated, galvanised corrugated steel or asbestos sheeting to the walls and roofs. Lighting and ventilation were obtained by incorporating steel windows or areas of patent glazing in walls and roofs. The offices and other ancillary buildings were usually combined on the main elevation of the station and the facing of brickwork and artificial stone on these buildings was usually extended to the whole of the main buildings on this elevation in order that a good external appearance was presented.

In the period following the Second World War, a number of power stations had turbine houses built with reinforced concrete frames and roofs. Precast, *in situ* concrete and composite construction was used and the use of concrete shell roofs with concrete columns and beams resulted in many excellent buildings.

During this period brickwork was adopted as the standard walling construction.

Until recently it was common practice to plan a power station in two halves, the first section often being commissioned before the second half was commenced. Problems presented by this form of construction included difficulties due to differential settlement of foundations owing to the settlement in the second half lagging behind that on the first. Relevelling of crane rails was necessary and provision had to be made for movement in all interconnected parts of the building and plant. A temporary gable end had to be provided to the completed first half and this usually took the form of a steel frame covered with cladding.

Several power stations have been built in the past using a semi-outdoor construction for the boilers. A structure was not provided to house the boilers with this arrangement although hutments and covered staircases, gantries and walkways were provided to protect operatives on the boilers. When these power stations were built, brick walls and concrete roofs were the accepted methods of construction for main buildings and some savings were effected. These savings would not be so apparent when outdoor boiler construction is compared with the more modern method of boiler house construction using lightweight cladding. Outdoor construction also has the disadvantage that many repairs have to be carried out in the open and extensive scaffolding may be necessary for work at the higher levels. There may also be objections to this type of construction on aesthetic grounds and it is unlikely to be used in this country in the future.

Although brick and steel chimneys have been used in the past, concrete is used for chimneys on projects now under construction. For chimneys in excess of 300 ft high concrete construction is cheaper than brick.

The first air cooling towers consisted of a timber frame built over concrete cooling ponds, the outside of the frames being covered with wood boarding and timber stacks were provided inside to ensure adequate cooling of the water. Steel was also used but corrosive conditions were so severe that in spite of maintenance painting the average life of a steel tower was 8 years. Concrete was then adopted for cooling tower construction and the first concrete towers were similar in shape to the timber towers. These towers were superseded by the hyperbolic concrete towers, which although smaller were of the same pattern as those now being constructed.

3.13. TURBINE AND BOILER HOUSE CONSTRUCTION

3.13.1. General

The main building superstructure for a 2000 MW power station is a building unique in size and associated loads when compared with other projects carried out by building and civil engineering contractors.

The design of the buildings is the result of co-ordinated efforts by civil engineers, architects, quantity surveyors, landscape architects, steelwork and reinforced concrete designers, services engineers and other specialists. Similarly the execution of the work on the site results from the combined efforts of steelwork, civil engineering and building contractors and also the many other specialist contractors and suppliers of materials.

All work on the superstructure of the main building is usually carried out under three main contracts, the structural steel contract including provision of lifts, a building and civil engineering contract for floors, brickwork, finishings, services, etc., and separate contracts for painting.

A small painting contract is usually let at a fairly early stage in order that all inaccessible surfaces can be painted, the remainder of the painting work being let as a single contract or as separate contracts for the buildings and the plant. Roofing and wall cladding has on occasions been carried out as a separate contract but it is more usual to include this work as a nominated sub-contract on either the steelwork or building contract.

3.13.2. Programme Requirements

The programme requirements for many types of buildings are simple from the clients' viewpoint, starting and finishing dates being the only limitations placed on the contractor. This type of programming is especially applicable where buildings are to be occupied by personnel who move in on completion, and can also be used for structures such as chimneys and cooling towers. The programming of contracts for the boiler and turbine houses is, however, not so simple for although the commencement and completion dates for the various contracts can be established from the overall programme for the project, intermediate completion dates must be established to ensure that foundations and cover are available for plant erection to proceed.

3.13.3. Structural Considerations

The total weight of structural steelwork required for the frame of a power station depends upon the plant layout, and other factors including the provision or omission of a basement and the use of reinforced concrete columns to the turbine house. The weight of steelwork, however, for a 2000 MW station expressed in terms of plant capacity varies from 12 to 14·3 tons/MW.

The basic dimensions of the boiler and turbine houses are decided by the plant layout and work cannot proceed on the design of the structures until this is established. When this stage is reached, development of the plant details and finalisation of the building design can develop together.

As previously stated, because of the long beam spans and the heavy loads on columns and beams, structural steel frames are normally used although composite construction in which reinforced concrete is employed for turbine house columns and also for the frames of associated structures has been used on 2000 MW power station buildings. Figure 3.13.3A shows the arrangement of the plant for No. 2 Unit at Thorpe Marsh power station and the steel frame for this unit in shown in Fig. 3.13.3B. Figure 3.13.3C shows a cross-section through the steel frame for No. 1 Unit at the same station. The longitudinal arrangement of the turbines at Thorpe Marsh power station resulted in the turbine house crane having a span of 80 ft.

In stations completed before 1960 boilers were slung from the main steelwork, the top beams of the large portals of the boiler house providing support for the boilers and roof whilst the columns formed the outer wall of the boiler house. The weight of a boiler in the newer 2000 MW power stations often exceeds 13,000 tons and the span of the boiler house is so large that intermediate columns are introduced to reduce the span of the steelwork carrying the boilers. The boilers are carried by slings from the overhead sling deck supported on columns which transfer the load to the foundation. For No. 1 Unit at Thorpe Marsh, the columns on lines D and E as shown in Fig. 3.13.3C are used to support the boilers and for No. 2 Unit the weight of the boilers is carried on the columns in lines C, H and J as shown in Fig. 3.13.3B.

It is only after consultation with the steelwork contractor that decisions on the type of members to be used can be made. Certain contractors associated with plate manufacturers favour box members whereas others can tender more economically on the basis of rolled sections. Compound members are used throughout in the frame for No. 1 Unit at Thorpe Marsh and although similar construction methods are used for the turbine house of No. 2 Unit, box members have been used in the boiler house construction. Lattice girders have been used to support the roof of the turbine house for both units.

Figure 3.13.3D. shows the cross-section through the structural steelwork for West Burton power station on the centre line of the boilers. This is a 2000 MW station with a longitudinal arrangement of the four 500 MW sets which gives a crane span in the turbine house of 126 ft. The boilers are supported from the columns on lines D, F and G, and all the columns in the boiler house and turbine house, together with the main beams in the boiler house are box members. A longitudinal arrangement of sets usually results in large spaces between boilers which in the case of West Burton has been used to accommodate the bunkers, and Fig. 3.13.3E shows the cross-section through the steelwork on the centre line of the bunkers. The roof beams used for the turbine house at West Burton are of triangular section and lattice construction. The compression members of the roof beams are at the apex of the triangle and the tension members at the bottom, which is level with the roof, the beams thus forming a triangular shaped pent house in which glazing and ventilation is provided.

Figure 3.13.3F shows the cross-section through the 2000 MW power station at Fawley. This station also has a longitudinal arrangement of sets and the span of the turbine house crane in the station is 160 ft. The main roof beams to the turbine house are box girders and the roof deck is level with the underside of the main beams.

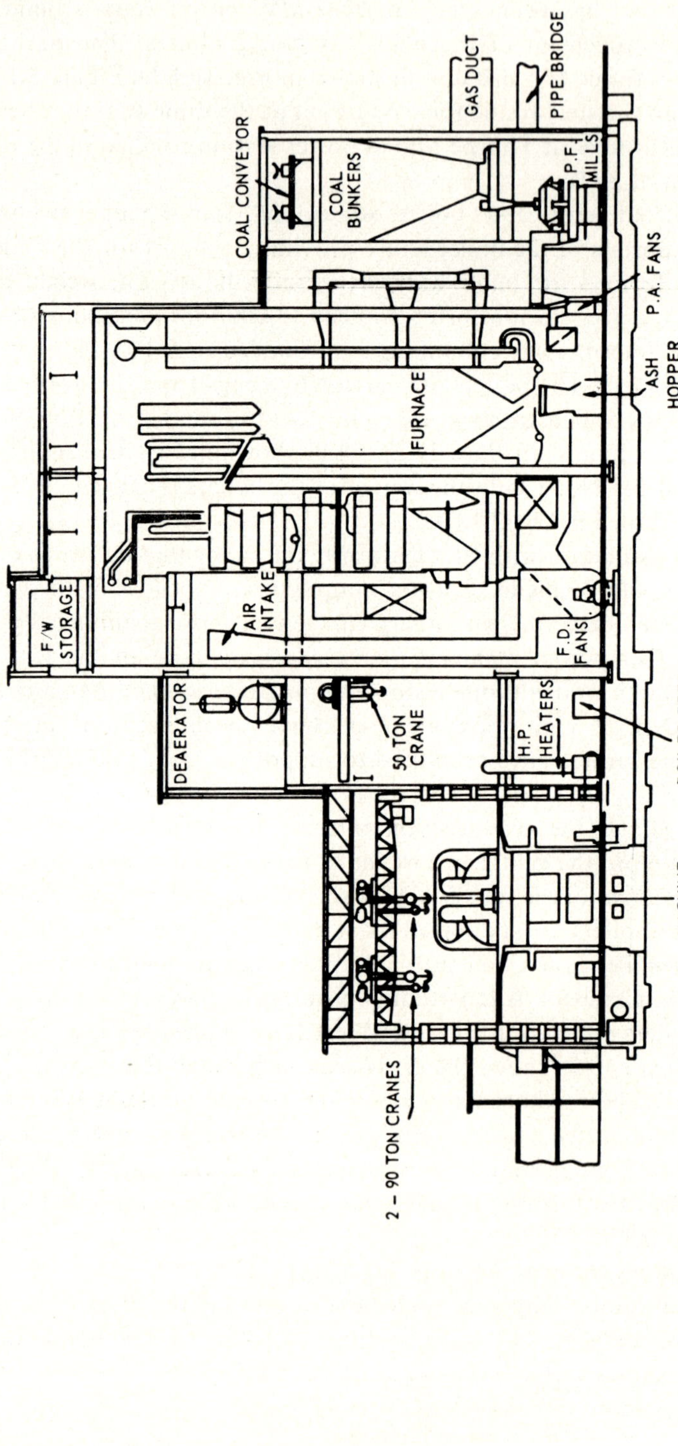

FIG. 3.13,3A. Section through Thorpe Marsh No. 2 Unit including plant.

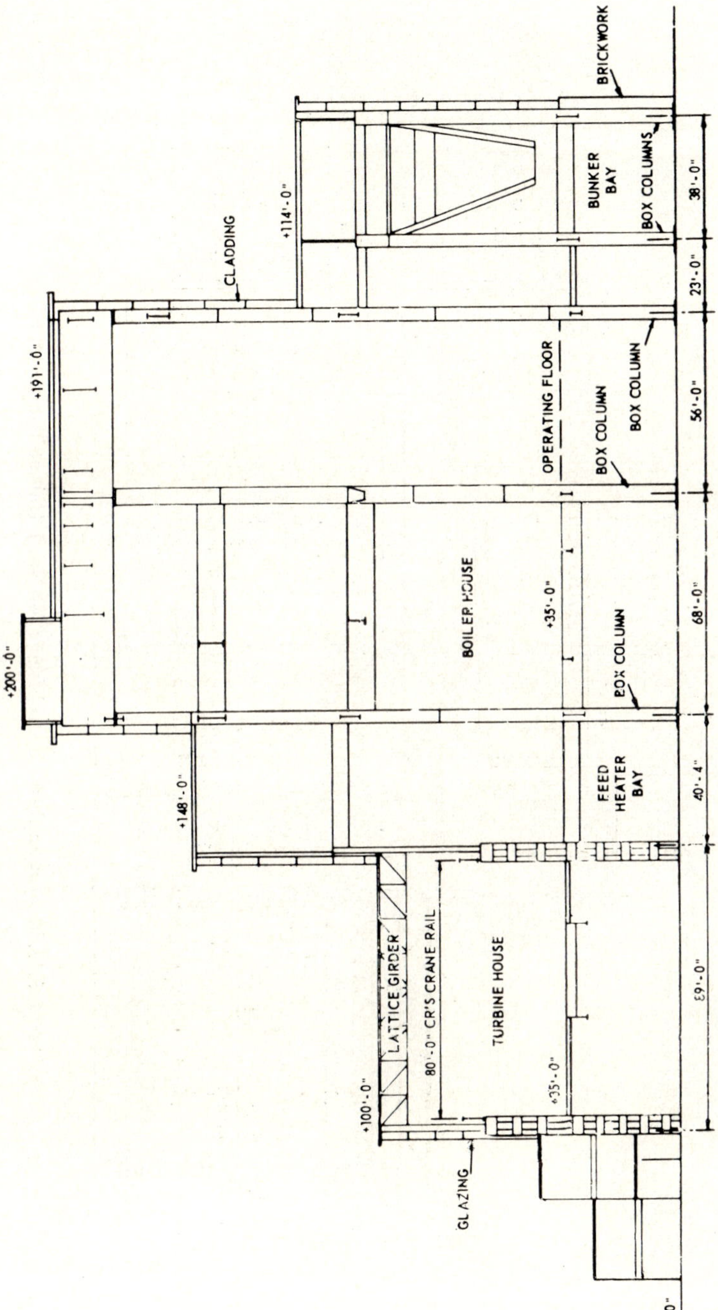

FIG. 3.13.3B. Section through steel frame for No. 2 Unit at Thorpe Marsh.

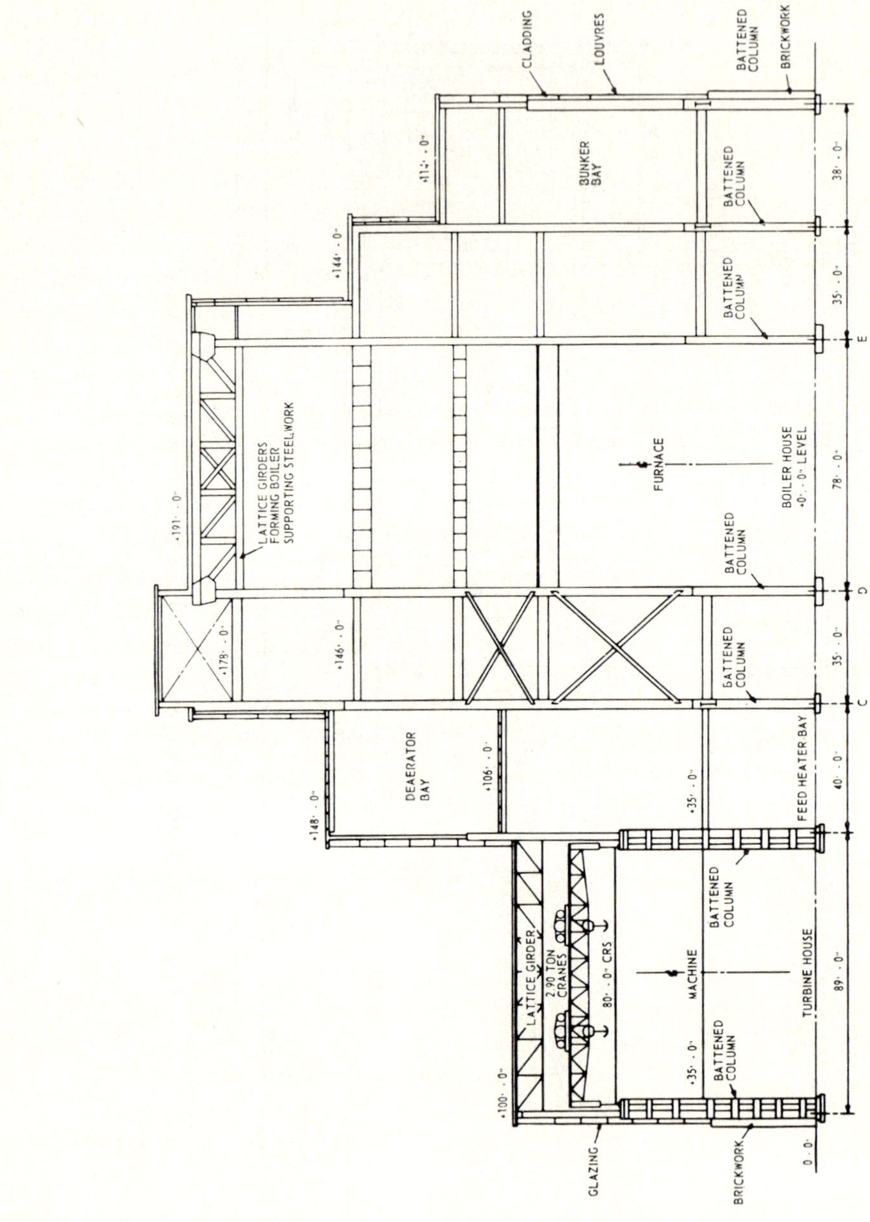

FIG. 3.13.3C. Section through steel frame for No. 1 Unit at Thorpe Marsh.

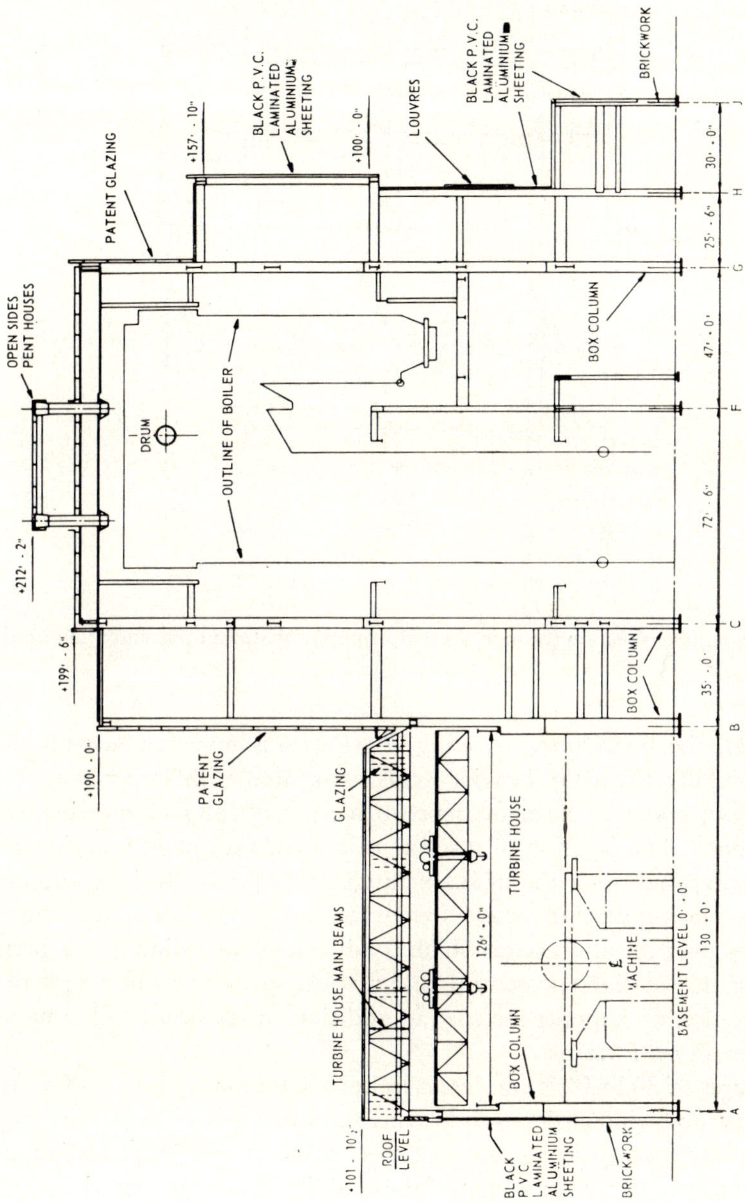

FIG. 3.13.3D. Cross-section through West Burton power station on centre line of boiler.

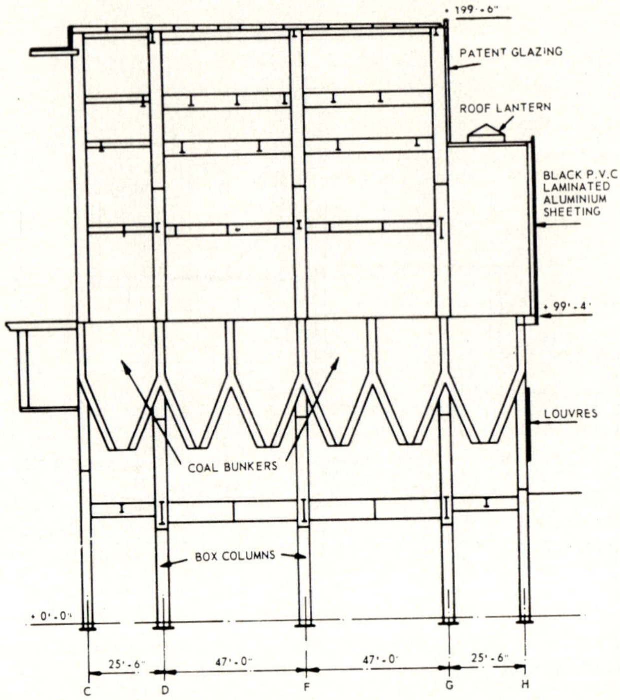

FIG. 3.13.3E. Cross-section through West Burton power station on centre line of bunkers.

The long spans, i.e. up to 200 ft, resulting from the transverse arrangement of sets, present problems, especially when the heavier loads associated with larger sets are considered and the use of two crane tracks involving a further line of columns carrying two crane rails is being developed. The position of this line of columns is not necessarily central and its actual location depends upon the lifting arrangements for the turbine and alternator. At Ferrybridge 'C' power station where this arrangement has been used, the smaller span crane is located over the alternators which is advantageous owing to the greater loads on this crane. On this station the central columns extend to the full height of the turbine house, and are used to support the roof in addition to the crane rails thus reducing the spans of the main roof beams.

Wind pressures of 30 lb/ft^2/h can result in horizontal loads up to 1500 tons on the face of a boiler house and wind girders and bracing is necessary to resist these forces. The main columns are often located at centres in excess of 40 ft and large section sheeting rails would be necessary to carry the weight of the cladding and resist the wind loading on such a span. It is usual, therefore, to introduce intermediate columns to give a sheeting rail span usually between 12 and 15 ft.

The main roof beams in the boiler and turbine house support the roof structures which normally consists of lightweight decking supported on rolled section purlins. Several of the more recent stations have been provided with a roof of precast concrete units to the boiler and these have been found to be not so susceptible to mechanical damage.

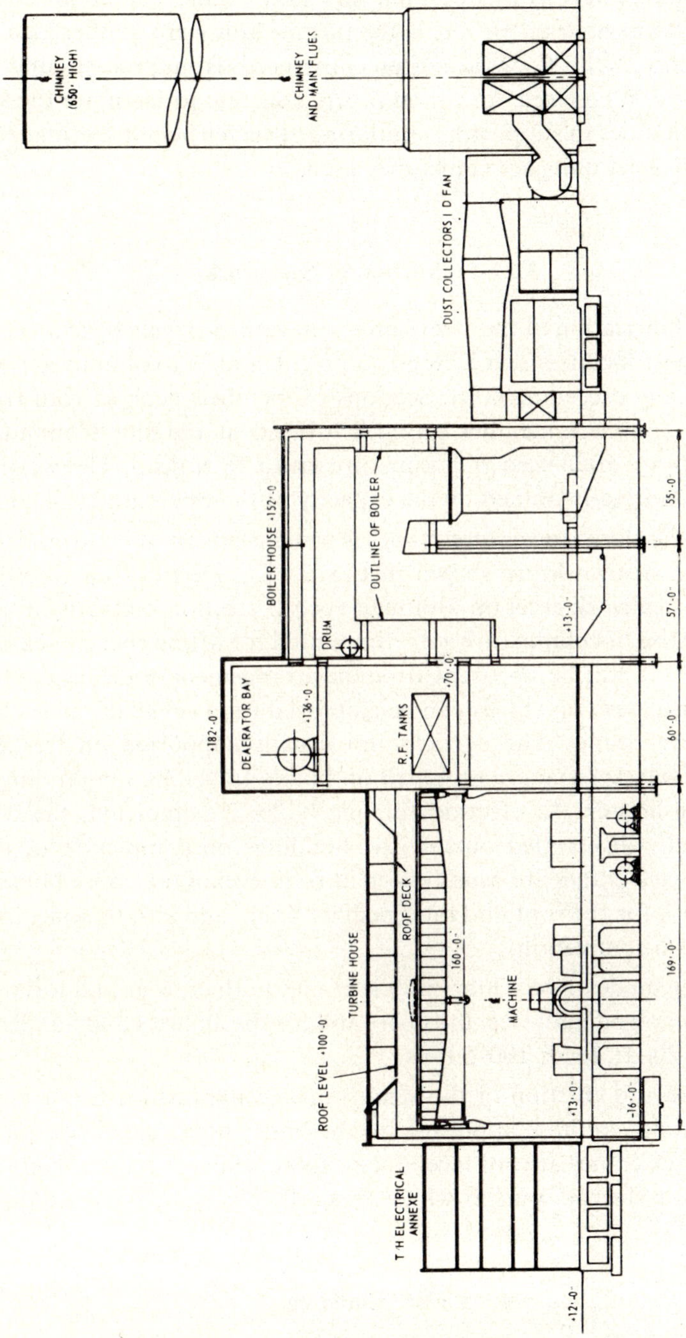

FIG. 3.13.3F. Cross-section through Fawley power station.

Beams and columns are used at intermediate levels to support floors, pipework and items of plant. Open grid flooring is normally used for floors, gantries and staircases above operating floor levels as this permits easy flow of air. The boiler house operating floor and also the floors around the sets in the turbine house are of filler joist or reinforced concrete construction. Solid floors are often constructed from precast concrete units or a combination of precast concrete units and *in situ* concrete to facilitate the speed of erection and avoid difficulties in supporting shuttering at such a height. Permanent shutterings of rigid galvanised sheet units are sometimes used.

3.13.4. **Erection of Steelwork**

The design and fabrication of the steel frame is governed largely by available fabrication techniques, transport facilities and lifting equipment, and box columns are manufactured in lengths not usually exceeding 40 ft. Sections of members such as roof trusses may be assembled into a complete member before it is lifted at the site. Columns of a cross-section 5 ft by 5 ft are often used and beams are up to 18 ft deep. The weight of sections delivered to site is normally limited by the capacity of the works and site cranes.

Although the use of large tower type cranes is being tried out, the normal crane arrangements for erecting steelwork are shown in Fig. 3.13.4. Derricks up to 50 tons capacity are now used for steelwork erection although special erection methods involving the use of catheads on the top of columns are sometimes used for lifting special beams in excess of the crane capacity. It will be apparent from the arrangement of the cranes shown in Fig. 3.13.4 that in many cases two cranes can be employed for a single lift as in the case of the turbine house main beams. The derricks are usually supported on trestles known as gabbards, and the whole structure mounted on bogies travelling longitudinally along the boiler and turbine houses. As erection for one boiler is completed, the derricks move away and are finally dismantled outside the buildings on completion of the steelwork erection. Smaller track cranes are usually used at roof level of the boiler house to complete the lighter steelwork for the roof and intermediate floors and also to assist in dismantling the main derricks on completion.

Use is now being made of very long mobile cranes both crawler and lorry mounted for erection of both main and infill steel members and for the annexes and auxiliary buildings. Jib lengths exceed 200 ft, loads 100–5 tons.

The manufacture and erection of the steel frame to the turbine house is governed by conditions very similar to those applicable to the boiler house steelwork except that it is not usual to use track cranes at roof level due to there being very little if any intermediate steelwork apart from the roof purlins which are easily placed by the main derrick.

3.13.5. **Cladding**

It would be extremely difficult to use an *in situ* concrete roof and brick walls for the cladding of a modern 2000 MW power station and also conform with the present pro-

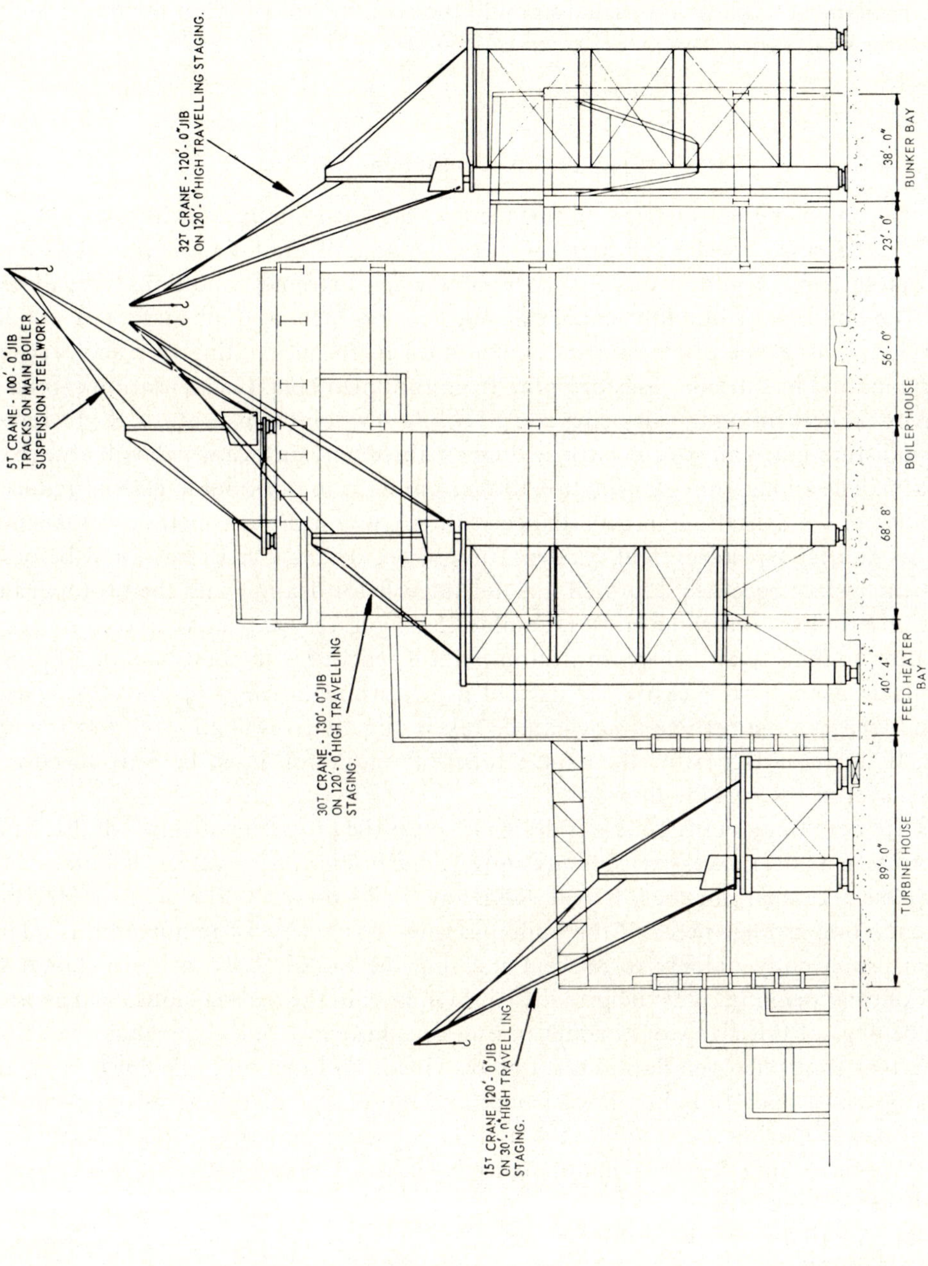

FIG. 3.13.4. Arrangement of erection cranes for power station steelwork.

gramme requirements. This together with other factors including the higher loadings which would result from the use of concrete and brickwork have led to the adoption of metal deck roof construction and lightweight metal wall cladding.

For architectural reasons and also to provide walls at a lower level which are more resistant to mechanical damage it is usual to build the bottom 20 to 40 ft of the boiler and turbine house walls using brickwork, exposed aggregate concrete slabs or other similar material.

3.13.6. **Ventilation**

Ventilation is no great problem in turbine houses. The amount of heat given off by the plant is not excessive when compared with the volume of the building and a reasonably comfortable temperature is maintained except when extreme conditions prevail externally. The build up of humidity can, however, be a problem and air at roof level cooled below dewpoint temperature causes condensation to form resulting in discoloration of decorations and in extreme cases dripping from roof members. Condensation can be prevented by (a) preventing the air being cooled below dewpoint temperature, (b) by removal of the moisture-laden air, or (c) by providing an absorbent surface which will absorb condensation. The application of insulation to metal and the use of double glazing reduce the possibility of condensation in accordance with (a), and although method (c) is suitable for areas subject to intermittent condensation where drying out is permitted before the absorbent surface becomes saturated it is not suitable for dealing with the prolonged condensation problem in turbine houses. Double glazing is very expensive and in most cases ventilation to remove air before saturation point is reached is the best method of preventing condensation. This is easily carried out if the turbine house is open to the deaerator bay and air from the turbine house can be taken directly from high level into the boiler house. If this is not possible then some form of ventilation must be introduced at the highest level in the turbine house.

A recent development in condenser design involves the provision of valves or diaphragms as protection against excessive pressure in the condensers. These valves or bursting discs are capable of discharging the full load steam flow which for a 500 MW set is 16,000 ft^3/sec. The ventilating arrangements of the buildings must be capable of removing this quantity of steam sufficiently quickly to prevent it filling the building. To achieve this at least 400 ft^2 of clear openings must be provided at high level in the turbine house and as permanent openings of this size would result in excessive heat loss during normal operation, it follows that controlled ventilation must be provided. To function in an emergency, these ventilators must open fully in a few seconds and motor operated louvres or opening sections of patent glazing are suitable, the motors being controlled by a push button in the control room. A high degree of maintenance is necessary if ventilators of this type are to be kept in working order.

Normal ventilation of the turbine house is usually provided by opening windows in the walls and roof lights.

The boiler intakes are normally located at the top levels in the boiler house and louvres are used at the lower levels in the walls, as shown for West Burton in Fig. 3.13.3D, to

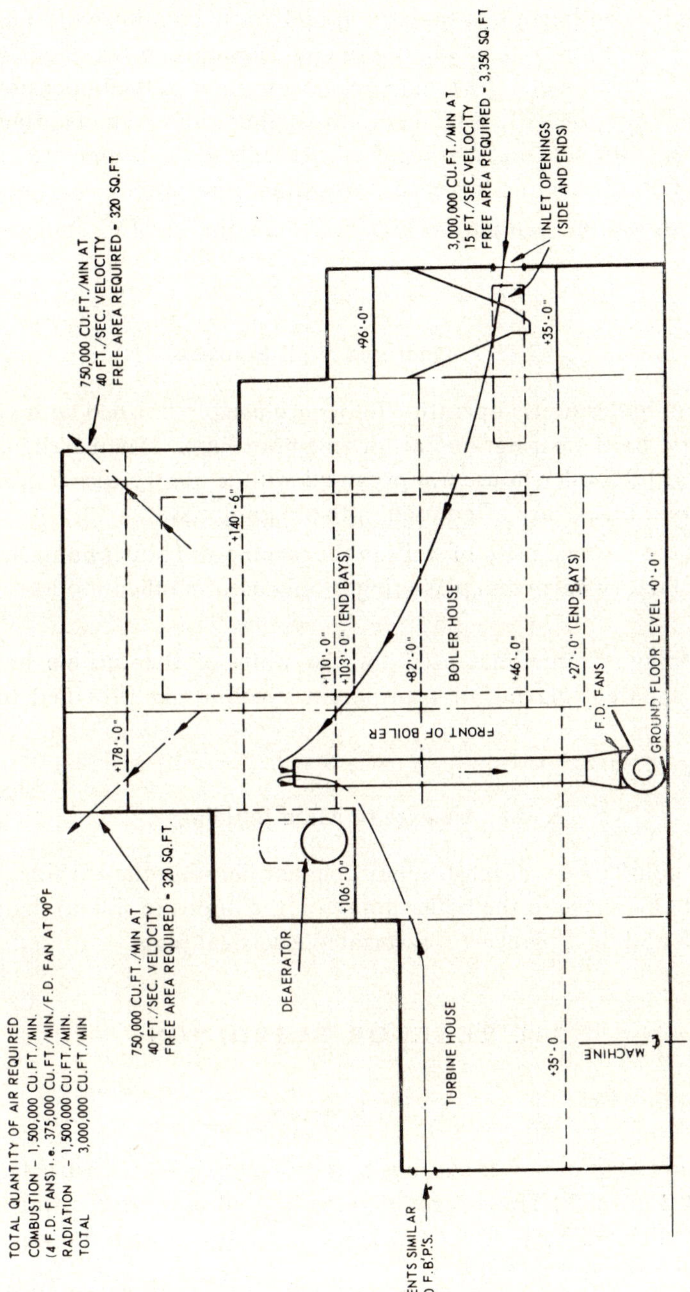

FIG. 3.13.6. Ventilation diagram for Thorpe Marsh No. 2 Unit.

provide the necessary air inlet. The pent houses to allow the escape of boiler radiation losses are also shown on this drawing. Approximately 3,600,000 ft^3 of air per minute is required for combustion in a 2000 MW station on full load. In addition to this air which is required for combustion, a similar amount must be allowed to pass through the boiler house to remove the boiler radiation losses making a total of 7,200,000 ft^3/min, of which half enters the boilers, the remainder pouring out of the boiler house at the high levels. Insulation is not normally provided to the boiler house wall cladding although it is usually used on the roof decking, mainly as an element in the composite construction to support the roof felting which in this case is its primary purpose.

Figure 3.13.6 shows the ventilation and boiler air intakes for Thorpe Marsh power station.

3.13.7. Floor and Wall Finishes

The turbine and boiler house operating floors are usually finished with vitreous tiles or other high quality, hard wearing, non absorbent floor finish. The basement floors to the turbine house and the boiler house are provided with a quarry tile or granolithic finish and end grain wood blocks are often used in the loading bays.

Large areas of the internal faces of walls in the turbine and boiler house are only pointed to provide a satisfactory finish and plastering is only used in offices, toilets and other similar locations.

Insulated cladding is sometimes used for the walls of the turbine house and large areas of windows, patent glazing or transparent sheeting are provided to give natural light.

3.13.8. Annexes to Main Buildings

These usually consist of a deaerator bay located between the turbine house and the boiler house and an annexe to the boiler house on the opposite side to the turbine house, which houses ash and dust plant with associated electrical plant.

3.14. REACTOR BUILDINGS

3.14.1. Reactors

The first reactors used in power stations had pressure vessels constructed from welded steel plate about 2 in thick. These reactors were located in a large cylinder formed with concrete biological shield walls about 6 ft thick and integral with the walls was the concrete forming the base and pile cap.

As the development of reactors proceeded it became apparent that if the concrete shield were increased in thickness and reinforced with prestressed cables, then the steel pressure vessel could be replaced by a comparatively light steel lining about $\frac{1}{2}$ in thick to the inner face of the concrete. This design allowed for the majority of the tension forces in the concrete due to gas pressure being resisted by the prestressing cables.

Models of pressure vessels of this type were made and tested, and from estimates and assessments of performance it became apparent that the advantages of concrete pressure vessels over steel made them economically and operationally attractive. Some of the more important advantages of containing reactors in concrete pressure vessels are listed below.

(a) A prestressed concrete pressure vessel is cheaper than an equivalent steel vessel with concrete biological shield.

(b) The concrete is subjected to its greatest compressive stresses during the prestressing operations, for when the reactor is operational, the gas pressure has the effect of reducing the compressive stresses induced in the concrete during the prestressing operations.

(c) The steel wires which carry the tensile forces induced by the gas pressure are protected from overheating by a good thickness of concrete.

(d) There is relative immunity to failure from local weakness in the tensile steel wire used for the cables.

(e) The demand for skilled labour on a prestressed concrete vessel is less than for a steel vessel.

(f) Using prestressed concrete construction, the size of the pressure vessel and the contained pressure can be increased without the fabrication difficulties associated with thick steel plates. It therefore becomes possible to introduce the boilers into the pressure vessel, eliminating the gas ducts.

(g) The load-carrying cables can be examined during the life of the vessel and if external or ungrauled cables are used, they can be replaced if necessary.

A concrete pressure vessel would not fail catastrophically.

Figure 3.14.1A shows the arrangement of a typical cylindrical prestressed concrete pressure vessel in which the ends and walls are reinforced with prestressed cables. There are no special requirements for the concrete used in concrete vessel construction, normal good quality concrete with a minimum crushing strength of 6000 lb/in^2 at 28 days being suitable. The arrangement of the cable ducts for the prestressing wires in the ends and side of the cylinder is such that the concrete will not be subject to cracking with gas pressures up to 1·65 times the working pressure. The precompression of the concrete due to the stressing of the cables result in compressive stresses of between 250 and 300 lb/in^2 in the concrete, and the loading of the cables by the jacks induces stresses up to 140,000 lb/in^2 in the steel. The main reason for the spiral layout of the cables in the walls of the cylinder is to ensure that with alternating spiral layers of cables in both directions there is no weakness at any of the penetrations in the walls for the entry of steam and feed pipes or control rods, and the arrangement of the stressing gallery for cables in the wall of a pressure vessel of this type is shown in Fig. 3.14.1B.

Figure 3.14.1C shows the arrangement of a spherical prestressed concrete reactor on which a large proportion of the prestressing cables are fixed externally.

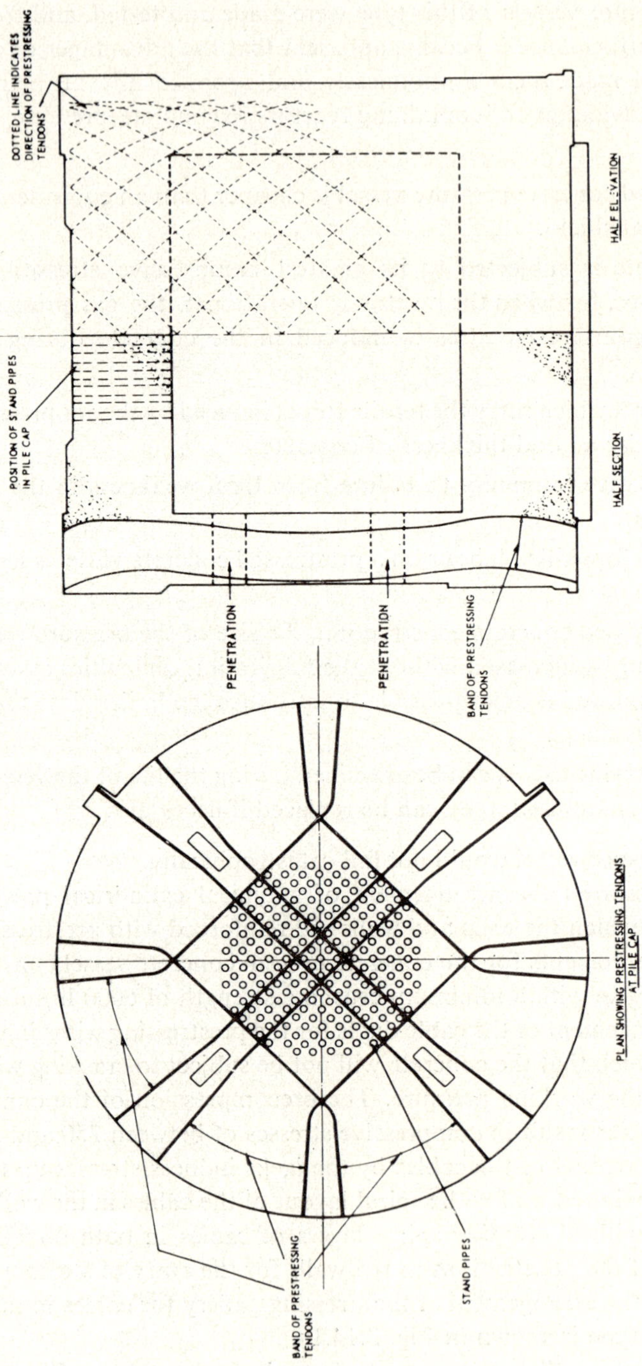

FIG. 3.14.1A. Details of cylindrical concrete pressure vessel.

FIG. 3.14.1B. Arrangement of stressing gallery for cylindrical concrete pressure vessel. (Reproduced by kind permission of the Nuclear Power Group.)

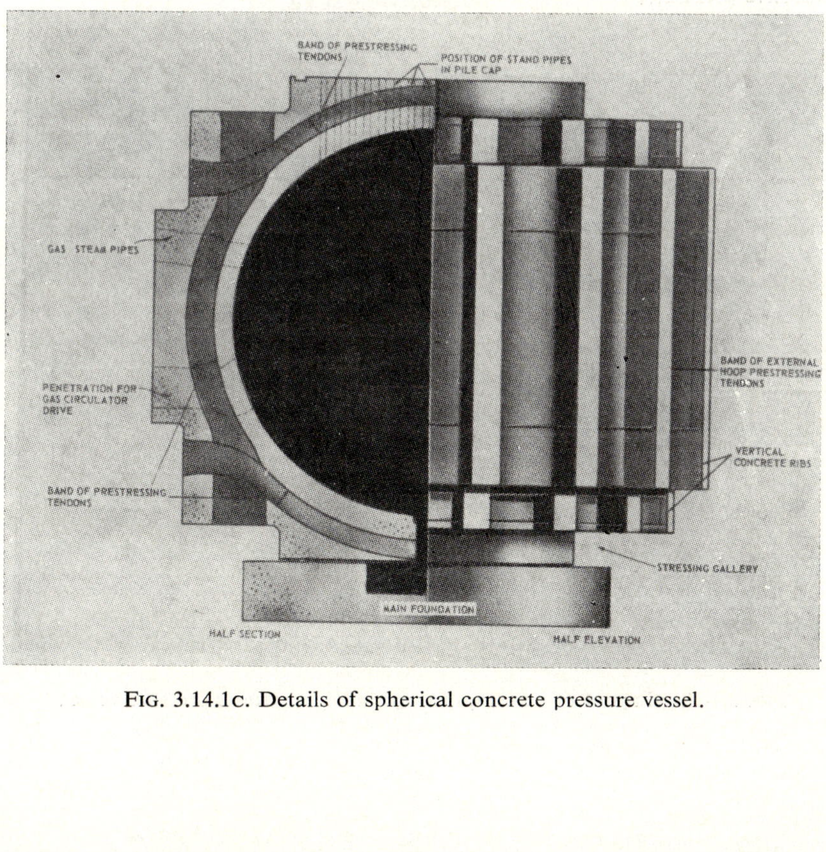

FIG. 3.14.1C. Details of spherical concrete pressure vessel.

3.14.2. **Reactor Buildings**

Layouts of the reactor buildings and turbine houses for stations using prestressed concrete cylindrical and spherical reactors are shown in Figs. 3.14.2A and B respectively. Generally turbine houses are very similar to those used for coal- and oil-fired stations. Although the first reactor buildings were constructed largely from concrete, structural steelwork being only used on the framing for the structures enclosing the boilers and the roof over the charge face, the more recent ones have used structural steelwork on an increasing scale for economic reasons. Figure 3.14.2C shows a perspective cutaway view of the reactor building for Wylfa power station.

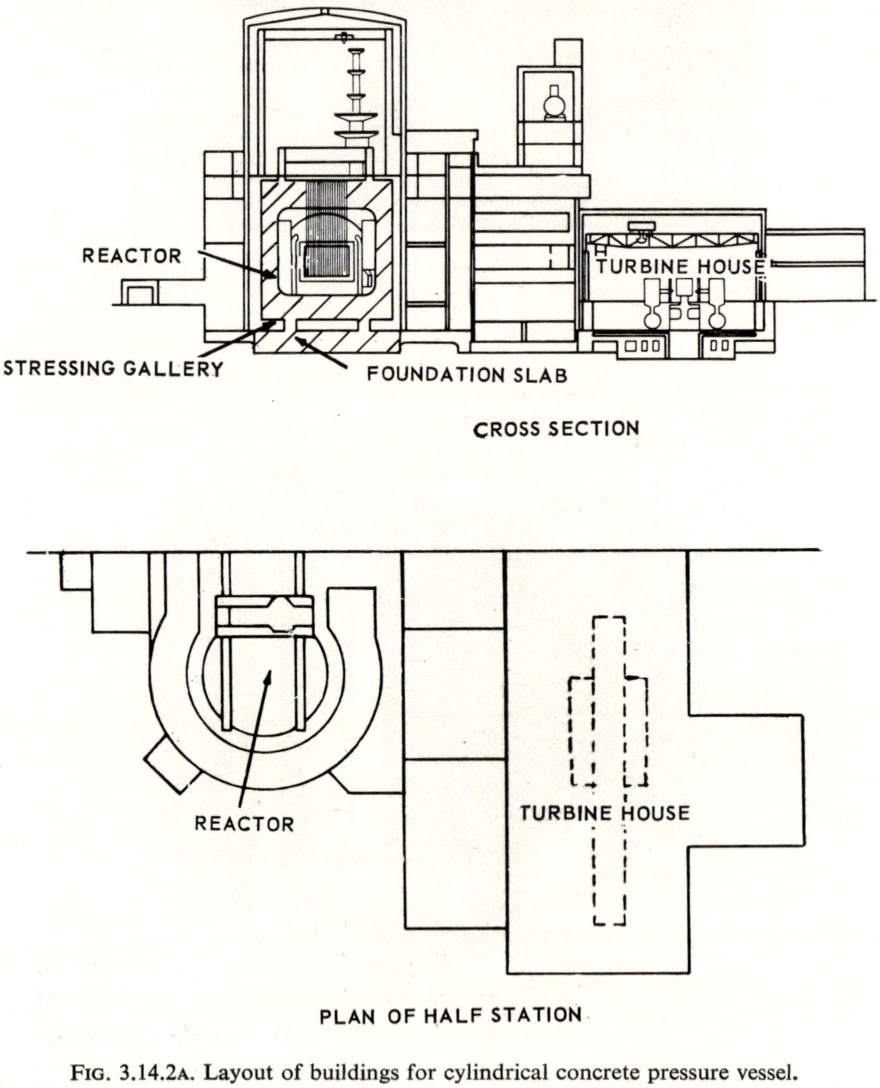

FIG. 3.14.2A. Layout of buildings for cylindrical concrete pressure vessel.

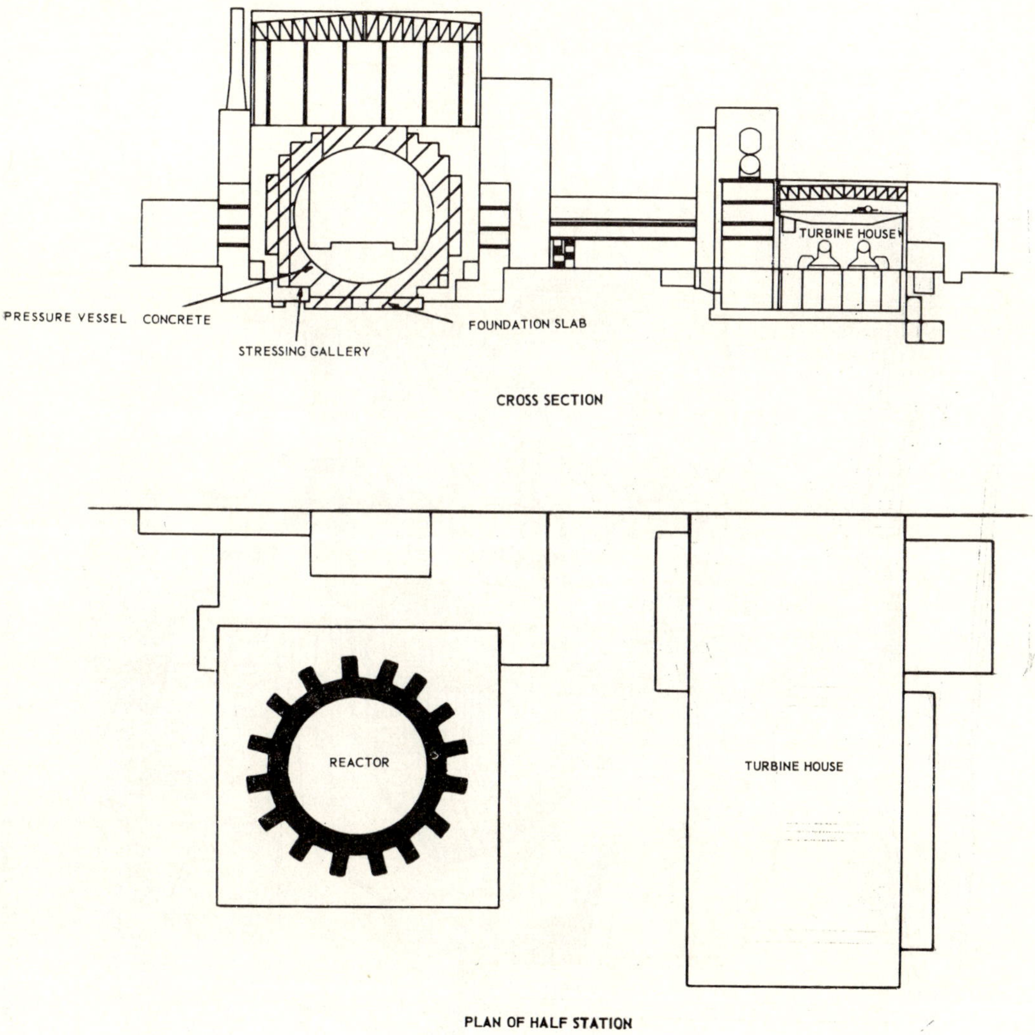

FIG. 3.14.2B. Layout of buildings for spherical concrete pressure vessel.

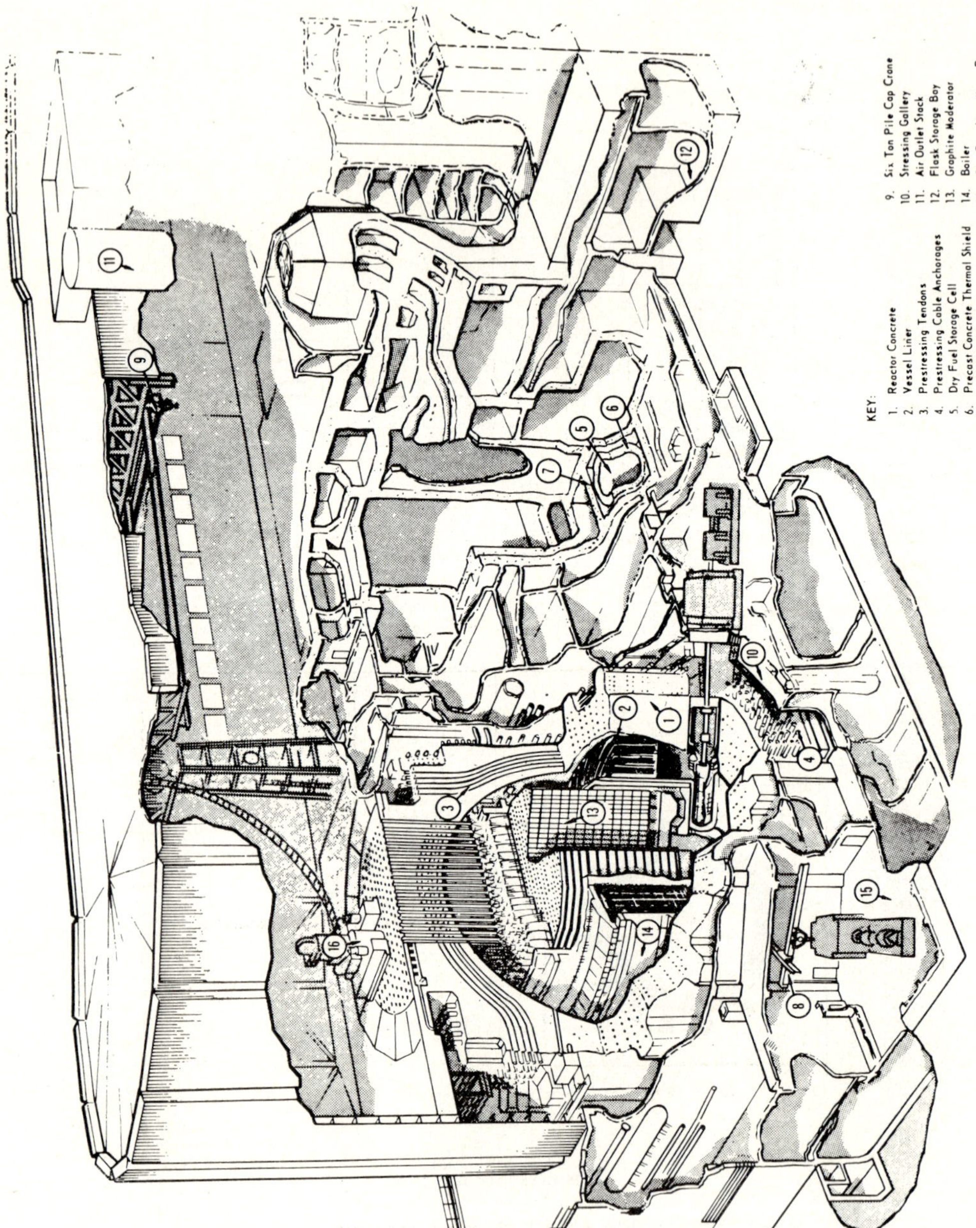

FIG. 3.14.2c. Cutaway view of Wylfa reactor building. (Reproduced by kind permission of English Electric–Babcock & Wilcox–Taylor Woodrow.)

3.15. ANCILLARY BUILDINGS

3.15.1. General

In the past there have been many views held on the desirability of including ancillary buildings as part of the main structure, grouping them in annexes to the main buildings or locating them away from the main building. The present trend is for buildings, especially those occupied by personnel for long periods, to be located away from the main buildings. This minimises noise nuisance and also enables the architect to obtain more natural lighting in the buildings. If, however, larger stations are developed in the future, it may be necessary to introduce toilet and ablution facilities, stores, workshops and other similar accommodation actually in the main buildings.

One big disadvantage with providing accommodation for ancillary functions in the main buildings or annexes is that the steelwork must be designed in conjunction with the main building steelwork and this takes place at a very early stage in the project. Due to their size, the time of construction for ancillary buildings is less than that for the main buildings and if separate buildings are used, commencement can be delayed allowing further time for design development. In many cases this results in more economic layouts as early assessments of requirements for annexe layouts may result in excessive accommodation being provided.

3.15.2. Control Building

The modern control rooms now house the control equipment for the complete operation of the power plant and the transmission switchgear. Some new stations have the control room centrally located in the annexe on the front of the turbine house, others have it located at the end of the turbine house nearest to the first set to be commissioned and others have it in a separate building either centrally located in front of or at the end of the turbine house. Moving the control building away from the turbine house has the effect of reducing noise transmitted from the main buildings and a favoured position is between the turbine house and switching compound or switch house. Another advantage of separate control building is that work can proceed at a very early stage and so give more time for the installation of the complex control equipment.

Although the control building is much smaller than the main building the programme must allow for its completion in time for commissioning the first set and boiler. Control buildings usually have a reinforced concrete or steel frame with brickwork, precast concrete or similar cladding. They are usually several storeys in height, cable basements being provided under the control room and switchgear rooms. Due to the large number of cable openings, trenches, plinths and similar arrangements, floors are usually of *in situ* concrete construction as is also the roof of the building.

The control building has finishings of a fairly high standard in the entrance hall, staircases and the control room itself. Terrazzo or similar floors are frequently used in entrance halls and staircases and rubber or similar noise absorbent floor finishes in the control room. Timber panelling or similar finishes are used on the walls in these rooms and suspended ceilings using acoustic tiles or other removable panels are used to conceal ventilating ducts and other services located at high level. Cable basements usually have granolithic floors, fair faced brick or concrete walls and concrete ceilings.

To keep the air free from dust and to maintain a low humidity a forced or balanced plenum air conditioning system is installed and doors are located in such a manner that air locks are provided to minimise heat loss. The necessity for accurate reading of instruments on the panels means that the lighting of the control room including the arrangement of windows requires special treatment. Diffused ceiling lighting usually gives good results.

3.15.3. **Gas Turbine House**

Special features to be considered during the design of this building are the large filters and ducts for combustion air to the sets and cooling air to and from the alternator and oil coolers, built in structural foundations. Preferably it should be independent of the main buildings and located in an area away from offices and similar buildings because of the noise factor. A steel or reinforced concrete frame is normally used as provision must be made for an overhead crane of about 15 tons capacity. To give good sound insulation $13\frac{1}{2}$ in or 18 in brick or similar construction walls are used together with a precast or *in situ* concrete roof. Natural lighting is best obtained by the use of high-level windows or preferably by roof lighting.

The ventilation of the gas turbine houses is usually by means of louvres and as the turbines operate for comparatively short periods and are not manned, the louvres can be electrically operated with control thermostats, or similar equipment.

The turbine house, filter room and other ancillary plant rooms usually have a granolithic floor finish with fair faced brick walls and concrete ceilings. The control rooms have plastered walls and ceilings together with thermoplastic tile or similar floor finishes. Heating of the control room is normally by means of electric tubular or similar heaters.

3.15.4. **CW Pumphouse**

The CW pumps on some power stations are located in open pits, unprotected from the weather or if outdoor equipment is not used they are provided with a low superstructure.

Other arrangements of pumphouses require superstructures up to 50 ft high, these having structural steel or reinforced concrete frames which provide support for overhead cranes and carry the walls and roofs. Lightweight metal deck roofing and lightweight metal wall panels are the usual forms of cladding.

Circular pumphouses have been constructed for hydraulic reasons and because they have certain structural advantages for the foundation design. They usually have a radial crane for handling pump components.

To prevent mechanical damage to the walls at low level it is usual to have a plinth of brick or precast concrete panels 10 to 12 ft high and this is normally finished with a fair face internally. The cladding is not usually lined and large areas of glazing are provided. Ventilation is by opening windows manually controlled and heating is not normally provided. Floors generally have a granolithic finish.

Should it be necessary to provide accommodation for personnel or control panels, then brick or similar materials are used for the walls to partition off the section required, the ceiling being of wood wool slabs or similar insulating materials and finishes are used to suit the requirements of the occupied area with heating and ventilation being provided accordingly.

3.15.5. **Coal Dust and Ash Plant Buildings**

These buildings, apart from the amenities for personnel operating the plant, usually consist of steel frames cladded in a similar manner to CW pumphouses. The cladding improves the appearance of the structures and acts as a weatherproof covering to protect plant and personnel. In addition to supporting the cladding the steel frame is used to support pipes, ductwork and plant items. It also supports cranes and lifting beams necessary for plant repair and maintenance.

Some of the smaller buildings such as ash pumphouses and the superstructures of transfer houses located mainly below ground level which do not warrant a steel frame are constructed from load bearing brickwork with *in situ* or precast concrete roofs.

These buildings are not normally heated and ventilation is by means of opening windows. Provision for personnel, control panels or similar equipment is also made in a similar manner to that used for CW pumphouses.

3.15.6. **Workshops and Stores**

The main workshops and stores are usually steel framed buildings with insulated lightweight cladding for walls and lightweight metal deck roofs. Overhead travelling cranes are provided in the main workshops and the heavy stores. Natural lighting is normally provided by roof lights and glazing in the walls. It is normal for walls to have a plinth 10 to 12 ft high of brickwork or precast concrete panels to help resist mechanical damage. A typical layout of stores and workshops is shown in Fig. 3.15.6.

Ventilation is usually by manually operated opening windows or roof ventilators in the workshops and stores. Hot water or steam unit heaters at high level are the most satis-

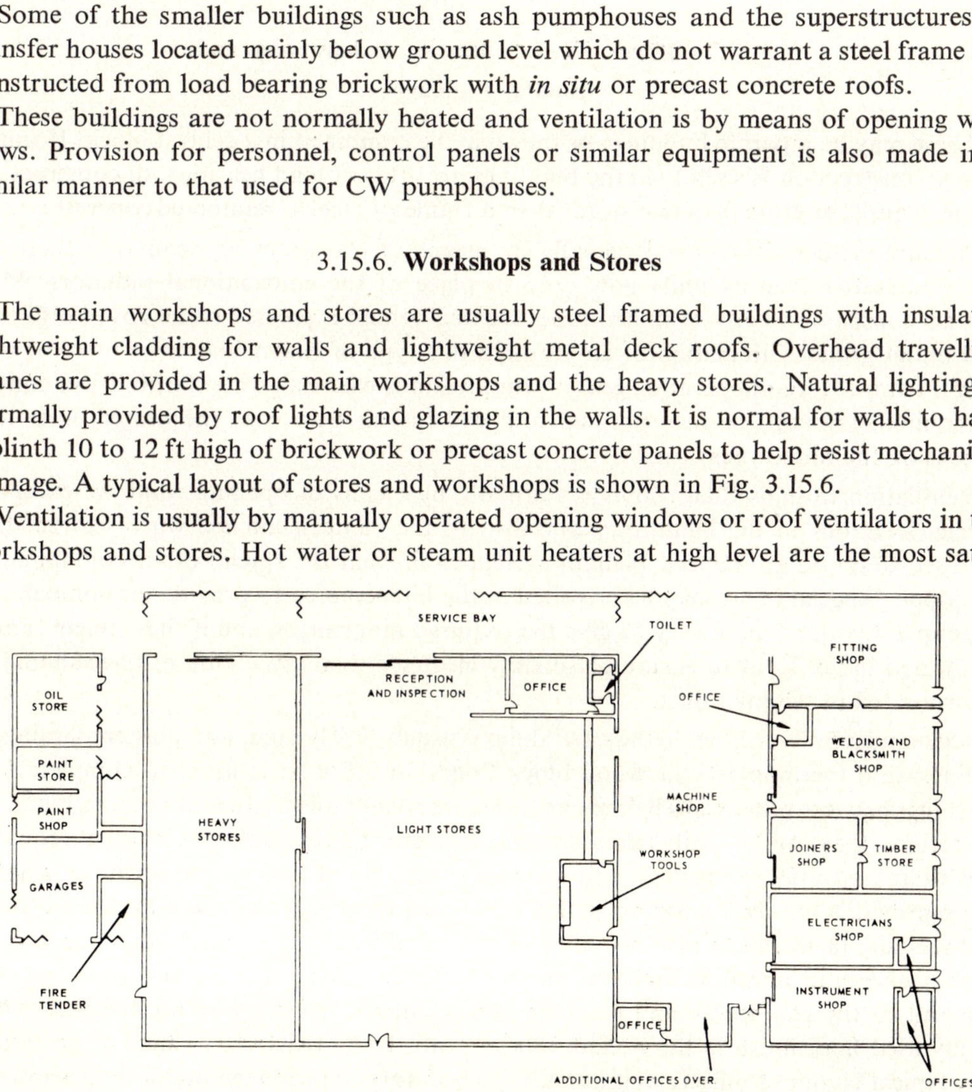

FIG. 3.15.6. Arrangement of stores and workshops.

factory form of heating these buildings. Heating of the stores office is usually done in the same manner as that used for the main office block. Walls in the workshops are usually fair faced brickwork to the plinth and floors have a granolithic finish except for the machine shop and other special areas where end grain wood block is sometimes used.

Special attention must be paid to garages, paint shops, joiners' shops, blacksmiths' shops and other similar locations where the fire risk is high. One satisfactory method of dealing with this problem is to locate them in brick buildings with concrete roofs, separate from, or as annexes to the workshops. Some form of extract ventilation is usually provided in the joiners' and paint shops, especially if woodworking machinery is used and spray painting is carried out.

3.15.7. **Offices, Welfare Blocks, Laboratories and Similar Buildings**

These may be separate buildings or they may be combined in a central block. If single-storey construction is used then the buildings are often of load bearing wall construction. If they consist of more than one storey then a frame of steel or reinforced concrete is used.

Heating of these buildings is usually by means of hot water or steam and there are many attractive heating units now used in place of the conventional radiators. Most heaters consist of a bank of finned tubes located behind a pressed metal cover with inlet and outlet grilles. Circulation of air through this type of heater may be controlled by a simple damper type of operating gear or, for locations where a greater flexibility of control is required, fans can be incorporated to speed the air flow over the heater tubes thus increasing the rate of heat transfer.

Ventilation to individual rooms is normally by means of opening windows except in special locations in the laboratory and also in the canteen kitchen where extract fans are used to create an induced plenum system to prevent the spread of air contaminated by odours. Special inlets may be provided in the kitchen area to provide the comparative large quantity of air necessary to give the required air changes, and if these inlets cannot be located in positions to ensure reasonably clean air then some kind of filter should be incorporated in the inlet duct.

The standard of finishings in these buildings is usually fairly high, with plastered walls and ceilings, and thermoplastic or wood block floors. In offices it is usual to complete floor finishings between outer walls before partitions are erected and as these are normally supplie in sections, they can be easily taken down and re-erected to a different layout if required.

Welfare amenities are provided as necessary throughout the station, separate facilities being provided for coal gangs and other personnel working in isolated locations. Toilet and washing facilities are also provided at central points in the station buildings but the main locker and ablution facilities are normally provided in a central block usually adjacent to the gate house and clocking arrangements. Quarry tiles are the most commonly used floor finish in the welfare blocks, walls being plastered or tiled as necessary.

A typical layout of offices, welfare block, laboratory and canteen buildings is shown in Fig. 3.15.7.

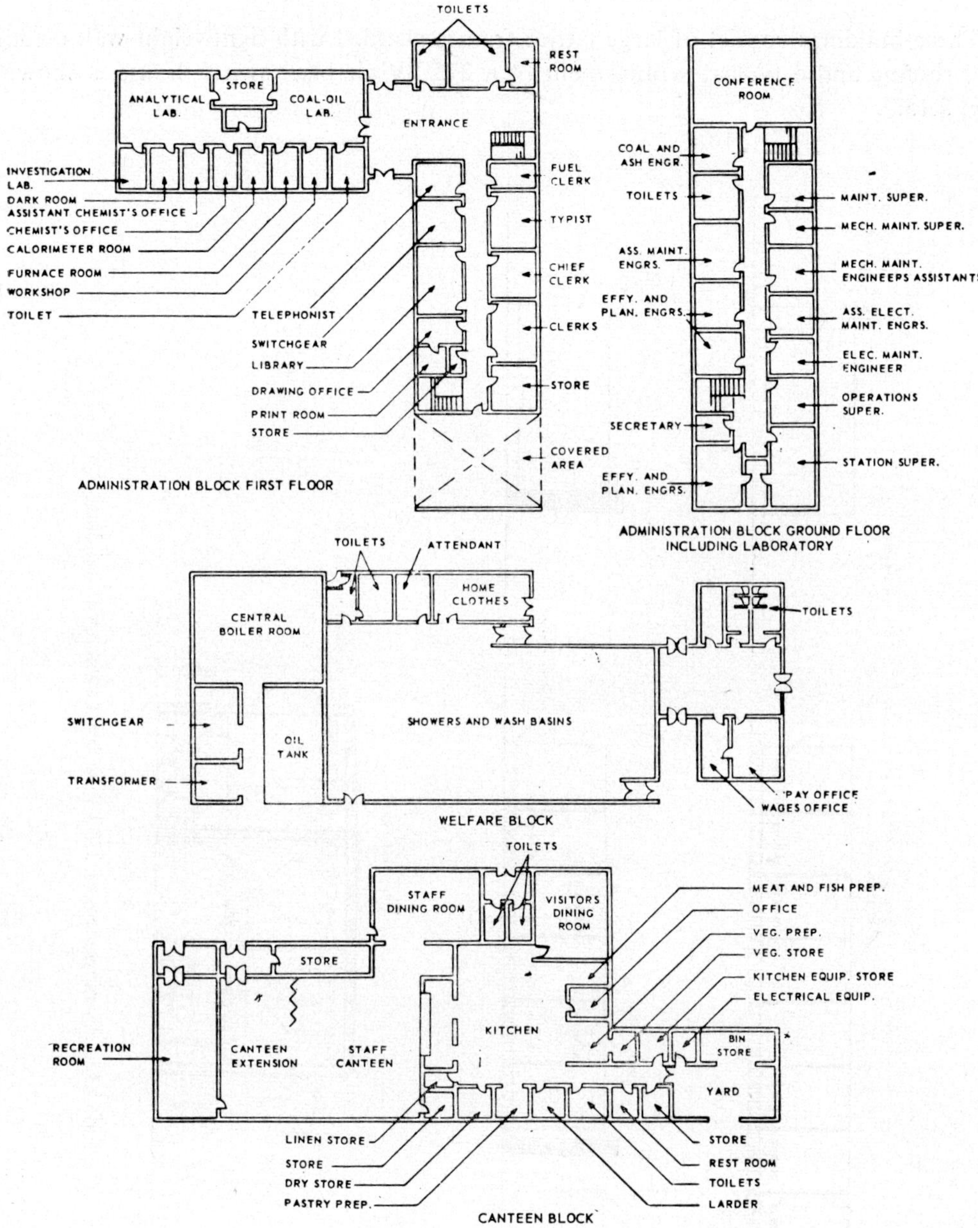

FIG. 3.15.7. Arrangement of offices, welfare block, laboratory and canteen buildings.

3.15.8. **Other Buildings**

There are many other buildings located in different parts of the station which require special treatment. Amongst these are the water treatment plant house where special consideration must be given to the floor and wall finishes to resist acids and alkalis.

3.15.9. **Air Blast Switch Houses**

These buildings consist of large lattice frames cladded with lightweight wall cladding and roofing and a typical arrangement of a 275 kV air blast switch house is shown in Fig. 3.15.9.

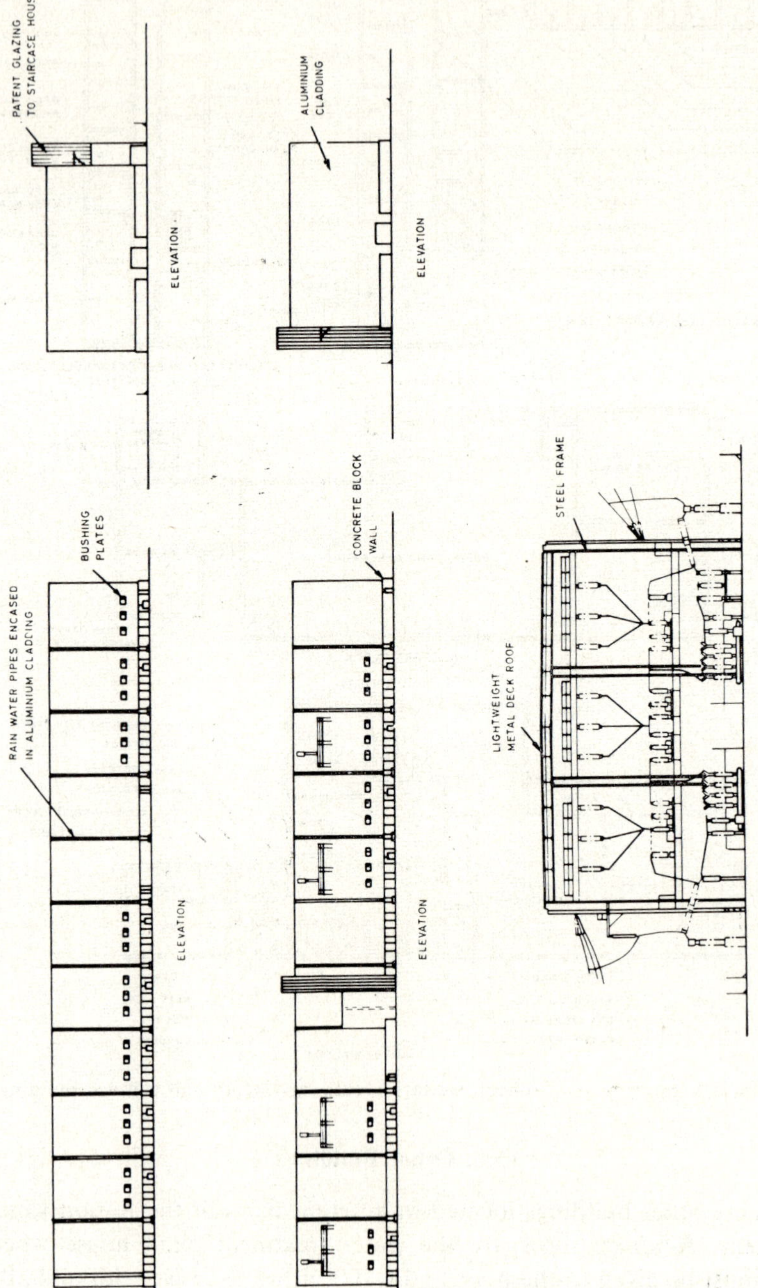

FIG. 3.15.9. Arrangement of 275 kV air blast switchgear building.

A brick plinth wall is normally provided and the floor finish is usually bitumen macadam.

The steelwork is usually designed to support electrical equipment in addition to the wind, snow and dead load of the structure.

Heating and ventilation is not normally provided in these buildings, and a small structure is usually constructed inside the building from load bearing brickwork with a concrete roof to house the office, stores and welfare facilities for the switching station.

3.16. CHIMNEYS, COOLING TOWERS AND PRECIPITATORS

3.16.1. General

As previously stated these civil engineering structures are items of plant, chimneys being extension of gas ducts to discharge combustion products at a high level, cooling towers being incorporated into the CW system to cool circulating water and precipitators forming the casings for the equipment to remove solid particles from the flue gases.

3.16.2. Chimneys

Although single flue chimneys have been used on 2000 MW power stations the present trends are to use multi-flue stacks for all future stations of this type. For power stations with four boilers, four flues, one to each boiler, are contained in a single concrete cylinder which acts as a windshield and Fig. 3.16.2A shows a typical arrangement of this type of chimney. Although this arrangement is suitable for 2000 MW stations, other arrangements of the flues in the windshield including discharging the flue gases from two boilers into one flue may be necessary for larger stations. The outer cylinder of concrete is constructed to withstand the greater part of the wind forces and its own dead load. The flues are built off the lowest intermediate floor or foundation slab and an acid resisting brickwork or other suitable lining is provided. There is no structural connection between the flues and the windshield, the flues being permitted to expand or contract freely according to temperature variation. The space between the windshield and the flues is accessible at all times, including the period when the boilers are operating. Access to the various levels is provided by a concrete or steel staircase and in some chimneys of this type, a lift is also provided.

Other types of multi-flue chimneys have windshields supporting floors at 30 ft intervals, the flues consisting of brick cylinders with external insulating built off each of these floors to form continuous gas ducts.

On the first type of chimney the windshield and flues are of reinforced concrete not normally exceeding 12 in thick apart from the lower section below the floor supporting the concrete flues. Steel or timber framework is usually used and to ensure continuous concreting operation, projections are reduced to a minimum. Brick linings to the flues are normally only half brick thick and are self-supporting for a maximum height of 30 ft. Concrete corbels, therefore, are cast on the inner face of the flue to carry each successive lift of brickwork. The use of corbels at 30 ft intervals has the advantage that in addition

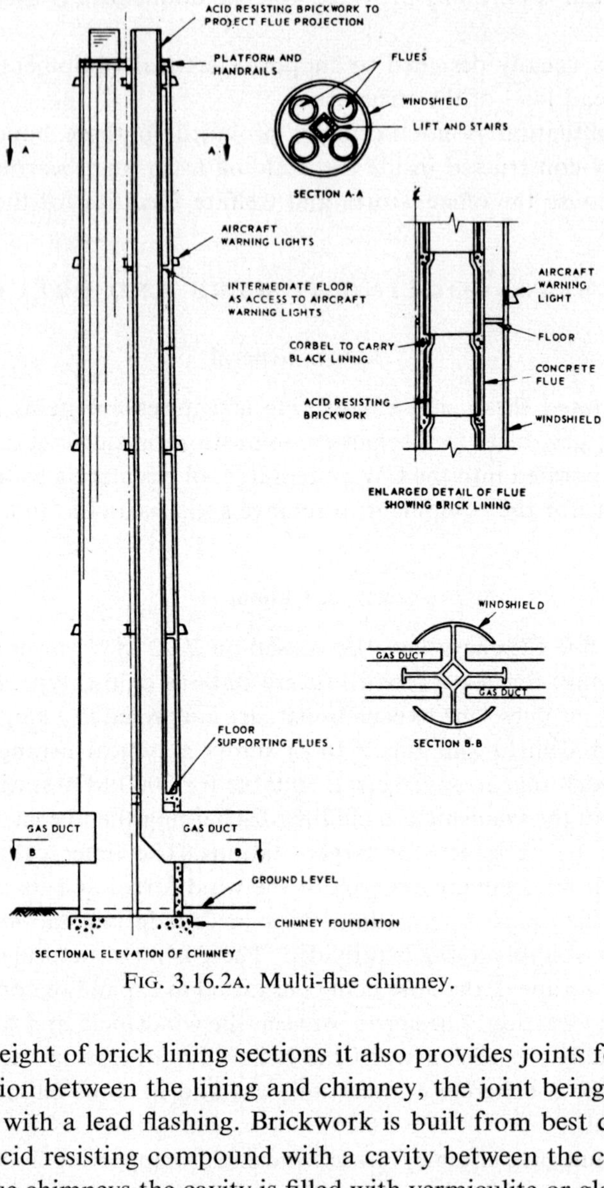

FIG. 3.16.2A. Multi-flue chimney.

to restricting the height of brick lining sections it also provides joints for accommodating differential expansion between the lining and chimney, the joint being packed with glass wool and covered with a lead flashing. Brickwork is built from best quality engineering bricks jointed in acid resisting compound with a cavity between the concrete and brickwork. On single flue chimneys the cavity is filled with vermiculite or glass wool insulation or is ventilated by holes in the concrete, and on multi-flue chimneys the cavities are either sealed or filled with insulation as for single flue chimneys.

Protection against acid attack can be given to the concrete at the top of chimneys by the use of engineering brickwork or protective coatings applied to the outer surface of the concrete windshield and flues. Quarry or similar tiles are used on the flat surface on the top of the windshield and cast iron segments are used on the top of the flues.

If flue gas is to be prevented from penetrating the chimney lining and entering the cavity between the lining and the concrete, a negative pressure must be maintained inside the chimney. The difference in pressure caused by the relative densities of the gases inside the chimney and the atmosphere must exceed the losses caused by the bend at the flue entry, the head loss due to friction inside the chimney and the head necessary to give the gas the required exit velocity of 75 ft/sec.

A negative pressure inside the flue is more important in chimneys where flues are built in brickwork off intermediate floors for a gas leakage in a chimney of this type means that gas would enter the occupied zone inside the windshield.

Chimneys must be designed to withstand the effects of wind load, change in temperatures and dead load. The method of design used is based on the American Concrete Institute Code No. 505.54 and stresses due to dead load, wind load and temperature differentials of reinforcing steel and concrete are considered.

In accordance with Ministry of Aviation requirements, chimneys are provided with aircraft warning lights and the usual arrangement consists of lights located at the top of the chimney and at 150 ft vertical intervals. Three fittings at 120° intervals or four fittings at 90° intervals depending upon the arrangement of flues are provided at each level. Fittings until recently have each had four tungsten bulbs with individual red shades, but a recently developed aircraft warning beacon has a discharge lamp in a fibreglass casing with a vertical transparent plastic front. This latter fitting has an anticipated lamp life several times greater than that for a tungsten fitting and all future chimneys will be equipped with discharge fittings. On multi-flue chimneys, the lights are usually fixed to doors in the outer face of the windshield in such a manner that when the door opens, the light is brought into the chimney where maintenance can be easily carried out. The floors provided inside the windshield are usually located at levels which coincide with the aircraft warning light doors. On chimneys without internal access, i.e. single-flue chimneys, lamp replacement as for maintenance must be carried out by steeplejacks and because of this duplication of fittings is required.

A lightning protection system is essential on such high structures as chimneys. Air terminals are provided on the top of the windshield and flues, and coronary bands are provided at various levels to intercept strikes below the top of the chimney. If a strike occurs, the down conductors are at a high potential and to avoid the risk of a flashover to other metal structures inside the chimney all reinforcement and steelwork are tied in to the down conductors to ensure no difference of potential occurs.

The Code of Practice C.P. 326, *Protection of Structures Against Lightning*, permits the use of steel reinforcement to concrete as down conductors provided the reinforcement cage is adequately earthed and tests on completion show that there is adequate continuity. It is probable that in the near future this system of protection will be adopted for all chimneys. Cast iron cappings, handrails or other air terminations will be bonded to the reinforcement.

Although access sockets in which steeplejacks may screw their ladder fixing hooks are provided on the outer face of single-flue chimneys, sockets are not provided on multi-flue chimneys.

It is also necessary to discharge at a high level the exhaust from gas turbines and as the rapid rise and fall and high working temperature of exhaust gases would have an adverse effect on the main chimney lining, a special chimney is normally provided. Developments in chimneys of this type are still proceeding and independent concrete or steel chimneys with various forms of insulation and protection have been used. Most chimneys have separate flues for each machine and the details of the steel tower and ducting forming a typical gas turbine chimney are shown in Fig. 3.16.2B.

3.16.3. **Cooling Towers**

If a piled foundation is used for a cooling tower, the column loads are transmitted direct to the pile caps and the pond wall serves only to retain water. However, if ground conditions permit, a strip footing is provided round the periphery of the pond and the wall provides a rib thus forming an inverted T beam. The shell is based on the hyperboloid and all the different designs are based on smooth reinforced concrete shells which present the most suitable shapes to the wind and are stable without requiring ribs, beams or buttresses. Figure 3.16.3 shows a typical hyperbolic cooling tower.

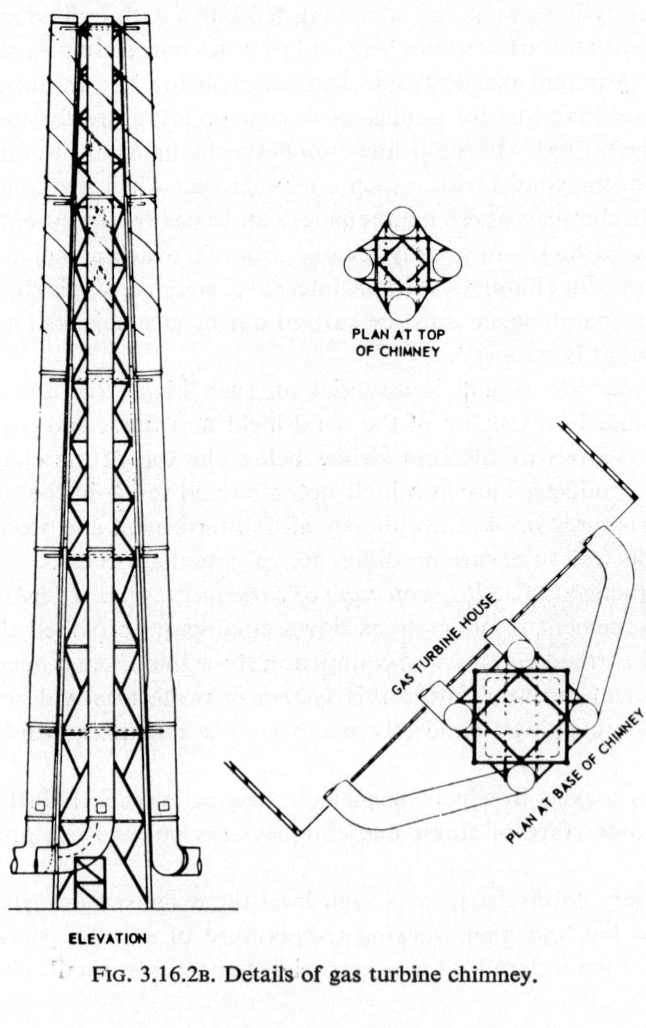

FIG. 3.16.2B. Details of gas turbine chimney.

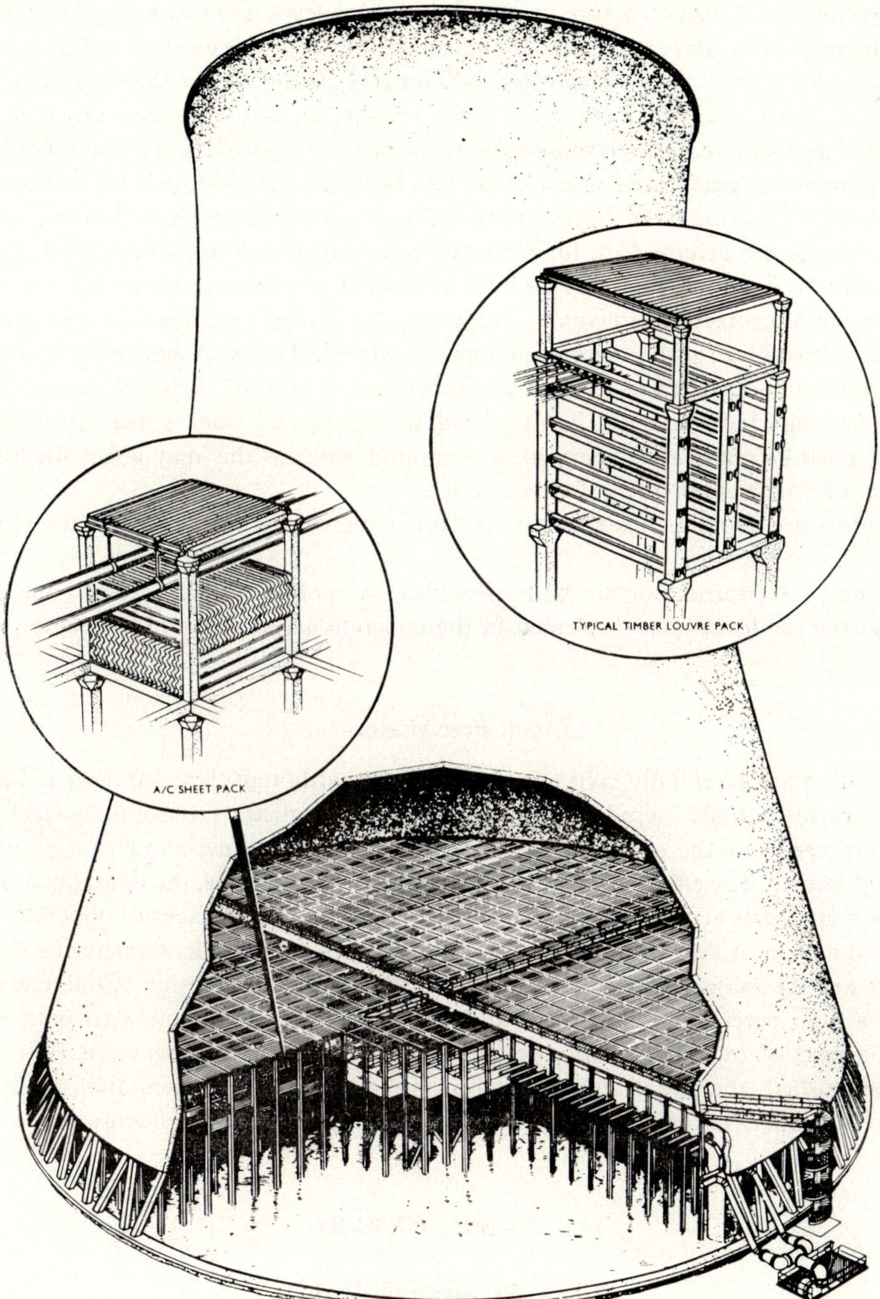

FIG. 3.16.3. Concrete cooling tower.

Within the tower from just above the open column area down to pond level is the "cascading" structure. The water from the condensers is pumped through the CW culverts up to the top of this structure or "stack". At this level it enters a large box culvert running diametrically across the top of the stack and into pipes connected at intervals along both sides of this culvert which feed the sprayers arranged over the top of the stack. The stack is constructed independently of the tower shell so as not to throw any load upon it in the event of differential settlement and is usually constructed from precast concrete columns supporting perforated precast concrete beams. Laced through the perforations of these beams are triangular section timber fillets at varying levels within the cascade structure to ensure a diverse path for the water falling from the spray nozzles. The damp warm conditions in which the timber is located make it very susceptible to wet rot attack. A high degree of protection is given by subjecting the timber to a pressure impregnation process of salt solution using mercury, copper or arsenical salts. Asbestos cement sheets are also used in cooling stacks instead of timber and in view of their anticipated longer life their use may be economic. Before deciding the type of stack some assessment of durability must be made after analysing the cooling water as this may affect the life and economics of a particular form of construction.

Eliminators are usually constructed in a similar manner to the cooling stack but at a higher level.

Lightning protection is not normally provided on cooling towers as the shape of the tower and the reinforcement contained in the concrete are considered adequate protection.

3.16.4. **Precipitators**

Precipitators are essentially civil engineering work, although they are designed by the boiler contractor and are included as part of the boiler contract. Occasionally steel precipitators are used but the common form of housing for the dust extraction plant is a concrete structure. The effects of high temperature on the concrete, the consequent differential thermal movement between hot and cold inner and outer faces of concrete walls, thermal expansion of the structure and the formation of acid condensates inside the precipitators are the major problems associated with precipitator design. Concrete which contains a high percentage of pulverised fuel ash is particularly suited to precipitator casing construction owing to its good resistance to acid and high temperature. Hair cracking of the finished concrete is also reduced owing to the slower hardening of this type of concrete. In some cases exterior insulation is employed to reduce condensation.

3.17. LANDSCAPING

3.17.1. **Preparatory Work**

The landscaping aspects of the sites around power stations must be considered in relation to the architectural aspects of the buildings themselves and also the nature of the surrounding developments and countryside. Landscaping must be considered at a very early stage and the advice of landscape architects taken on the initial layouts. Although most of the tree planting, seeding and other similar works are carried out towards the end of a project, a large amount of the preparatory work such as contouring, forming mounds and storing top soil must be carried out during the early stages of construction work in order that the expense of double handling vast quantities of excavated soil can be avoided. If trees and shrubs are planted as soon as possible after work commences on the site, several years growth can be established before commissioning. It is not possible, however, to do this in the close proximity of buildings due to construction works and contractor's storage areas, because of the likelihood of resultant damage to the planted areas.

3.17.2. **Layout**

Landscaping does not consist merely of forming earth mounds and belts of trees to hide unsightly structures together with lawn and flowerbeds around the main buildings. These features can usefully be utilised in the overall scheme but each site requires individual treatment depending upon location, arrangement of buildings and other special features. The economics of a layout must also be considered for although the planting of large mature trees and extensive use of turf give excellent results very quickly, their high cost when compared with planting saplings and sowing of grass limits their use to the areas of importance. To avoid contamination of top soil during construction, the sites of the buildings and surrounding working and storage areas are stripped of top soil which is then stored in dumps. Any parts of the site which can be shaped to its final profile and is outside the working areas are then soiled and grassed or planted as necessary. Finally on completion of the building and civil engineering works the remaining areas are contoured and finished to suit the requirements of the scheme.

Trees and grassed areas are not the only considerations in a landscaping scheme. The Board have established standards for types of fencing to be used in different locations bearing in mind the degree of security required. Boundary walls, paving and other items also receive similar consideration.

The final appearance of a completed power station must be considered from several aspects. There is the appearance of the station when viewed from local vantage points outside the confines of the station and also the amenities of the people employed in the station whose views when inside the buildings or moving round the site must also be considered. The Fine-Arts Commission are consulted on such matters in the early stage of design.

Finally any scheme adopted must be maintained and as the site for a 2000 MW power station may be several hundred acres of which only a half may be occupied by buildings, coal store, cooling towers and similar items, then obviously the cost of maintenance must be borne in mind when planning the landscape scheme.

3.18. REGULATIONS

3.18.1. Government Instruments

The Government Regulations which apply to power station construction and operation are:

(a) The Factories Act, 1961, which applies to health, safety and welfare of the personnel working in the power station when commissioned.

(i) The Building (Safety, Health and Welfare) Regulations, 1948.
(ii) The Construction (General Provisions) Regulation, 1961.
(iii) The Construction (Lifting Operations) Regulation, 1961.

(b) The Offices, Shops and Railway Premises Act, 1963, which lays down the minimum standards for office accommodation.

3.18.2. Factories Act

Only the aspects of this legislation which relate to the layout of buildings are dealt with here.

The Factories Act, 1961, applies to power stations when commissioned and clauses relating to the following topics must be borne in mind in the design construction stage.

3.18.2.1. Safe Means of Access to Place of Work

This covers items such as ladders, walkways, doorways, lifts, etc., and during the commissioning of a power station the Inspector of Factories will inspect these items. The Act requires adequate access and fencing in all working platforms and footways and the Inspector of Factories will state his requirements regarding the measures to be adopted.

It would be very difficult to advise the Factory Inspector of the developments in design and to obtain his requirements at an early stage in the project. For this reason and also to obviate any tendency to remove the responsibility for implementing the Act from the employer to the inspector, inspections do not normally take place until construction is well advanced.

3.18.2.2. Means of Escape in Case of Fire

Sections of the Act lay down the provisions which must be made for exit doors, their marking, warning systems, etc. The provision must be certified by the local Fire Authority as reasonable.

3.18.2.3. Fixed Lifting Equipment

All lifting tackle, blocks, cranes, etc. are tested by the manufacturers, marked with the safe working load and test certificates are issued to the Generating Board when they are taken over. There are, however, many lifting installations which rely on the attachment to and strength of building frames in addition to the lifting equipment and runway beams or crane rails and these can only be tested after installation. Proof loads in excess of the working load must be traversed over the whole operating area of the crane before the certificate is issued and the lifting equipment marked with the safe working load.

3.18.2.4. Health (General Provisions)

This part of the Act sets out the minimum requirements relating to drinking water supplies, washing facilities, sanitary accommodation, accommodation for clothing, first aid, ventilation and heating of working places, notification of accidents and other similar matters. The Board has issued its own Design Memoranda and Standards which provide accommodation with facilities in excess of the minimum required by the Act.

3.18.3. Buildings (Safety, Health and Welfare) and Construction Regulations

These Regulations set out codes of practice for safety, health and welfare arrangements to be followed in building and construction work.

The Regulations deal with the safety aspects of means of access such as ladders, gangways, scaffolding, etc. Lifting equipment such as runways, cranes, etc., are dealt with by giving the limitations on the use of the different types of equipment and stating how it shall be inspected and maintained. Falling and falling objects are the greatest hazards in construction work. To help guard against these sources of injury the Regulations lay down that all working platforms shall be close boarded, fitted with handrails to a minimum standard and shall be provided with toe boards to prevent materials or equipment being kicked off the platform. All holes in platforms and in locations on the works subject to foot traffic shall be securely fenced except when the progress of the work makes this not possible. Holes of this nature shall remain unfenced for the shortest possible time consistent with the work in progress and whilst the fences are removed all reasonable precautions shall be taken to prevent accidents.

All scaffolding, shoring, lifting, equipment, etc., has to be inspected and tested at prescribed intervals.

Under the section "Health and Welfare" stress is laid on the provision for prompt first aid treatment for injury. Although contractors are required to make all provisions to conform with the Regulations, the Board on large projects endeavour to persuade contractors to combine in financing and providing central first aid facilities which are available to all employees employed by contractors on the site. This avoids duplicating facilities and in certain cases ensures that better facilities and treatment can be obtained for with this arrangement it is usually possible to employ a full-time first aid man.

The Regulations also require that a minimum provision of shelters be made for personnel employed on the site and also that canteen facilities be provided. On most projects the Board provide a temporary canteen which is available for the use of all personnel on the site and in certain isolated locations it has been found necessary to provide a labour camp to accommodate contractors' employees during the currency of the works.

3.18.4. **Offices, Shops and Railway Premises Act**

The parts of this Act which affect the design of power stations are those relating to the standard required for office accommodation.

Premises must be registered and the Act lays down that a minimum standard of cleanliness must be maintained. Although it is sometimes misleading to quote requirements out of context, a minimum area of 40 ft^2 must be allotted for each person in a room or if the ceiling is lower than 10 ft then a volume of 400 ft^3 must be substituted. For accurate information the Act should be consulted as circumstances can affect interpretation. It also lays down that a minimum temperature of 20° C must be maintained and that adequate lighting and ventilation shall be provided. Further Regulations are to be published regarding minimum provision of sanitary conveniences and washing facilities and adequate supplies of drinking water must be available. Arrangements must be made for accommodating and where necessary drying clothes. The Act also covers seating requirements, eating facilities and also the construction of floors, passages and stairs, especially finishes and handrails.

The part of the Act relating to fire precautions covers provision of escape routes, fire fighting appliances and fire alarms. It also covers escape drills for employees and the procedure for issuing fire certificates.

The Act also covers procedures for notification of accidents and the division of responsibilities if the occupier is not the owner of the office building.

EXERCISES

1. List the various factors which must be considered when deciding whether an *in situ* reinforced concrete or structural steel frame is to be used for a building. When considering these factors list the advantages and disadvantages of each type of frame.

2. Before selecting the material to be used for cladding a building, various requirements such as heat insulation must be considered. List these requirements and state briefly how the use of the buildings affects their importance when deciding the type of cladding to be used.

Explain how consideration of these requirements would affect the selection of the wall cladding for a boiler house and also a canteen building.

3. Concrete may be placed *in situ* or precast on or off the site. Give a brief account of the advantages and disadvantages and typical applications suitable to each method.

4. A concrete column is 18 ft high and is properly fixed in position and direction at ends. Given the following information:

(i) Size of column	= 12 in×12 in
(ii) Main reinforcement	= 4 No. 1 in diameter bars
(iii) Permissible compressive stress in concrete	= 760 lb/in^2
(iv) Permissible compressive stress in steel	= 18,000 lb/in^2

calculate the maximum axial load which may be placed on the top of the column and sketch a typical arrangement of reinforcement.

5. Describe the various means which are available for connecting structural steel members, listing the advantages and disadvantages of each method.

6. Various factors affect the conditions which are responsible for the comfort of human beings inside a building. State what these factors are and how these conditions are affected by various parts of the structure.

7. On a multi-flue concrete chimney state:

(a) How the brick lining to flues is normally constructed.
(b) The reason that there is no continuity between the concrete and reinforcement forming the wind-shield and floors and the concrete and reinforcement in the flues.
(c) What precautions are taken at the top of the flues to prevent damage to the concrete by flue gases.

8. What are the principal considerations when selecting a floor finish for a particular location? How do these considerations affect the choice of floor finish for the following locations?

(a) Turbine house operating floor.
(b) Heavy workshop.
(c) Canteen kitchen.

CHAPTER 4

MAIN PLANT LAYOUT

4.1. THE OVERALL CONCEPTION

Plant layout is concerned with the logical and economic use of space and the relationship of one piece of plant with another. Subtle factors of prejudice and taste play their part and sometimes have a powerful influence particularly in the early stages when the relationship between the boiler and the turbine is fixed.

The generation of electricity depends on a flow process which can be shown very simply as in Fig. 4.1. This shows the three basic streams.

(a) Energy flow.

(b) Regenerative flow.

(c) Waste flow.

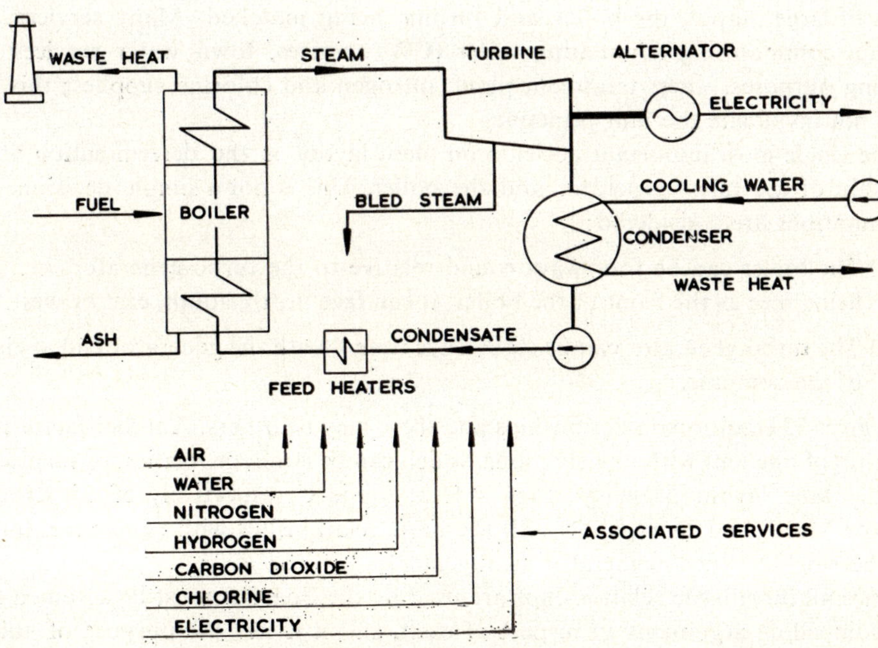

FIG. 4.1. Power station flow principle.

An efficient layout of plant minimises losses and hence running costs in each stream but, vitally, in the energy flow stream. Ideally plant items should be close together and connected to each other in short straight paths. This approach to design could be termed the "Datum Philosophy" and is used to assess the practical achievement with its necessary deviations and compromises, to evaluate new ideas and to reject what is bad or unnecessary.

There is unlimited scope for fresh ideas in plant layout, some current overall layouts for big machines are discussed in the next section.

4.1.1. **Which Way Round?**

The three previous chapters have discussed the siting of the main building within the land available, its orientation with respect to fuel delivery and storage and cooling water supplies. Considering both fuel and cooling water, the datum philosophy must be applied with minimum compromise and the policy continued within the station building to determine how the products of a large number of manufacturers and contractors can be related and welded into a complete unit.

The word unit in the power station sense needs to be defined. It is the association of a single turbo-generator and boiler, together with auxiliary services, to form a complete operating unit of virtually self-contained design.

Many engineers will know of or be working at what is known as a "Range" station where the boilers steam into a common steam receiver and feed water is returned in a common range. This gives a very flexible system with two or more boilers per turbo-generator set. In modern terms this kind of provision to maintain availability is uneconomic and inefficient and since 1950 it has been the Board's policy to build completely integrated units of large output, the boiler and turbine being matched. Many services, however, remain common, e.g. circulating water (CW) services, town water services for other cooling purposes, water treatment plant, nitrogen and chlorine supplies; this, however, does not invalidate the unit principle.

The single most important decision on plant layout is the determination of the relationship of the turbo-generator and the boiler. This is not a simple decision, when the permutations are considered:

(a) The boiler can be four ways round relative to the turbo-generator, i.e. taking the firing face as the front of the boiler, it can face north, south, east or west.

(b) The turbo-generator can be eight ways round, with the generator end at eight points of the compass.

This gives 32 traditional relationships and there may be others. Another factor is the relationship of one unit with its neighbour, which can be a mirror relation or identical.

The three layout drawings, Figs. 4.1.1A, B and C respectively of Littlebrook 'D',* Thorpe Marsh and Drax, show well the basic relationship with transverse, longitudinal and herringbone turbo-generator arrangements.

Since all three basic relationships are used by the Board it must be assumed that there are compelling arguments in respect of each, and it is not the purpose of this chapter

* Littlebrook 'D' not yet proceeded with.

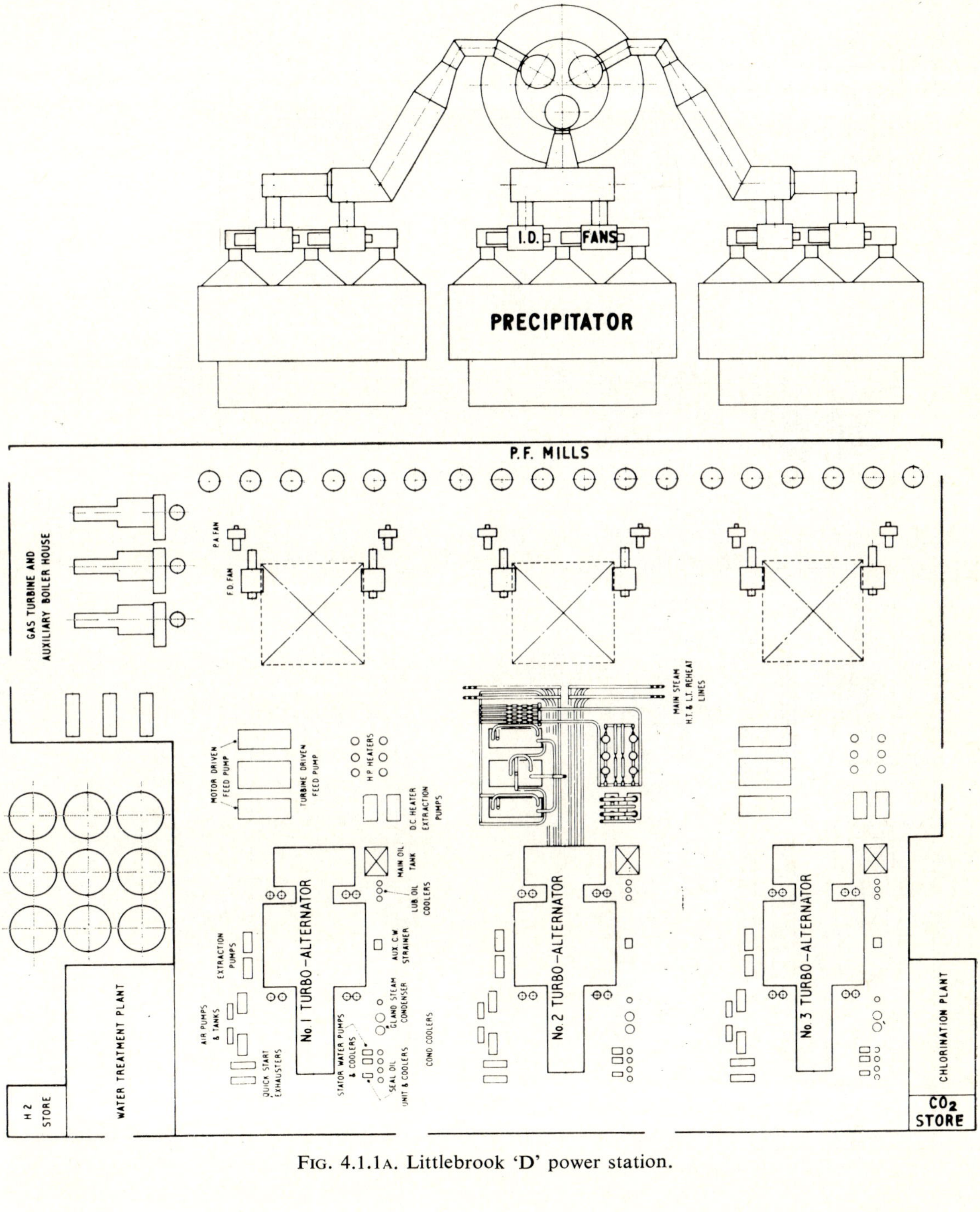

FIG. 4.1.1A. Littlebrook 'D' power station.

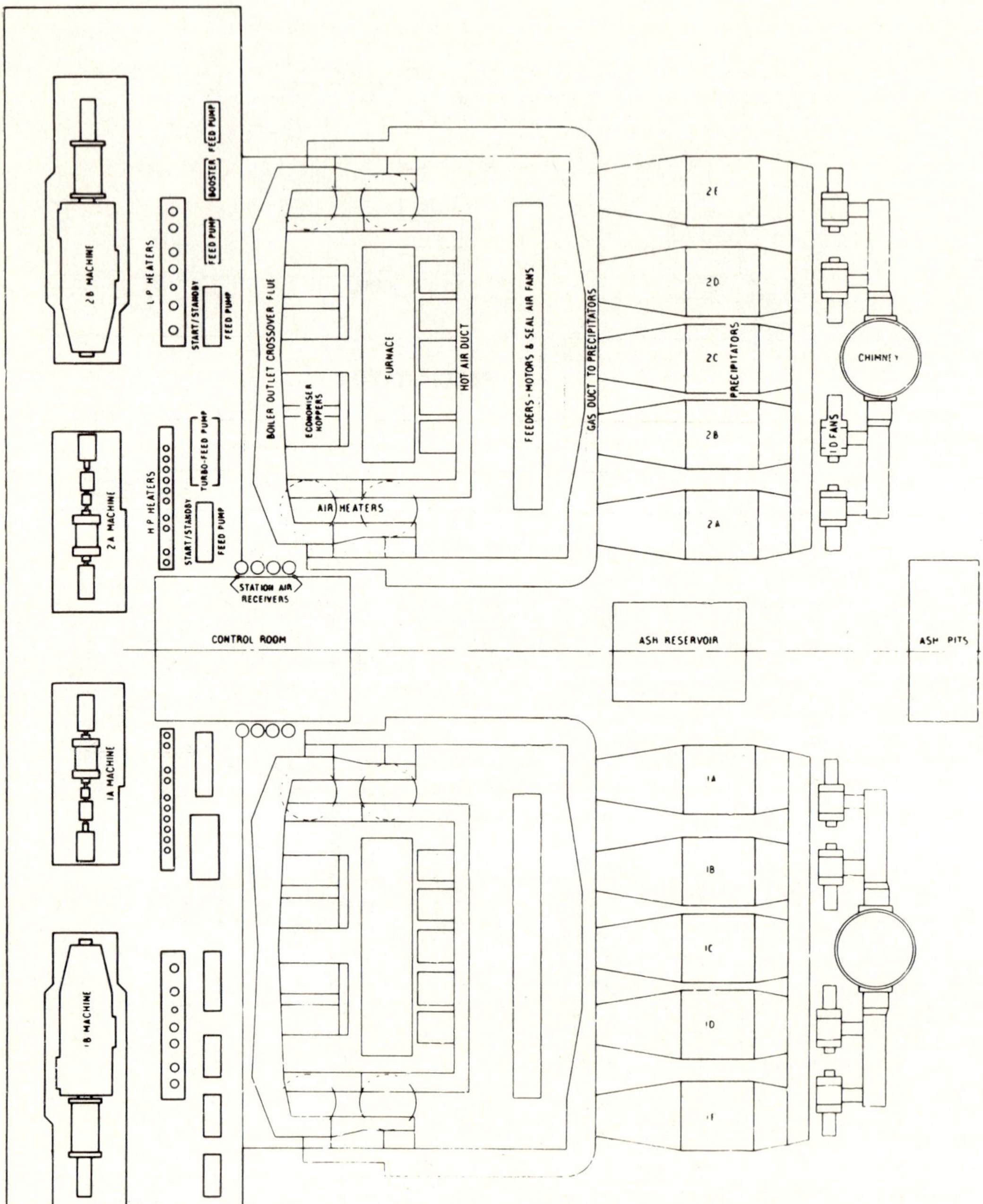

FIG. 4.1.1B. Thorpe Marsh power station.

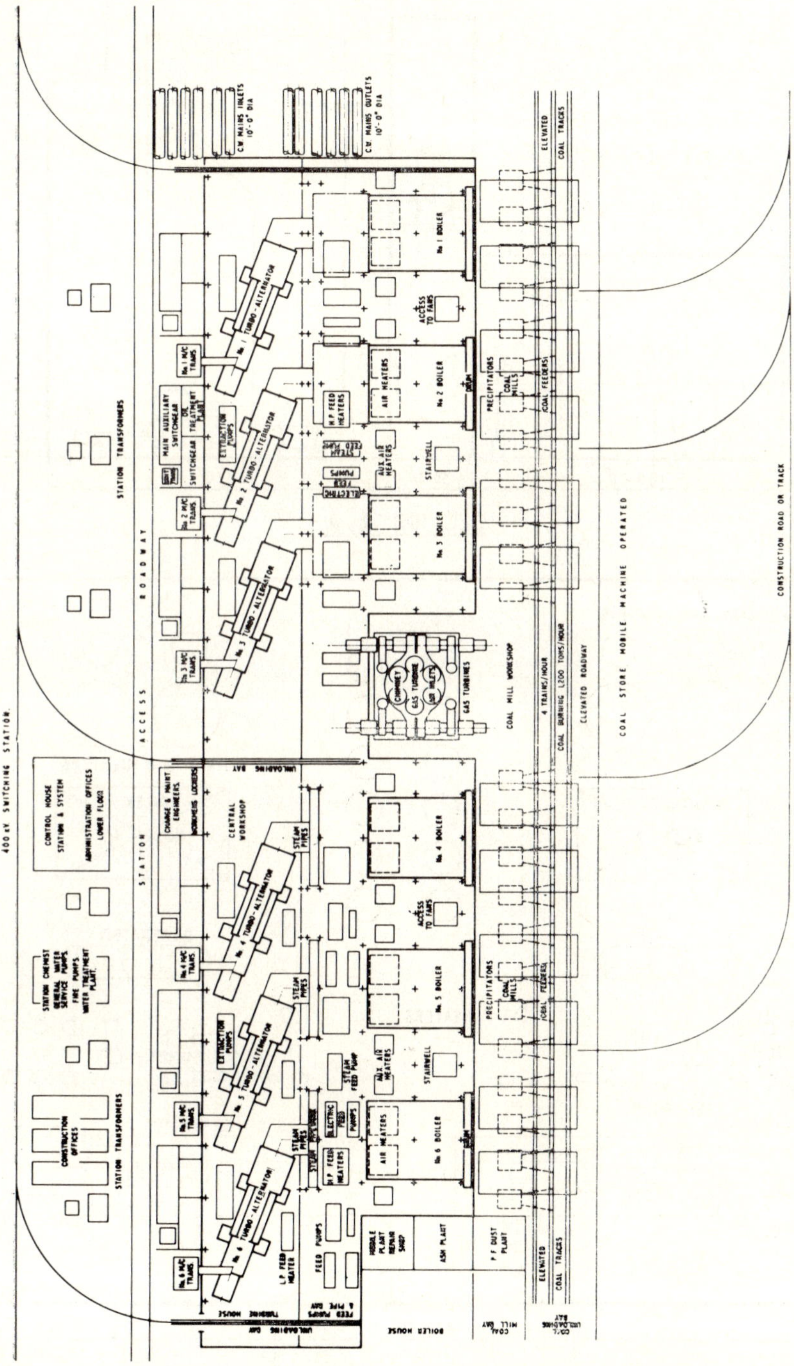

FIG. 4.1.1c. Schematic layout for Drax power station.

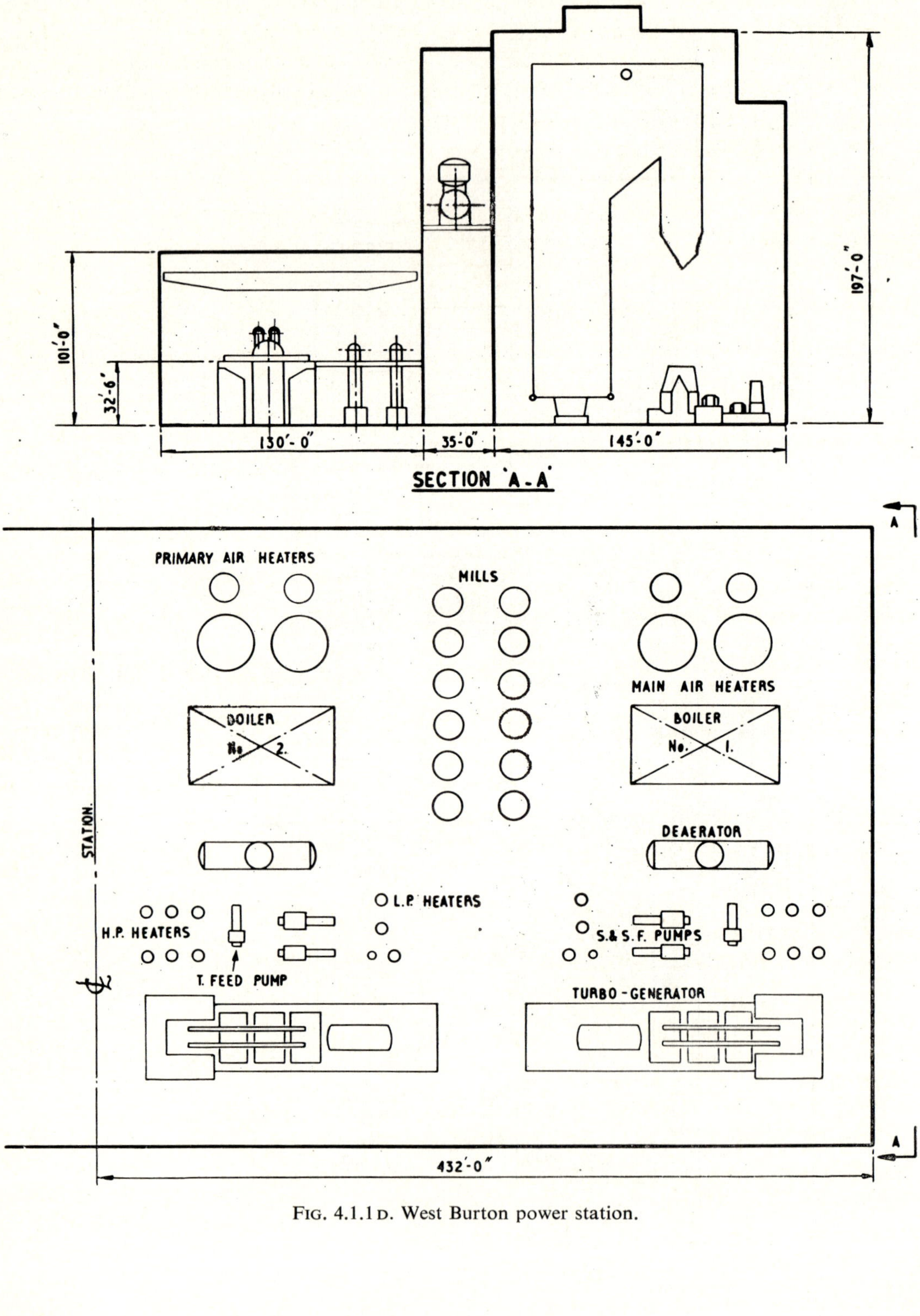

FIG. 4.1.1D. West Burton power station.

to prejudice thought in respect of one particular method. However, it must be realised that only one method will give the most economic layout for a given make of boiler and turbo-generator, taking into account feed heating systems and condensers.

The decision must rest on logic and economics, although the engineer's personal likes and dislikes may effect the final outcome.

It is useful at this stage to consider a number of current modern stations, laid out in the three main modes already mentioned and to see how these depart from the datum philosophy with respect to energy flow. Later sections deal in specific terms with the local environmental needs of the main plant items.

4.1.2. **Longitudinal Turbine Arrangement**

Historically this is the most common arrangement in the Board's existing stations and the reasons are quite logical in relation to "Range" stations having two boilers per generating set, since the set length did not exceed the length taken up by two boilers side by side and the main turbine house overhead crane could have a relatively short span.

The energy flow stream passes through two right angles in plan, i.e. into the turbine and out of the generator, but it can be argued that physically this does not necessarily cause losses, provided the steam piping and generator copperwork are run economically.

This arrangement has been adopted at a number of stations equipped with 500 MW units, typically Fawley, Pembroke and West Burton (see Fig. 4.1.1D). Of these, West Burton is coal-fired and space must be provided for P.F. mills, bunkers and dust-arresting plant.

From Fig. 4.1.1D it can be seen that the generating set is a great deal longer than the boiler width, consequently the sets determine the length of the power station, and there is plenty of useful space between the boilers. By installing two adjacent sets in mirror image to each other and placing the turbine steam chests opposite boiler centre lines, a bay is produced between boilers in which the mills for each pair of boilers can be housed, with feeders, bunkers and coal conveyors vertically above. A disadvantage of the "mirror" layout is that the number of layout drawings is doubled, with increased margin for error in left- and right-handed plant.

This layout gives a relatively long turbine and boiler house with total width reduced by the amount that longitudinally arranged coal plant would take up plus the minimum possible width of turbine house. This method has less effect in oil-fired stations where the width of the turbine house is the only real gain and this is also true of the Thorpe Marsh layout shown in Figs. 4.1.1B and E.

It is concluded that the *shape* of the site available, taken with the grid sub-station and access requirements will in the future be the main factor leading to the adoption of longitudinal sets.

The Thorpe Marsh layout shown in Figs. 4.1.1B and E is an interesting special case since the unit size is nominally 550 MW, but uses two cross-compounded turbines each driving a 275 MW generator. The layout problem was, therefore, one of two separate turbine generators served by a single boiler, and involved rather complex pipework connections.

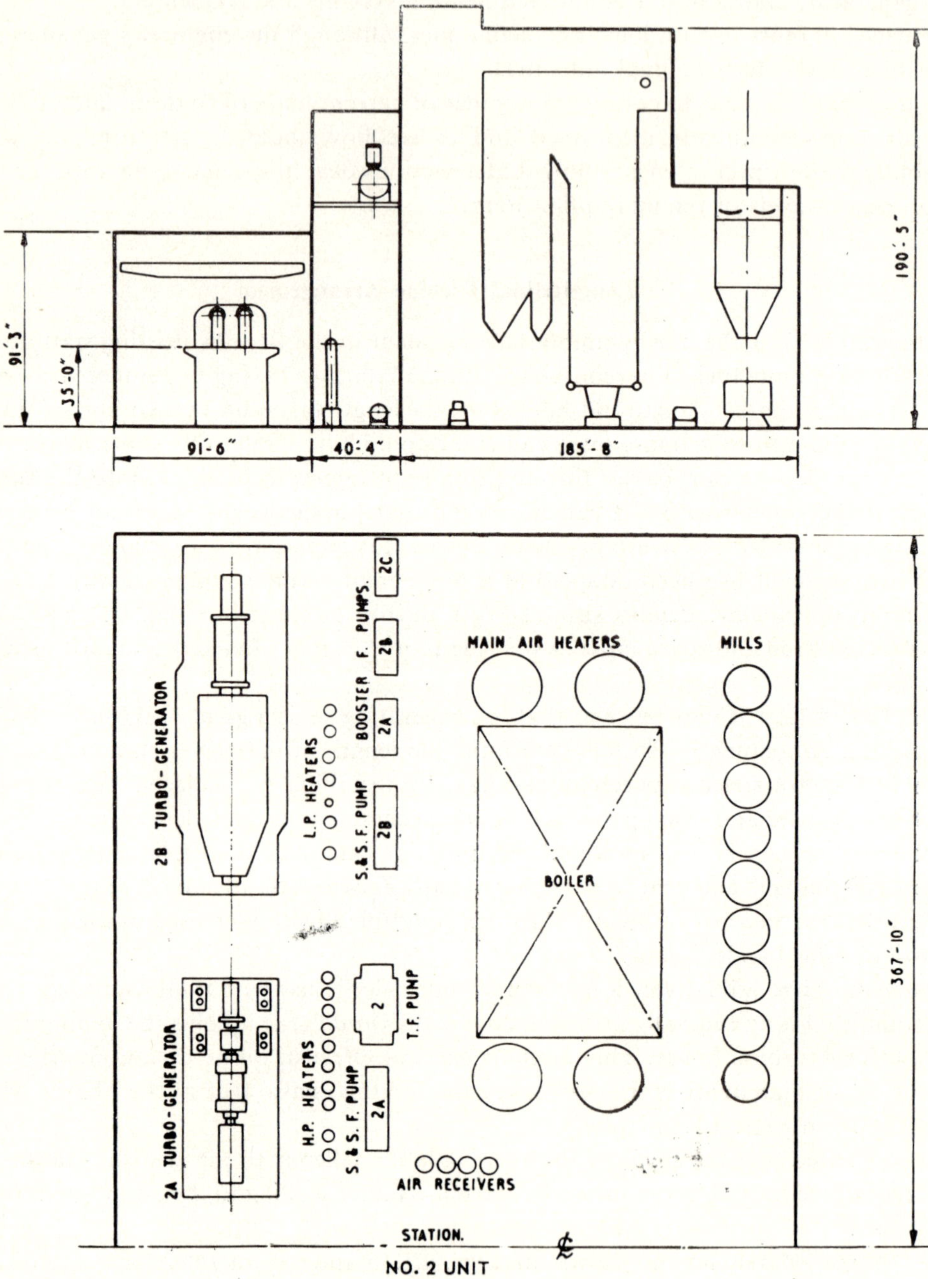

FIG. 4.1.1E. Thorpe Marsh power station.

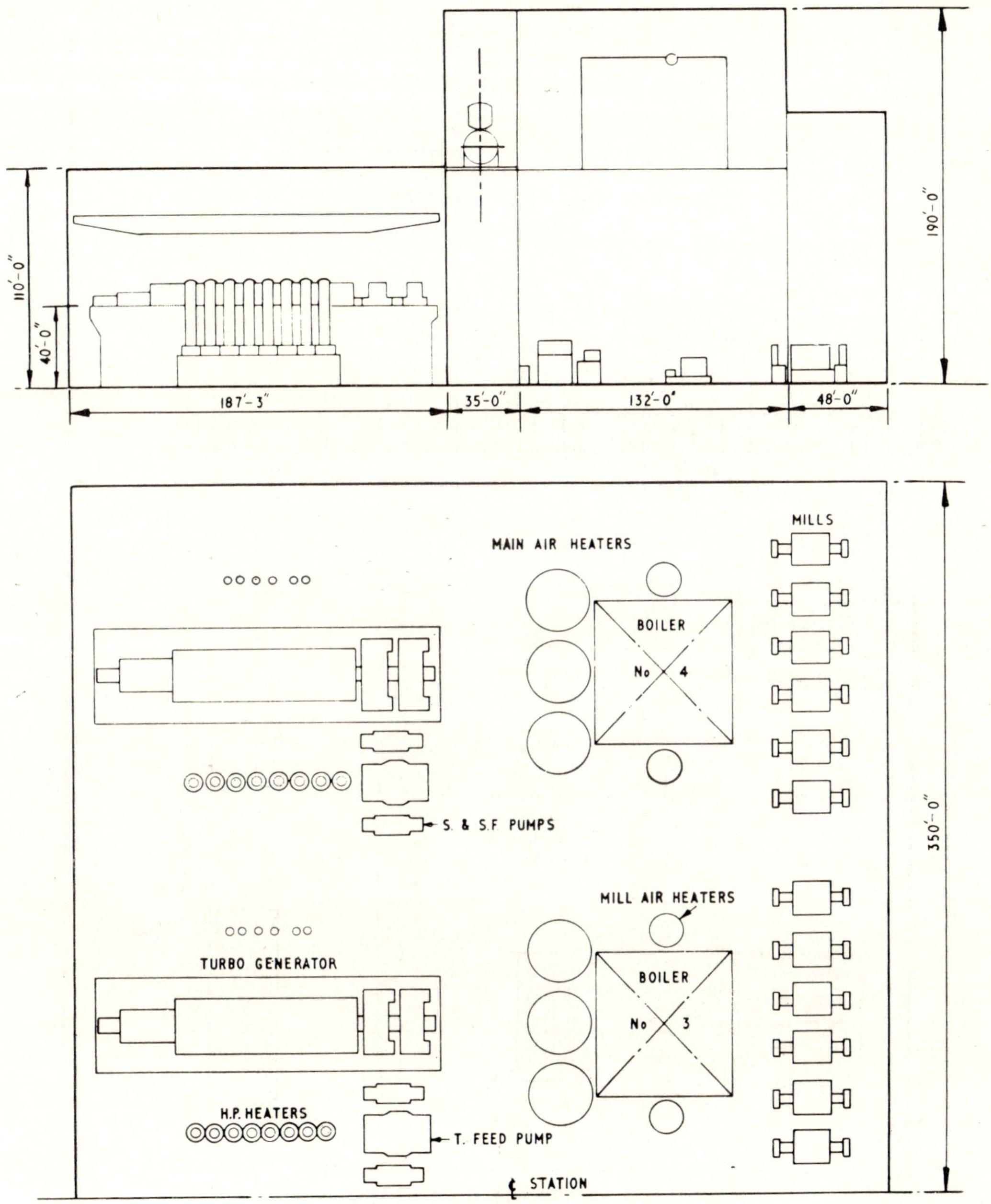

FIG. 4.1.1F. Eggborough power station.

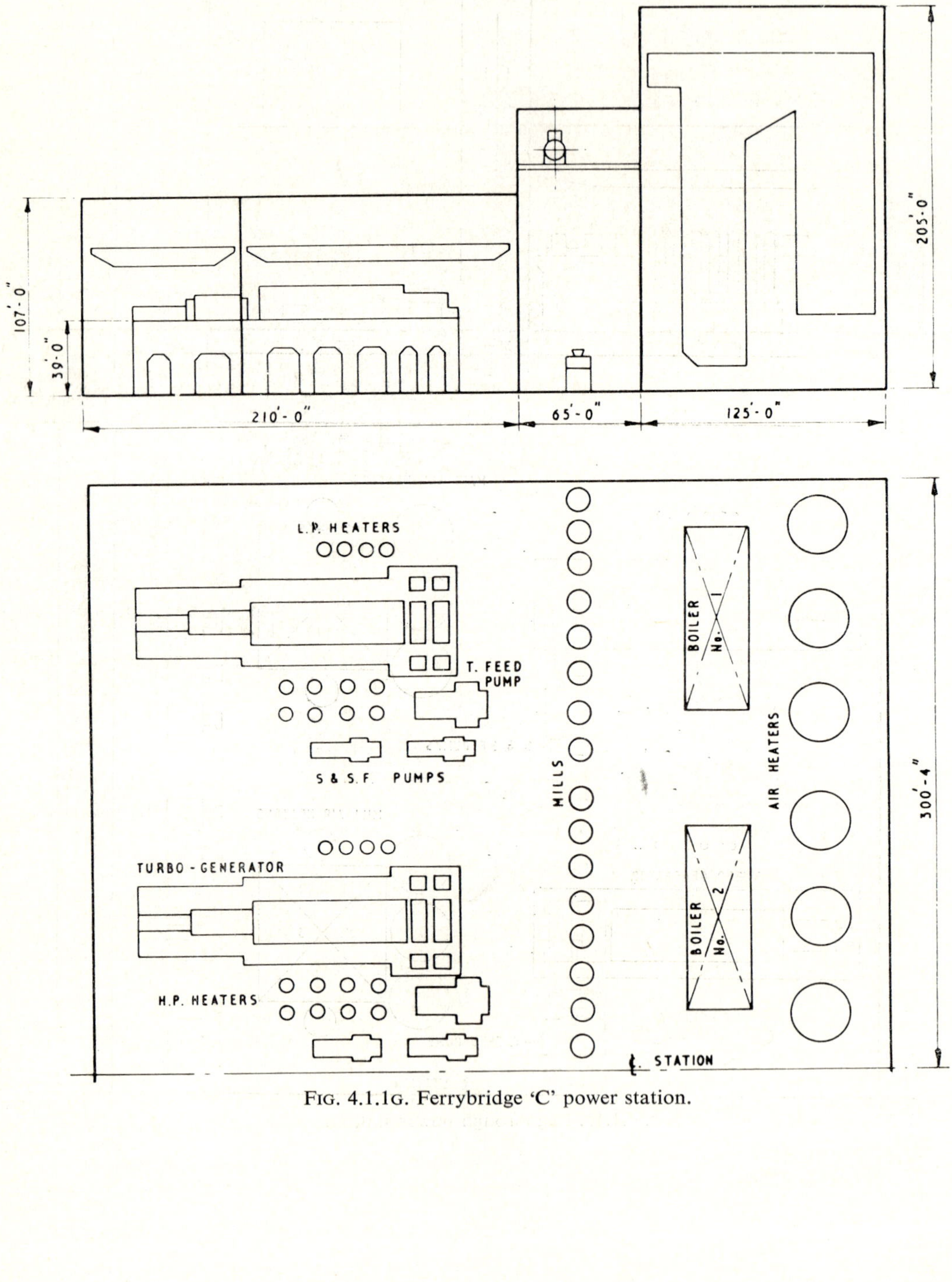

FIG. 4.1.1G. Ferrybridge 'C' power station.

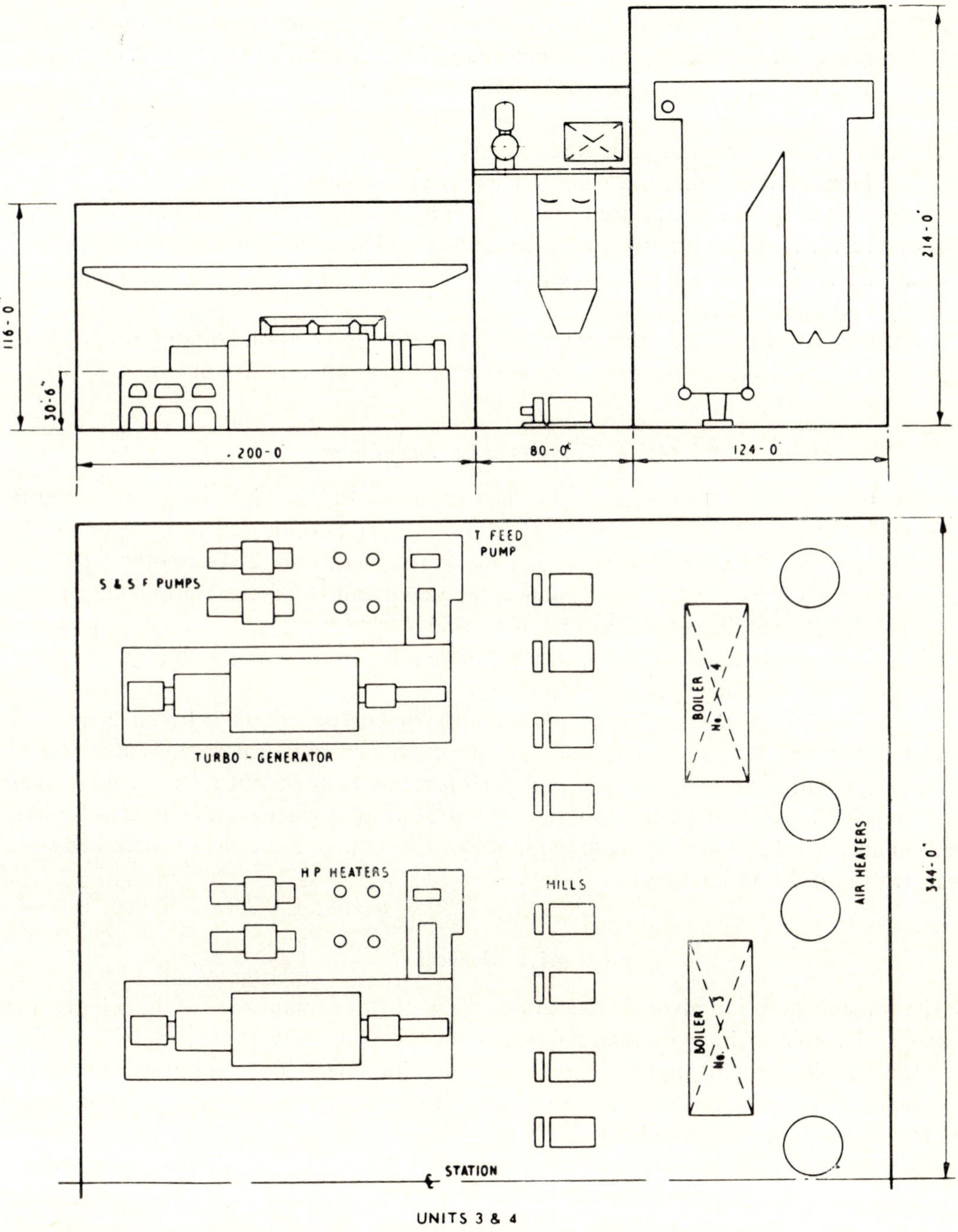

FIG. 4.1.1H. Cottam power station.

4.1.3. **Herringbone Arrangement of Sets**

Figure 4.1.1C shows the arrangement adopted for Drax power station (660 MW units). A very compact arrangement has been achieved giving a narrow turbine house and reducing the overall length to that dictated by boiler spacing, because the sets overlap each other.

The energy flow stream in plan is closer to a straight path and short connections are obtained between boiler and set, generator and transformer.

Floor space between sets has been reduced to a minimum and this could lead to congestion when a set is stripped down for maintenance, and since there are no clearly defined trucking ways within each block of three sets, movement of heavy plant items may have to be by overhead crane.

It must be concluded that this arrangement offers the greatest economy in terms of floor area, but subjectively the operators may find it a confined environment.

4.1.4. **Transverse Layout of Sets**

Some typical block drawings of this layout are shown in Figs. 4.1.1A, E, G and H. At present there appears to be a trend towards common adoption of this layout for the large 500 and 660 MW sets, since (unless the shape of the site dictates otherwise) this arrangement presents the best environment for the layout of associated auxiliary plant. The flow streams can in every detail be more nearly approximated to a straight path and each unit made into a standard entity, since there is no overlapping with adjacent units, nor does the mirror image pairing of units arise.

Comparison of Figs. 4.1.1A, E, G and H shows that good access between generating sets is a common feature of each of these layouts. For practical reasons this should be generous with transverse layouts and the operating staff benefit from the more open environment. There is scope for variation of positions of auxiliaries, mainly feed heaters, feed pumps, coal handling and milling plant and the relative positions of these items are dealt with in detail in later sections.

4.1.5. **General**

The foregoing section gives a brief overall view of three main ways of associating the boiler and turbine with particular emphasis on the plan view. There must, however, be considerable thought given to the third dimension. Here it is necessary to think in specific terms of plant height and maintenance lifting and unlike the plan problem, there are few alternatives of principle available to the layout engineer.

Not many years ago it was customary to measure the success or otherwise of a specific layout in terms of enclosed building volume per MW produced. This is unsatisfactory, because it is very unusual to be able to compare like with like, i.e. identical turbo-generating sets, feed heating plant and boilers. Over the years a trend appears of continually reducing volume per MW produced, but this is better shown by comparing the physical size of main plant items as cycle pressures and temperatures have increased, e.g. a 500 MW

	Agecroft 'C' 2×120 MW	Padiham 'B' Number 1 120 MW	Padiham 'B' Number 120 M
Volume of boiler house per unit (ft^3/kW)	14·6	12·52	
Suspended weight of one boiler (tons)	2200	2330	3250
Capacity of coal bunkers per unit (tons)	150	1500	1500
Number of superheater furnaces per boiler Number of reheat furnaces per boiler	Common furnace	Common furnace	Common furnace
Number of chimneys per station	One	One	
Number of P.F. mills per unit	Five	Three	Five
Number of exhausters per unit	Five	—	Five
Number of P.A. fans per unit	—	—	—
Number of F.D. fans per unit	Two	Two	Two
Number of I.D. fans per unit	Two	Two	Two
Number of gas recirculating fans per unit	Two	One	—
Number of precipitators per unit	Two	Two	Two
Total length of main steam pipes per unit	—	Two pipes total length 504 ft	Two pipes length 64
Total length of reheat pipes per unit	—	Four pipes total length 837 ft	Four pipes length 11
Total length of feed piping to economiser per unit	—	One pipe 213 ft	One pipe 242 ft
Number and diameter of P.F. pipes at mills	5–27 in bore	9–$14\frac{1}{4}$ in bore	5–24 in b
Number and diameter of P.F. pipes at burners	20–12 in bore	18–10 in bore	24–11 in b
Method of handling P.F. mill parts	Runway beams	Special table	Runway b
Method of handling F.D. fan parts	Runway beams	Runway beams	Runway b
Method of handling exhauster of P.A. fan parts	Runway beams	Runway beams	Runway b
Method of handling air heater parts	Runway beams	Runway beams	Runway b
Types of main air heaters	Plate type	Howden rotary	Howden ro
Method of access for furnace inspection	Special access equipment	Special access equipment	Special ac equipmen
Are P.F. mill foundations insulated from main boiler house foundations?	No	No	No
Method of transport of components to workshop	Trolley	Trolley	Trolley

TABLE 2.

2	High Marnham 5×200 MW	West Burton 4×500 MW	Thorpe Marsh Number 1 550 MW	Thorpe Marsh Number 2 550 MW	Eggborough 4×500 MW
	16·37	12·22	12·95	12·85	11·98
	3640	7090	9102	10108	9700
	2350	3140	4400	4400	2750
	One	Common furnace with dividing water wall	One	One	Common furnace
	One		One	One	
	Two	Two	Two		One
	Six	Six	Eight	Nine	Six
	Six	—	—	—	—
	—	Four	Eight	Nine	Two
	Two	Two	Four	Three	Two
	Two	Two	Four	Four	Two
	Boilers 1–3 Two; Boilers 4 and 5 provision for two future	Provision for two future	Provision for two future	Provision for two future	Provision for two future
	Two	Two	Five	Five	Three
otal ft	Two pipes total length 647 ft	Four pipes total length 1490 ft	Four pipes total length 1600 ft	Four pipes total length 1260 ft	Four pipes total length 1520 ft
total 2 ft	Four pipes total length 1531 ft	Eight pipes total length 3920 ft	Eight pipes total length 3467 ft	Eight pipes total length 2576 ft	Eight pipes total length 2800 ft
	Two pipes total length 978 ft	Four pipes total length 1320 ft	Four pipes total length 1930 ft	Four pipes total length 1190 ft	Four pipes total length 1340 ft
re	6–36 in bore	6–38 in bore	8–42 in bore	18–26 in bore	12–23 in bore
re	32–16 in bore	48–14 in bore	32–15 in bore	54–15$\frac{1}{4}$ in bore	24–14$\frac{1}{2}$ in bore
ams	Runway beams	Special travelling crane	Runway beams	Runway beams	Runway beams
ams	8 ton E.O.T. crane	Runway beams	Runway beams	Runway beams	Runway beams
ams	Runway beams	Runway beams	Runway beams	Runway beams	Runway beams
ams	12 ton travelling crane for 5 boilers	Twin runway beams with special crab	Single runway beam with electrically operated block		Runway beams
tary	Howden rotary	Howden rotary	Howden rotary	Howden rotary	Rothemuhle rotary
cess	Special access equipment	Not yet decided	Special access equipment	Special access equipment	Special access equipment
	Yes	No	No	No	No
	Trolley	Trolley	Trolley	Trolley	Trolley

turbine is less than three times the length of a 30 MW turbine with nearly 17 times the output.

Figure 4.1.1J shows clearly the reducing volume in terms of ft^3/kW. It can be seen that the trend is hyperbolic and extrapolation shows little further gain as set sizes rise above 550 MW. The really big economies came in the change from 30 MW to 200 MW unit sizes. Tables 1 and 2 show comparative data relating to various unit and total station sizes and again the economies gained in going to large unit sizes are well shown. Figures 4.1.1K, L and M show various layouts for 4×500 MW sets with typical main dimensions.

The layout engineer's task is not to meet some MW/sp. vol. or any other similar volume criteria, but to provide an efficient layout at minimum capital and running costs within the available space and the best possible environment for control, operation and maintenance.

4.1.6. **Control, Operation and Maintenance**

The ability of station staff to perform these three functions efficiently and economically is governed primarily by the selected layout of plant. The main lines of communication throughout the plant are part of the layout and must be established at an early stage of overall conception.

These are best assessed from the position of main assembly points relative to each unit such as:

Administrative offices.
Maintenance and stores centre.
Staff amenities centre.
Operations control centre.
Road and rail traffic.

All of these centres must serve all units equally well in terms of direct communication, but since construction of units progresses usually over five or six years from one end of the station, truly central facilities are not easy to plan because the lines of withdrawal of construction equipment and personnel must be secured.

As a consequence a large number of existing stations have the maintenance workshops, stores and staff amenities at one end with a separate block of administrative offices.

Because of its special nature as the centre of control, the control room position is of high importance and a close study of this both as a construction problem and from the operator's point of view has been made. The control room may be at the centre of the station at operating floor level, or offset either on the generator transformer (or electrical annexe) side of the turbine house or on the boiler house side in the mechanical annexe or between boilers.

The control room should always be free from plant noise and must remain secure in the event of catastrophic failure of adjacent plant and this suggests that the best position is off-set on the electrical annexe side of the turbine house. This position, however, gives a problem in construction since the control room must be cabled forward from the first unit to be commissioned and electronic equipment erected under clean conditions in advance of plant construction, but concurrent with major civil engineering work. This

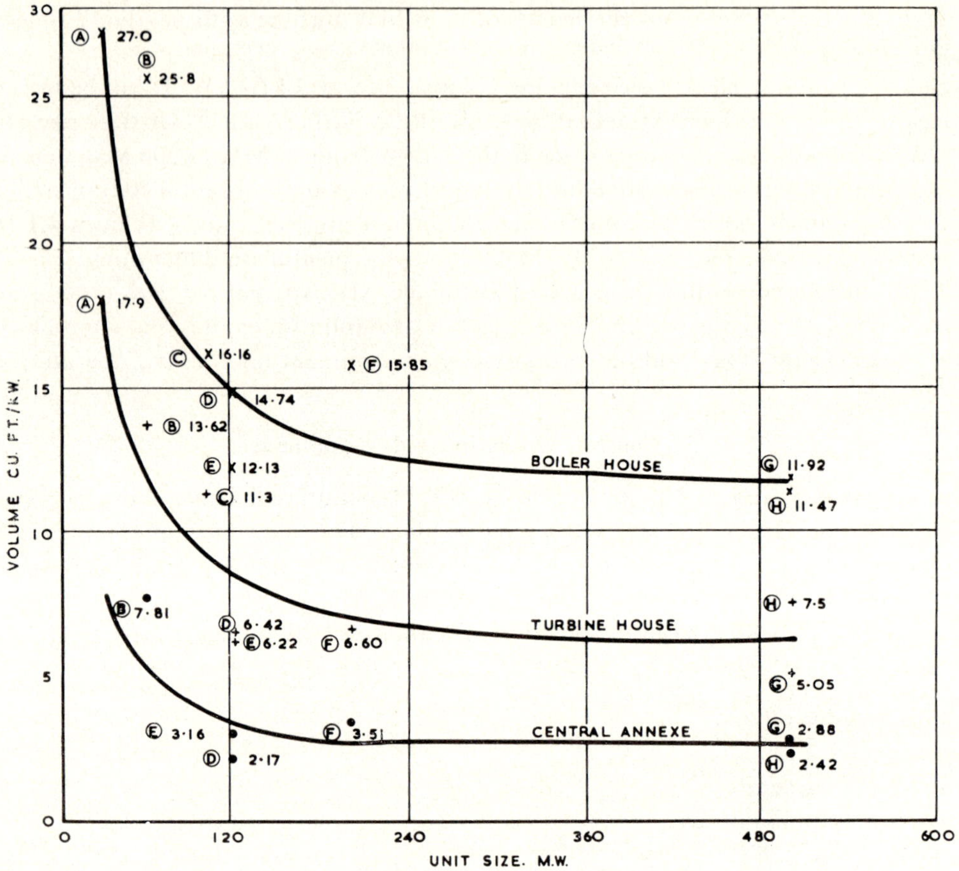

FIG. 4.1.1J. Curves showing relative volumes of main buildings.

problem is overcome by careful planning at site. The charge engineer's office suite should be located adjacent to the control room for obvious reasons.

The lines of communication from these areas must form a natural pattern linking with the units and the basis is very easily formed in plan, but vertical communication may not be given proper consideration resulting in haphazard positioning of lifts, not always in the best position or linking with horizontal access to plant.

For large modern power stations, fast efficient lifts available for use by all staff are required in each unit area. The lift car must be, as a minimum, large enough to carry a stretcher and the men handling it. Indication of the position of the car must be signalled at each landing.

Detail access to plant items depends on the nature and form of each item and must be seen in this relationship. Coupled with this there must always be consideration of main access for men and equipment for maintenance purposes, and for the essential duties of operators. It should be remembered that even the smallest valve must be accessible for inspection and lubrication as must also remotely operated valves.

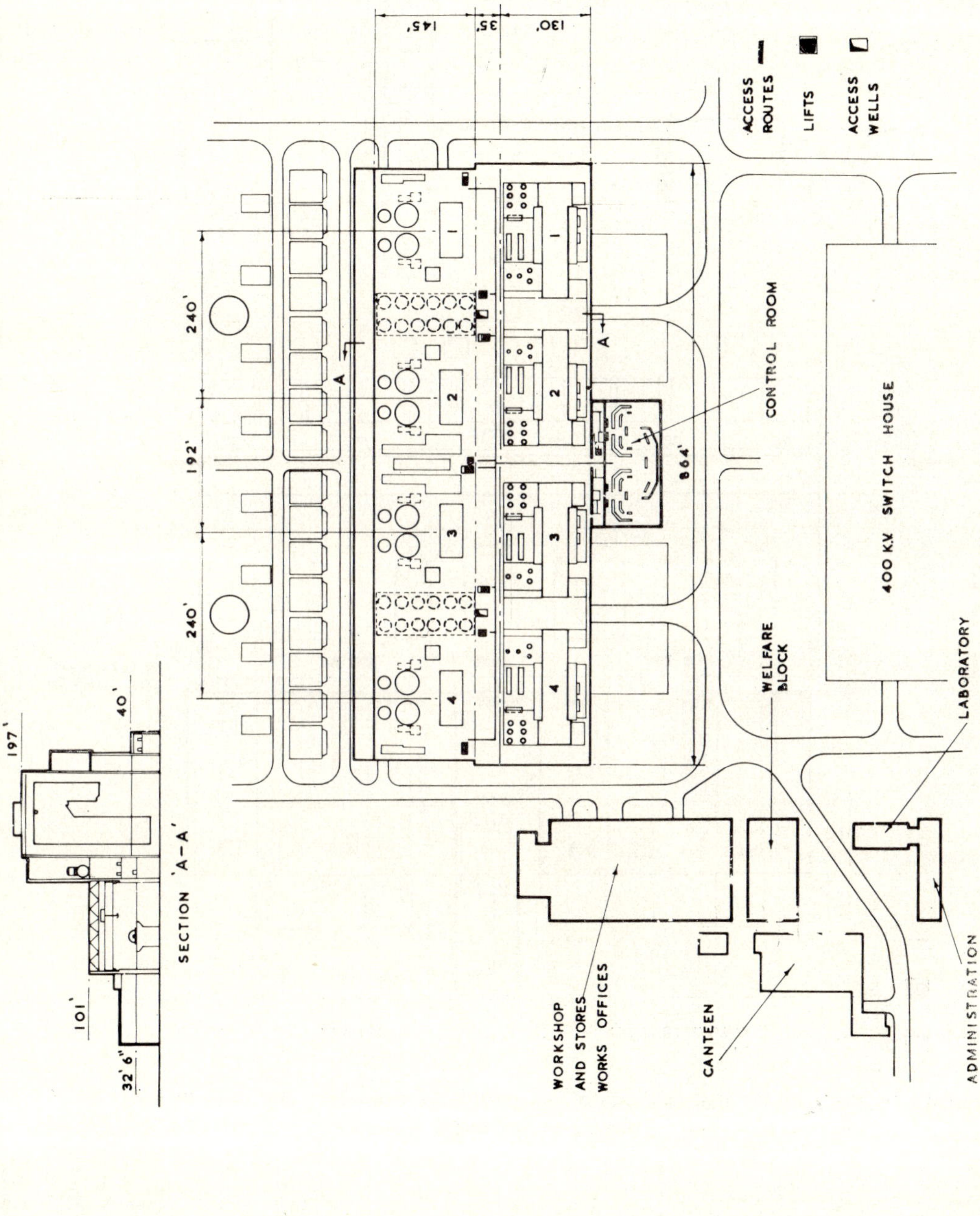

FIG. 4.1.1K. Typical main dimensions for 4×500 MW units with sets longitudinal.

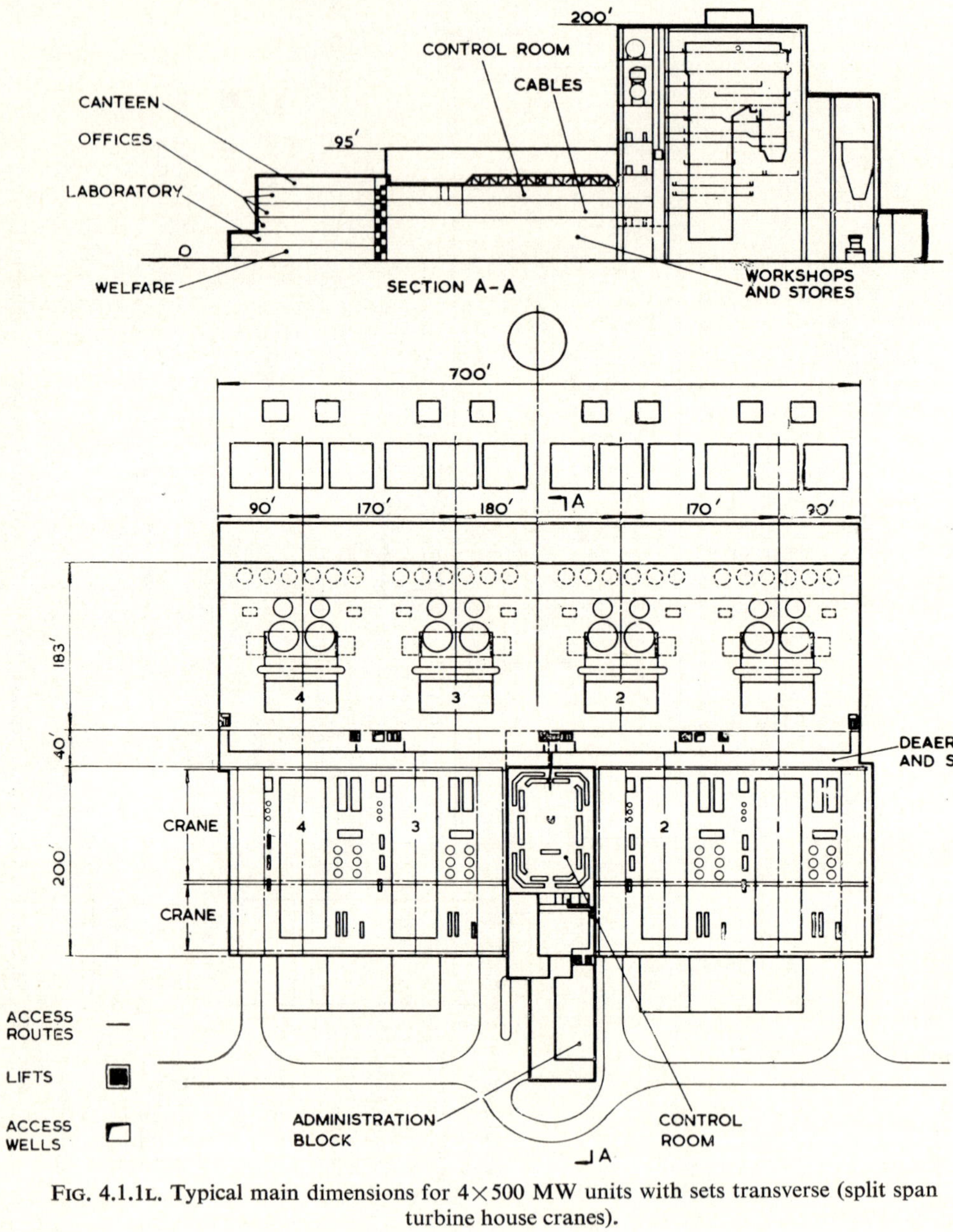

FIG. 4.1.1L. Typical main dimensions for 4×500 MW units with sets transverse (split span turbine house cranes).

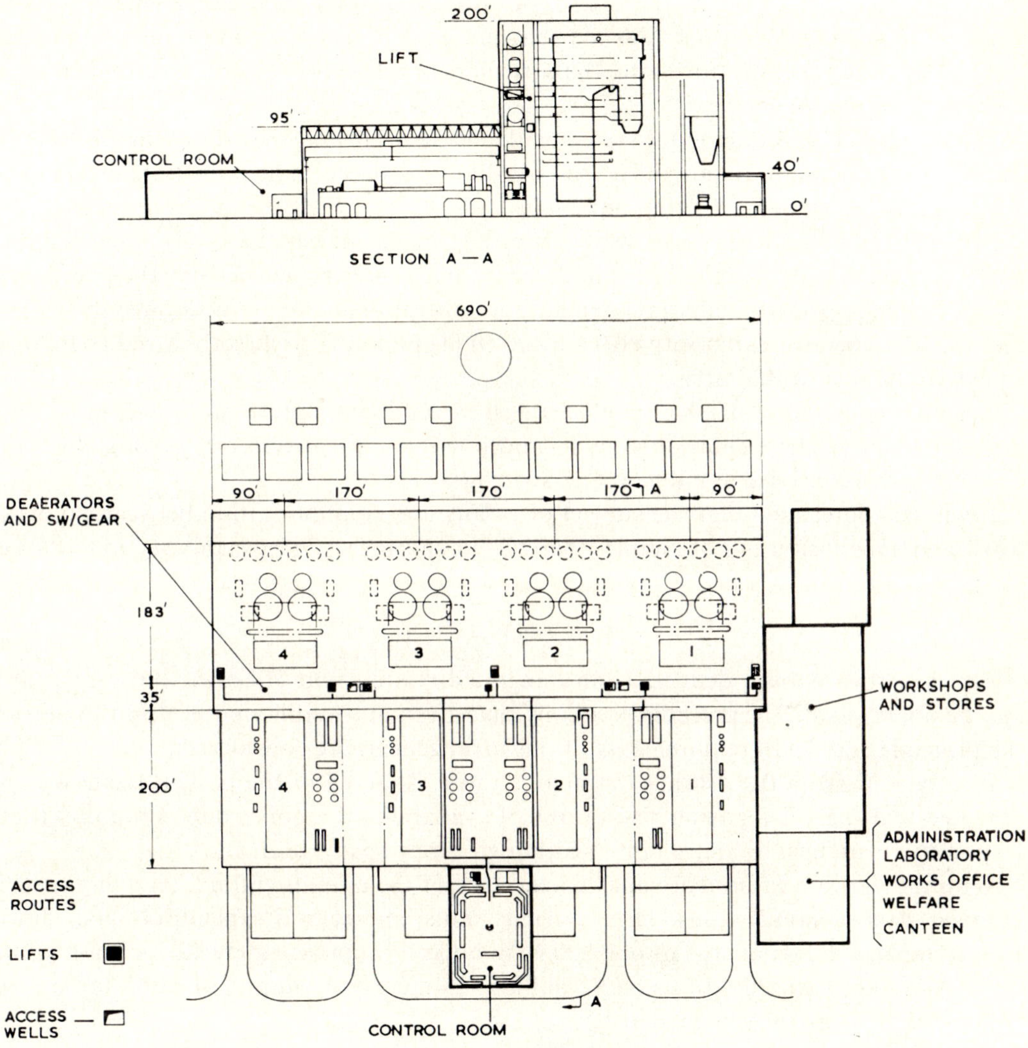

FIG. 4.1.1M. Typical main dimensions for 4×500 MW units with transverse arrangement (single span in turbine house).

4.2. COMMUNICATION

In addition to good access to plant from central areas, it is essential to provide good staff location and calling systems centralised on to the control room which is the focal point for operation.

A survey in August 1966 by the Board, of operator's views, including eighteen stations of more than 200 MW output, showed that present provisions in most stations are not considered to be generally adequate. In particular the reliability of certain equipment gave cause for complaint.

The standard P.A.X. telephone system with adequate coverage satisfies general administration communication requirements but staff locating lamp indication systems are criticised mainly on account of siting difficulties.

Personal radio calling systems were, according to the survey, generally preferred, but in a power station there are many "blind spots" and people forget to carry their receivers.

The "roving operator" concepts demands an operational communication system, such that the unit operator can positively call staff in his unit area to directly wired telephones which are unique to that area.

General maintenance of plant could be well served by miniature radio telephones, but for control and instrumentation systems and power electrical systems, jacking field telephone points are needed at cubicles and switchgear areas.

It will be appreciated that the subject of calling and communication between points of work and supervision is not simple, there is no shortage of ideas, this section sets out the problems.

4.3. THE UNIT

This has already been defined as an integrated relationship of boiler, turbo-generator and auxiliary plant. The preceding sections have given an overall view and in this section the principle and detail requirements of this arrangement are considered.

In departing from the "Range" system, so well favoured in the past, it is known and accepted that there is a greater risk of loss of availability if a boiler fails, since on a range station there are usually spare boilers although rarely spare turbines.

With the increase in pressure and temperature of the operating fluid, reducing specific volumes, it is economic to use large capacity units, and capital expenditure goes down. For example, the capital cost of one 200 MW boiler is approximately £2·7 million whilst a 300 MW boiler would cost say £3·2 million, an increased output of 50% for an increased capital cost of less than 19%.

With a unit station each unit must be seen as a separate producer operating on a simple flow process given in Fig. 4.1 from the arrival of fuel, via a common service to all units, to the delivery of power to the Grid System. Each unit consumes about 4% of the power it produces to drive its auxiliaries and this figure must be kept to a minimum by careful design and *layout*. This applies particularly to the choice of static heights over which condensate and condenser cooling water must be pumped, together with the pressure drop sustained through piping services.

4.3.1. **Unit Levels**

Basement level is always taken as datum and its precise level relative to ordnance datum should be assessed for economic reasons. For a river or sea water-cooled station, pumping costs over the life of the plant at its expected average load factor must be set against civil engineering costs in piling and excavations, etc. This is not an important factor in a cooling tower station unless the lie of the land involves excavation to bring down the relative static height of the cooling towers.

The conduits below basement level should be limited to:

(1) Cooling water culverts.

(2) The ash sluiceway.

(3) The main drains.

Many existing stations have considerable tunnelling for cables, pipe trenches and other services and the excavation for and building of these is expensive. Further disadvantages are the risk of flooding and difficulty of maintenance. To overcome these disadvantages, it is convenient to have cables and piping above basement level but below a mezzanine or semi-basement floor at say 9 ft to 10 ft above basement level.

The basement floor must be laid with adequate falls to the main drains to ensure quick removal of water resulting from washing down, maintenance work and plant drainage (see Fig. 4.3.1A). The main drain is laid below the turbo-generator, mainly because this is where a large amount of drainage occurs particularly from condensers.

The semi-basement floor should be between 9 ft and 10 ft above the main basement level. It need not be continuous but will provide operating and maintenance access to turbine auxiliaries such as pumps, coolers, filters, etc., and suspension for pipes to and from the auxiliaries which are run in the interspace which replaces trenching, as indicated on Fig. 4.3.1B.

The boiler and turbine operating floors may not be at the same level on modern plant since with the use of integral or pannier type condensers, the height of the turbine horizontal centre line above basement level can be greatly reduced by comparison with older layouts. The limitation on the boiler operating floor level is the size of the bottom ash hopper. If, however, the lie of the land is suitable, it may be possible to step the basement level down with minimum excavation to obtain a continuous operating level.

It will be found that all levels above operating floor level will follow plant requirements. The roof levels will be dictated in the turbine house by the position of the main crane, for the heater bay and mechanical annexe by the deaerator or first direct contact feed heater, both of which must be served by an overhead crane. The boiler house will of course be fixed by the height of the boiler.

4.3.2. **Cranes and Lifting Equipment**

Major permanent travelling cranes are required in the following places:

(1) Turbine house.

(2) Feed heater bay.

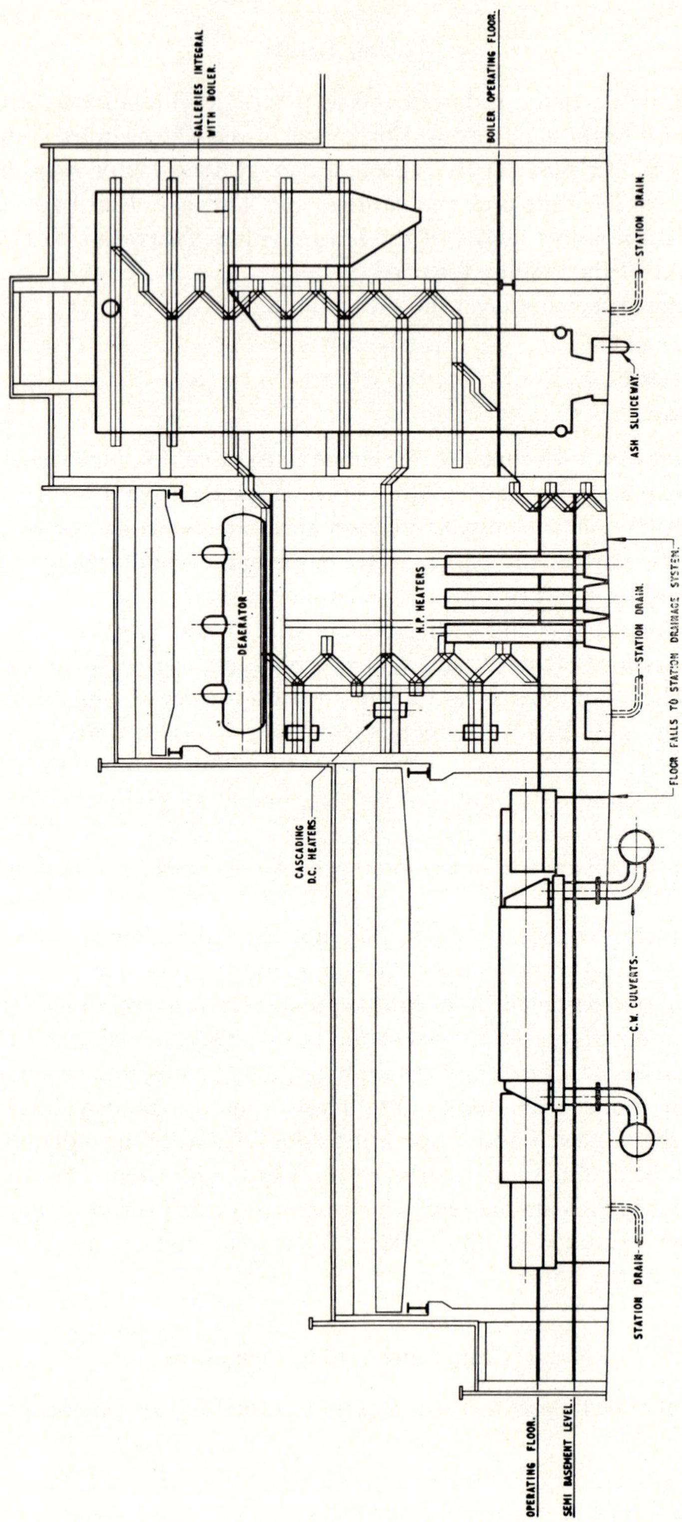

FIG. 4.3.1A. Outline floor level sketch.

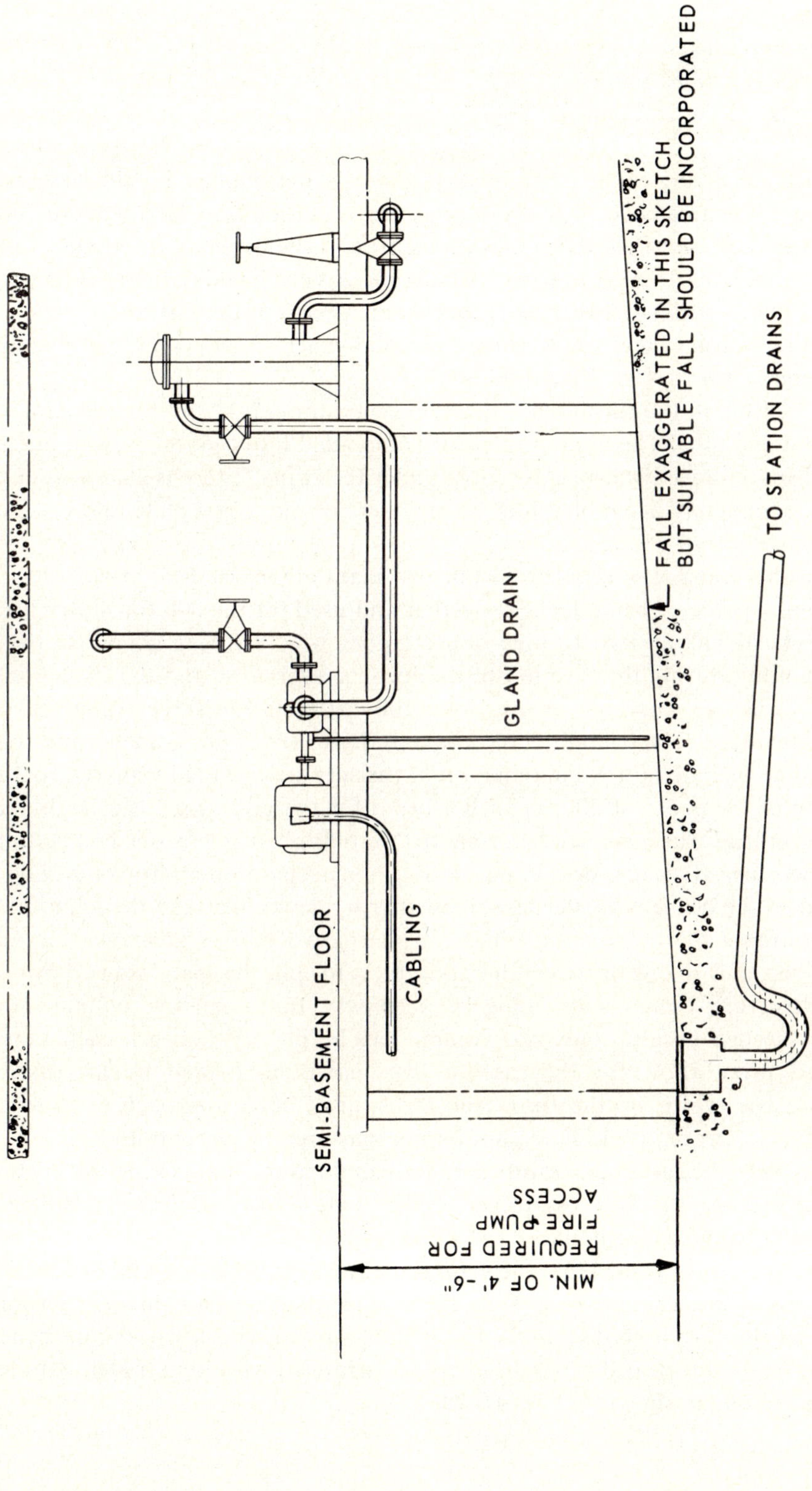

FIG. 4.3.1B. Auxiliary plant on semi-basement floor.

(3) Workshop.
(4) Stores.
(5) CW pumphouse.
(6) Furnace ash sales point.
(7) Gas turbine-generator house.

The maximum lifting capacity of each of these is determined by the heaviest single lift expected in the area covered, both during construction of the station and maintenance during the operational life of the station. This lift and the live load it represents when the crane is travelling, has to be carried by the building structure and is, therefore, a major item among the data required at a very early time in the design of the station.

The turbine house cranes are foremost in importance since no turbine erection work can commence until they are functioning.

The heaviest lift in the turbine house is the generator stator and for 660 MW sets this will be about 300 tons. The alternator rotor is threaded into the stator from one end and withdrawal space must always be available under the crane. There is also a trend towards bringing in completely assembled turbine cylinders on their bed plate and some of these assemblies may weigh 150 tons.

Based on this trend it is usual to install two main cranes each of about 170–180 tons capacity, but capable of being locked together and used for the 300-ton stator lift.

Advantages of having two turbine house cranes with a lifting beam are the greater crane availability during the erection of the turbine alternator and the reduction in cost of the structural steel and civil works. For a 2000 MW station three cranes have a considerable advantage. The two crane technique requires that both cranes travel longitudinally along the turbine house, consequently if the sets are installed transversely the crane span will be of the order of 200 ft. Such a beam for the given load can be designed but very long overhead travellers can be prone to "crabbing", which is one end getting in advance of the other. This can be minimised by having separate long travel driving motors at each end of the crane with electrical load sharing device between them and by careful parallel mounting of the running rails. An alternative used a great deal in American power stations, and in one or two in this country, is to split the span into two by providing a row of structure columns founding between sets, but a massive column and beam structure is needed because only one column can be pitched between each set and the design load must allow for the case of two maximum loaded hooks adjacent (see Fig. 4.3.2A). The width of the beam and the limit of hook approach to the end of the crane beam also creates a blind area not plumb-mastered by either hook.

At Drax where the turbo-generators are at an angle to the main crane traverses, a separate motion is fitted to the cross travel so that long and cross travel combine, to give angled motion to the crane hook.

In Section 4.3.3 it is argued that the main generator transformer could be located at the generator end of the turbine block. With such an arrangement a loading bay running the full length of the turbine house could be available, fitted with a workshop crane, which itself might be used to transfer overhead cranes arranged with long travel, parallel to the turbine main axis, as shown in Fig. 4.3.2B.

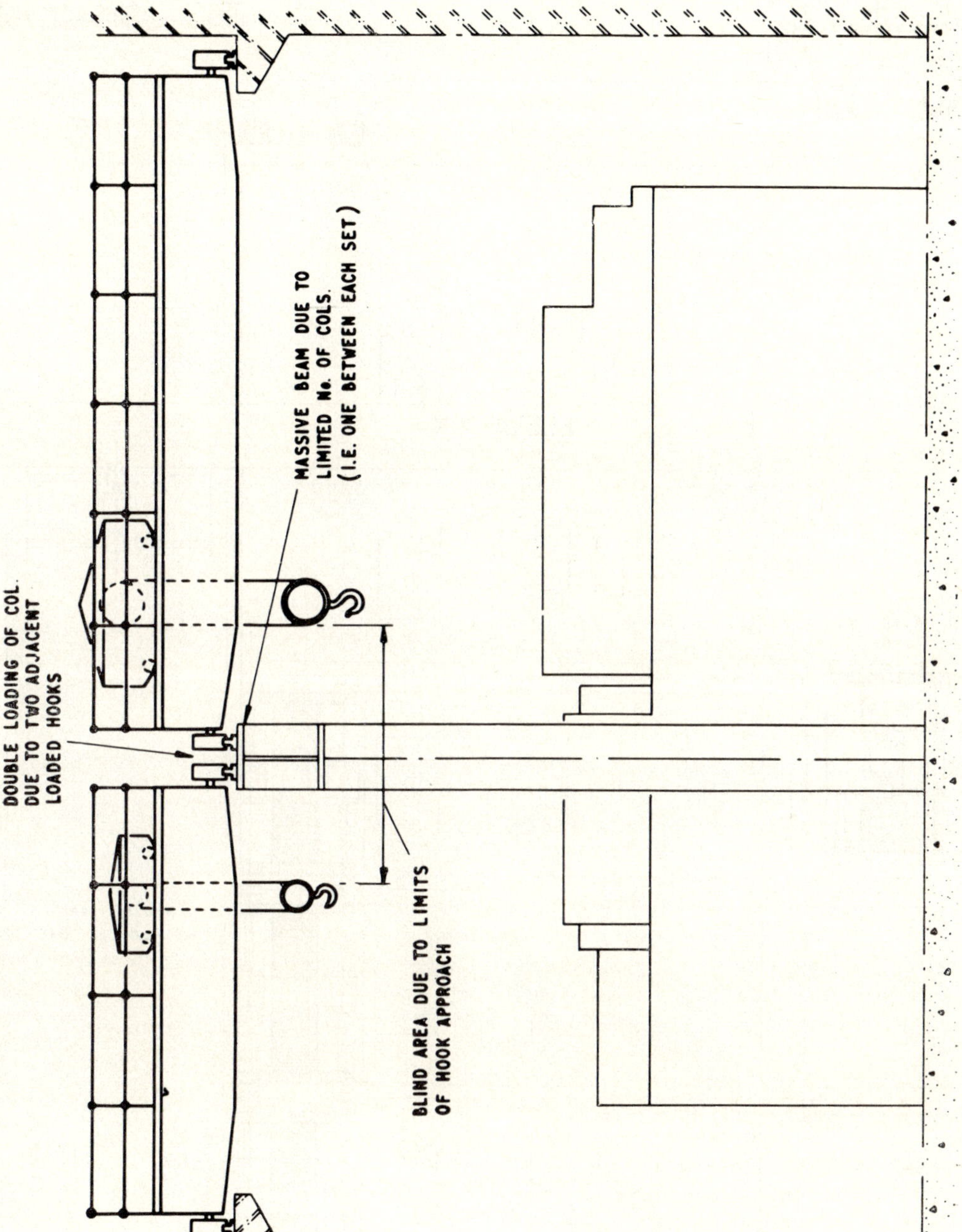

FIG. 4.3.2A. Split span turbine house cranes.

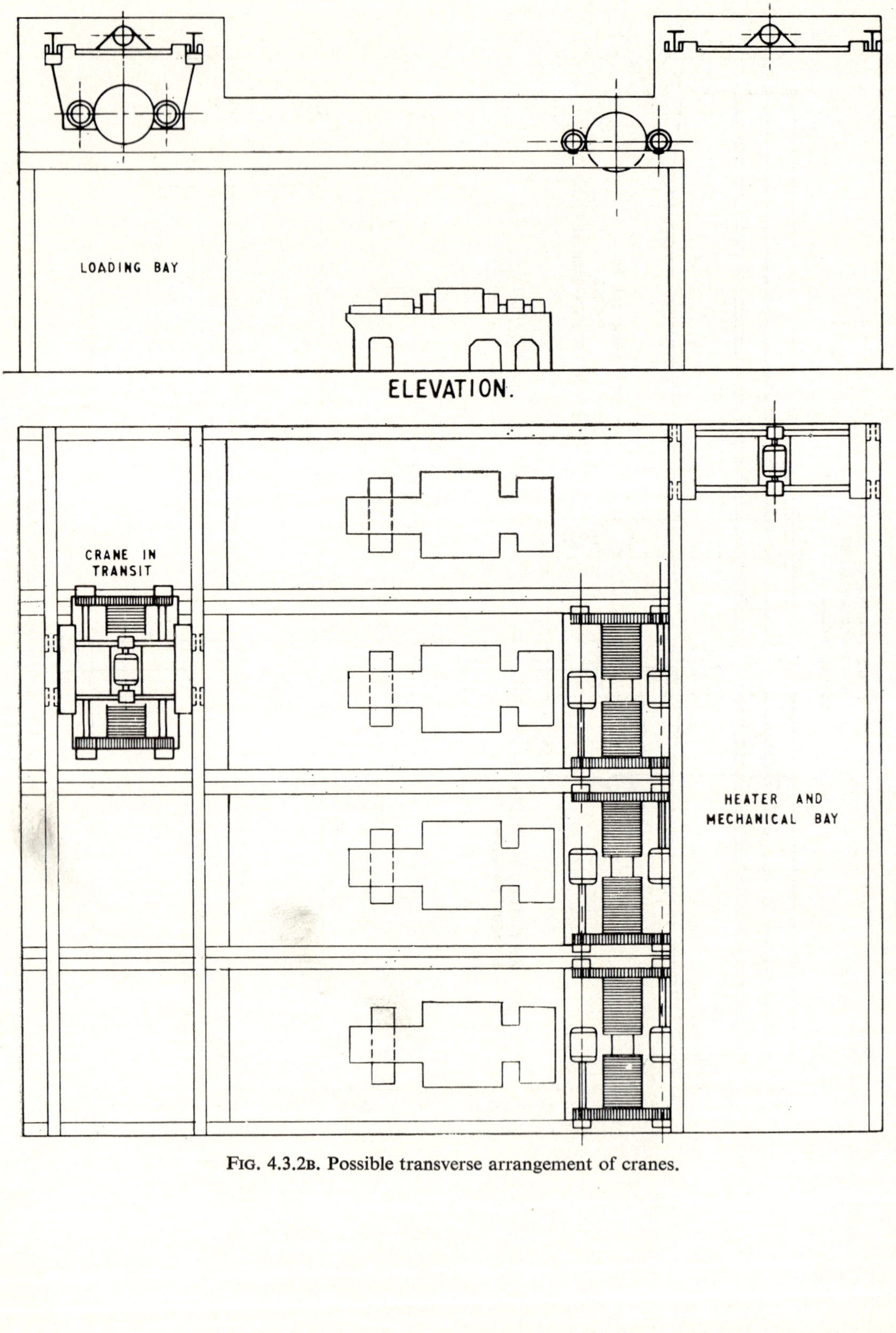

FIG. 4.3.2B. Possible transverse arrangement of cranes.

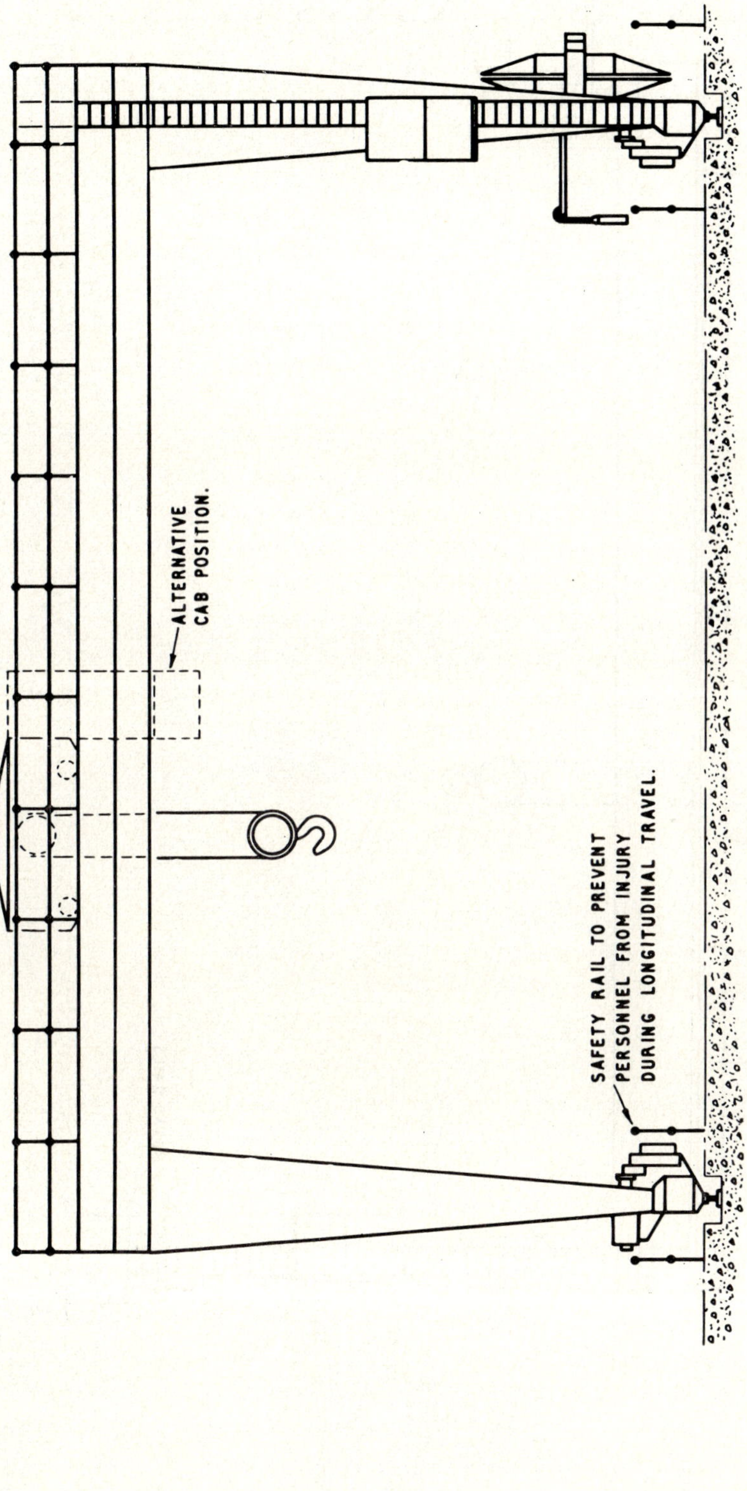

FIG. 4.3.2C. Typical Goliath type of crane.

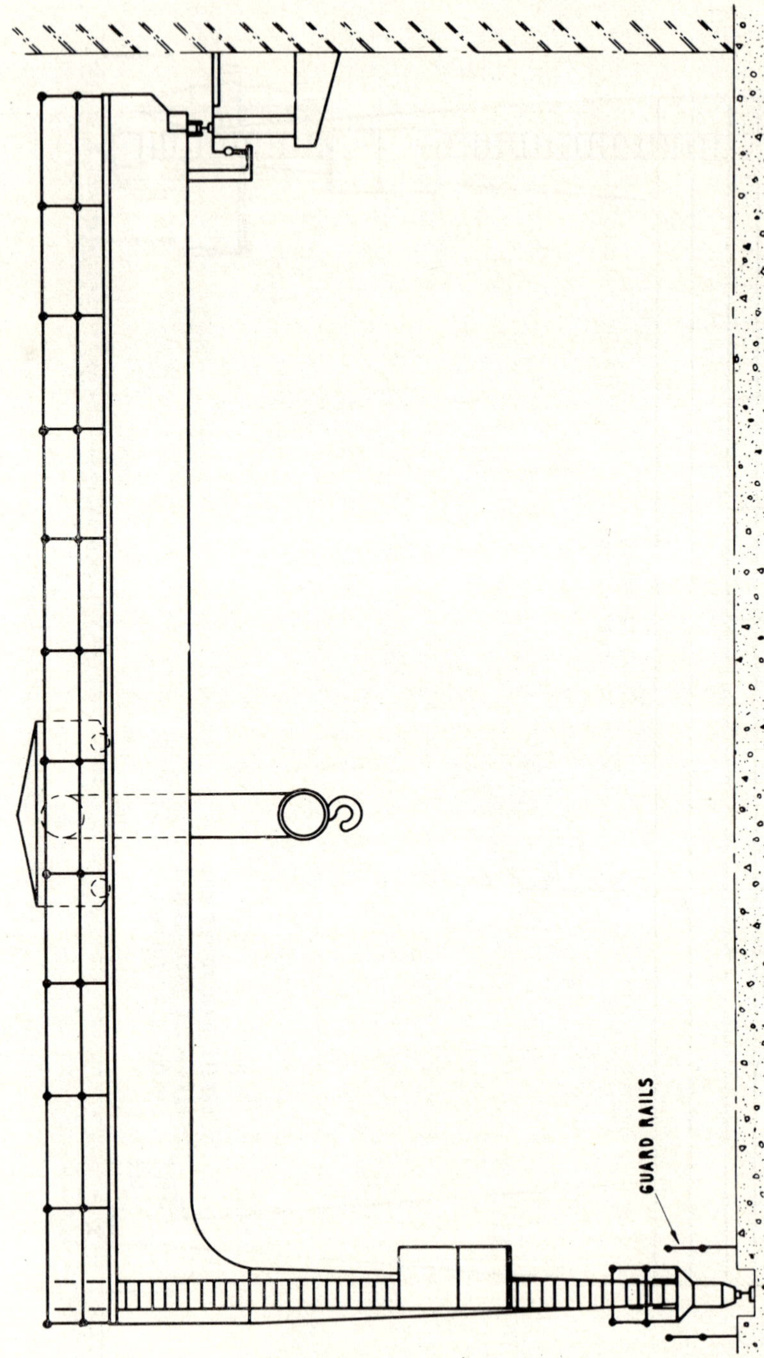

FIG. 4.3.2D. Semi-Goliath type of crane.

The Goliath type of crane shown in Fig. 4.3.2C has attractions for the layout engineer since it carries its vertical supporting columns with it and there is a large reduction in the weight and size of building and structure columns. The semi-Goliath type shown in Fig. 4.3.2D is also a useful compromise.

Since the wheel tracks of Goliath cranes are often on or near access areas it is necessary to rail off securely the whole run of track as indicated in the drawings.

In addition to the main cranes, hoists on runway beams are essential over items of plant not accessible to main crane hooks, important items are pulverising mills, large fans, pumps, boiler air heaters, etc. Where a model is used in the design stages (see the last section of this chapter) the maintenance staff should be given an early opportunity to check and carry out lifting exercises on the model to determine the most satisfactory arrangements.

4.3.3. **Transformers**

When considering plant layout for a power station the relationship of transformers to the associated plant is important since, as a general rule, the low-voltage cable runs from the transformers must be as short as possible to save copper and minimise voltage drop.

The most important transformer in the power station is the generator transformer, which steps up the generated voltage to the switching voltage and must be located as near the generator as possible, usually in a cell just outside the turbine house. The generator is connected to the transformer by copper bus-bars air insulated, and carried in trunking arranged to give the necessary electrical clearance. This trunking is of about 8 ft square section on large units, but this method of enclosure is now being superseded by a much more compact design in which each phase is suspended by insulators in individual circular ducts of about 48 in outside diameter. The latter design saves space and better control of cooling the bus-bars can be achieved by air or inert gases.

Apart from designing the bus-bar system to a minimum length, changes of direction must also be minimised since, under short circuit conditions, high stresses are set up in the bars and mechanical failure could occur at the weakest points, i.e. bends and sets, etc.

All bus-bars are flexibly connected at or near the generator terminals to prevent transmission of vibration and work hardening of the copper.

Bus-bar arrangements determine to a certain extent the layout of the turbine generator, and Figs. 4.3.3A and B show alternative layouts for transverse and longitudinal machines.

With a transverse layout of machine, a simpler probably shorter route can be obtained as illustrated. The bus-bars on this layout should be run below the generator since an overhead arrangement would require dismantling of the copper work each time the generator rotor was withdrawn. The neutral point is made and earthed under the alternator also incorporating generator protection C.T.s. As shown on the sketch the unit transformer is teed off of the main bus-bars just before the connection to the low-voltage side of the generator transformer. With this arrangement of bus-bars the winding connections are quite simple since the three phases are arranged across the generator frame.

A further point in favour of the transverse machine layout with respect to bus-bars is that usually very little plant is located at the end of the machine at basement level whereas

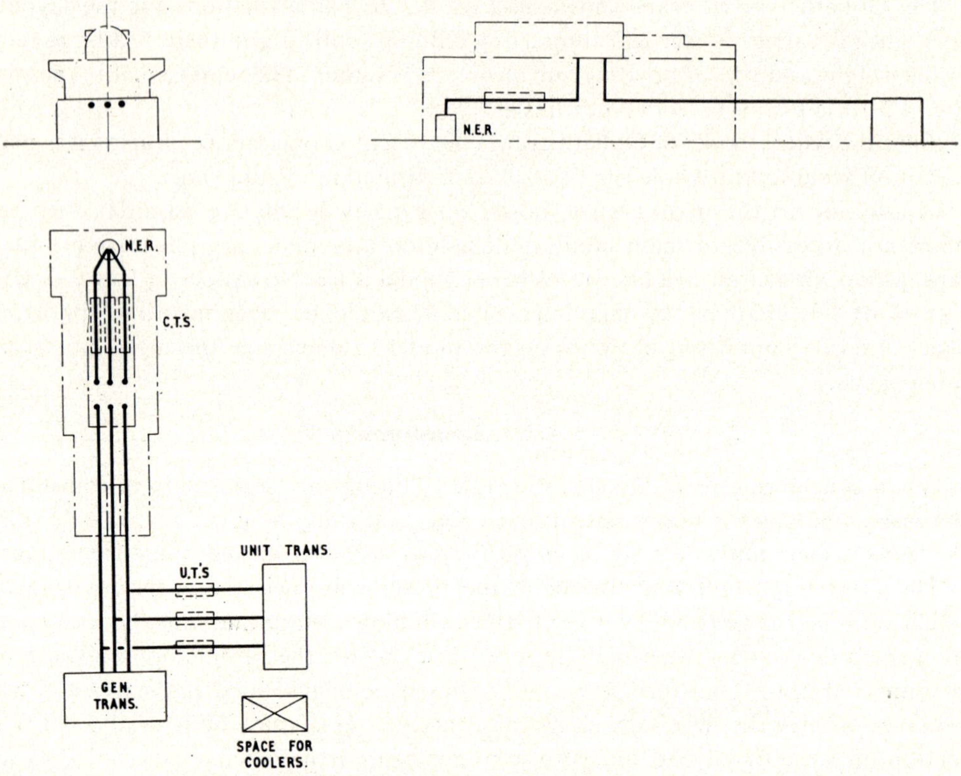

FIG. 4.3.3A. Bus-bars for transverse layout.

with the longitudinal arrangement, the basement area is sterilised to a certain extent since the bus-bars pass overhead.

The second illustration shows bus-bar arrangements for a longitudinal arrangement of machine and as can be seen this causes complications to the arrangement when the connections are brought from under the machine. With top connected bus-bars, the arrangement is much simpler although the star point is still brought out under the generator belly thus requiring two sets of three-phase terminals. With this latter arrangement the bus-bars and enclosure can be vulnerable to damage from loads carried by the overhead crane.

To take the connections from a side of a longitudinally arranged machine would produce difficulties in arranging the stator hydrogen cooling system since the tubes carrying the coolant are generally arranged down the sides of the stator thus blanking off the windings. This arrangement would also block access alongside the generator on the one side where the connections are taken out, creating unnecessary access problems around the machine.

Probably the next most important transformer to be located is the generator unit transformer which is supplied off the main bus-bars, usually at right angles (see Fig. 4.3.3A), and here again, to save on copper work, the bus-bars are kept as short as possible. This transformer should be placed in a cell adjacent to the generator transformer.

A most important transformer to be accommodated near the main buildings is the

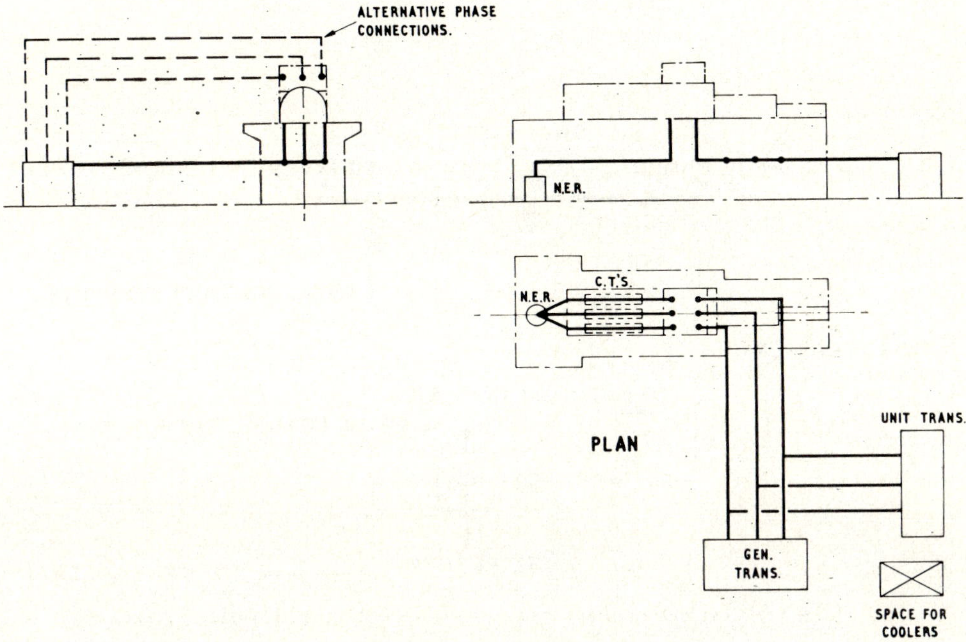

FIG. 4.3.3B. Bus-bars for longitudinal layout.

station transformer. There are usually at least two of these and they step down the grid sub-station voltage to the highest plant auxiliary voltage required in the station. This means that these transformers are wound for operating at 400 kV/11 kV or 275 kV/11 kV on modern power stations. These transformers must be sited so that they are available for commissioning the first plant in a new power station.

The remaining transformers, i.e. turbine unit, boiler unit, coal plant, ash plant, etc., should all be located as near the associated lower voltage switchgear as possible to reduce copper costs and losses. A new design of low-voltage switchgear (415 V) now incorporates a transformer built into the suite of panels, but problems associated with size limitation and heat dissipation arise.

Oil-filled transformers are a fire hazard and all must be separately protected with an emulsifying water system which is initiated by temperature rise in the transformer cells, hence transformers must be separated by blast walls. Large volumes of water and oil can be released into the compound and adequate arrangements must be made for drainage to a large sump which is fitted with oil separators (in case there is a failure on the oi circuit of the transformer). The water can be pumped away to the station drains but the oil must be retained and disposed of by road tanker or other means.

Cooling of oil-filled auxiliary transformers is usually by radiators, but the generator and unit transformers are now so large and the heat generation is such, that the heat in the oil is dissipated by cooling water in the heat exchanger. The limiting factor on transformer cooling, by other than natural means, is the hydraulic strength of the transformer tank. The head on the tank is limited to the level of the conservator and these are kept as low as possible to minimise tank pressures. The water-cooling pressure must

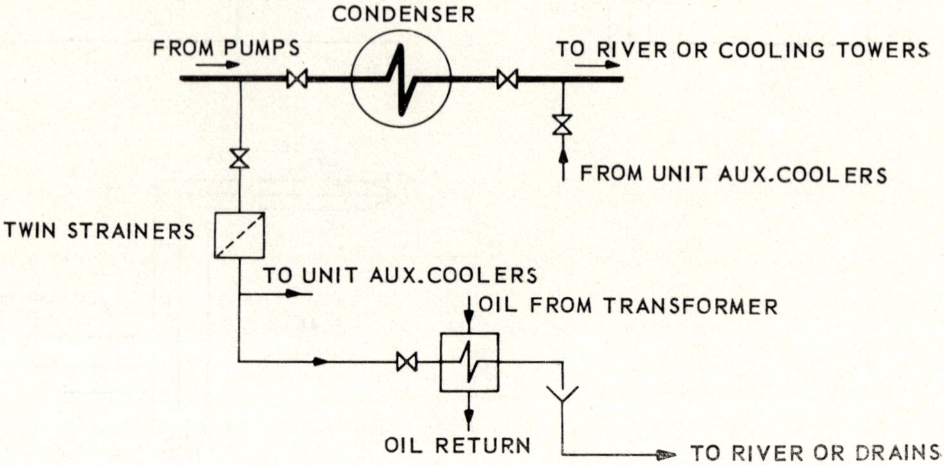

FIG. 4.3.3C. Typical auxiliary transformer oil-cooling system (single stage).

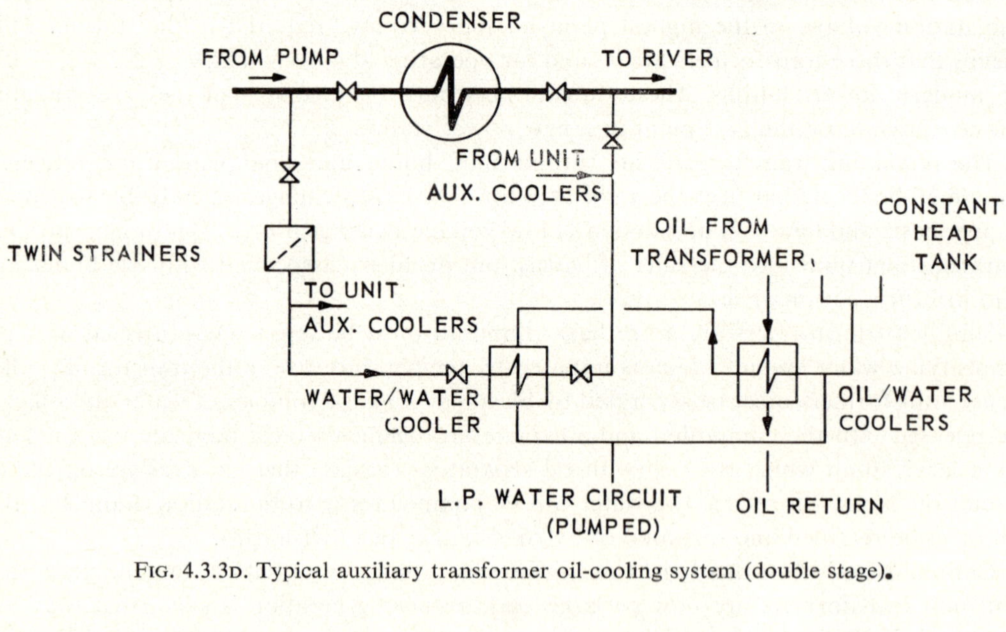

FIG. 4.3.3D. Typical auxiliary transformer oil-cooling system (double stage).

be kept below the oil pressure to ensure that should leakage between the fluids occur the flow will be into the coolant, thus preserving the insulation resistance of the oil.

The coolant is normally drawn from the turbo-generator unit auxiliary cooling water system and on high-head CW systems, i.e. associated with cooling towers, and because of the water/oil pressure differential, individual valveless atmospheric discharges are arranged on the water/oil coolers. The cooling water is led back either to the river or to the station drains. On siphonic CW systems a similar arrangement to the above is sometimes used since closed valve conditions could arise with the coolers shunted across the main condensers, but this is wasteful of power if the coolant is returned to pumped station drainage system.

An alternative scheme adopted on some siphonic CW systems uses interstage cooling between the oil coolers and the river water coolers. In this method the oil is cooled by water at a lower pressure than the oil in the oil/water coolers and in turn the water in this l.p. circuit is cooled by river water in a water/water cooler. Figures 4.3.3C and D illustrate the alternative transformer cooling arrangements on modern power stations.

When using interstage cooling the same l.p. cooling water circuit can be utilised and run in parallel for both generator and unit transformers. The siting of the coolers is a problem to the layout engineer since they must be placed local to the unit in the case of a shunt arrangement with the coolers across the condenser and also within an area close to the transformer for security of the oil lines. Lifting beams must be provided in the cooler cell for handling the heat exchange tube nests.

Figure 4.3.3E shows how the generator transformer could be installed immediately against the generator thus reducing the copper connections to a minimum, this method is used at Clarence Dock power station. It is attractive to layout engineers, since a loading bay running the full length of the turbine house can be arranged with the 275 kV cables run in trenches below loading bay floor level to the switching compound. In addition to improving layout and access there is less possibility of inducing currents in concrete reinforcing rods and local structure steelwork from the field effects of long copper runs.

The main objection to this arrangement is the fire hazard although, judging from past experience, there is little evidence of risk and with modern fire fighting techniques any transformer oil fire can be extinguished in less than 10 sec. A noise nuisance may be created in the turbine house from transformer hum, but this can be dealt with by screening techniques and is not a major problem.

4.3.4. **Cabling and Switchgear**

The cabling required for a unit may be regarded as the spinal cord of the unit; on a modern unit with fully automatic and remote control it is a very considerable installation as the following table shows:

Unit control cables	310 miles
Unit power cables	200 miles

The quantities shown represent the major cable installation for one 300 MW unit and are roughly representative of most modern installations.

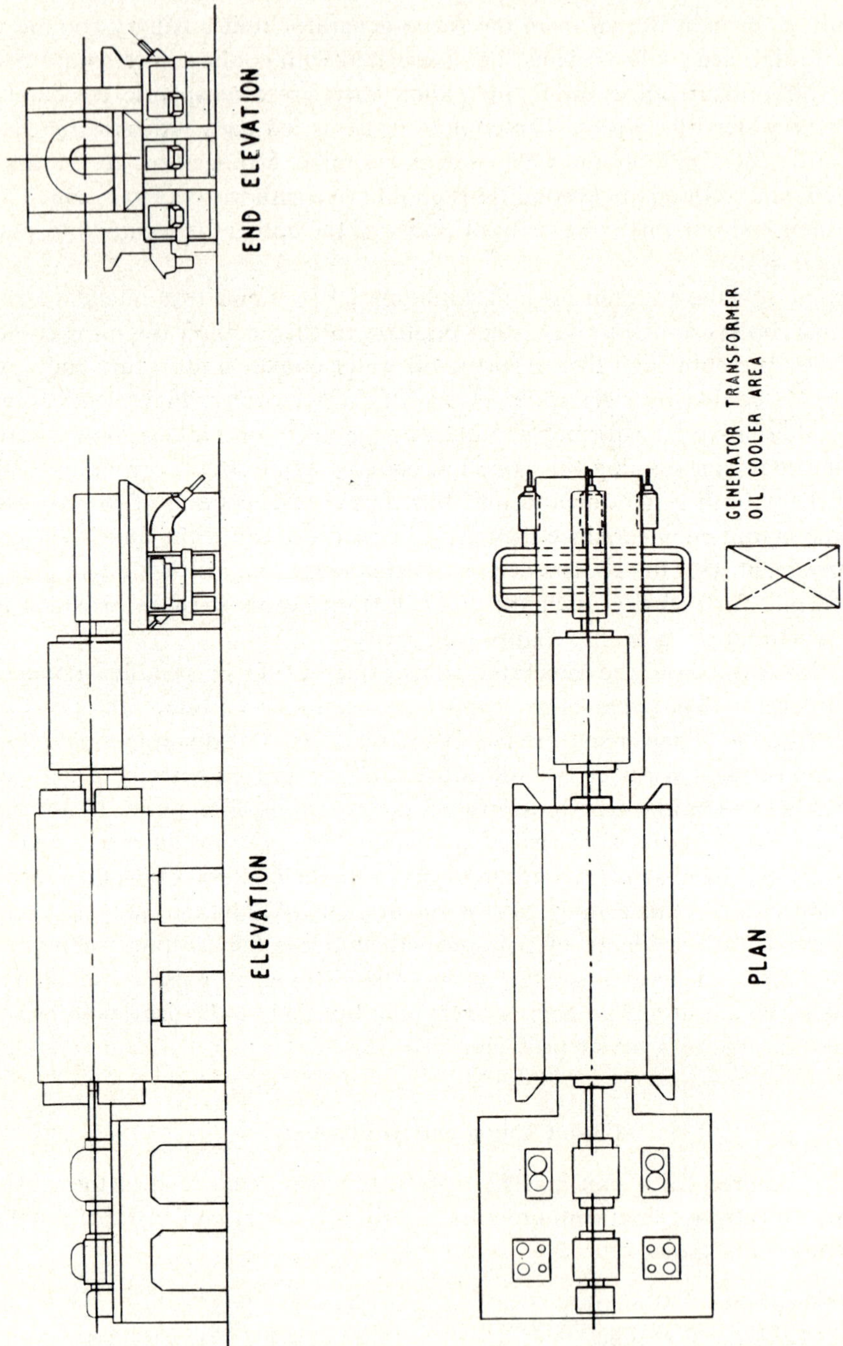

FIG. 4.3.3E. Arrangement of generator transformer in turbine block.

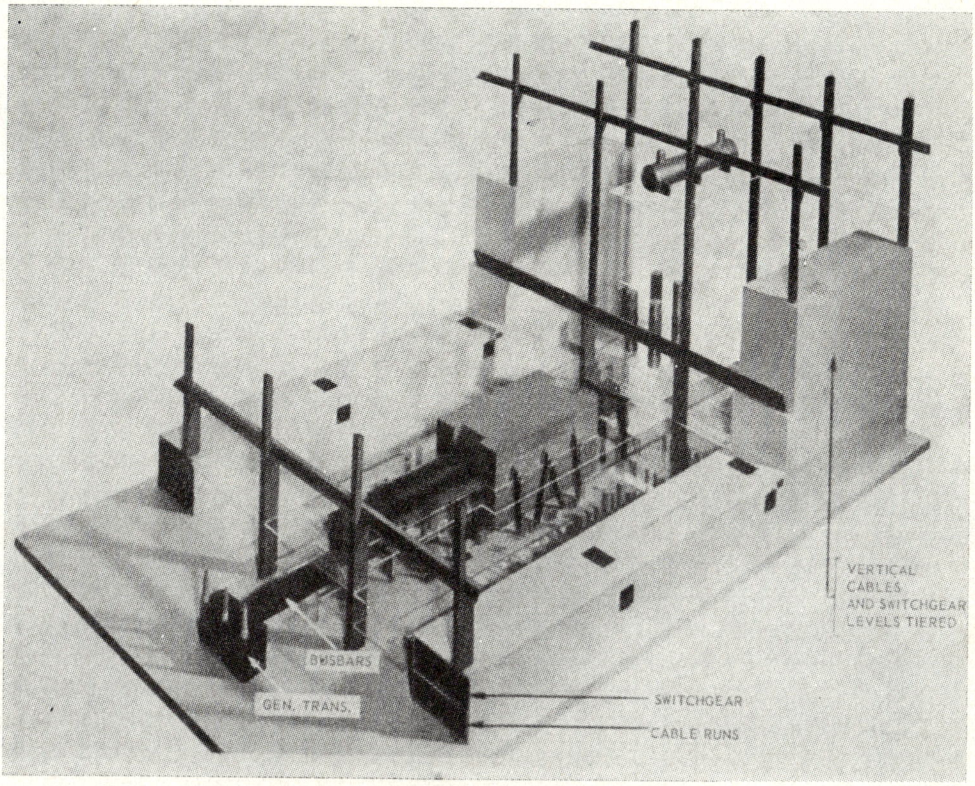

FIG. 4.3.4A. Model of Littlebrook 'D' (500 MW units) showing arrangement for cabling on spinal cord principle.

The arrangements for cabling proposed for Littlebrook 'D' (500 MW units) express the spinal cord principle quite well as shown in Figs. 4.3.4A and B which are photographs of the $\frac{1}{8}$ in to 1 ft scale model made to investigate preliminary layouts. Here it was decided not to have the traditional cable tunnels below basement floor level because they are liable to flooding and are always limited in cross sectional area by civil engineering and cost considerations. Instead, roomy two storey buildings run the width of the turbine house with cables below and switchgear immediately above and the roof is to be made strong enough to be a useful "lay-down" area for turbine parts during construction and maintenance overhaul.

The structure becomes vertical in the mechanical annexe to convenient levels housing boiler auxiliary boards and cables. Depending on the control room position the structure can terminate at either end with relay equipment from which telephone type cables are run to the control room. Since the control room is best situated off-centre on the electrical annexe side, the relay equipment should be placed at this end and cross-cabling between units run in the electrical annexe.

This arrangement reduces to a minimum the lengths of connections from the main cable runs to switchgear and plant, provided that the switchgear units are logically placed in relation to the plant they serve.

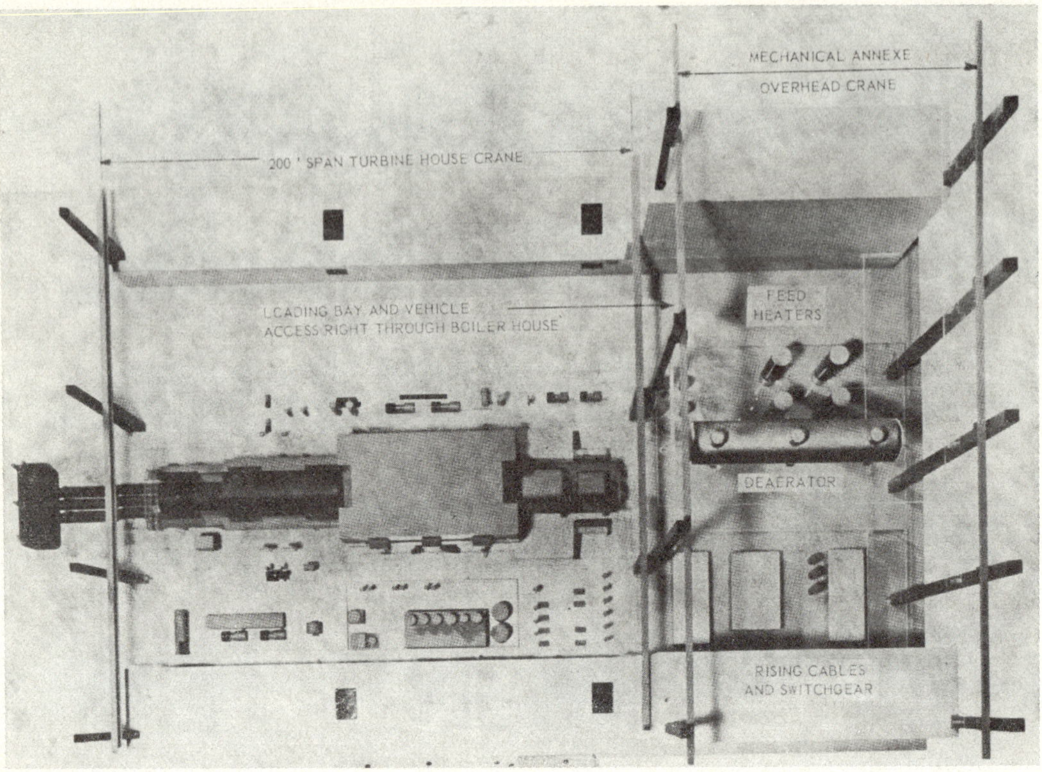

FIG. 4.3.4B. Model of Littlebrook 'D' (500 MW units) showing arrangement for cabling on spinal cord principle.

All cabling must be protected from mechanical damage, initially by the sheathing selected and specified but also by the method of installation. It should not be possible to walk on a cable, run a truck over it or for it to be hit by a load swinging from a crane.

Not so obvious is the need for cleanliness. Most cables now being installed do not catch fire easily under fault conditions, since the insulants used do not support combustion of themselves and cable protection from overload and consequent heating is taken care of in modern installations. The main danger comes from rubbish, such as timber and oily rags catching fire in the vicinity. Only good housekeeping prevents this, and the layout engineer has a prime responsibility to ensure that there is easy access for cleaning.

4.4. STEAM RAISING PLANT

4.4.1. Boiler Layout

The boiler is the largest single item of plant in the power station and in conjunction with its auxiliary items, e.g. mills, fans, ducting, fuel- and ash-handling equipment, presents an access problem to the layout and maintenance engineers.

The placing of the main mass relative to the turbine has been discussed in general terms; and it is necessary to look now at the local activity necessary on both construction and operation to establish economic communication and control.

A modern boiler consists of approximately 10,000 tons of steel which arrives at the site over a period of about two years. Consequently, the initial siting of the boiler must provide for secure lines of access for heavy lifting and withdrawal from the area. A modern feature of large boiler construction is the use of prefabricated membrane furnace walls and typical sections are shown in Figs. 4.4.1A and B. These are delivered in large sections up to 60 ft long and 7 ft wide and lifted directly from transport into position and welded-in as shown in Fig. 4.4.1C. This technique is likely to be of growing importance and eventually there may be an economic case for the provision of a permanent overhead gantry crane above boiler roof level for all heavy lifting and maintenance work afterwards.

Some 500 people will be continuously at work on the construction of one boiler for at least two years and their productivity can be severely limited by their environment. Consequently, the initial siting should take their welfare into account. The following list gives a guide to their needs.

(a) *Simple amenities*

(1) Lockers.
(2) Messing.
(3) Lavatories.

(b) *Fast communication*

(1) Central supervision. } Assembly points for instructions
(2) Local supervision. }
(3) Short journey times between amenities, supervision, material stocks and work points.

Consideration of these needs can affect the final plant layout and if permanent station facilities are used by contractors staff, decisions are required regarding amenity and final finish and fittings. Some thought will be given to the early construction of the whole boiler house with passenger lifts and central amenities, before maximum effort starts on boiler construction, although convenient lines of communication for movement of heavy plant would have to be preserved.

4.4.2. **Operation**

Many operational activities on a boiler will concentrate on the burner area, including on-load maintenance of burners and attention must be given to fire prevention and means of escape for personnel. Fuel oil at high pressure and temperature, often 350 lb/in^2 and 205° C, and hot P.F. supply lines are brought into this area, and it is here that most automatic control functions for air/fuel ratio and flame monitors are installed.

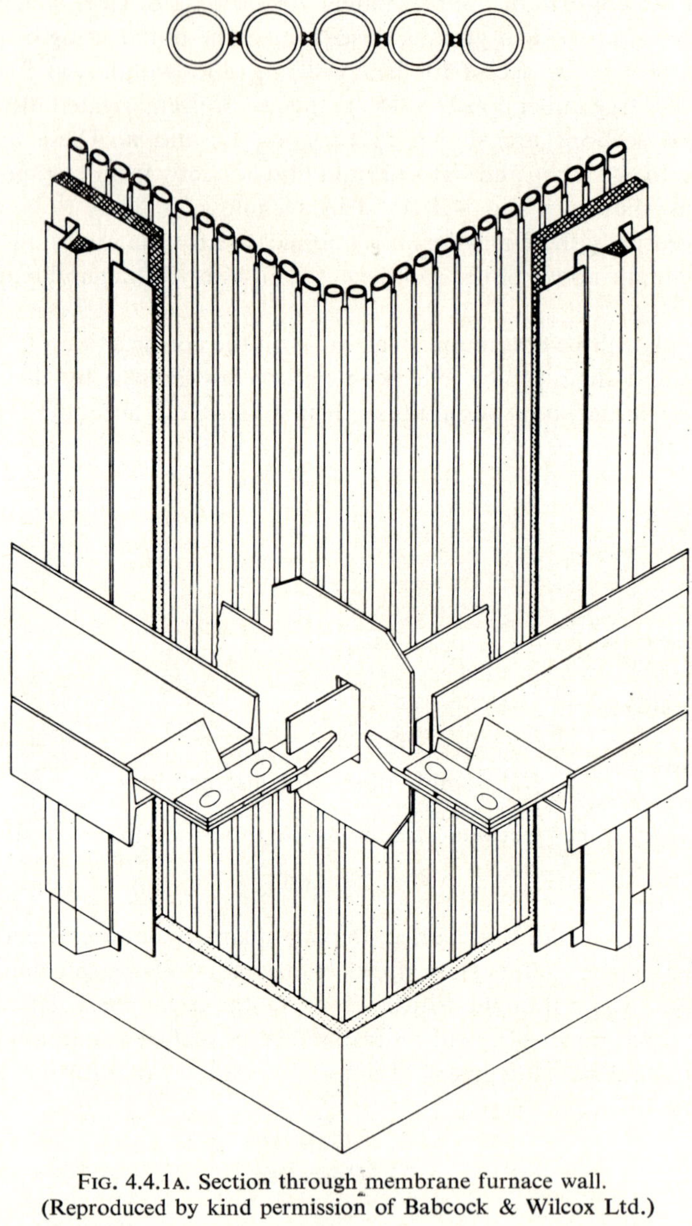

FIG. 4.4.1A. Section through membrane furnace wall.
(Reproduced by kind permission of Babcock & Wilcox Ltd.)

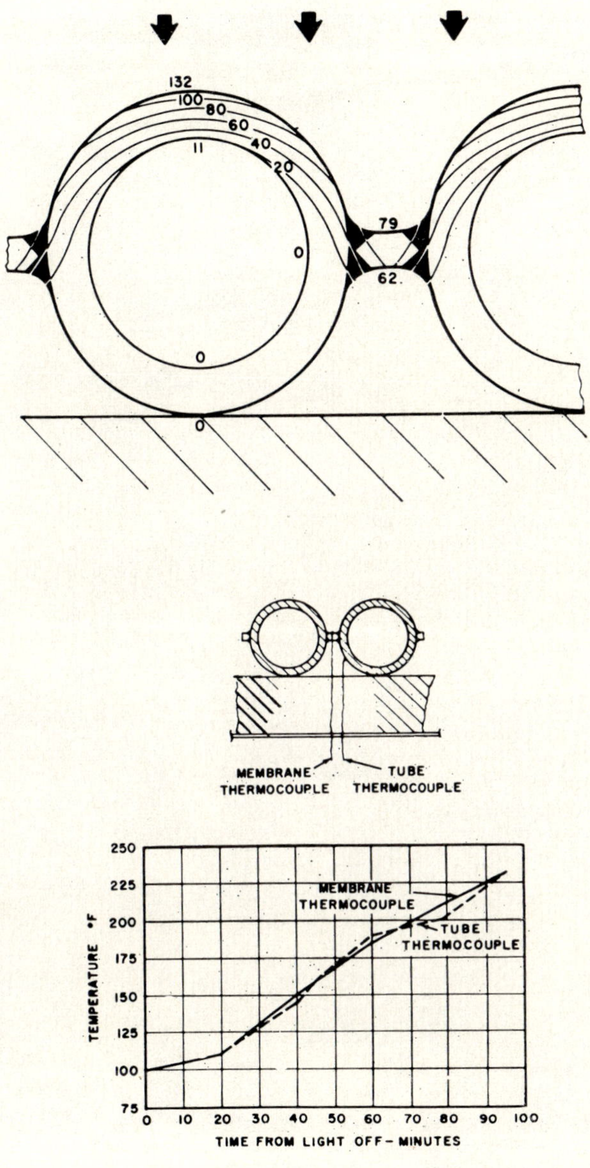

FIG. 4.4.1B. Temperature gradient through furnace wall.
(Reproduced by kind permission of Babcock & Wilcox Ltd.)

FIG. 4.4.1C. Completed panel of membrane wall.
(Reproduced by kind permission of Babcock & Wilcox Ltd.)

There should be ample platform area at each burner, wide enough to permit easy withdrawal of oil burner tubes and for fitters to work comfortably. The layout and drainage provisions on the working platforms should be designed so that it is a simple matter to keep the area clean and especially free from accumulated fuel oil drippings and P.F. coal dust, oily rags and other rubbish. Open grille flooring is not satisfactory as cleaning will discharge dirt to the floor below. Artificial lighting is always very important in these areas, since a great deal of work is carried out at night and the lighting level must be up to workshop standard.

Other places of activity requiring permanent access are:

All inspection windows.
All main access doors.
Sootblowers.
All valves, actuators, dampers, instrument sensing heads, transmitters and electrical equipment.

The spread of this equipment over the boiler demands a comprehensive series of galleries right round the boiler casing at suitable levels, all the way from firing floor to drum level, and the importance of designing these correctly and safely cannot be overemphasised.

4.4.3. **Boiler Relationship with Other Plant**

All modern boilers are top slung from an independent steel structure. From the beams a series of slings take up the loads represented by the drum, furnace tubes and casing, superheaters and reheaters (see Fig. 4.4.3A); the economiser is usually remotely slung below the primary superheater casing.

When the boiler is fired to full temperature conditions the whole mass expands downwards and horizontally in both directions. The downward expansion is greatest and can be up to 15 in and all connected equipment statically mounted, i.e. ducting, air heaters, precipitators, etc., must have expansion or articulated joints to provide complete freedom of movement for the boiler.

For the Board's latest 500 MW and larger units where coal is the main fuel, there is general agreement that the firing face and coal mills and bunkers should be remote from the turbine. The strongest technical reason for this is that the main steam and reheat pipe runs are considerably shorter (by the width of the mill and bunker bay). A further advantage is that coaling is confined to the back of the station. These factors do not necessarily apply to an oil-fired boiler, but by reference to the original simple flow process it is clear that the remote firing face is appropriate.

One boiler maker, however, employs corner firing, i.e. the burners are placed vertically above each other in the four corners of the furnace and a better arrangement of mills is between boilers. This is also likely to apply to the projected 660 MW boilers which will almost certainly employ front and back firing of the furnace and there will be two separate sets of burner galleries, one set being between the furnace and the convection pass casing to the economiser.

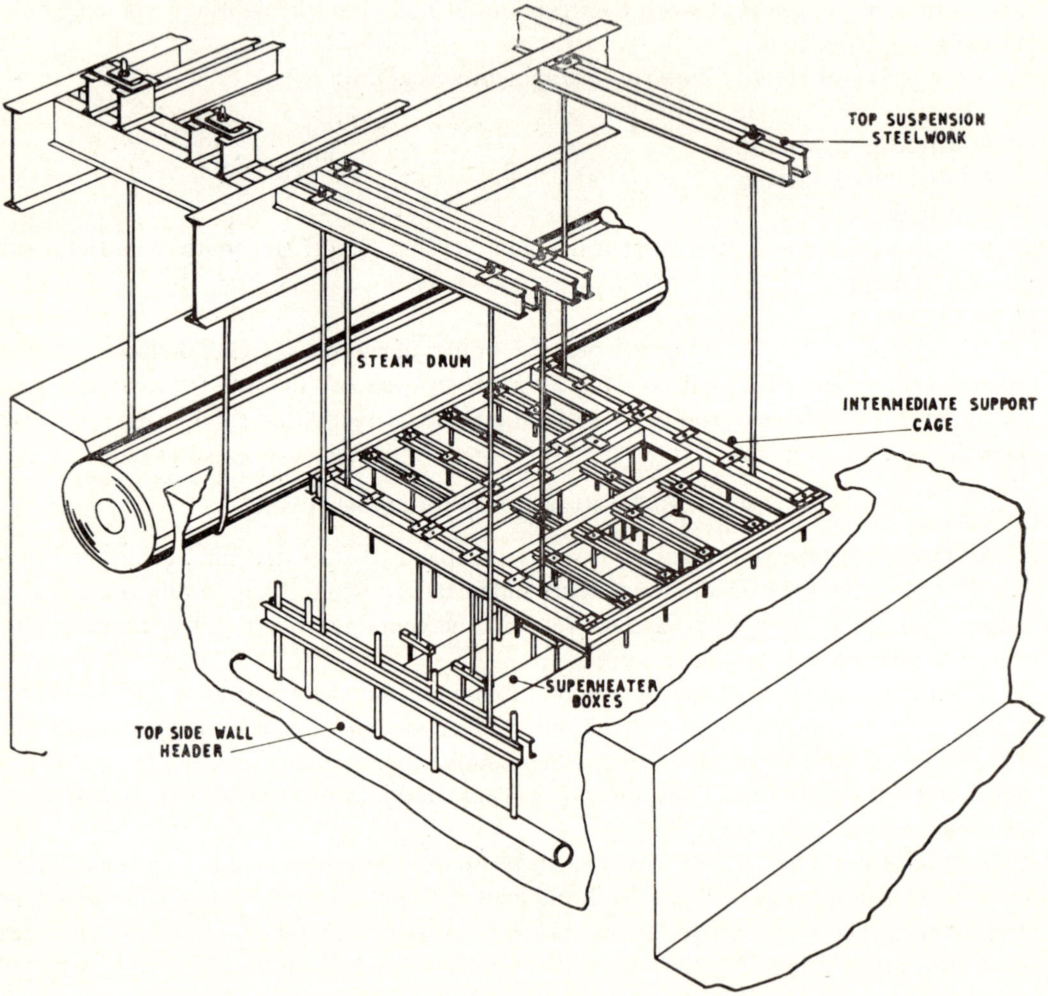

FIG. 4.4.3A. Top slung boiler unit with sub support frame.

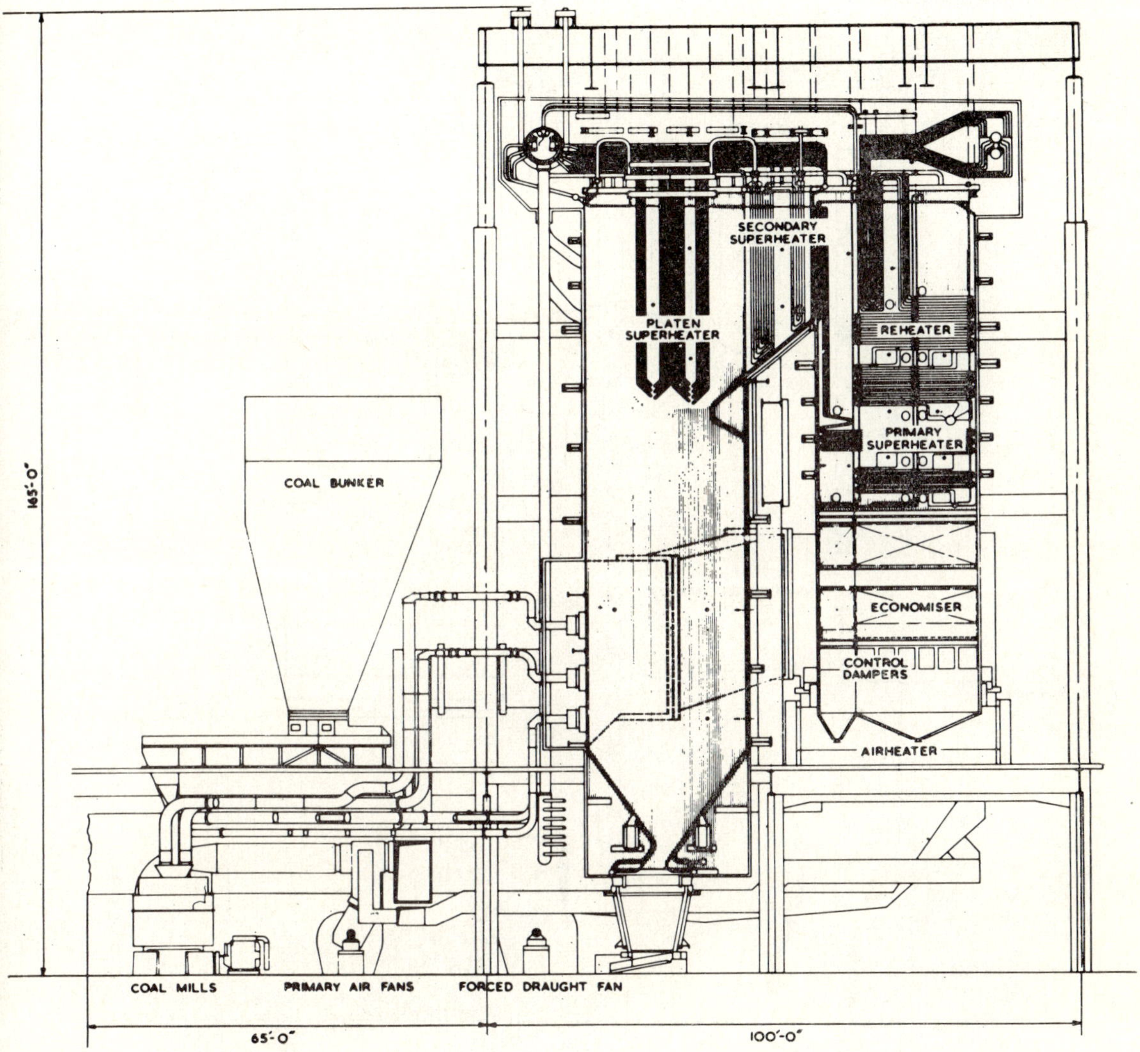

FIG. 4.4.3B. 300 MW boiler—West Thurrock. (Reproduced by kind permission of Babcock & Wilcox Ltd.)

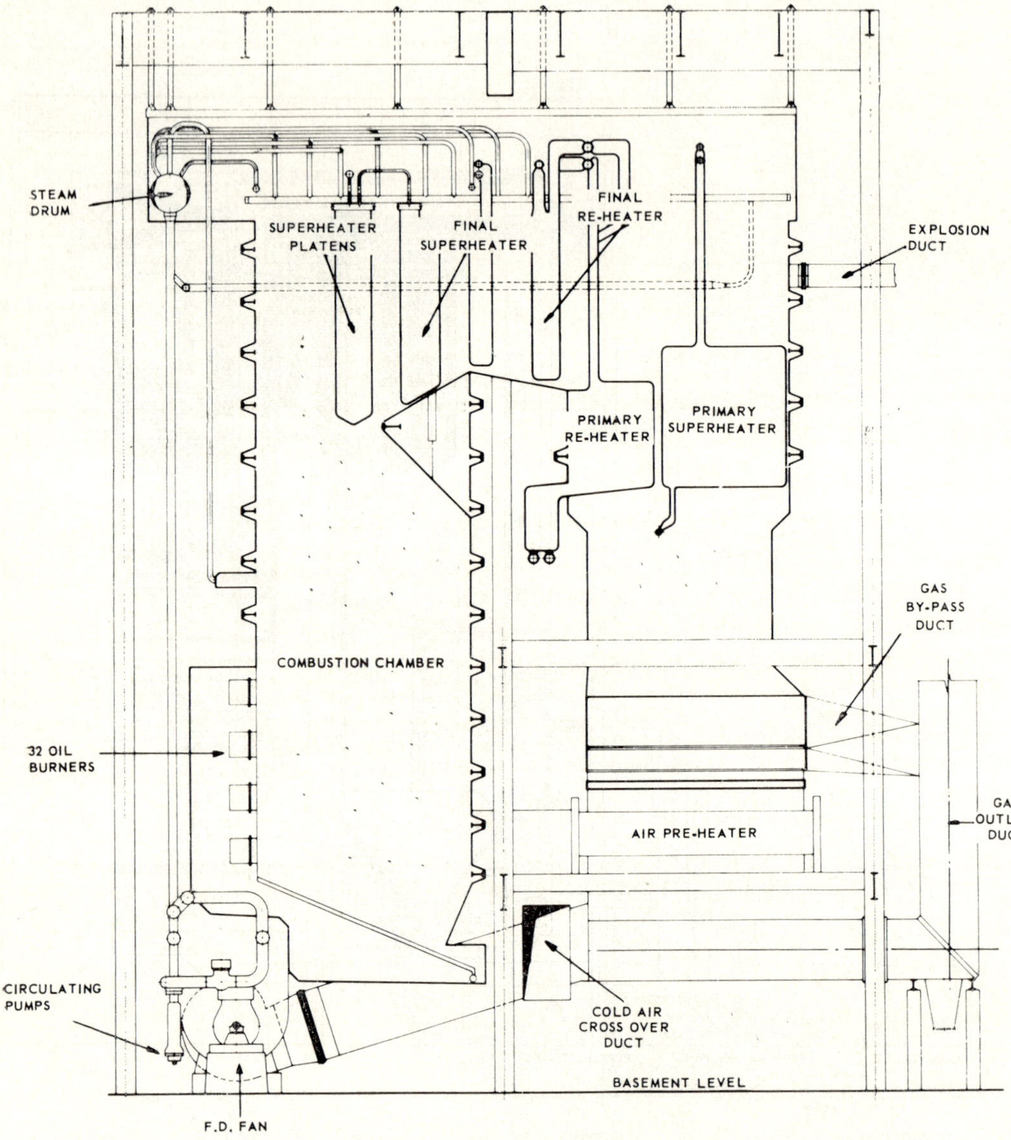

FIG. 4.4.3C. 500 MW boiler—Fawley.
(Reproduced by kind permission of John Thompson Ltd. - Clarke Chapman Ltd.)

Figure 4.4.3B shows a typical cross-sectional elevation through a 300 MW unit at West Thurrock. The boilers are of the open air type and the coal bunkers are completely separate in their own structure.

Figure 4.4.3C is a cross-sectional elevation of a 500 MW unit at Fawley power station which is oil-fired.

Determination of the distance between boilers is, as a minimum, dictated by boiler auxiliary plant. It is particularly affected by long lance sootblowers which are about 40 ft long and occupy this width when withdrawn from the furnace. Similarly air heaters, fans and ducting invariably come outside the net furnace width. Provision should also be made for a lift and hoisting well between each pair of boilers.

Obviously boiler centres increase if the milling plant is placed between boilers, but this is interdependent with arrangement of the set as previously discussed.

4.4.4. Boiler Fittings and Mountings

These are mostly externally mounted items many of which form terminal connections for other plant. The most important are:

Main steam stop valves and boiler integral pipework.
Reheater inlet and outlet connections.
Feed water inlet connections to the economiser.
Blowdown and drain valves.
Sootblowers plus their steam and drain piping.
P.F. and oil fuel connections.
Steam safety valves on the drum, superheater and reheater together with exhaust pipework and silencers.
Steam and water sampling equipment.
Chemical injection equipment.
Air release valves.
Drum water level gauges.
Remote water level indication and drum level television.
High and low water alarms.
Auto-control sensing devices, transmitters and cabling.

It is the layout engineer's responsibility to see that each item is so positioned that there is ample maintenance access and that the safety of each connection is assured.

4.4.5. Boiler Lagging

In general the boiler casing, particularly when the boiler is of skin case construction, requires a thickness of about 4 in of lagging to bring the outer surface temperature down to a maximum of 49°C and reduce radiation losses.

It is also necessary to surface this lagging with a cleading not only to avoid damage to the rather friable lagging material, but also to present a reasonable finished surface on the boiler. A number of cleading systems are available and an isometric sketch of one of these

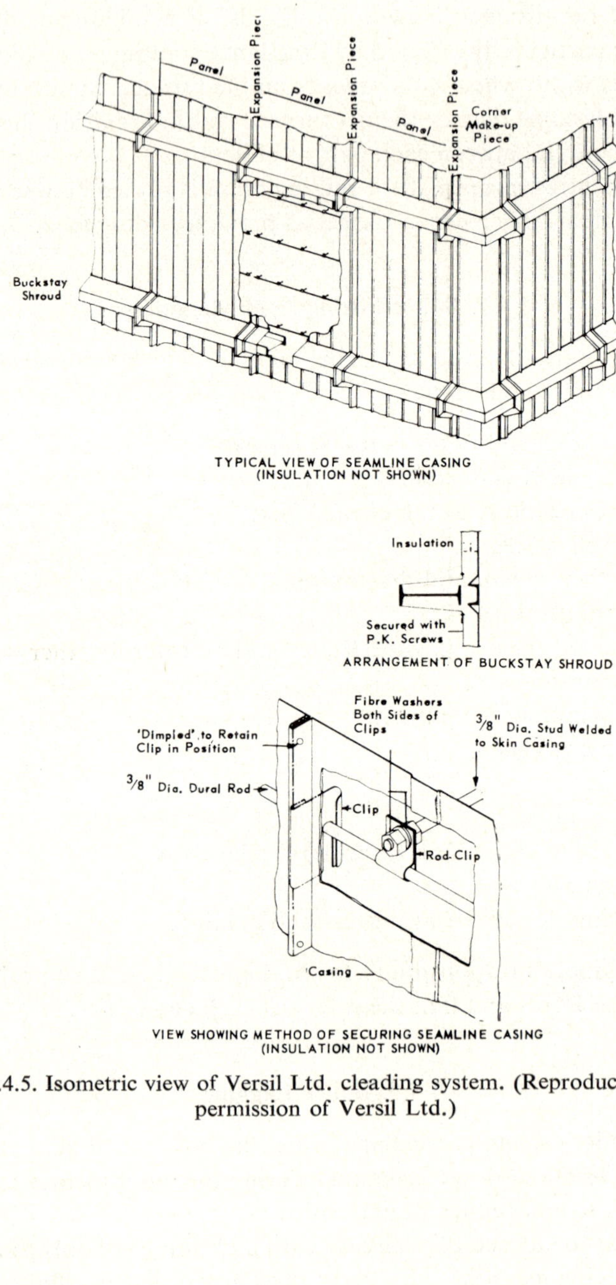

FIG. 4.4.5. Isometric view of Versil Ltd. cleading system. (Reproduced by kind permission of Versil Ltd.)

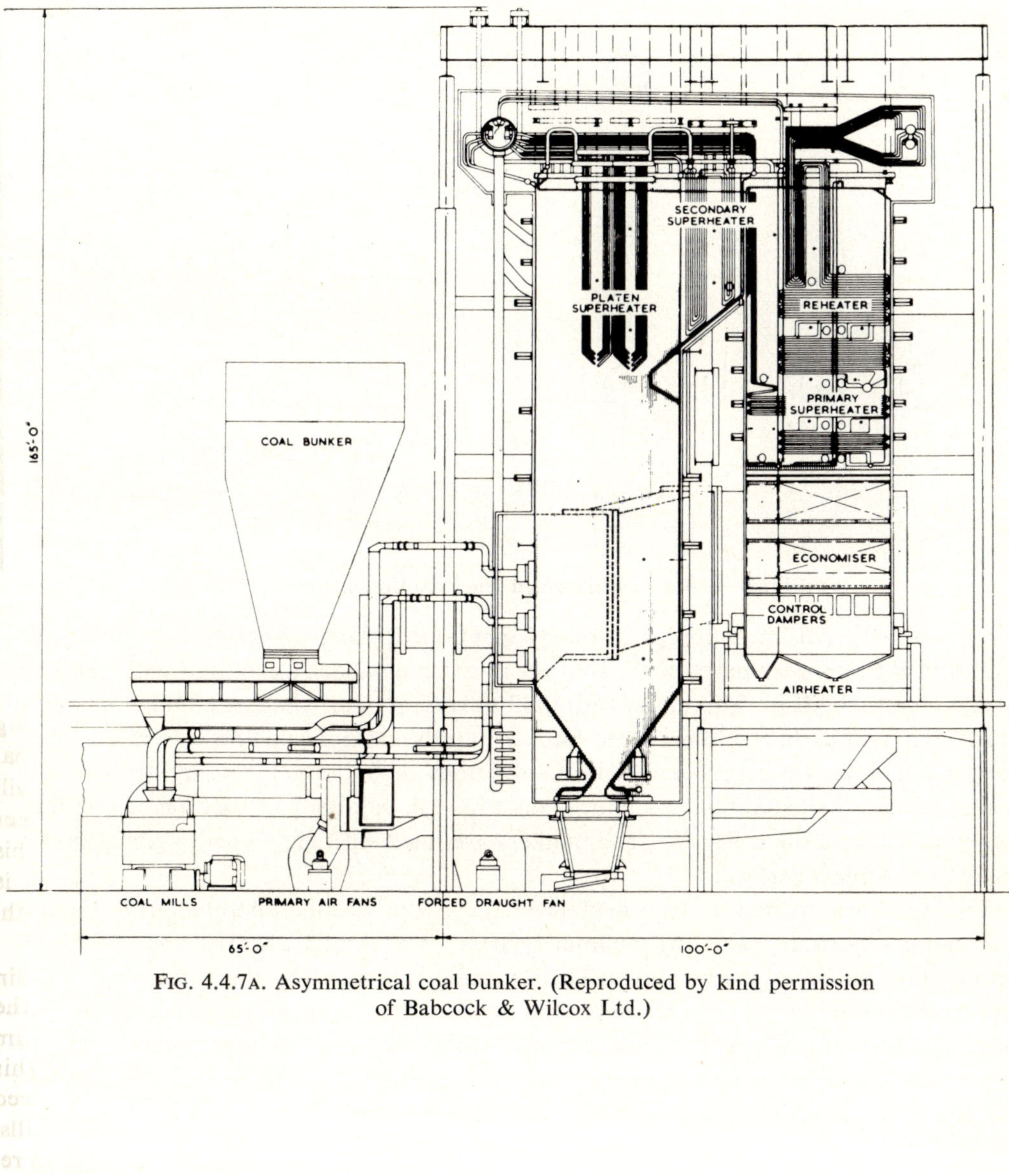

FIG. 4.4.7A. Asymmetrical coal bunker. (Reproduced by kind permission of Babcock & Wilcox Ltd.)

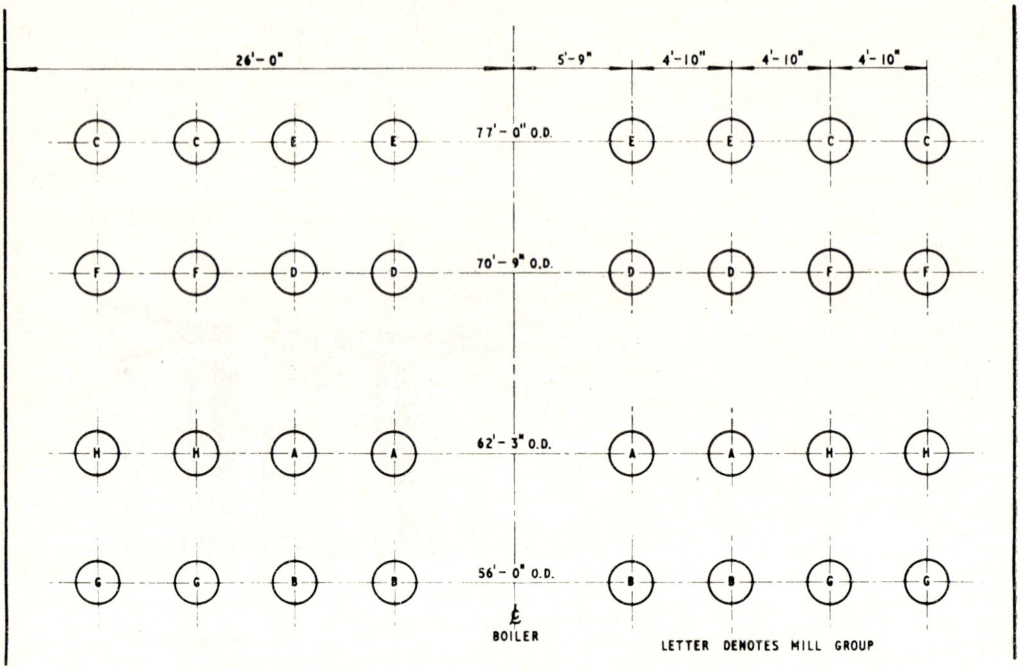

FIG. 4.4.7B. P.F. burner layout.

4.4.7. **Pulverised Fuel Equipment**

Based on the datum philosophy already mentioned, the bunkers, feeders, pulverising mills and P.F. delivery pipes to the burners must be compactly arranged with good coal flow principles built in. This starts with the bunkers; a badly designed bunker shape will cause coal to stick in the bunker and bunker mouth and hold up flow. The latest bunker designs are asymmetric, i.e. one side is at a different angle to the other, and typical of this design are the 300 MW bunkers shown on Fig. 4.4.7A where in each plane one side is sloped at 70° and the other 75°. The bunkers are lined with glass tiles to give a smooth, almost frictionless surface.

The mills are erected at basement level on a foundation separated from the main foundations by some insulating medium such as cork, which absorbs and isolates the heavy vibration from the equipment. Their spacing is dictated by this insulating medium and maintenance access requirements. The mills usually lie in a straight line since this suits coal bunker and coal conveyor design, but this requirement, together with the need to keep the outer mills as close to the burners as possible leads to fewer but larger mills. On 300 MW units there are five mills, four of which may carry full boiler load fuel requirements.

The layout of the pulverised fuel piping from the mills is complex. Consider a 200 MW case, there are 32 burners and each mill has four pulverised fuel pipes. The burners are in four horizontal rows of eight. Coal is fed from each of the eight mills by two $16\frac{1}{4}$ in fuel pipes, each pipe bifurcating into two 11 in bore pipes to supply four burners. However, in

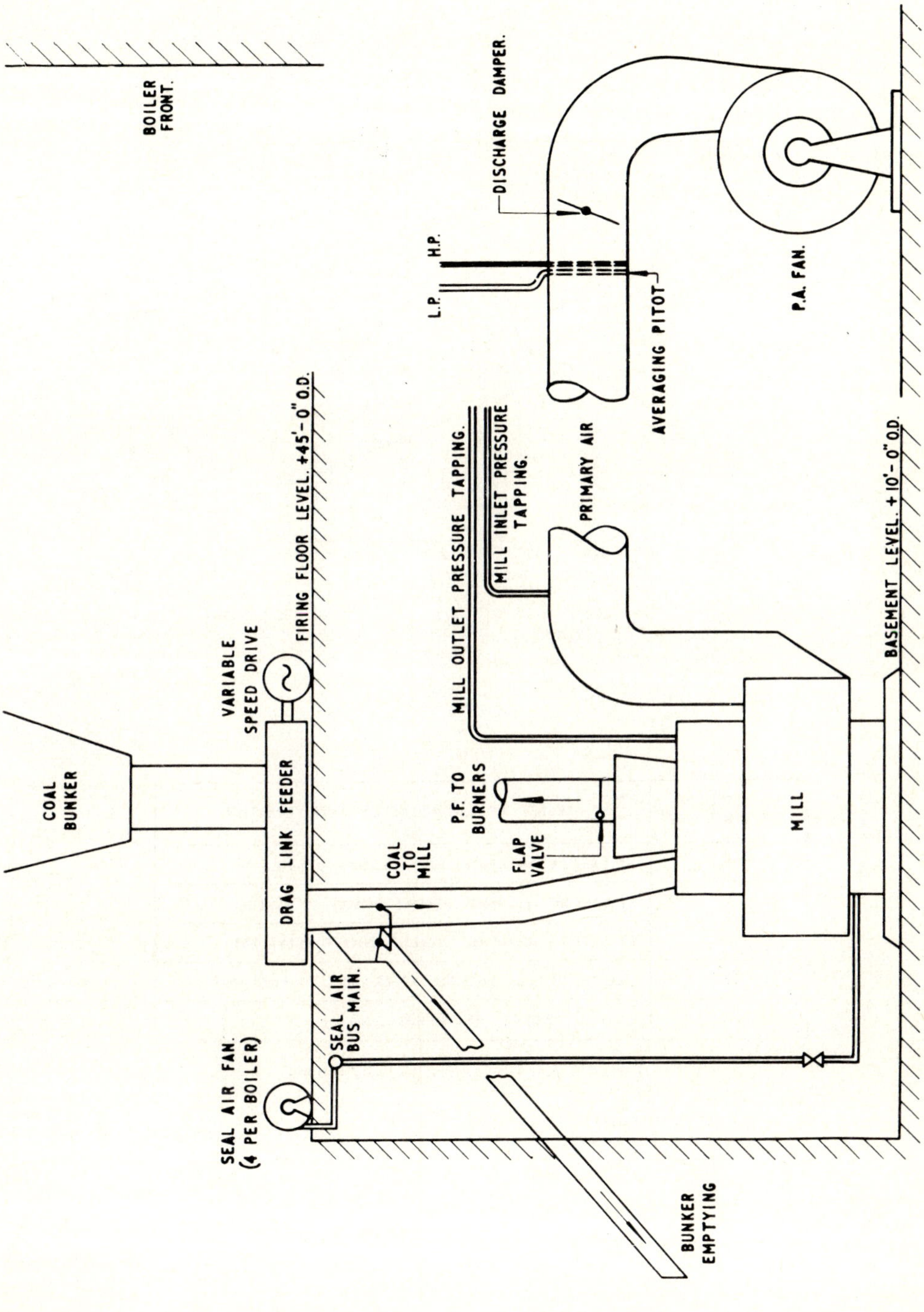

FIG. 4.4.7C. Mill layout.

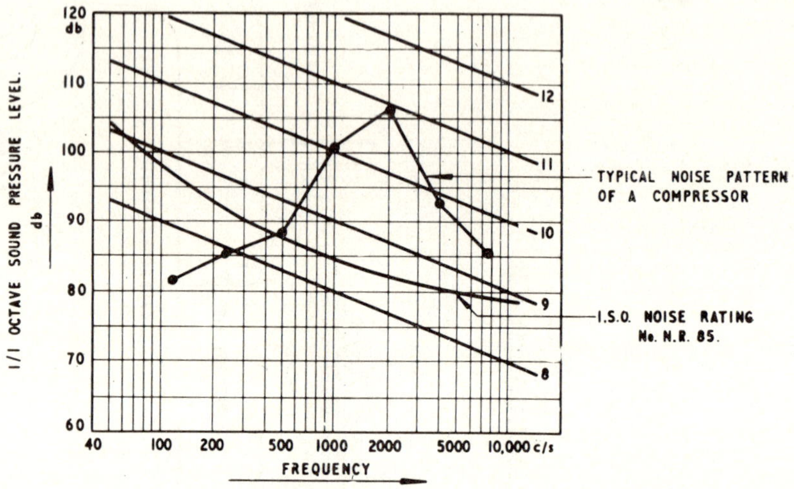

BY EXCEEDING THE 'LIMIT LINE.'	DEGREE OF DAMAGE AFTER GIVEN DURATION.
8	BY LASTING INFLUENCE SOME HEARING DAMAGE.
9	BY LASTING INFLUENCE HEARING DAMAGE IN NUMEROUS CASES.
10	BY LASTING INFLUENCE HEALTH DAMAGE PREVALENT.
11	HEALTH DAMAGE POSSIBLE AFTER ONLY A FEW HOURS.
12	HEALTH DAMAGE UNAVOIDABLE.

FIG. 4.4.7D. I.S.O. noise rating curve.

order to get flexibility of burner/mill operation and to maintain a good flame shape under all conditions, the destination burner for each pipe is as shown in Fig. 4.4.7B. In this figure each burner is lettered according to the mill feeding it, the mills being A, B, C, D, E, F, G, H from south to north. This will indicate the crossing and interweaving required for the pulverised fuel piping which must be properly routed for minimum pressure drop and supported on spring supports to cater for expansion.

Figure 4.4.7C shows diagrammatically a typical pressure mill layout with primary air fan and seal air fan. Note the emergency bunker emptying chute which is fitted to facilitate emptying of a bunker in the event of a bunker fire or for maintenance purposes.

Pulverising mills require frequent maintenance involving heavy lifts and the layout must provide for this. Normally each mill should be provided with its own heavy duty runway beam, and it is now a specification requirement that the mill supplier must also provide a suitable electrically operated hoist and demonstrate a suitable mill maintenance technique.

Each mill has a reject door arrangement which will eject on to the floor, pyrites or stones, etc., rejected by the mill classifier, and provision is necessary to gather these up for dumping possibly with a mobile shovel which will require manoeuvring space.

Milling bays are often very noisy and it is becoming essential to deal with this problem. This is a specialist's job and there are several methods of noise suppression. In a milling bay the most likely process would be to line the walls and ceiling with a sound absorbent material such as slag wool or spray limpet material and to face-off with perforated metal or plaster board. Local portable screen walls can also be used to screen off directional sound through doorways and other openings.

The noise spectrum should be checked with a sound level meter. If the I.S.O. noise rating curve No. 85 is exceeded and staff are in the area for a full shift, then more silencing measures should be adopted.

4.4.8. **Oil-firing Equipment**

An oil-fired boiler is attractive from a space utilisation point of view since mills, bunkers, etc., are not required. The oil store is a tank farm containing two or more large oil tanks often 180 ft diameter $\times$ 60 ft high. A simple flow process must be devised and piping routes from ship to shore and from stock to burners must be economically planned for good routing, lack of complication, and low-pressure drop.

Heavy residual oil is normally used as the main fuel for an oil-fired power station. This oil has a high viscosity at ordinary ambient temperatures and it is heated to about 80°C to facilitate pumping and flow, and reduce pumping power, when handling into stock and out to the pumping and heating units.

The oil often comes direct from the oil refinery (Fawley, Pembroke, Isle of Grain are examples) by pipeline and it is usual to seek a guaranteed temperature of fuel delivered. From the stock tanks, transfer pumps pressurise a ring main system which circulates the oil constantly back to stock through a pressure-sustaining valve (or valves), with each boiler taking oil from the ring main as required and at the pressure maintained by the ring main sustaining valve (normally 25 to 50 lb/in^2). Since there will be some heat loss from the fuel the piping system is trace heated and the stock tanks provided with heating coils.

Heating is usually by means of mineral insulated metal sheathed cable but where auxiliary boilers are provided consideration should be given to steam or hot water trace heating. Design formula are given in Volume 3, Chapter 5.

At the boiler, pumping and heating units are required. These take oil from the ring main at 25 to 50 lb/in^2 at 80°C and heat and pressurise the oil to the levels required by the burner design. This varies depending upon the manner in which the oil is atomised, and can involve pumping the oil at high pressure (350–500 lb/in^2) in which case a high oil temperature is necessary (100°C). Alternatively burners using air or steam blast atomising, work at lower oil pressure although it is usual to keep high temperatures.

The oil passes from the pumping and heating units to another ring main around the boiler firing face, each burner tapping off its supply through a control valve. This ring main spills back to the pumping and heating units.

In layout, the needs of operation and maintenance must be considered, in particular:

(a) Facilities to keep all firing equipment and its environment clean.
(b) To permit expansion.
(c) Provide clearances and access to valves.

Where P.F. burners are employed a supply of oil for lighting up is required although town gas or butane gas are now being considered for this purpose. The requirements for oil lighting up are similar to straight oil-firing, but on a smaller scale since the lighting up burners are sized to carry only 10% of the full boiler load.

4.4.9. **Ash- and Dust-handling Plant**

A modern coal-fired 2000 MW station will make about 250 tons of ash and dust per hour and this poses a tremendous disposal problem. The plant required is dealt with in detail in Volume 2 of this course, but there are many practical considerations for the layout engineer. It is best to consider the requirements of ash handling (by this we mean furnace bottom ash and clinker, etc.) separately from P.F. dust handling since two separate systems are involved. Of the total "make" about 25% will be furnace bottom ash and the balance P.F. dust.

Furnace bottom ash is collected in a hopper usually of welded steel construction firebrick lined. The hopper is static and has a water trough around its top periphery into which a seal plate welded around the furnace bottom outlet dips to prevent air ingress into the furnace. The water in this trough evaporates and must have a float operated make-up supply from river or town water. Some ash hoppers are full of water so that the ash is quenched as soon as it drops from the furnace, but more commonly the ash falls into a dry hopper which is fitted with access doors and high pressure water jets. The collected ash is jet-sluiced from the hopper into a sluiceway and is carried by the water into a settling tank outside the power station building where it can be drained, grabbed and sold. Alternatively the ash is crushed in roll crushers and pumped in water to a disposal area.

The size and hence capacity of the sluiceway will depend upon the operational technique adopted. Until recently it has been the view that all ashing should be carried out in one shift per day. With the volume of ash make at the latest stations it appears to be more economic to ash over two or even three shifts so that ash plant volumes can be reduced.

The ashing plant required in addition to the sluiceway and settling tank are high pressure water pumps for jet sluicing, which usually pump about 3000 gal/min of water at 375 ft head.

The ash, having settled out in the settling tank, has to be grabbed out by means of a crane and placed on a drainage apron for drying. After drying it is picked up again by the grabbing crane and loaded directly into purchasers' lorries or into a concrete silo of up to 500 tons capacity from which it may be discharged into purchasers' lorries.

The cycle of operations must be carefully worked out, starting from the total ash make per ashing operation, settling time in the tank and drying out time on the apron. The storage should ensure that purchasers can be served on a continuous basis during the day shift period.

On a 2000 MW station it is usual to provide two drying aprons (one on each side of the settling tank) for alternate loading, draining and pick up cycles, and two travelling overhead cranes each with a grab capacity of 10 tons are required for handling purposes.

Where ash is made surplus to sales, provision is made to crush the ash in roll crushers so that it can readily be pumped in water to a disposal area which may be at any reasonable distance from the station. The design power of the ash disposal pumps will consequently depend on the distances involved.

Generally the sluice pumps, roll crushers and ash disposal pumps are housed in a chamber below ground level and adjacent to the settling tank. These chambers can be very liable to flooding and should be provided with sump pumps having capability equal to or better than the flow capacity of the largest pipe in the chamber which could fail, for example, in freezing winter conditions.

P.F. dust handling plant is associated with the ash plant but must be considered as a separate layout problem. The dust is handled dry throughout the plant by pneumatic "air slides". These are described in detail in Volume 2, Chapter 5.

The following places require permanent dust removing connections:

(a) The boiler economiser hoppers.
(b) All low trap points on boiler gas ducting.
(c) All mechanical grit arrestor and precipitator hoppers.
(d) All points of gas ducting where interconnection is made into a common flue, and where an isolating damper is kept closed during operation.
(e) Common crossover ducts, particularly at the entrance to a precipitator where facilities are provided to isolate one channel of the precipitator, the load being carried on the remainder.

These points must be established at the design stage and facilities provided for the operators to remove the accumulation.

The pneumatic air slides connect the dust removal points on hoppers and ducts with specially designed dust pumps (Fuller Kinyon—see Volume 2, Chapter 5) which are provided on a unit basis in a central position so that the air slides can "fall" continuously at a minimum slope of 1 in 3 to the pump (see Fig. 4.4.9A) which then delivers the dust

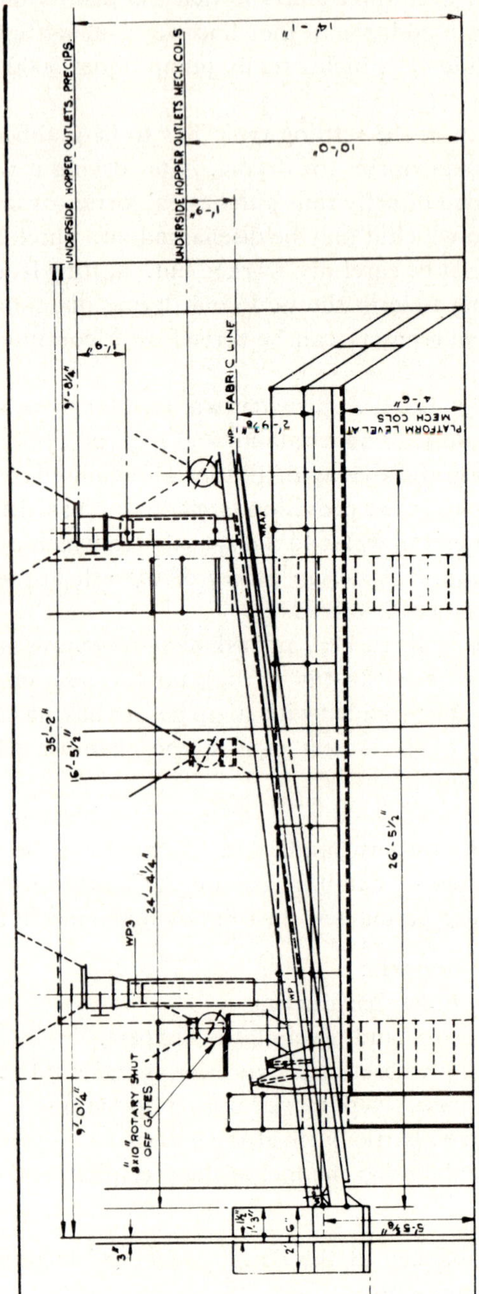

FIG. 4.4.9A. Typical arrangement of air slides.

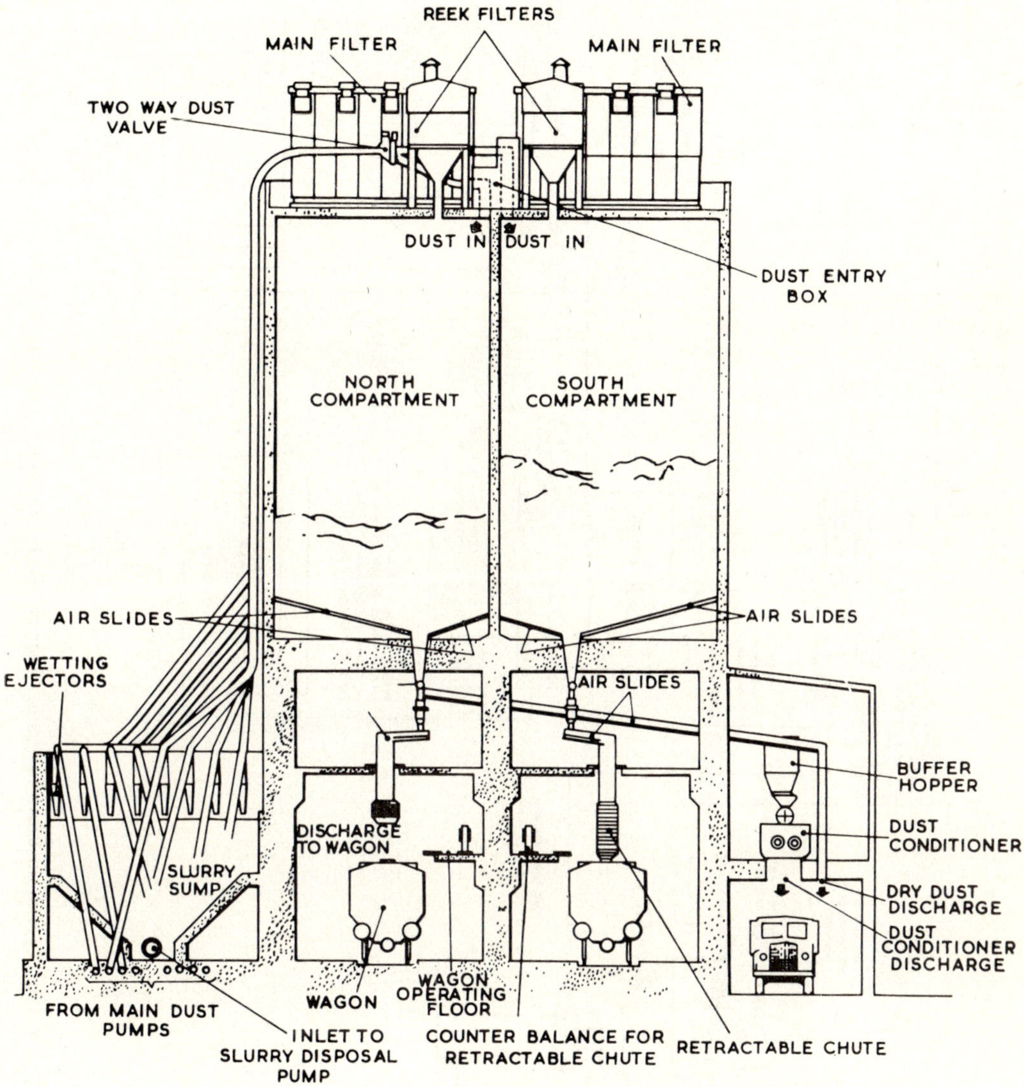

FIG. 4.4.9B. Typical dust bunker. (Reproduced by kind permission of Babcock & Wilcox Ltd.)

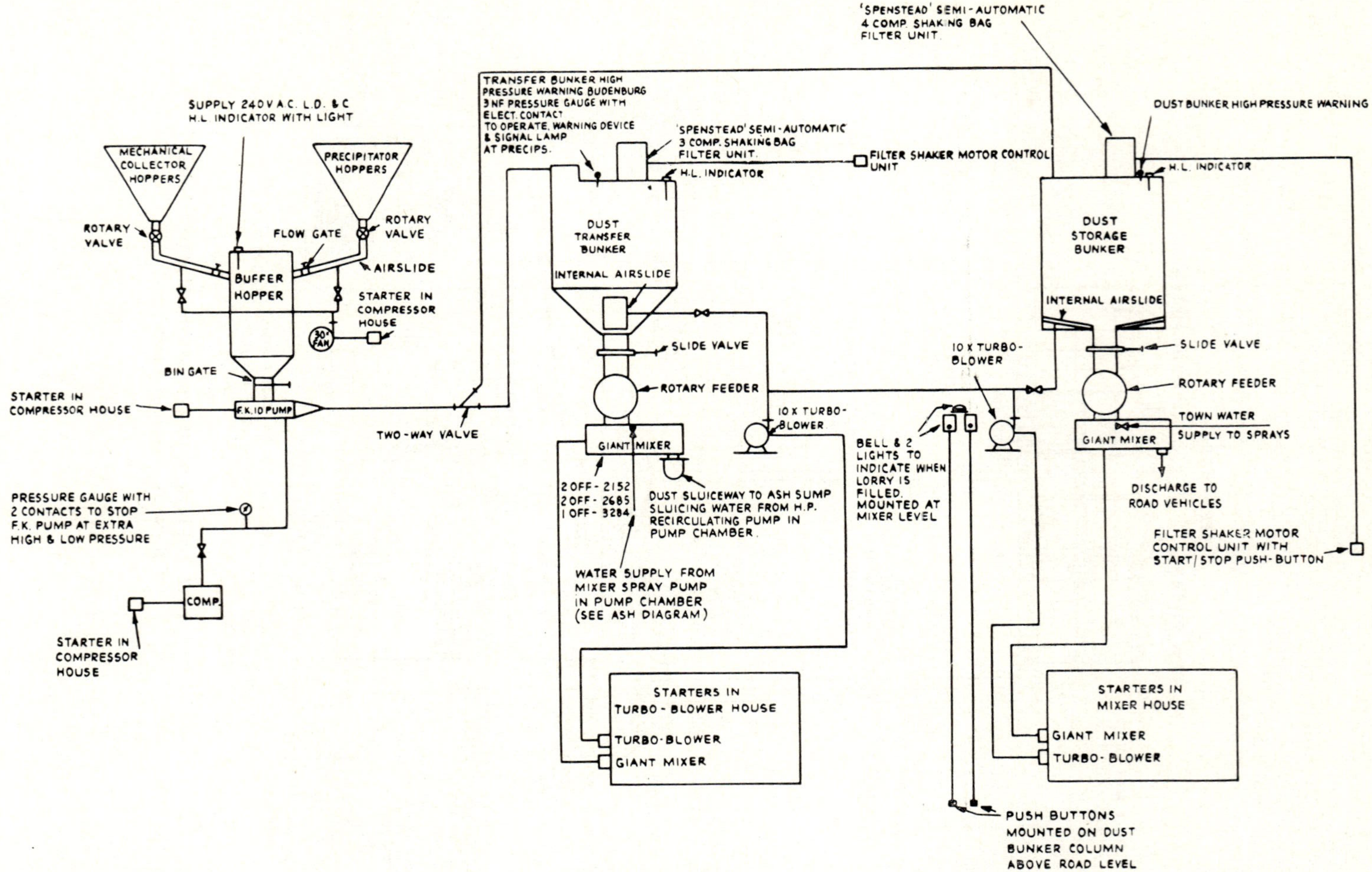

FIG. 4.4.9C. Typical dust flow diagram. (Reproduced by kind of permission Babcock & Wilcox Ltd.)

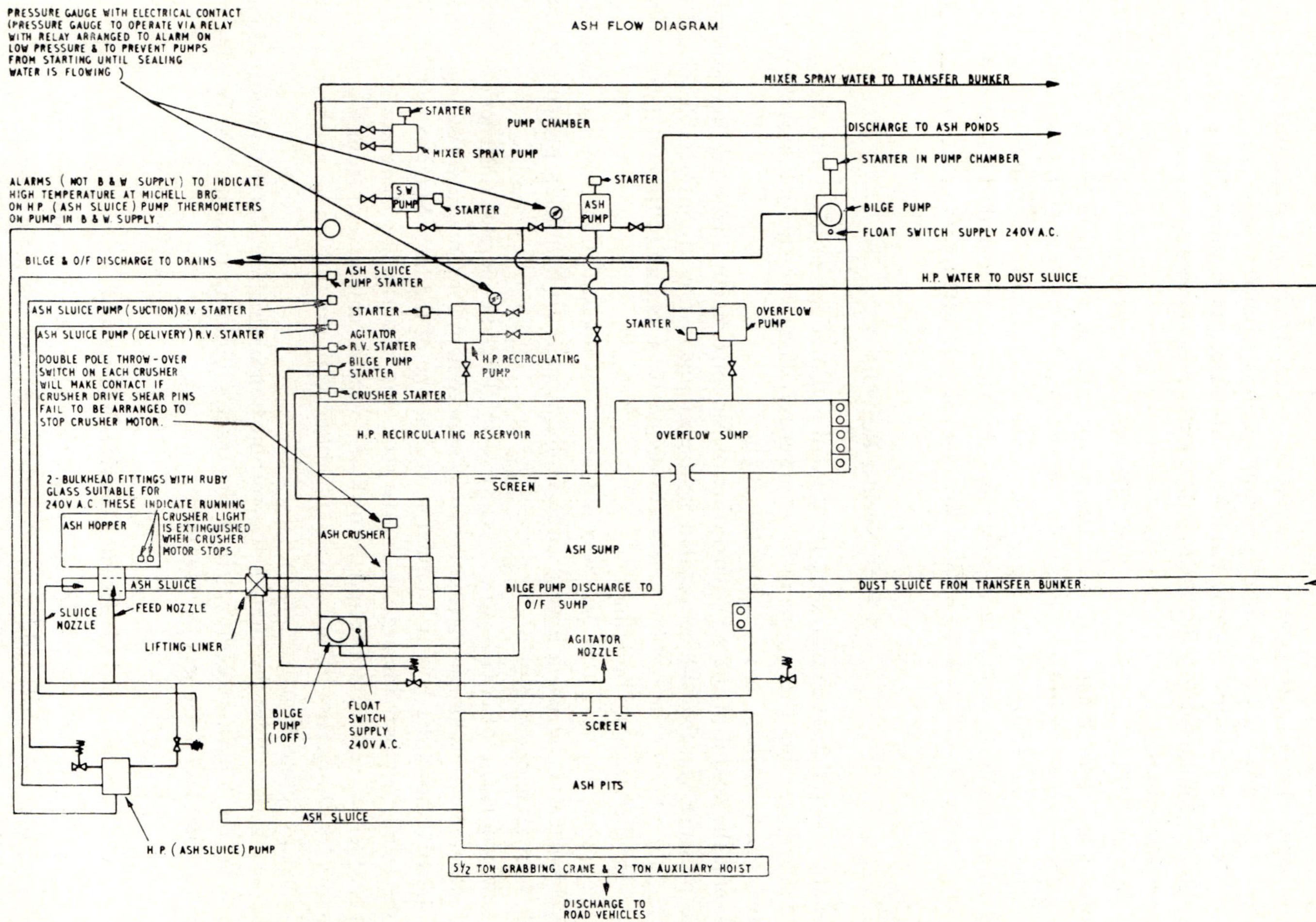

FIG. 4.4.9D. Typical ash flow diagram. (Reproduced by kind permission of Babcock & Wilcox Ltd.)

to reinforced concrete silos (or bunkers). These are usually of about 1000 tons capacity and at least two are provided, but the total storage capacity must be assessed on an economic basis related to demand for the dust by customers of any given station.

Figure 4.4.9B shows a dust bunker providing double compartment storage for dust. It has provision for conditioning by damping with water, as required by customers for building block manufacture, or dry dust loading into tanker lorries or railcars, as required by cement manufacturers.

Busy dust sales bring a continuous flow of large diesel lorries into the station and the road approaches to both ash and dust sales points must be adequate both in design for heavy loads and capacity in terms of traffic. The lorries should be able to drive under the discharge points and straight out after loading, without having to reverse or manoeuvre.

The loading operation should be as rapid and clean as possible, and for station security the whole activity should be kept outside the main station security fence.

4.4.10. Nuclear Reactors

These are very different in character from fossil fuel fired boiler plant and the services required are unique, but the argument for short energy flow paths to and from the turbo-generator remains exactly as for conventional plant.

In general the layout of the first generation of Magnox reactors ignored this principle and the pressure vessels with associated heat exchangers are a considerable distance (200 ft and more) from the turbine house. Consequently a feature of these stations is the piped services between the buildings.

The decision to build in this way was not based on nuclear hazard but to facilitate reactor construction by using a very large Goliath crane to handle in pressure vessel sections, heat exchangers, etc.

The new generation of A.G.R. are basically of reinforced concrete construction, there are no separate heat exchangers and there is no requirement for a major travelling crane. Figure 4.4.10A shows the layout adopted at Wylfa, one of the later Magnox stations, in which the trend to closer integration can be seen, but nevertheless there are long pipe bridges between the reactor house and the turbine house. As can be seen from Fig. 4.4.10B the reactors for Dungeness 'B' are closely integrated with the turbine house, becoming a boiler house in conventional terms.

This drawing also shows the long narrow turbine house which results from longitudinal arrangement, the length exceeding that of the reactor building by some 300 ft. Similar comment applies to Dungeness 'B' layout.

4.4.11. Reactor Services and Layout

The A.G.R. is now the focus of attention, but Figs. 4.4.11A and B relating to the Magnox reactors at Trawsfynydd are included to highlight the main changes.

Figure 4.4.11A is a plan of half of one reactor (the other half is a mirror image in layout terms). There are three heat exchangers (boilers) on each side of the reactor each served by an electrically driven gas circulator which circulates the CO_2 used to cool the reactor

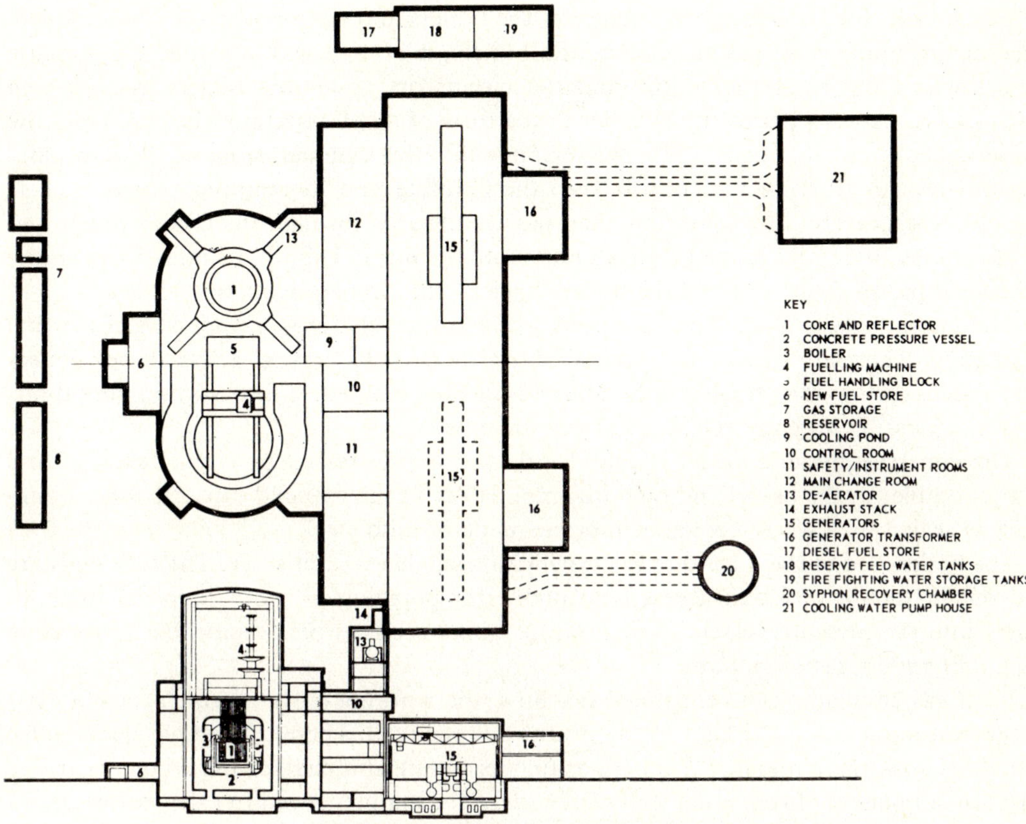

FIG. 4.4.10B. Section through Dungeness 'B' (A.G.R.).

core. Cooling fans for the biological shield are also necessary and are housed at the charge face level as shown in Fig. 4.4.11B. This drawing is a sectional elevation showing the local arrangement of one heat exchanger (repeated six times).

The main CO_2 inlet ducting is shown in full and the large, flexible expansion pieces can be seen. The vertical U-tube arrangement of this ducting is undesirable from the point of view of pressure drop on the gas side, but the whole plant is very tightly arranged and the duct arrangement provides articulation about the anchors on each bend at the top of the U.

Because of the very tight arrangement of the plant, the turbo-blower, flywheel and main bearing pedestal are below the main support ring for the heat exchanger and a hybrid crane arrangement is necessary in the circulator house. The main crane is a normal 22-ton overhead crane which has a 15-ton extension beam, lining up with a static beam above the flywheel of each circulator.

Ventilation openings and access doors are provided with V-shaped louvres, to prevent radiation "streaming". Radiation from a radioactive source always travels in a straight line and V formation of louvres prevents direct rays from passing out of the contained area.

The A.G.R. for Dungeness 'B' integrates the boilers and gas circulators. The complete reactor, including core, gas circulators and boilers, is housed within a concrete pressure vessel, with the core at the centre, and the circulators under the boilers as shown in Fig. 4.4.11C. An inner pressure cylinder in the form of a bell separates the core from the boilers. The circulators deliver into the region within this cylinder, some of the flow passing downwards through the core to keep the graphite and its restraint structure close to inlet temperature. The total flow then passes upwards through the fuel channels and charge tubes, and is discharged through ports into the plenum above the pressure cylinder, whence it passes down through the boilers back to the circulator inlets.

The internal walls of the concrete pressure vessel are clad with stainless steel thermal insulation. There are shields around and above the core to prevent activation of boilers and circulators so that these can be reached and maintained. Under certain conditions it is also possible to enter the space above the core.

The concrete pressure vessel is cylindrical, and is prestressed by circumferential and vertical cables in the barrel and by horizontal cables in the two end caps. Its inner boundary is sealed by a water-cooled membrane made of mild steel.

The boilers are of the once-through type with a single reheater stage. The tube ends are led out in such a way that access is available for plugging or other purposes, without entry into the pressure vessel. There are four boiler sections per reactor, each operating as a unit with a gas circulator.

Each gas circulator is a centrifugal machine, the working parts of which form a cartridge unit mounted at the base of the pressure vessel wall. There is a simple sleeve valve which isolates the circulator from the main gas circuit and enables the whole cartridge unit to be removed through the wall of the vessel for maintenance after depressurisation, but without filling the pressure vessel with air. If necessary, the other three circulators can be brought back to load with the fourth penetration blanked off. Each circulator is driven by a synchronous motor through a fluid coupling which allows the circulator speed to be controlled.

The uranium dioxide fuel is contained in stainless steel cans. These are arranged in three-ring clusters of 36 pins within graphite sleeves, to form 40 in-long fuel elements. The multi-pin A.G.R. fuel elements are entirely different from the fuel elements used in the Magnox stations.

Individual standpipes through the pressure vessel are used for access to every channel in the core. Eight fuel elements are made up into a stringer, which is attached to a plug unit located in the standpipe and the whole forms a composite assembly which is handled by the fuelling machine and loaded into the reactor as one unit. The plug units incorporate motor driven gags which can be used to provide adjustment of channel flow from the control room under normal operating conditions: this feature achieves maximum output while avoiding complex working on the charge face. The control rods are attached to similar plug units containing the drive motors and mechanisms and are interchangeable with fuel assemblies.

A single fuelling machine serves both reactors. It is mounted on a gantry which runs across the charge face and can be connected to any of the standpipes on either reactor or to the service holes in the central fuel handling block.

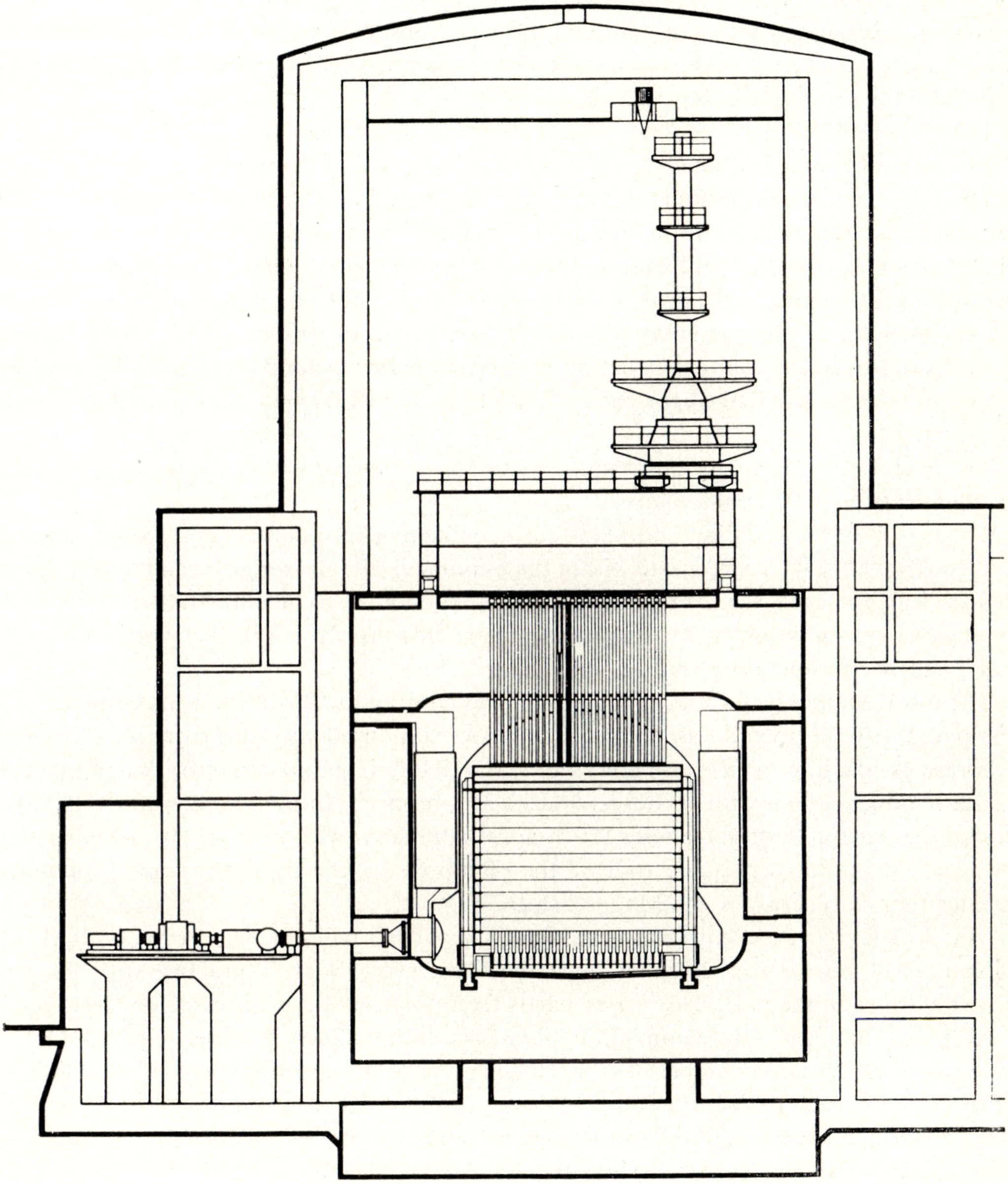

FIG. 4.4.11C. Section through Dungeness 'B' reactor.

The fuelling machine is designed to operate with the reactor at full load, but neither grabs nor any other fuelling machine components need to penetrate the pressure envelope below charge face level for this purpose. All handling operations over the reactor are under the jurisdiction of the control room.

New and irradiated fuel-handling facilities are housed in a central block located between the two reactors and this area also contains servicing facilities for the fuelling machine and the standpipe components.

The two 660 MW turbo-generators operate at the same steam conditions as in modern coal-fired power stations. The latest arrangement of pannier condensers is adopted.

The two reactors are housed, together with all the other plant and equipment for the station, in a single composite building. This building comprises a reactor unit, housing the two reactors and the fuel-handling block on the landward side, a turbine unit housing the turbo-generators and their auxiliaries on the seaward side, and a services unit containing the switchgear, instrumentation, change-rooms and other common services in between. Reinforced concrete construction is adopted below ground level and where heavy shielding is required; elsewhere the construction is steel-framed with protective metal cladding.

The operation of all major plant in the station is controlled from a single control room located centrally in the building.

The layout of the building and plant has been arranged to control contamination and to dispose of all solid active waste within the building itself, allowing all roads to be classified as clean and access to the main entrance of the building to be unrestricted. With these arrangements, radiation and contamination are adequately controlled with minimum inconvenience to operating staff or visitors.

The most striking fact about the layout of the Dungeness 'B' station is its compactness compared with the previous British nuclear power stations. This stems from the increased fuel ratings, the improved steam conditions, the tightly integrated reactor design and the single building complex for turbo-generators and reactors. In the Dungeness 'B' A.G.R. design the reactor centres are only 160 ft apart compared with over 400 ft at Dungeness 'A', and the output per square foot of the plan area for the station is some four times greater than is general for the Magnox stations.

The A.G.R. has excellent safety characteristics which will enable it to be built closer to centres of population than has been the Generating Board's practice with nuclear power stations in the past. This arises partly from the use of a prestressed concrete pressure vessel to contain the complete coolant gas circuit. Such a vessel depends for its integrity on a very large number of separate high tensile steel prestressing cables and many of these would have to fail simultaneously before the vessel would show signs of distress. As a result sudden loss of coolant is inconceivable and the worst accident of this type which needs to be considered is a slow depressurisation through an auxiliary circuit which would not cause overheating of the fuel.

This high integrity coolant circuit would also limit the maximum release of activity to an acceptable value if for some unforeseen reason a channel of fuel within the reactor became grossly overheated.

The safety of the reactor is also enhanced by the use of ceramic (uranium dioxide) fuel in stainless steel cans. Both these materials can stand quite severe over-temperature without distress and even at temperatures very much higher than normal, are chemically inert in the carbon dioxide coolant gas.

The A.G.R. represents another stage (after Magnox reactors) in the development of gas-cooled reactors. As in all such developments, engineering improvements in the design of the plant can be anticipated in A.G.R. stations constructed after Dungeness 'B'.

Simultaneously with engineering developments it appears, from development work now proceeding, that it should be possible to advance the working conditions of later A.G.R.s. An increase in coolant gas pressure to 600 lb/in^2 from the Dungeness 'B' level of 450 and substantial advances in fuel operating parameters and in fuel management schemes are anticipated. A reduction in the cost of electricity generated approaching 10% would be the target.

If the fuel element were made up of 60×0·4 in diameter pins instead of 36×0·57 in, the fuel rating could be approximately doubled. The fuel inventory could thus be markedly reduced, and the initial investment cut by around 7%. Increases in enrichment and in fuel fabrication costs could largely nullify the effect of this capital saving in generating costs. However, high fuel rating designs could be of interest if emphasis were placed on a low capital investment and could be considered for later A.G.R. stations.

The largest single source of possible saving for later A.G.R.s appears to be the increase of reactor size. Substantiation for the feasibility of A.G.R.s having about double the Dungeness 'B' output was obtained in a detailed design study carried out by the U.K.A.E.A. during 1964. The target saving for a 2×1200 MW station (with four turbo-generators) would be 10% in the cost of electricity generated compared with a 2×600 MW station.

Looking further ahead, there are other possible developments in gas-cooled reactors, including the use of "dispersed" fuels, i.e. small uranium carbide spheres embedded in highly refractory silicon carbide, instead of being canned in stainless steel.

4.5. TURBINE LAYOUT

4.5.1. Turbo-generator

The space required for a turbo-generator plus its close auxiliaries will be for a 500 MW unit approximately 225 ft×120 ft. (This excludes the heaters and feed pumps now almost always in a separate bay between the boiler and the turbine houses.)

Once the orientation of the turbine has been settled the main crane must be fixed, since it will determine the width and height of the turbine house roof. The height is fixed by checking out the maximum lift as shown in Fig. 4.5.1A. The width of the crane rail centres is established by the hook approach limit at each end of the crane; the hook must plumb particular loads, e.g. the alternator rotor in a transverse layout beyond the end of the turbine.

The height of the turbine house is also fixed by the height of the machine above basement level, and it is becoming apparent that all future large machines will be fitted with

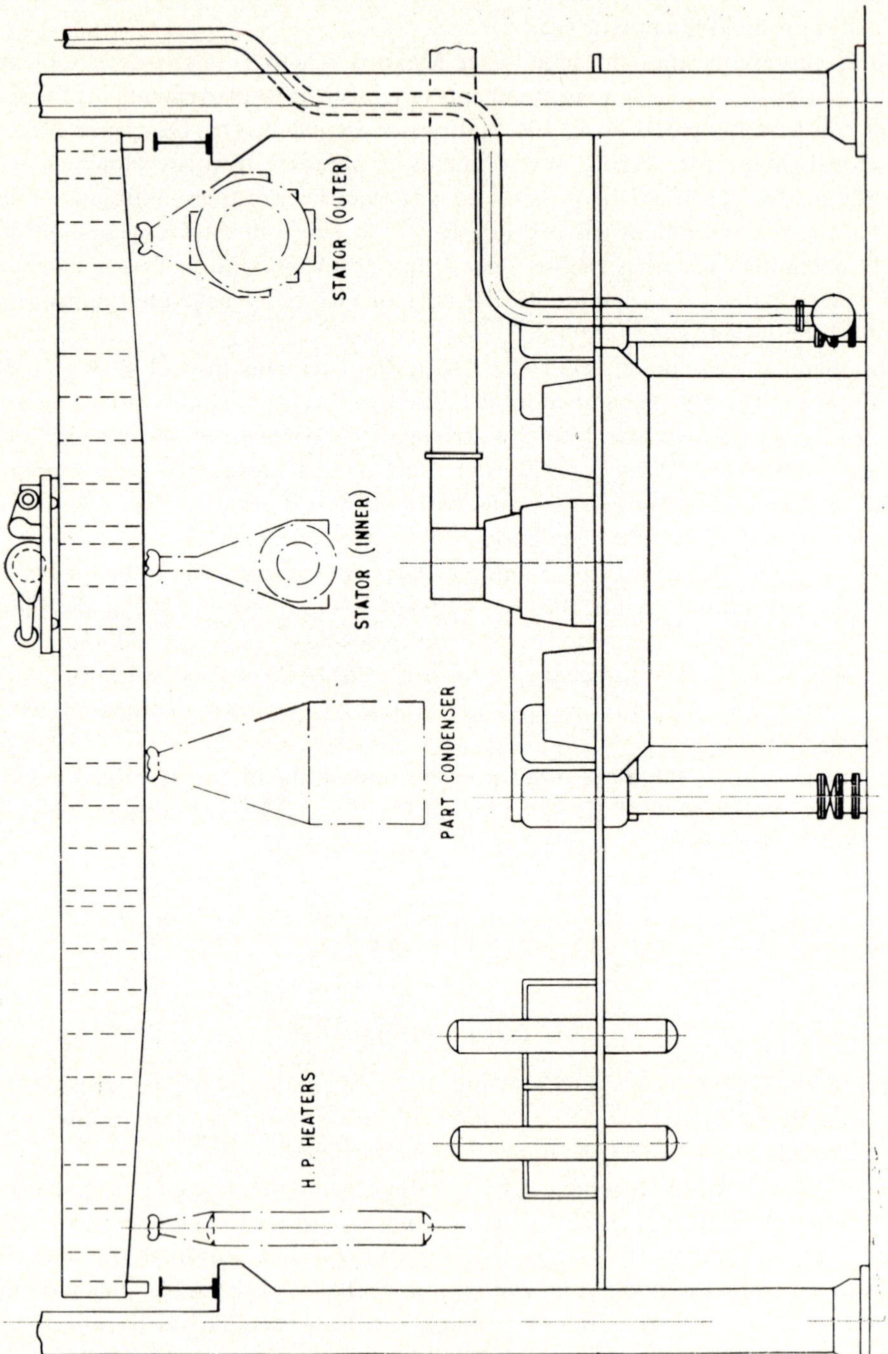

FIG. 4.5.1A. Crane hook approach limitation.

either pannier or integral condensers which allow the turbine operating floor level to be only 18 ft from basement level.

A loading bay must be provided for receiving parts delivered by road or rail and this must be covered by the crane. The arrangement and allocation of these bays vary considerably. In older stations it was general practice to have a loading bay at one end of the turbine house and later one at each end. With the larger stations it is necessary to be more liberal. Littlebrook 'D' will have two loading bays, one between each pair of sets, and Drax will have three for a six-unit station.

Although a central loading bay increases the length of the turbine house this is a logical position because maintenance workshops, amenities, water treatment plant, etc., can be concentrated in this area on the centre of gravity of the station.

Loading bays serve a dual purpose, they are primarily an area below the main crane for the specific purpose of loading or unloading heavy plant items, but they are also a useful standing area for top half cylinders, pipes and other large pieces dismantled from the turbine when it is undergoing a main overhaul. These areas must be liberal for a modern turbine, the Littlebrook 'D' layout referred to earlier provides specifically that the long continuous roof of the switchgear house between sets is load-bearing, and dismantled parts will be stored on this large convenient area.

There is no high level of operational activity around a modern turbine with remote control, but there is a peak maintenance activity during regular planned overhaul, nowadays assumed to be necessary every 2 years and this must be planned for maximum manpower over a minimum time period. In layout terms this means that good access between the set and the maintenance centre is essential and unimpeded clear wide access all round the set at operating level must be provided. This access area should be in concrete with a good non slip tiling and have a high level of lighting, although this latter may involve provision for temporary erection of lamp posts with plugs for the electricity supply.

Looking now at the turbine in more detail, there will be auxiliary equipment for the oil systems, generator cooling, generator connections, etc., and the turbine maker will have set ideas about the layout of these. Most of these items have natural positions in relation to the main plant and present a simple problem of co-ordination and efficient use of space.

Early positioning of the turbine main oil tank is important as it can have a decided influence on positioning of adjacent plant both below and above operating level. Filling and emptying connections are also required and these connections should be piped to a convenient unloading point for road tankers supplying the oil. The remainder of the oil system comprises flushing and jacking pumps, purifiers, coolers, piping and valves, all of which are arranged around the turbine end adjacent to the oil tank.

The oil system is a potential fire hazard in that a leak on a high pressure relay oil line may squirt on to a hot steam vessel, soak the lagging and ignite; it is important to minimise the number of joints and to avoid pipe-joints such as compression couplings which are not intrinsically safe. As a second defence, all lagged surfaces in the vicinity of oil equipment must be close cleaded with mild steel sheeting or shielded with mild steel plates. At Thorpe Marsh power station it was decided, in view of the fire risk, to site

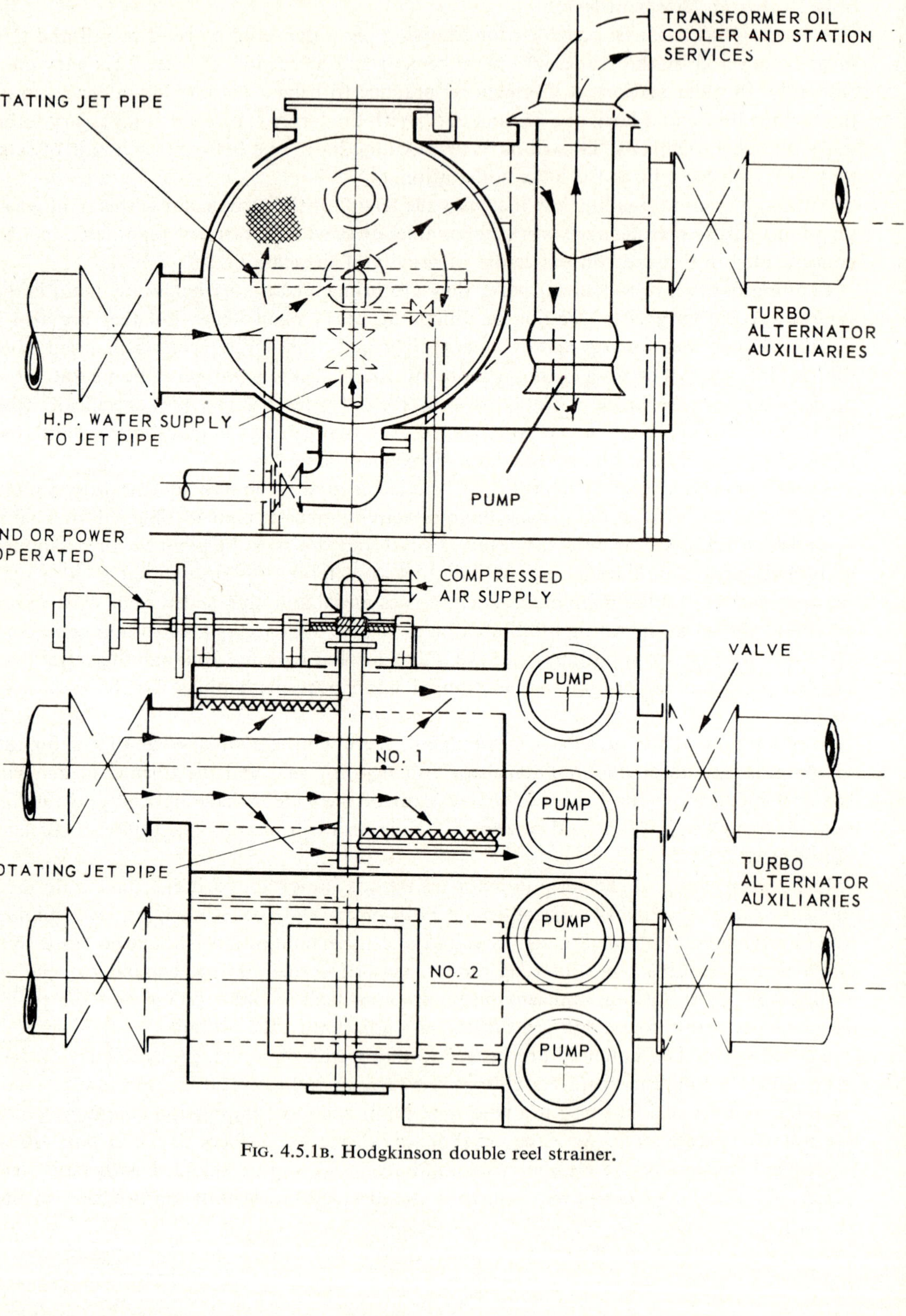

FIG. 4.5.1B. Hodgkinson double reel strainer.

the turbine oil tank, together with pumps and coolers, etc., into a separate annexe with a fire screen adjacent to the turbine. The ability to do this depends to some extent upon the layout of the turbines. For instance, on a transversely arranged set with the steam end nearest the boiler, unduly long oil service pipes would be needed to a tank and associated equipment in the electrical annexe. The lubricating and relay oil service piping should be well protected with a fixed emulsifying spray extinguishing system. This, however, cannot be automatically initiated since it is pointless to hazard generation by spraying the area with water when the fire may be local and easily dealt with by hand held hoses, which should always be available.

The generator end of the set will accommodate the equipment required for cooling the generator windings slip-rings, etc. These are fans, coolers, ventilation ducting, and so on. With modern hydrogen-cooled machines, space is required for seal oil equipment, hydrogen control panel, CO_2 purging equipment, etc., together with racks for hydrogen and CO_2 bottled supplies, although for larger units it is becoming customary to provide separate hydrogen and CO_2 bottle stores with a lorry unloading platform.

Hydrogen is a highly explosive gas when mixed in certain proportions with air and piping services in the vicinity must all be bonded electrically down to earth at the joints, since sparking, set-off by static electricity or currents induced from the field of the generator connections, could be disastrous.

A major source of cooling water is required for coolers associated with the turbine, generator, and general services. On large units this will be via a 40 in diameter pipe taken off the inlet CW culverts through, typically, a "Hodgkinson double reel strainer" (see Fig. 4.5.1B).

This cooling water supply, having passed through the coolers, must re-enter the main outlet culvert and consequently by-passes the main condenser which normally operates on a syphon (for sea or river water cooled stations). Under these circumstances the auxiliary cooling water may have too high a head available and flow is restrained through throttling valves on the cooler outlet. There may, however, be cases where the auxiliary CW head available is insufficient in which case booster pumps are necessary. These can now be integrated with the Hodgkinson strainer in a neat installation taking up little more room than the regular strainer.

Figure 4.5.1B gives a general arrangement of these.

4.5.2. **Condensing Plant and Auxiliaries**

The modern condenser is now virtually part of the turbine since it is so closely integrated, either as pannier mounted on each side of the l.p. cylinders or completely surrounding the l.p. end as shown in Figure 4.5.2A. This arrangement has considerably reduced the height requirement of the turbine house.

The major space requirement is for tube withdrawal and insertion; nowadays this is always along the length of the turbine towards the generator and advantage is taken of the wide access platform now required all round the machine.

Extreme care must be taken with cooling water and condensate extraction pipework connections to the condenser. Integral and pannier type condensers do not move very

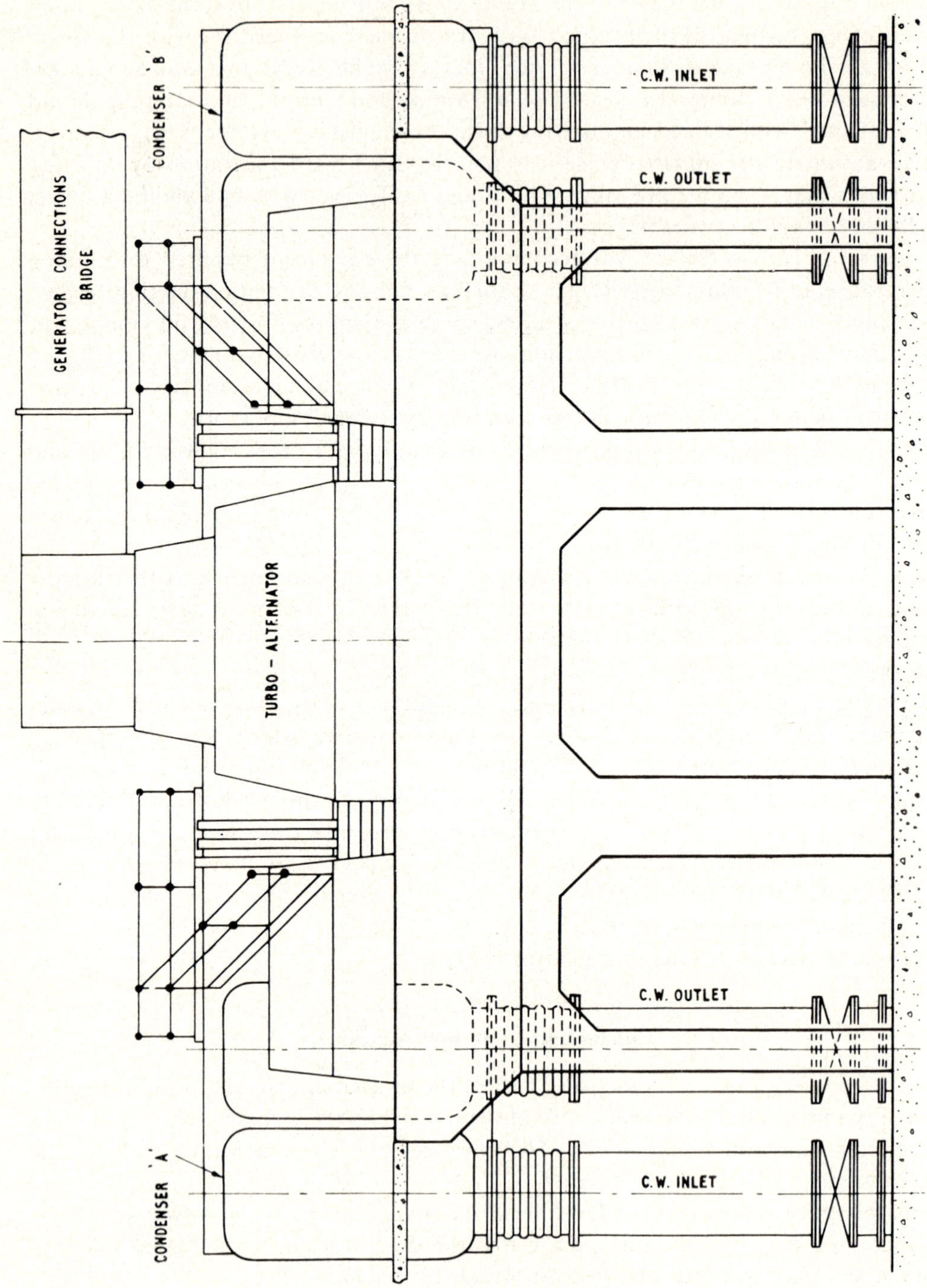

FIG. 4.5.2A. 500 MW turbine with pannier condensers.

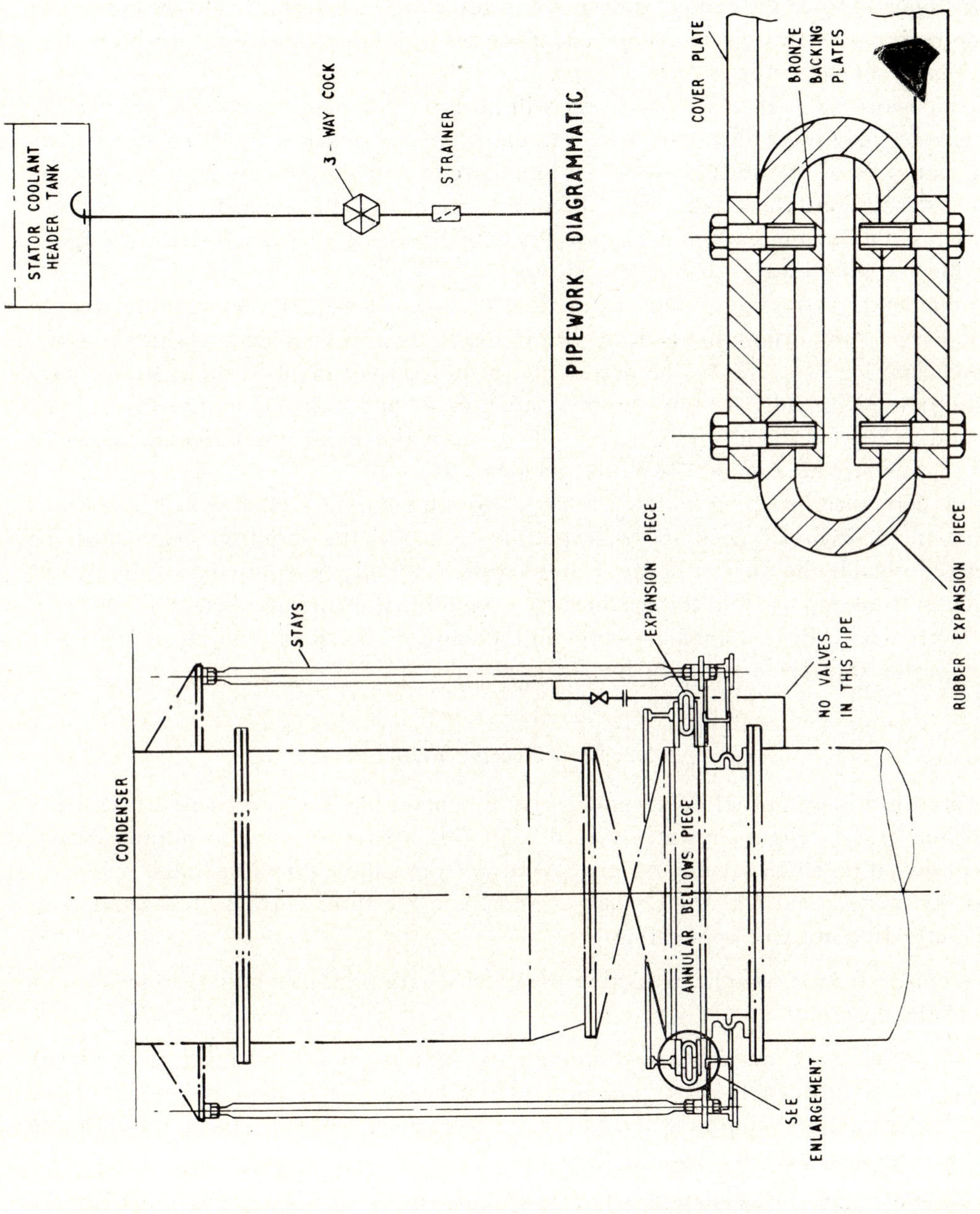

FIG. 4.5.2B. Self compensating expansion joint.

much with change in temperature since they are about the main static point of the machine, but there will be sufficient terminal point movement to make compensation of pipework connections essential. Bellows type expansion fittings are usually fitted to accommodate such movement, but internal pressure acting on the differential area of the bellows tends to force the bellows open and this force can be transmitted to the condenser causing undesirable turning moments. An ingenious self-compensated joint which eliminates this is shown in Fig. 4.5.2B.

CW pipework to a 500 MW condenser will be 120 in diameter for single water entry although this may be divided into twin 84 in diameter pipes each with an isolating valve, which should be of the butterfly type. On smaller condensers there are large wedge gate valves for isolation; these take up considerable room and the butterfly valve proportionally takes up much less space because the long operating gear associated with wedge gate valves is unnecessary for butterfly valves.

The condenser extraction pumps are sited at basement level and their level relative to the condenser is important and is dealt with in the section on feed heaters which follows. It is sufficient here to refer to the need to have these heavy plant items and the rotary air pumps used to pull vacuum on the condenser immediately below the main crane hook and with good maintenance access. These are major plant items and are shown in Fig. 4.5.2C as provided on 300 MW unit condensers.

All modern turbine l.p. cylinders are provided with bursting discs, which rupture if pressure in the cylinder goes above a specific value and the resulting steam must be discharged outside the turbine house. Considerable ingenuity is required to route two or three large trunk mains from the machine to a point outside and at a height above basement level which will not embarrass lifts on the main crane. Such routing is probably simpler with a transverse arrangement of the set.

4.5.3. **Feed Heating Plant**

The feed heating plant, although not an item of main plant is nevertheless essentially part of the process chain shown in Fig. 4.1. In this area it is vital to appreciate the effect of height on efficiency and of any deviation from simple flow principles.

The feed system starts at the condenser and there are three current arrangements of condenser extraction and l.p. heating.

(a) Figure 4.5.3A shows the extraction pumps discharging through l.p. surface heaters to the deaerator.

(b) Figure 4.5.3B shows the extraction pumps discharging first to the generator coolers and then through the low pressure surface heaters. It is possible with a multi-stage extraction pump to connect the generator coolers between stages, thus limiting the water pressure in these coolers.

(c) Figure 4.5.3C shows the different arrangement required when using direct contact heaters. Note that the first direct contact heater may have its own extraction pump or use the last stage of the condenser extraction pump. The other heaters cascade to an independent deaerator feed pump which is backed up by an emergency

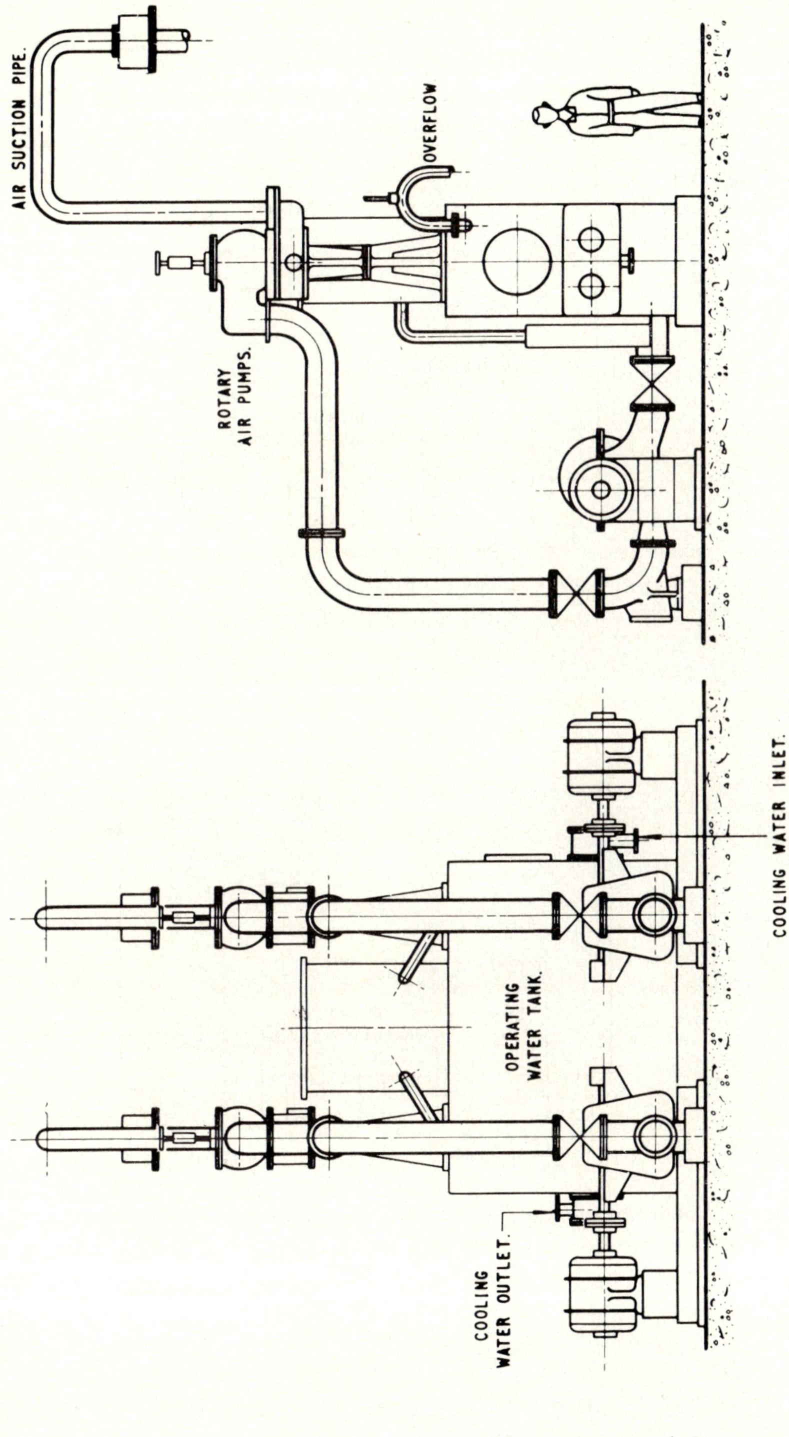

FIG. 4.5.2C. Rotary air extraction pumps.

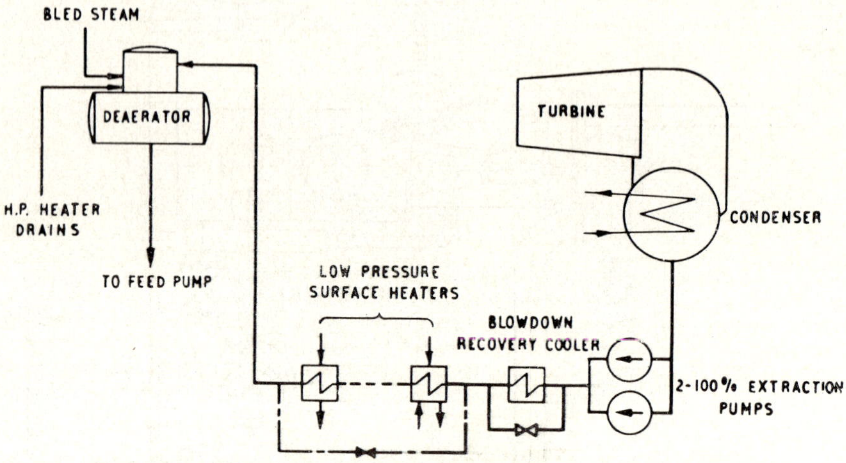

FIG. 4.5.3A. Condensate system — l.p. surface heaters and deaerator.

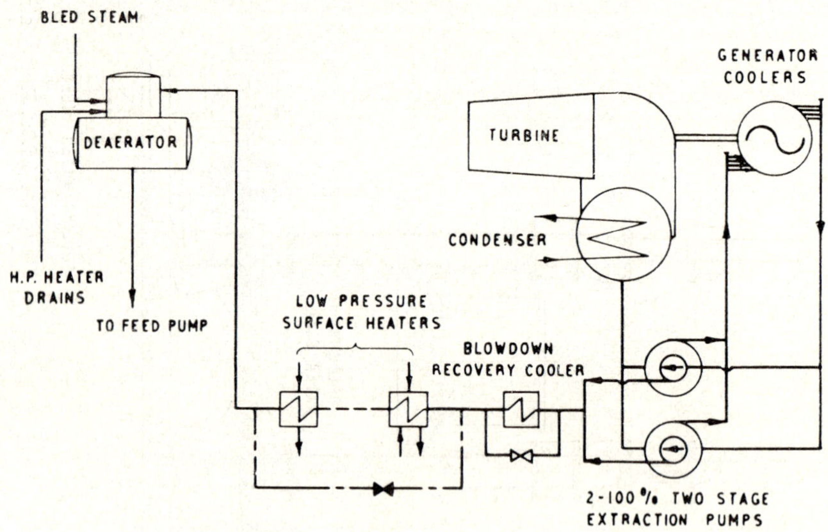

FIG. 4.5.3B. Condensate system — generator, coolers, l.p. heaters and deaerator.

pump by-passing the heaters. The layout of the plant is a function of economic pumping to minimise cycle energy consumption. The height of the deaerator is fixed by feed pump suction requirements (dealt with later) and the problem is one of apportioning positions for heaters and pumps and of employing the most efficient pipe routing and pipe diameters, to reduce friction loss which is in the ultimate the running cost of the pipeline.

The assessment of total head on the extraction pump would be made along the following lines for the system shown in Fig. 4.5.3A:

System characteristics	Feet head at turbine MCR flow + 1% make-up
(1) Net static pressure	
(a) Deaerator pressure at full load	
$33{\cdot}3\ \text{lb. in}^2 \times \dfrac{1}{0{\cdot}434}$	77
(b) Condenser outlet pressure generally assumed to be 33 ft	33
(c) Deaerator inlet to condenser extraction pump suction level	130
	240
(2) Total frictional pressure	
(a) Generator coolers	13
(b) Water extraction condensers	10
(c) L.P. heaters	64
(d) Deaerator vent condenser and sprays	17
(e) Piping:	
Extraction pump suction piping	1
Piping from pump discharge flange to l.p. heaters	3
Piping from l.p. heater outlet flange to deaerator	5
Total	353
Margin 10%	35
Design head	388

Note: The 10% margin is intended to cover deterioration of pump in service and possible inaccuracies in head assessment.

The head duty of heater extraction pumps is of course determined in a similar manner.

The deaerator is the highest point in the feed system and is a feed heater in its own right; its static height is carefully calculated to give a minimum net positive suction head (NPSH) at the feed pump. The feed water leaves the deaerator at saturation temperature necessitating a careful study of conditions likely to occur at the feed pump suction. The total head on the suction at full load is the sum of the static head and the steam pressure in the deaerator, less the friction and velocity losses in the piping, valves and strainers.

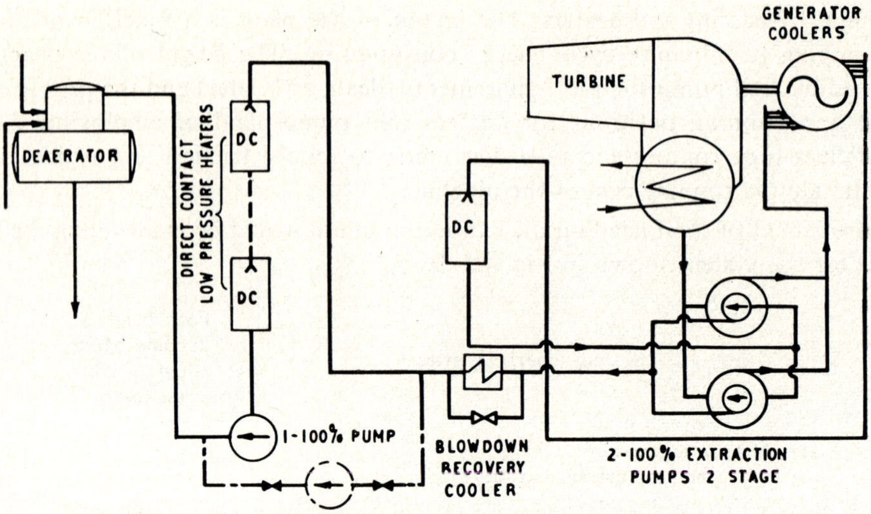

FIG. 4.5.3C. Condensate system — generator, coolers, d.c. heaters and deaerator.

There are several conditions which result in a reduced net positive suction head (NPSH), the most serious occurring when full load is suddenly thrown off. The turbine governor closes the steam inlet, thus reducing immediately the pressure of the bled steam supply to the deaerator and the deaerator pressure falls in a period of minutes from its full load pressure towards condenser vacuum. The water in the feed pump suction piping is still at the saturation temperature associated with the pressure at full load and will remain so until the pump has passed a considerable quantity of water.

The calculation of the head required to prevent flashing in the impeller eye under load throw off conditions consists in finding the best compromise between the transient loss of NPSH and the frictional losses in the suction pipe, for although a large suction pipe reduces the loss of suction head due to friction it also increases the storage capacity of the high temperature water.

A method of calculating the NPSH of the system and arriving at the optimum pipe size follows. This is derived from a paper by A. Gage entitled "Criterion for the Suction Conditions of Feed Pumps".

TYPICAL CALCULATIONS OF SUCTION PIPE DIAMETER AND STATIC HEAD REQUIRED

Symbols

H = head in feet,
D = pipe diameter in inches,
L = length of pipe in feet,
Q = quantity in pounds per hour,
V = velocity in feet per second,
α = ratio of weight of water in suction pipe to total equivalent weight of water in the deaerator.

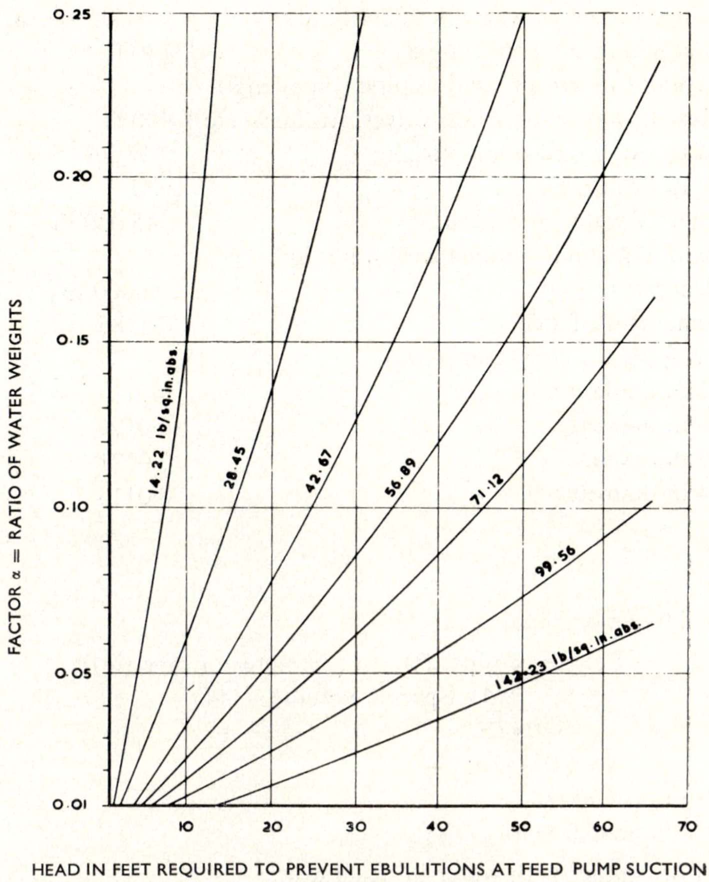

FIG. 4.5.3D. Deaerator and feed pump suction calculations.

Total suction head required by feed pumps

$$H = H_1 + H_2 + H_3 + H_4$$

where

H_1 = head required to prevent ebullitions at pump suction due to difference in temperature of water at pump suction and deaerator storage vessel during transient load conditions (see Fig. 4.5.3D),

H_2 = normal static head required by pump (NPSH required),

H_3 = loss of head in suction pipe due to friction,

H_4 = other losses independent of pipe diameter, e.g. suction strainer, flow orifice, etc.

Data

Absolute minimum NPSH over vapour pressure	35 ft
Normal operation NPSH over vapour pressure	60 ft
Feed pump full load capacity	$3{\cdot}52 \times 10^6$ lb/h
Deaerator bled steam pressure	42·54 lb/in²
Equivalent saturation temperature	132·8°C (271°F)

Specific volume of water at 132·8°C (271°F)	0·01718 ft^3/lb
Actual length of suction pipe	140 ft
Equivalent length of suction pipe (pipe length plus allowance for bends, valves, strainers, etc.)	400 ft
Loss through feed pump strainer	3 lb/in^2
Loss in orifice, etc.	6 ft
Weight of water in deaerator	545,000 lb
Weight of metal in contact with water in deaerator	90,000 lb
Specific heat of steel	0·1185
Friction factor (for steel pipes)	
12 in diameter	0·013
16 in diameter	0·0129
20 in diameter	0·0121
24 in diameter	0·0118

Calculations

To find H_1

Weight of water in suction pipe

$$= \frac{\text{Length of pipe (ft)} \times \pi\,(\text{diameter of pipe (in)})^2}{4 \times \text{Specific volume} \times 144}$$

$$= \frac{140\pi D^2}{4 \times 0{\cdot}01718 \times 144}$$

Weight of water in deaerator

$$= 545{,}000 \text{ lb}$$

Weight of metal in contact with water in deaerator

$$= 90{,}000 \text{ lb}$$

Equivalent weight of water

$$= 90{,}000 \times \text{Specific heat of steei}$$
$$= 90{,}000 \times 0{\cdot}1185$$
$$= 10{,}665 \text{ lb}$$

Total equivalent weight of water in deaerator

$$= 545{,}000 + 10{,}665$$
$$= 555{,}665 \text{ lb}$$

$$\alpha = \frac{\text{Weight of water in suction pipe}}{\text{Total equivalent weight of water in deaerator}}$$

$$= \frac{140 \times \pi D^2 \text{ (in)}}{4 \times 0{\cdot}01718 \times 144 \times 555{,}665}$$

$$= 0{\cdot}00008 D^2$$

If D = 12 in, α = 0·01152.

If deaerator bled steam pressure = 42·67 lb/in^2 abs. then from Fig. 4.5.3D, H_1 = 5 ft.

To find H_2

This will be specified by the pump manufacturer.

To find H_3

Velocity in suction pipe

$$= \frac{Q(\text{lb/hr}) \times \text{Specific volume} \times 4 \times 144}{3600 \times \pi D^2\ (\text{in})}$$

$$= \frac{3{,}520{,}000 \times 0{\cdot}01718 \times 4 \times 144}{3600 \times \pi D^2}$$

$$= \frac{3080}{D^2}$$

Friction factor for 12 in bore pipe = 0·013.

H_3 = Head loss due to friction

$$= \frac{\text{Friction factor} \times \text{Equivalent length of pipe (ft)} \times \text{Velocity}^2 \times 12}{2 \times g \times \text{dia. of pipe (in)}}$$

$$= \frac{0{\cdot}013 \times 400 \times \left(\frac{3080}{D^2}\right)^2 \times 12}{2 \times g \times D}$$

$$= \frac{9{\cdot}192 \times 10^6}{D^5}$$

When $D = 12$ in, $D^5 = 2{\cdot}488 \times 10^5$

and $H_3 = \dfrac{9{\cdot}192 \times 10^6}{2{\cdot}488 \times 10^5} = 36{\cdot}95$ ft., say 37 ft.

Similar calculations for 16 in, 20 in, and 24 in pipes give the following compared results:

TABLE 3.

D	12	16	20	24
D^2	144	256	400	576
D^5	248,832	1,048,576	3,200,000	7,962,624
α	0·01152	0·02048	0·032	0·04608
H_1	4	7	10	14
H_2	60	60	60	60
H_3	37	8·7	2·7	1·05
H_4	6	6	6	6
H	107	81·7	78·7	81·05
Static head	90	90	90	90
Margin	−17	+8·3	+11·3	+8·95

The data given on page 331 shows that a margin of 25 ft is available between normal and absolute minimum NPSH required by the pump, use of a 12 in pipe in this case would reduce this to (25−17) = 8 ft, which is too small to cater for transient changes. The choice, therefore, is between 16 in and 20 in pipe and cost difference between them; alternatively, it may be possible to raise the static head and use a 12 in pipe, the cost difference again will be a factor.

This particular problem was resolved by using the 20 in bore pipe to obtain maximum margin with small extra cost.

If the static head was raised in order to reduce the piping cost, additional revenue costs would be incurred in pumping from condenser to deaerator over the life of the plant.

4.5.4. Feed Pumps

Feed pumps are the heart of the power cycle. Their intrinsic design is highly specialised but the layout engineer determines their arrangement within the plant and their rated duty. He reviews the suction and delivery arrangements, obtains the pressure drop characteristics of feed heaters and the boiler and finally coordinates the calculation of system resistance at varying flows.

4.5.4.1. Layout of Pumps

A modern feed pump will weigh around 50–60 tons, but the heaviest single lift is likely to be the electric motor, around 25 tons for a 100% electrically driven 200 MW feeder. This will require the pumps to be situated immediately below an overhead crane, usually in the mechanical annexe. The pumps must stand on solid foundations brought up from basement level (see Figs. 4.15.4D and E, Fawley model) so that each pump is accessible from the semi-basement floor and has complete maintenance access. Provision should also be made for a portable miniature lifting frame with a runway beam and hoist, which can stand over the pump and take long time holding lifts whilst dismantling and assembly is going on. This releases the main crane for heavier work.

The precise placing of the pumps relative to other plant should permit economic pipe routing, and initially the grouping should be such that the common suction pipe from the deaerator can drop vertically with little or no deviation from a straight line. At some point around operating floor level the suction pipe must be split into (usually) three supplies, one to each pump suction filter, as indicated in Fig. 4.5.4.1. It is most important that this trifurcation is hydraulically satisfactory; an apparently satisfactory arrangement in layout terms may, in practice (due to balance of pressure drops), starve one pump and cause severe cavitation in the pump impeller. A hydraulic model can be made up quickly in glass tubing (in a station chemist's laboratory) and the flow conditions studied for any permutation of pumps working and idle, by using an appropriate Reynolds number scale relationship. This is an essential practical experiment for the development of the design.

This important consideration taken with the positioning of feed heaters which must have an economic hydraulic relationship with each other, means that the pumps and the turbine tend to fix the layout in general terms and the balance of work becomes a piping layout problem which is dealt with in detail later.

4.5.4.2. Disposition of Pumps in the Feed Heating Train

Figure 4.5.4.2. illustrates diagrammatically some of the systems that have been used on modern power stations:

(a) Consists of low- and medium-pressure heaters and 100% duty main and standby high-temperature feed pumps, each with its own booster pump.

(b) A similar system to (a) but with the 100% duty main feed pump driven by a bled steam turbine or alternatively from the main turbo-generator through a variable speed slip coupling, with 2×50% starting and standby electrically driven pumps and 3×50% duty booster pumps.

(c) 100% duty main feed pump driven by a bled steam turbine and situated before the heaters, together with 2×50% starting and standby pumps.

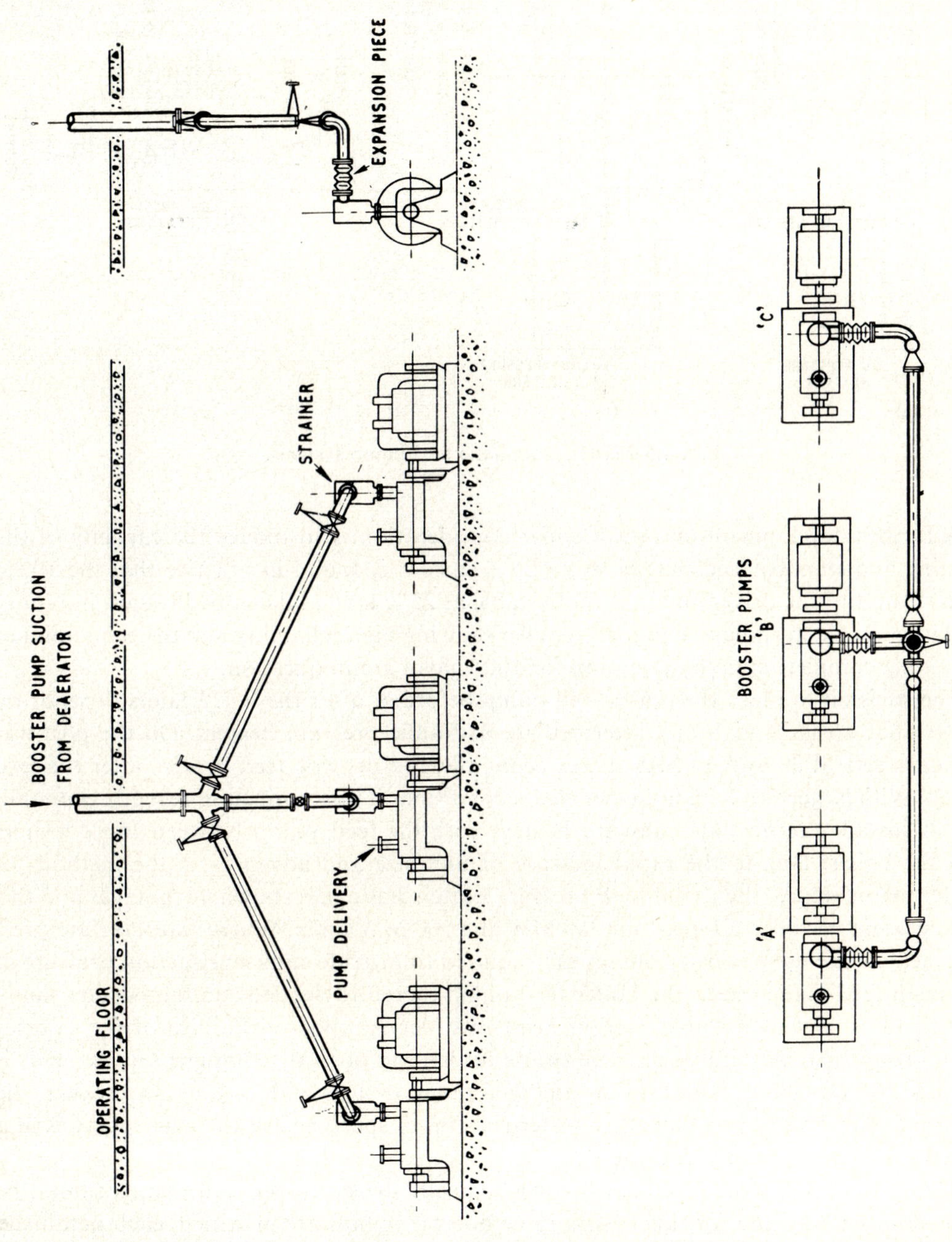

FIG. 4.5.4.1. Feed pump suction pipework arrangement.

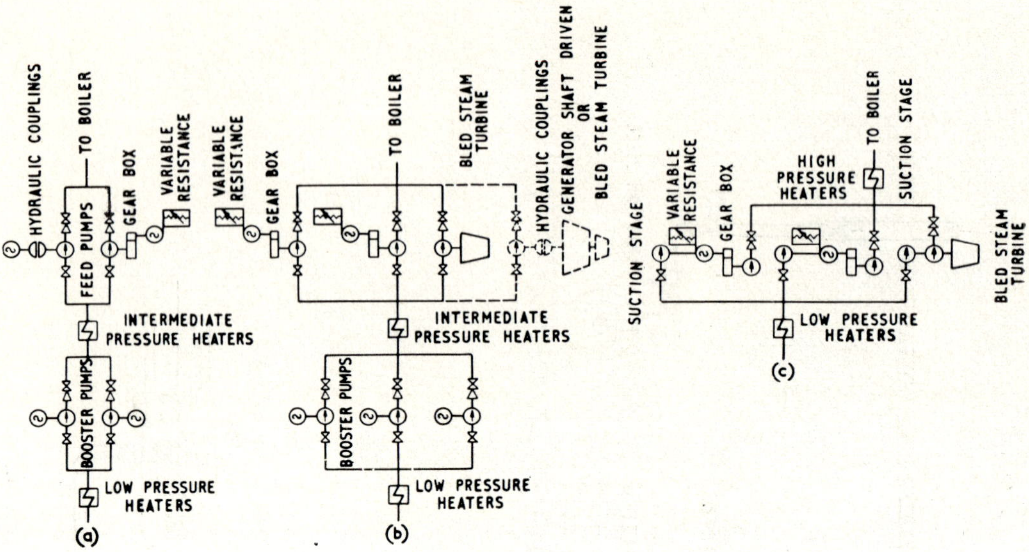

FIG. 4.5.4.2. Diagrammatic feed pump systems.

A further development currently under consideration is to reduce the capacity of the starting and standby feed pumps to 2×25% since it is found in practice that the 100% bled steam turbine driven pump is quite stable at 25% load and can be brought into duty at this level. The turbine and pumps are also showing high reliability and the economics of providing complete standby pumping for 100% load are in question.

The decision to place the boiler feed pump before or after the h.p. heaters depends on the relative costs of high and intermediate hydraulic pressure heaters. On the post-war standard sets of 30 and 60 MW it was economical to provide feed heaters after the feed pump. With larger sets having higher steam conditions it was until quite recently economical to install intermediate pressure heaters with the feed pump between those heaters and the boiler. Due to the rapid increase of unit size and advances in the methods of heater manufacture, the use of high pressure heaters has again shown advantages and this arrangement has been adopted for 350 MW and 500 MW units. Where intermediate pressure heaters are used booster pumps are required to provide sufficient suction pressure at the main feed pump inlet. The boiler feed pump therefore depends on the booster pump to provide sufficient positive suction head in order to operate. Failure of the booster would result in considerable damage to the feeder and possibly complete seizure. This is avoided by controlling both booster and feed pump motors with one circuit breaker, the motors being designed so that the booster runs up to speed before the main pump is at a speed that can cause any damage.

The booster pump arrangement for mechanically driven boiler feed pumps cannot be treated in this way and for this reason three booster pumps are provided, each nominally rated for 50% duty, but capable of supplying individually sufficient pressure and quantity to prevent flashing in the eye of the main pump under normal operation. Two boosters would normally be working together, with the third on automatic standby.

4.5.4.3. Speed Variation

In the section of this chapter dealing with h.p. feed piping it will be shown that the rated capacity and pressure of a feed pump should be a little in excess of the unit design steam conditions in order to ensure a sufficient feed flow under all conditions. A feed pump output may be reduced from the rated value to that required under normal conditions, either by throttling the flow with the pump running at constant speed, or by reducing the speed of the pump.

The pressure to be broken down by the feed regulating valves on a constant speed pump is the difference between the head generated by the pump at the required flow and the system resistance at this flow. A typical system resistance curve compared with a pump characteristic curve is given in Fig. 4.6.2D. These two curves cross each other at the rated duty point of the pump and diverge as flow reduces.

By providing speed reducing equipment the pump speed can be automatically controlled and the head generated matched exactly to the system resistance. The least complicated control method is to take the pressure differential across the feed regulating valve and maintain this at a constant value, say 50 lb/in^2 by varying the pump speed. The opening of the feed regulating valve would be independently varied by the boiler three-element control system.

At present, electrically driven feed pump speed variation is most economical with a slip ring motor and liquid resistance type speed controller. There are other methods available, in particular variable speed gear boxes, which could be attractive where there is an initial requirement for a gear box between the motor and the pump, since the speed control element is quite simply added.

A liquid resistance controller requires a lot of space basically because of the electrolyte tank and the provision of electrolyte cooling arrangements consisting of $2\times100\%$ duty pumps and CW coolers. These items can be physically quite remote from the feed pump motor and installation becomes a problem of fitting them in with the general plant arrangement. Due to space consideration the electrolyte tanks are often installed at deaerator floor level, with the cooling pumps at basement level. In this case special attention must be paid to the pump glands since the electrolyte can be wasted by leakage at these points.

4.5.4.4. Suction Strainers

Because of the presence of foreign material in the feed water, particularly during the commissioning period when temperature variation releases pipe scale, etc., the fine clearances of the pump must be protected. This is usually achieved by strainers and by acid cleaning and high pressure flushing to remove most of the debris. Ordinary basket strainers tend to choke rapidly and strainers of the type illustrated in Fig. 4.5.4.4 have been developed for permanent installation on all feed pump suctions.

The strainer elements rotate against a scraper and the debris drops to a flushing connection leading to a convenient dumping vessel such as the boiler blowdown tank.

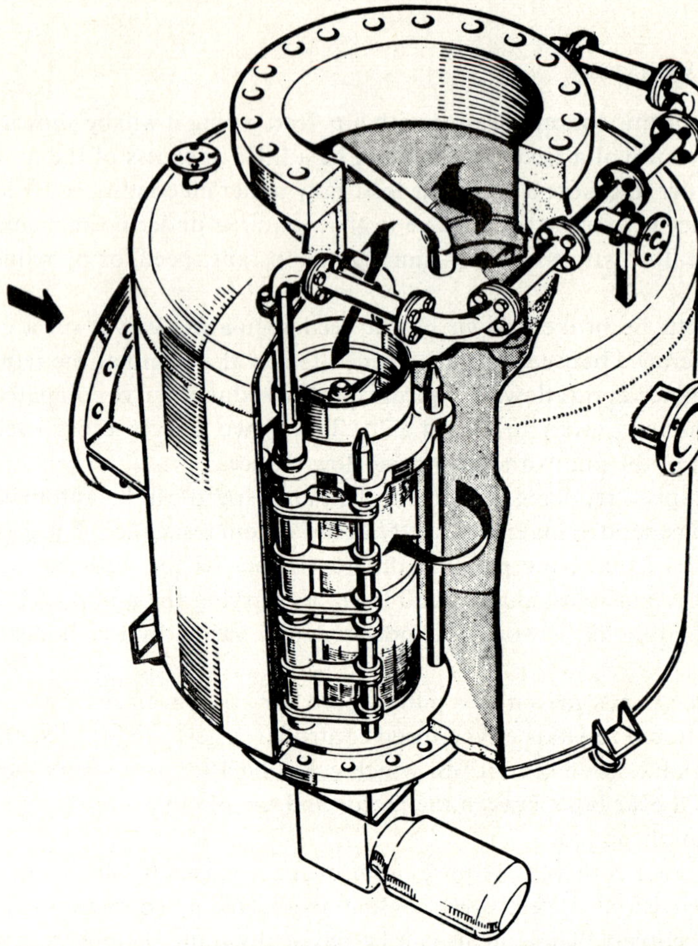

FIG. 4.5.4.4. Self-cleaning suction strainer. (Reproduced by kind permission of Auto—Klean Ltd.)

4.5.4.5. Feed Pump Services

Modern feed pumps are normally fitted with leak in/leak out labyrinth glands which are sealed with a high pressure water supply direct from the booster pump delivery or from an interstage connection on an integrated booster/feed pump. The seal water flows into the pump and outward towards atmosphere. A second injection point receives water from the extraction pump delivery via a strainer; part of this flow goes inwards and mixes with water from the first injection, the balance flowing outwards.

The first exit point for seal water is situated between the first and second injection points and is connected to the deaerator. The second exit point is situated on the outward end of the gland and passes the outward flow from the second injection point to a suitable collecting tank where (provided there is no oil or silica contamination) the water can be restored to the unit make-up system.

The pipework for these supplies and the drain should be well supported and arranged compactly about the feed pumps. Since the gland seal, gland quench and both the drain

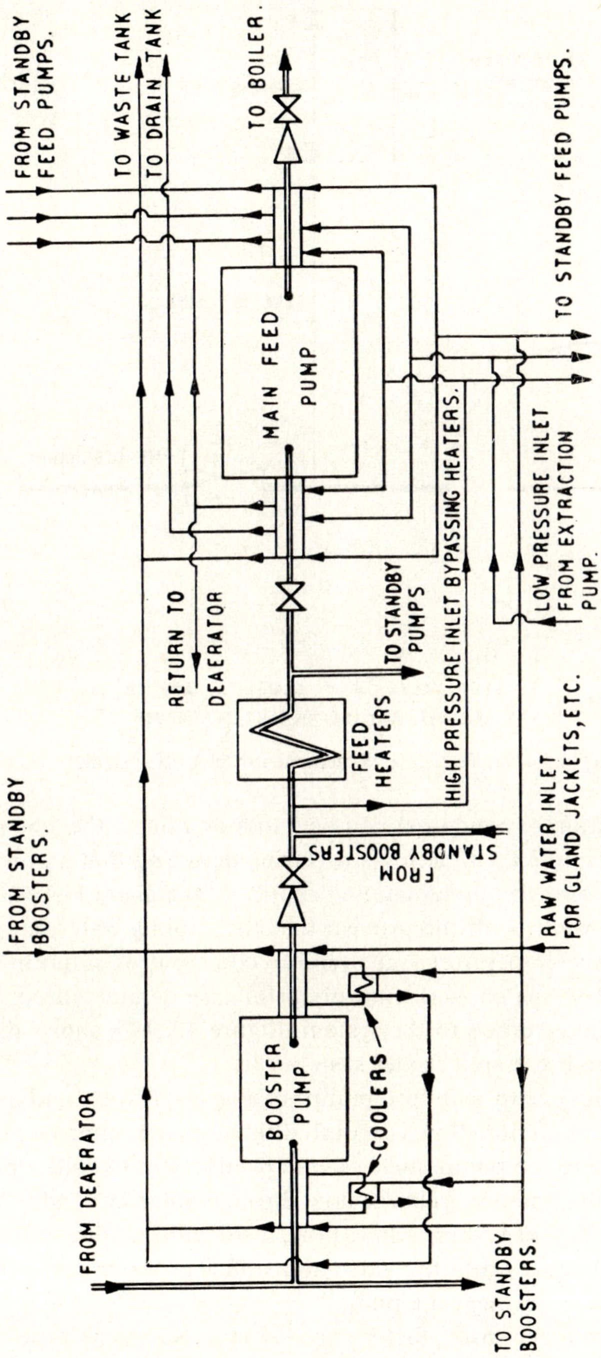

FIG. 4.5.4.5A. Gland cooling and sealing arrangement.

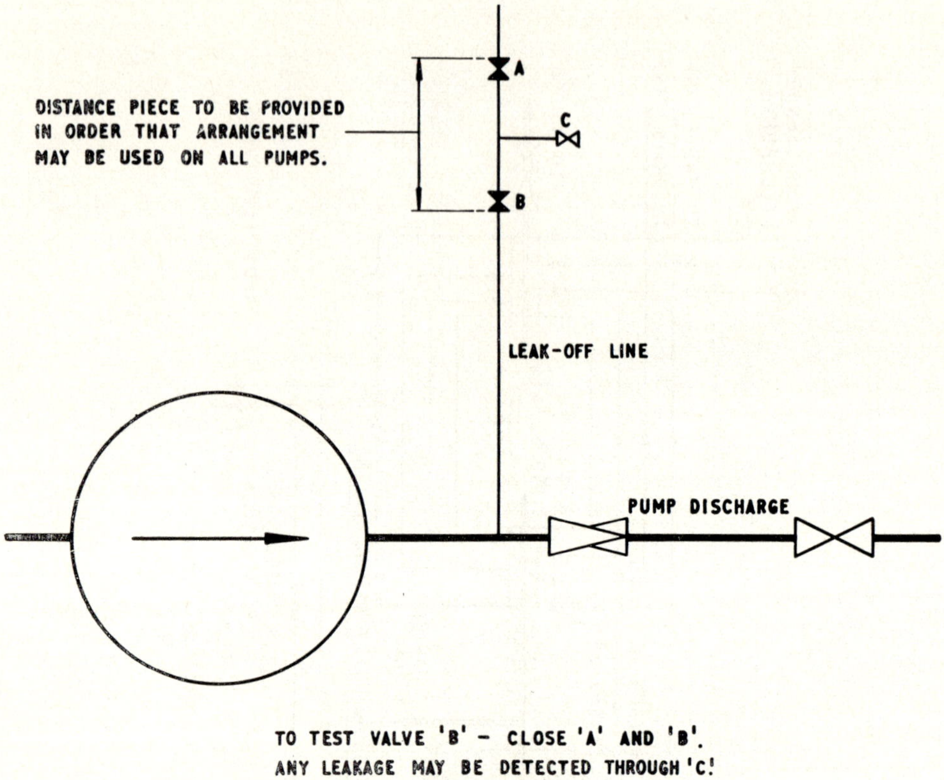

FIG. 4.5.4.5B. Testing arrangement for leak-off system.

to the deaerator and drain to condenser can be a loss or gain to the feed flow system it is necessary for each to be fitted with a flow measuring device so that a precise feed balance can be deduced for turbine testing. A cooling supply to the gland jacket is also required; this can be clean raw water, with the proviso that the cooling water temperature cannot exceed 60°C. Above this temperature, salts tend to come out of solution and the cooling jacket annulus could become choked. In a doubtful case demineralised water should be used and conserved and returned to the system. Figure 4.5.4.5A shows diagrammatically the connections required to a split feeder system.

Another important service to each feed pump is the leak-off valve and piping. Whenever the pump is running, a minimum flow is required through the pump to prevent the possibility of severe damage to the pump due to steam production. On older plant this protection is afforded by reading pump discharge flow through an orifice and, when it drops to a level usually equal to 25% of turbine MCR flow, a solenoid holding a valve closed is de-energised, the water pressure opens the valve and sufficient water passes to the deaerator to ensure a safe flow level through the pump.

This system has proved very unsatisfactory on later large pumps because of the large quantity of leak-off required and the sudden loading on pipework caused by a snap-open valve.

It is now customary to open the valve slowly by using an electrically or air driven actuator and an even better system is to use a modulating leak-off valve which operates via a transmitter reading pressure drop across an orifice plate; this system will always maintain smoothly a minimum desired value (DV) of flow through the pump irrespective of boiler demand.

The disposal of leak-off water is normally to the deaerator but where a blowdown recovery system is installed there is a good case for disposal into this system.

Since the temperature of the leak-off water may be considerably above 100° C (the saturation temperature at atmospheric pressure) the flow in the leak-off pipe must be maintained at some pressure slightly above the appropriate saturation pressure; the best way to achieve this is to place the leak-off valve at the disposal vessel and to fit a pressure diffusing pipe inside the vessel.

Two isolating valves, normally locked open, and a test valve must be provided at the pump connection to leak-off so that the test engineer can prevent this potential loss of water flowing to the boiler during efficiency testing. The arrangement is shown diagrammatically on Fig. 4.5.4.5B.

4.6. H.P. PIPEWORK AND VALVES

4.6.1. Main Steam and Reheat Piping

The design basis for calculation of pipe wall thickness, bore and pressure drop is dealt with fully in Volume 3, Chapter 5. However, it is vital to have this background firmly in mind when looking into layout and routing of the extremely heavy main steam and reheat pipes, since on piping every design factor is related such that improvement in one respect can have an adverse effect in another way.

Consider the main relationships and the constants. The constants are:

Steam pressure.
Steam temperature.
Steam flow.
Expansion/unit length on warming up to working temperature.

The variable relationships are:

Total pressure drop varies with total route length, changes in direction, changes in pipe bore and velocity of flow. On a modern unit, pressure drop from boiler stop valve to turbine stop valve is limited to no more than 100 lb/in^2, and the total reheat circuit pressure drop to no more than 10% of the exhaust pressure at the h.p. cylinder.

Flow velocity will obviously vary with the bore diameter but it is not so obvious that pipe wall thickness is directly proportional to pipe bore diameter. Thus wall thickness can be kept to a minimum by designing for high velocity of flow and although total pressure drop is increased flexibility is improved. Flow velocities of between 180 and 200 ft/sec have been consistently used on recent 500 MW designs for main steam and high temperature reheat mains, with rather less, around 160 ft/sec, for low-temperature reheat mains.

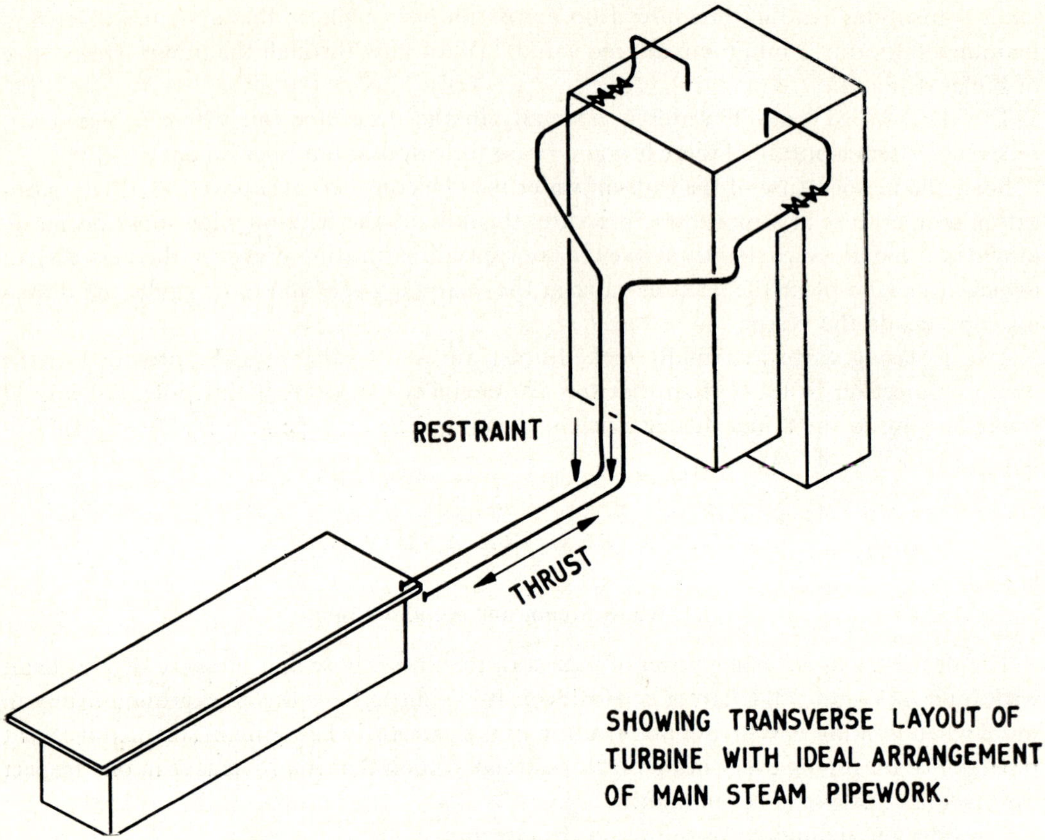

FIG. 4.6.1A. Main steam pipework — transverse turbine.

The first principles in laying out any steam main are to provide for flexibility and to minimise total piping run; both of these are very much a function of all the variables quoted above and the resulting pipe run will be a compromise on an economic basis.

Flexibility is a major consideration as total expansion of the pipe from the cold to the hot condition is considerable, about 9 in per 100 ft length from 16°C to 538°C. This is normally taken up in the cold condition by cold pull-up equal to the total expansion, thus when the pipe is heated and fully expanded there is no stress acting due to longitudinal expansion. If less than 100% cold pull-up is used, say 60%, the residual 40% will remain as a stress loading in the hot condition, possibly leading to relaxation due to creep and loss of cold pull-up.

The load due to cold pull-up is translated into thrusts and turning moments at the boiler and turbine terminal points. These are most difficult to calculate manually although there are manual solutions available, normally of the graphical type. The modern method is to use a computer program based on mathematical analysis. This gives not only the forces and moments at the anchor points, but also the point of maximum stress and the point of maximum bending moment. The results can then be compared with the design working stress for the pipe material.

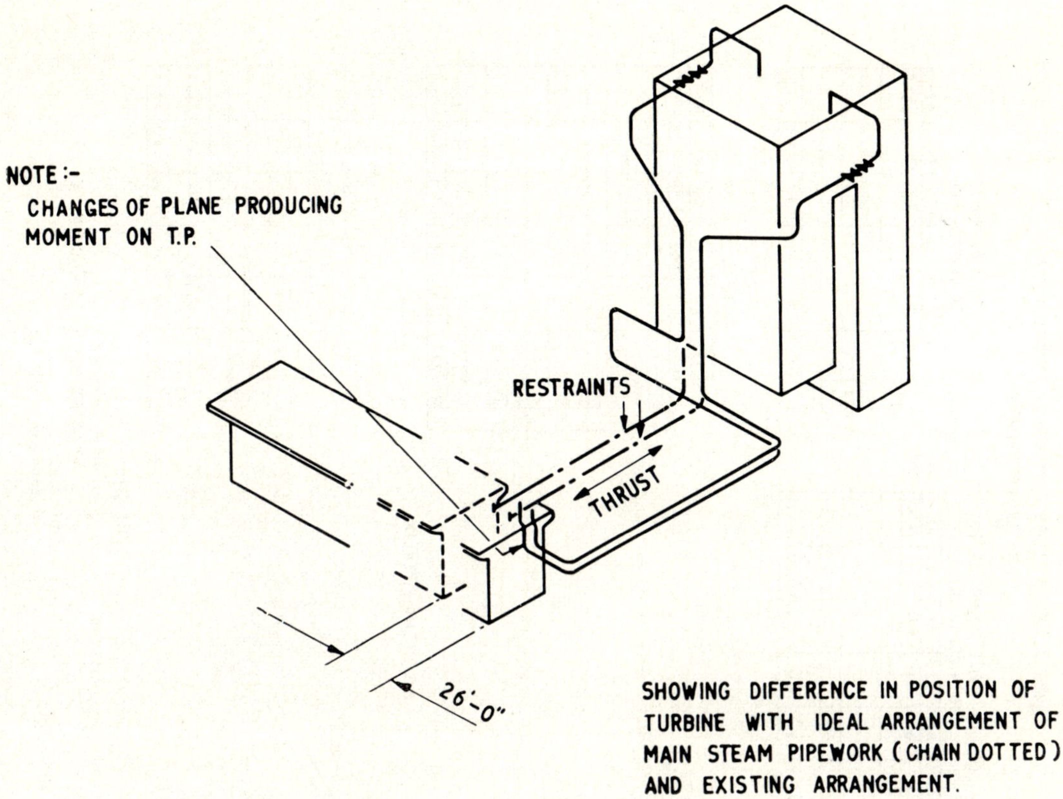

FIG. 4.6.1B. Main steam pipework – longitudinal turbine.

The position of the set relative to the boiler, as previously discussed, has a marked effect on the production of an economic steam pipe layout. The most logical arrangement is with a transversely installed turbine on the centre line of the boiler, with symmetrical pipe routes from each side of the boiler, converging in a flexible manner to a point below firing floor level. Here the last change in plane will be made and restraints located to anchor against forces which would, if unrestrained, set up a moment on the turbine connections through the long lever arm represented by the pipe run from this position to the turbine, as shown in Fig. 4.6.1A.

Assuming a long straight run to the turbine from this point, the forces at the turbine terminal can be resolved in a single thrust in the horizontal plane, provided that the turbine steam chests are arranged with connections in the same plane as the incoming pipe and have balanced flexible loop pipes to the turbine nozzles.

The low temperature reheat pipes connect at the turbine end direct on to the h.p. cylinder. With the longitudinal machine arrangement, the cold pull-up develops a sideways force on the machine and much lower loadings are necessary than with the transverse arrangement where the pipe force is in the same plane and much higher loadings can be tolerated. However, it should be noted that expansion bellows can be provided which will allow full compensation of low temperature reheat connections.

LOADS ACTING ON GUIDES (G) AND AT TERMINAL POINTS (T)

L.H.S. STEAM PIPEWORK

ORIGINAL ROUTE (CHAIN DOTTED)
100 % COLD PULL

	FX	FY	FZ	MXY	MYZ	MZX
	LBS	LBS	LBS	LB. FT.	LB. FT.	LB. FT.
G_2	-848		495			
G_4	1,988					
G_6			2,875			
G_8	1,924	-1,380				
G_{10}			-3,597			
T	-2,297	-1,815	228	-135,393	-87,267	32,450

EXISTING ROUTE
100 % COLD PULL

	FX	FY	FZ	MXY	MYZ	MZX
	LBS	LBS	LBS	LB. FT	LB. FT	LB. FT
G_2	171 –		129 –			
G_4	1					
G_6			2084 –			
G_8	1,981 –	315 –				
G_{10}			2,779			
T	1,461	1,080	566 –	72,921	70,992	2,407 –

EXISTING ROUTE
72 % COLD PULL

		FX	FY	FZ	MXY	MYZ	MZX
		LBS	LBS	LBS	LB. FT.	LB. FT.	LB. FT
G_2	HOT	37		28			
	COLD	123 –		93 –			
G_4	HOT	NEG[BLE]					
	COLD	"					
G_6	HOT			456			
	COLD			1,500 –			
G_8	HOT	433	69				
	COLD	1,426 –	227 –				
G_{10}	HOT			607 –			
	COLD			2,001			
T	HOT	319 –	236 –	124	15,939 –	15,517 –	526
	COLD	1,052	778	408 –	52,503	51,114	1,733 –

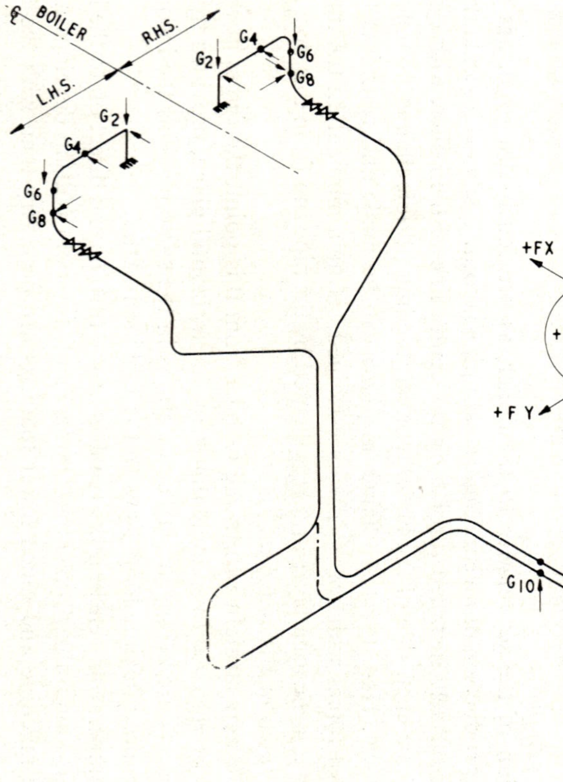

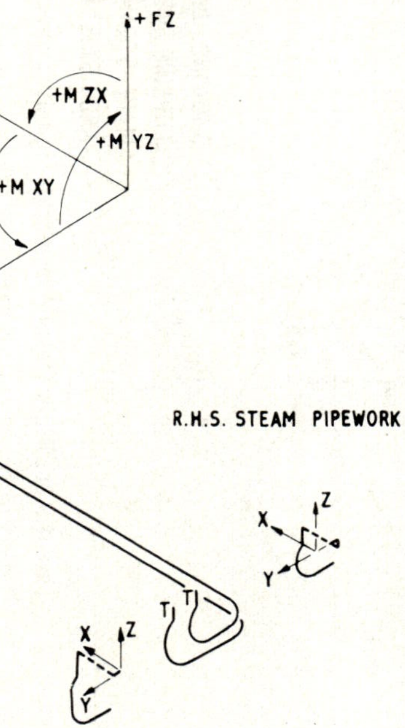

EXISTING ROUTE
100 % COLD PULL

	FX	FY	FZ	MXY	MYZ	MZX
	LBS	LBS	LBS	LB. FT	LB. FT	LB. FT
G_2	-1,249		767			
G_4	6,313					
G_6			4,062			
G_8	-3,324	6,260				
G_{10}			-5,672			
T	-2,652	-359	842	-37,160	-71,858	14,045

PIPE SIZE:
15" O.D. X 2·94" TH
10 1/16" CONTROLLED BORE ± 1/8" (Cr. Mo. Va.)
CONDITIONS:
2480 psi. 1055° F.

FIG. 4.6.1C. Loads acting on main steam pipes.

Another effect of having the turbine longitudinally arranged is the unsymmetrical piping arrangement shown in Fig. 4.6.1B. If the set was displaced by 26 ft off the centre line of the boiler a symmetrical layout could be achieved.

The particular layout above is given in more detail in Fig. 4.6.1C which also details the forces and moments at several anchors and restraints. In practice the moment about the turbine terminal in the *xy* plane, set up by the left-hand main steam lead and generated by 100% cold pull-up, could not be accepted by the turbine designer. This moment necessitated the introduction of a "relieving loop" as shown and reduction of cold pull-up to 72% in order to reduce the moment to an acceptable magnitude. The introduction of the relieving loop increases the pressure drop marginally as would be expected and it is essential to remember *all* piping design parameters so that when a change is made in one respect the effect is assessed in all respects.

The same applies to the reheat steam circuits. In this case the whole circuit must be optimised for flexibility and pressure drop. The h.p. cylinder exhaust pressure is known, also the design pressure at the intercept valves on the i.p. cylinder, so that the smallest pipe size can be obtained quite critically for maximum conditions.

The key to any piping layout is the initial provision of a line diagram and this must be the first drawing produced. It can be drawn without reference to physical layout in the early design stages and used to show valve requirements, instrument tapping points and much other information. The diagram shown in Fig. 4.6.1D is typical for a 500 MW unit and it should be noted that for this unit size and above, the total number of mains is double that of smaller units, in order to handle the fluid flow in pipe sizes which are practical to manufacture from a given billet size.

4.6.2. **Feed Piping**

Figures 4.6.2A and B are typical feed flow diagrams showing both split pumping as used on 300 MW units and single pumping using h.p. heaters as proposed for 500 MW units. Note that the diagrams also indicate all instrument tapping points.

Most of the comment given under main steam piping also applies to feed piping which must basically satisfy the following requirements:

(a) Be of sufficient bore diameter to carry the required quantity of water at an acceptable flow velocity.
(b) Be able to sustain the internal pressure exerted by the feed water over the range of operating temperature.
(c) Possess sufficient flexibility to accommodate cold pull-up without imposing excessive loads on terminal plant.

Until quite recently design feed flow velocities have been limited to 10 ft/sec, but 15 ft/sec was adopted on 300 MW units to reduce pipe bore and hence wall thickness for greater flexibility; no operational difficulties have been caused. Indeed, for the particular layout considered, it was shown, by comparing the additional capitalised pumping cost with the saving in piping costs, that it would have been economically justifiable to use flow velocities of 22 ft/sec for a unit load factor of 65%. Experience shows that a protec-

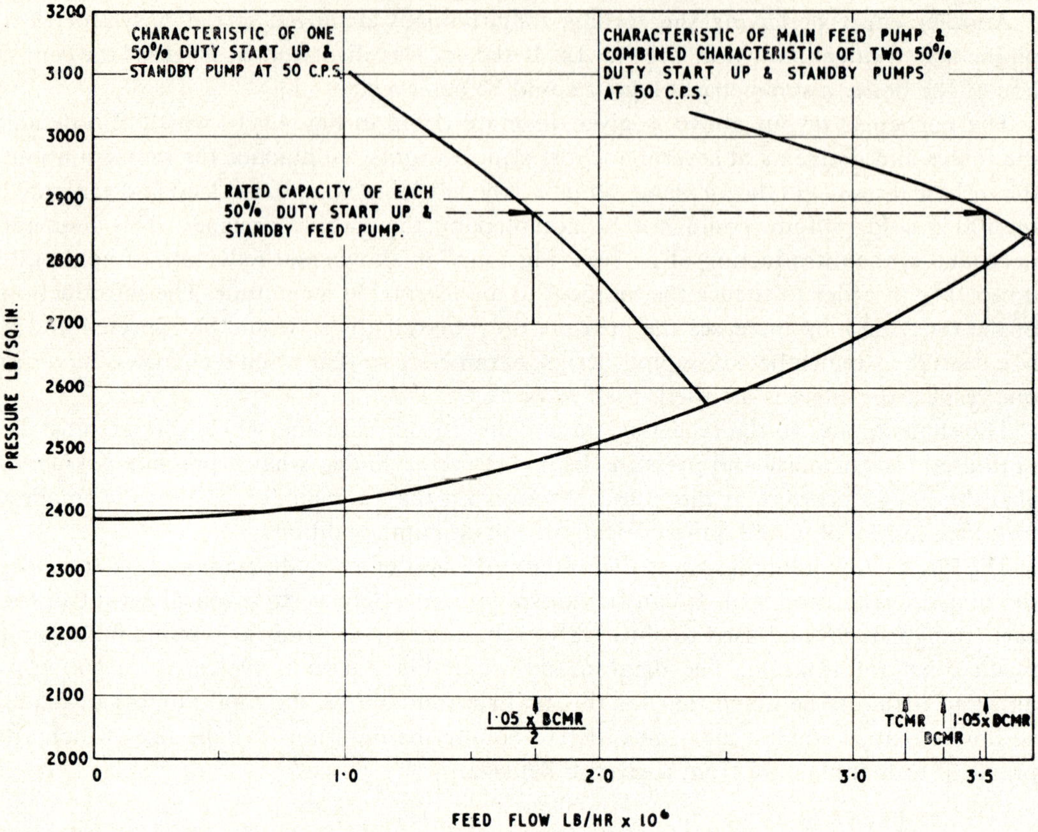

FIG. 4.6.2C. Typical feed pump and feed system characteristics.

tive magnetite coating persists in pump flow passages where water velocities of 50 ft/sec are normal, which suggests that erosion is not likely to be a problem. Consequently on any new design the calculated economic maximum velocity should be derived and used to obtain minimum bore size.

There is a big difference between the maximum pressure a feed pump will generate against a closed valve and boiler design pressure. On 300 MW units the feeder no-load pressure was 3900 lb/in^2 and the boiler design pressure was 3000 lb/in^2. By placing the boiler feed regulating valves immediately after the feed pump discharge isolating valve and ensuring that no further valves were installed downstream, the bulk of the feed piping could be designed for the lower pressure with considerable saving in cost and weight of metal and increased flexibility. On the later 500 MW units with single pumping the h.p. feed heaters are placed after the feed regulating valves (see Fig. 4.6.2B) and since each bank of heaters must be provided with isolating valves for maintenance purposes a relief valve on a by-pass is provided to prevent over-pressurising should both heater banks be inadvertently isolated.

The design flow upon which these considerations must be based is taken as the turbine M.C.R., shared equally by all feed lines to the boiler, although the feed pump itself is always designed to pump a maximum of about 105% boiler MCR, which is itself 5% more than turbine MCR. This is important to the layout engineer in determining the system resistance. It determines the pump duty and the character of speed reduction required to match pump output pressure with the system resistance. At the pump duty point (or pump MCR) a margin of 3% is normally added to cover the possibilities of:

(a) The feed pump failing to generate the pressure for which it was designed.
(b) The system resistance itself being greater than calculated.
(c) Loss of performance of the feed pump due to wear in service.

The following example is typical of the system resistance for a 500 MW unit with the feed pumps arranged upstream of the feed heaters (as in Fig. 4.6.1D).

SYSTEM CHARACTERISTIC

	lb/in^2 at boiler MCR flow
1. *Net static pressure*	
Control point pressure (i.e. superheater outlet pressure)	+2415
Drum water level to economiser inlet	+24
Economiser inlet to feed pump centre line	+34
Deaerator steam pressure at water level	−49
Deaerator water level to pump centre line	−40
Total	+2384
2. *Total frictional pressure*	
Secondary superheater	100
Primary superheater	60
Desuperheater and associated piping	35
Drum to economiser inlet	35
Feed piping:	
Feed pump discharge to economiser inlet (excluding h.p. heaters and feed regulating valve)	30
H.P. feed heaters (inlet to outlet)	97
Piping between deaerator and pump inlet	2
Feed pump suction strainers	3
Total	362
3. *Feed regulating valves controlled pressure drop*	50

(N.B. Positive items represent positive pumping head)

The summation of these three values represents the total pressure rise required to be generated between the feed pump suction and the delivery of the discharge non-return valve.

The total frictional pressure (section 2 of preceding list) must be calculated for 25%, 50%, 75%, 100% and 105% boiler MCR taking into account the change in feed temperature which occurs at each different flow rate. Table 4 is typical and one should be prepared for each type of feed pump (50% and 100% duty).

TABLE 4.

% boiler MCR	0	25	50	75	100	105
Boiler MCR $\times 10^6$ lb/h	0	0·84	1·675	2·51	3·35	3·5175
Net static pressure, lb/in²	2384	2384	2384	2384	2384	2384
Total friction pressure, lb/in²	0	24	94	201	362	400
Controlled p.d. across feed regulating valve, lb/in²	0	50	50	50	50	50
Total system resistance, lb/in²	2384	2458	2528	2635	2796	2834

The top and bottom lines are the horizontal and vertical coordinates of the system resistance curve, which should be plotted across the pump characteristic curves as shown in Fig. 4.6.2D.

It can be argued that with varying flow rate the true suction pressure from the deaerator will vary since bled steam pressure and frictional losses in the suction pipework will change with reduced flow. It is a useful exercise to see what the order of change is by repeating the calculation given in the section on feed pumps for, say, 50% flow basically to ensure that net positive suction head is maintained above minimum in all cases.

The general layout of feed piping is subject to space considerations and the need to maintain access above each pump for maintenance lifting. A typical modern layout is shown in Fig. 4.6.2E; this shows the complete integration of the 100% and 50% pumps and h.p. heaters at Fawley power station (4×500 MW units).

4.6.3. **Drains and Air Releases**

All steam piping must be fitted with ample drains for removal of condensate when warming up. This is helped by running horizontal pipes with a fall preferably in the same direction as the steam flow. The Board has standardised on $1\frac{1}{4}$ in and $\frac{3}{4}$ in drains for main steam piping and at each drain take-off point a thermocouple is fitted with indication in the control room. When the drain pocket is above saturation temperature the operator can close the drain valve remotely and prevent the passage of superheated steam to the blowdown vessel.

With the reheat steam circuit, the drains are taken to the condenser flash box so that vacuum is not lost via the i.p. cylinder connection during warming up, but the condenser

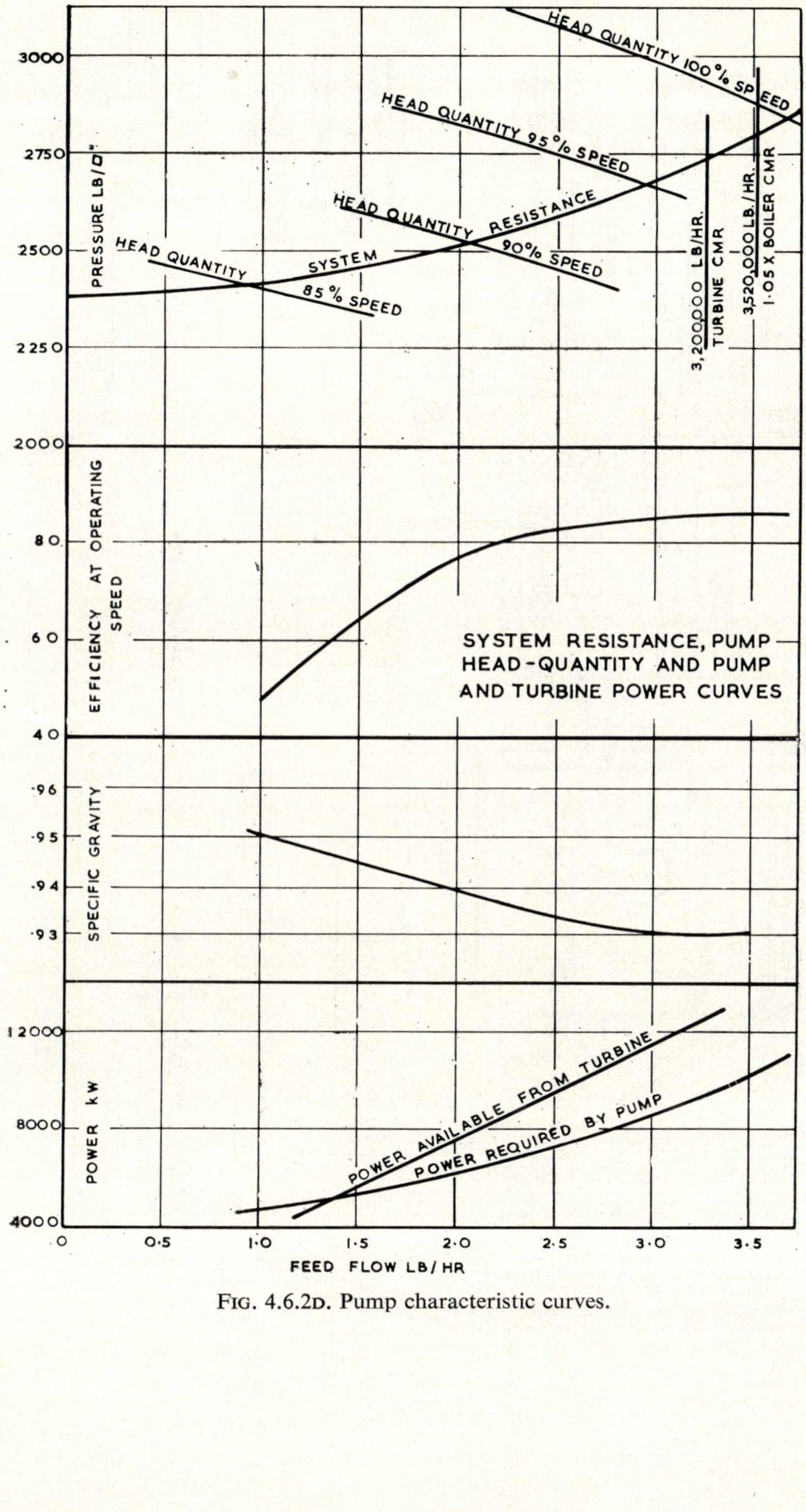

FIG. 4.6.2D. Pump characteristic curves.

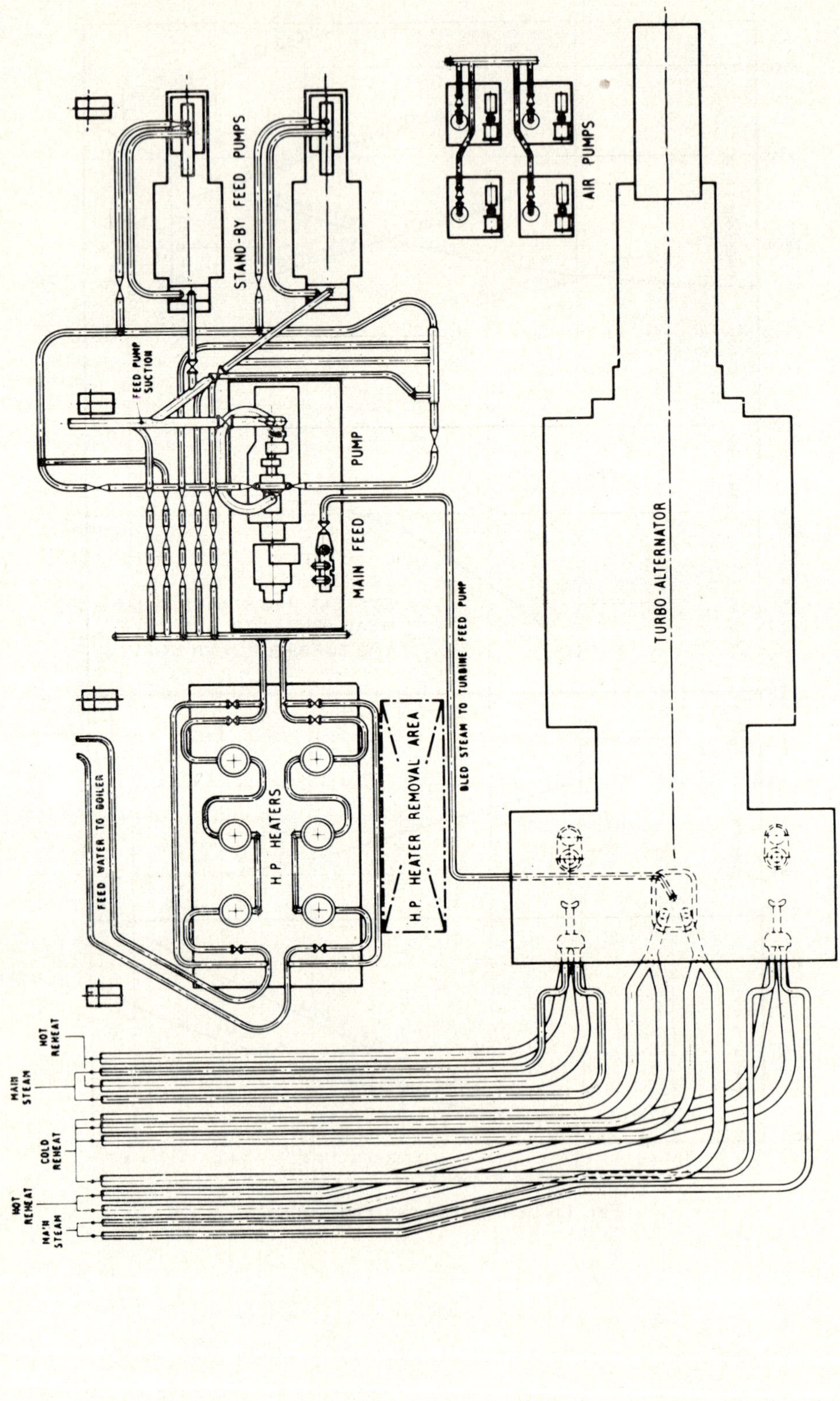

FIG. 4.6.2E. Typical layout of feed piping.

manufacturer will require the drain capacity to be strictly limited by means of an orifice plate at the flash-box nozzle. This limited drain flow rate cannot be specified because it is dependent on flash vessel designs, drainage from feed heaters and other plant. It is necessary, therefore, to review the total expected drainage under cold starting conditions and to ensure ample capacity so that start up time is not extended or flooding back could not take place. Account must be taken of the superior reheat steam pressure that could build up in the drain system.

All reheat boilers are provided with spray attemperators on the low temperature reheat inlet to the boiler and it is possible to flood the low temperature reheat steam pipe from this source. To overcome this an automatic trap drain must be provided which by-passes the drain valve and any water build up is automatically removed when the set is on load.

There is rarely any need to fit air release valves on main steam and reheat pipes because of the way they are routed and the ease with which air can be removed via the condenser air pumps. However, auxiliary and bled steam supply pipes must always be examined for the odd rising loop where air could be trapped and, where necessary, they must be vented to a suitable vessel.

The venting and drainage of feed piping is quite critical since whenever a pump is opened up for maintenance the immediate piping and the pump itself must be drained. Every length of pipe between valves must also be vented and drained, since the water between valves will be at high pressure and well above atmospheric saturation temperature. The procedure will be to open drains and drop pressure and then open the air releases for final emptying. When refilling, air vents are opened until water appears and indicates that all mains are properly primed.

4.6.4. **Supporting**

Pipes must be well supported and secure but at the same time the supports must permit expansion and cold pull, consequently heavy spring supports are used. These are set to carry a particular load and there is full coverage of the various approved types in C.E.G.B. standards. These standards cover constant load type supports which are spring loaded but have compensating geometric linkage which maintains a constant supporting pull on the pipe, however it moves due to expansion (see Volume 3, Chapter 5).

Hanger supports must provide for the rods to swing as the pipe expands, a typical support of this type, using so-called dummy branches as swivels, is shown in Fig. 4.6.4. Note that with very hot pipes the pipe clip must either be protected by a hard asbestos ring or alternatively the clip and hanger rod and all nuts, bolts and washers must be in a heat resistant alloy.

Swivelling hanger rods are always erected so that when the pipe is steaming the rod is in the true vertical position and working in the most efficient plane.

4.6.5. **Lagging**

On a main steam pipe this is likely to be 6 in or more thick, thus a steam main of 14 in o.d. will finish at about 26 in o.d. and a 20 in reheat steam main will finish at 32 in diameter. This fact must not be forgotten in laying out pipes which move under the

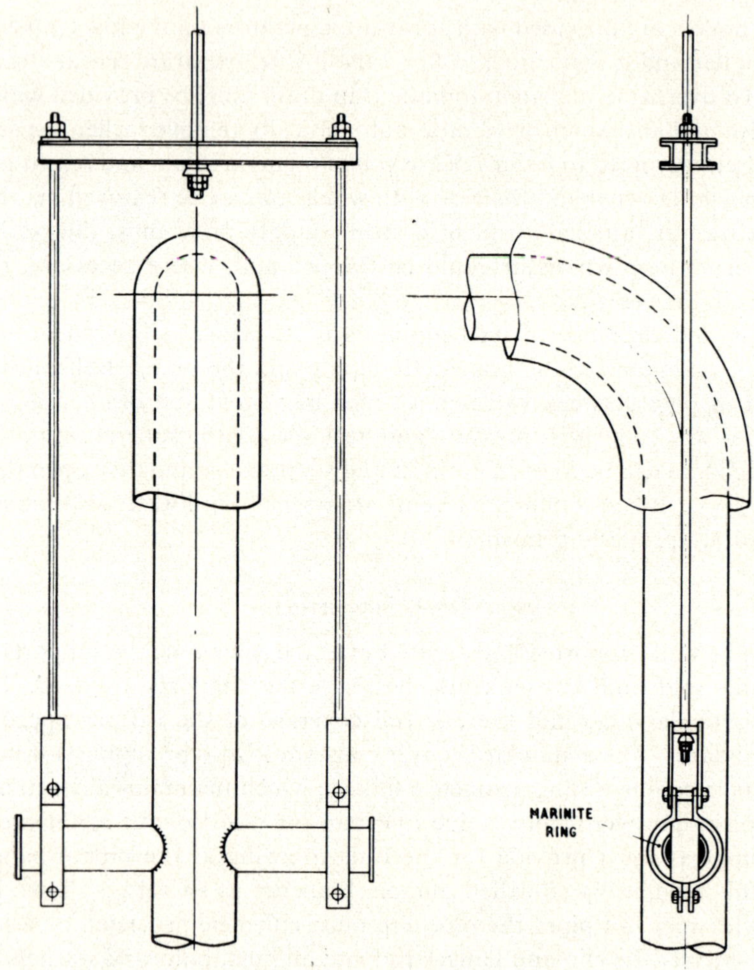

FIG. 4.6.4. Typical trunnion type support.

influence of temperature and the total clearances with other plant, holes through the floors and walls, must be carefully arranged.

For the future, lagging on pipework will consist of rigid preformed sections in double layer with the joints staggered between sections. These sections will be wired or bonded on and close cleaded with galvanised steel plate. The cleading surface temperature must not exceed 49°C for personnel safety and a general heat loss level of 85 Btu/ft^2 of lagged surface satisfies the economic cost equation. The graphs given in Figs. 4.6.5A to G give lagging thicknesses for most hot face conditions for a heat loss of 85 Btu/ft^2 and will enable the layout engineer to establish all clearances required.

4.7. L.P. PIPEWORK AND VALVES

The detailed requirements are dealt with in Volume 3, Chapter 5 but space must be provided for the following services:

Town water.
River water.
Treated (demineralised) water.
Chlorine solution dosing.
Reserve feed water.
Blowdown water recovery.
Fire hydrant system.
Fixed water spray protection.
Ash sluicing supplies.
Nitrogen supplies.
Potable water supplies (fit for drinking).
Compressed air.

Many of these mains lend themselves well to trunk main routing through the length of the station with branch supplies out to each unit and an ample route reservation is essential, probably in the heater bay between the sets and the boilers.

In the early stages of design it is essential that diagrams are prepared for all l.p. services and a pattern established with an approximate idea of valve and tube requirements obtained. The diagrams are then constantly developed as information becomes available about services required by main plant items.

At the earliest stage possible piping should be run on the model and from this isometric drawings prepared. Very few rules can be given, since the l.p. system has to be very flexibly planned; but as with all piping problems the route and pressure drop should be optimised for complete efficiency.

Piping should not be run in trenches or small ducts hidden from sight, but with its associated valves should be visible and accessible for maintenance. There must be proper drainage falls with air release valves fitted where necessary, and the pipework must be well supported and secure. Pipe joints are usually the weakest link in a piping system and demand the highest standards of materials and workmanship. The number of joints should be minimised by the use of welding.

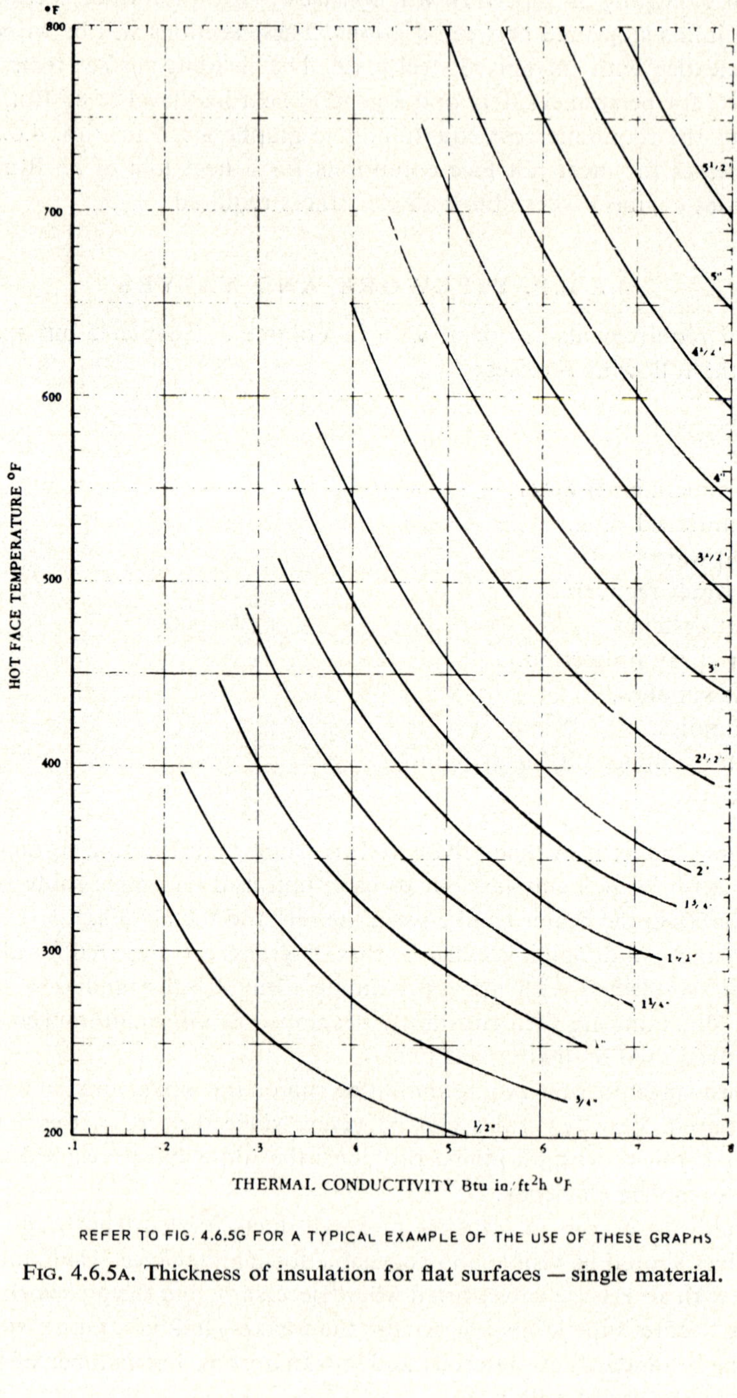

REFER TO FIG. 4.6.5G FOR A TYPICAL EXAMPLE OF THE USE OF THESE GRAPHS

FIG. 4.6.5A. Thickness of insulation for flat surfaces — single material.

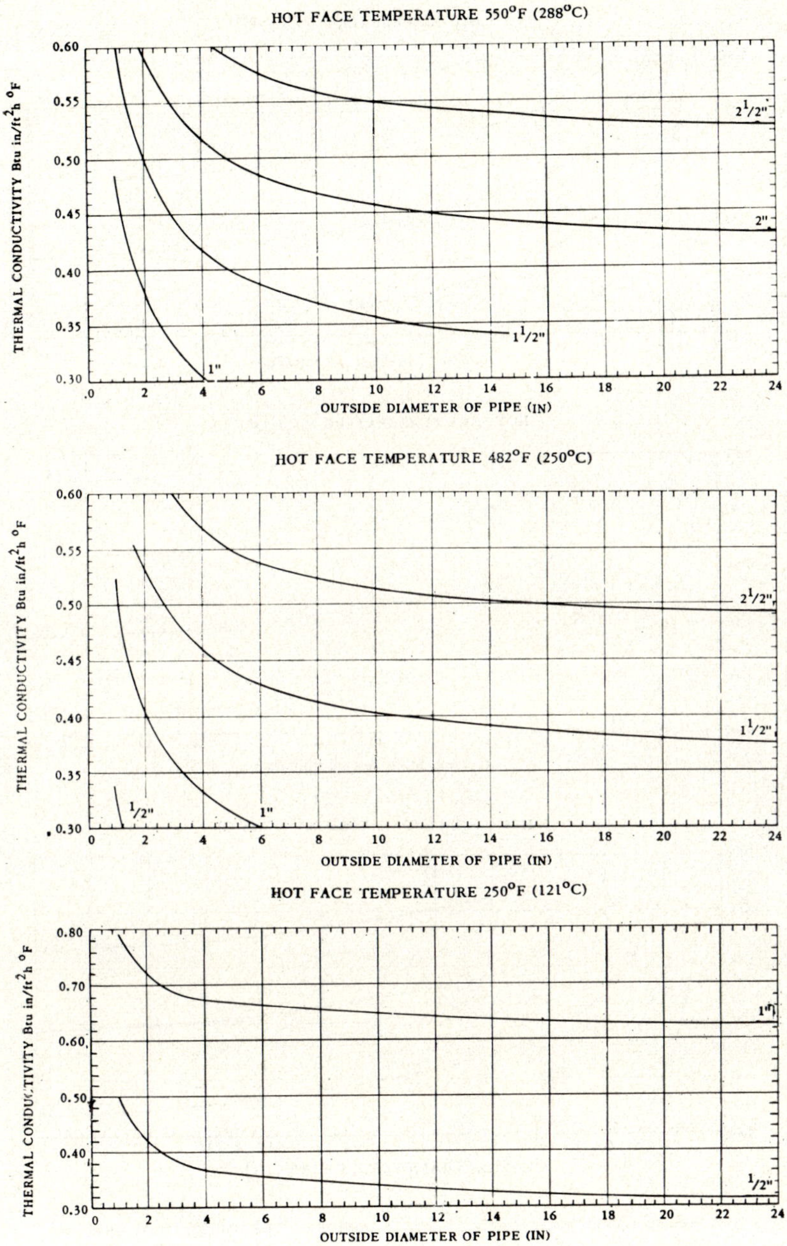

REFER TO FIG. 4.6.5G FOR A TYPICAL EXAMPLE OF THE USE OF THESE GRAPHS

FIG. 4.6.5B. Thickness of insulation for curved surfaces – single material.

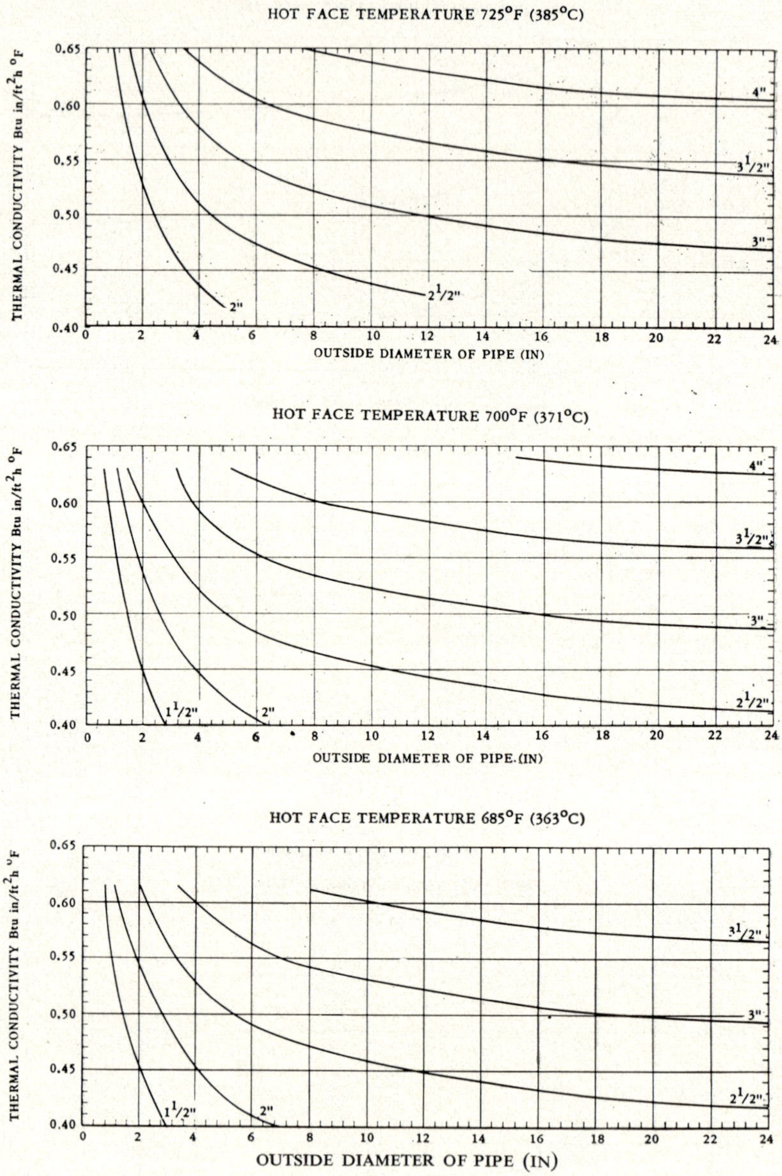

REFER TO FIG. 4.6.5G FOR A TYPICAL EXAMPLE OF THE USE OF THESE GRAPHS

FIG. 4.6.5C. Thickness of insulation for curved surfaces – single material.

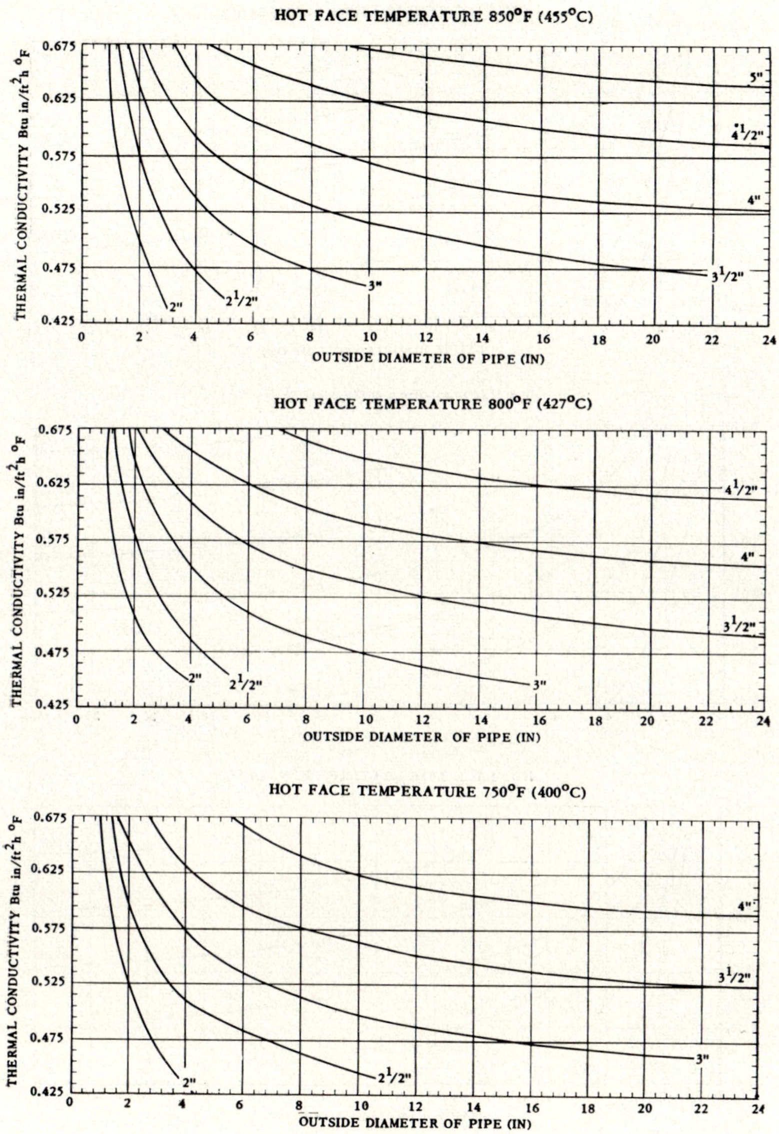

REFER TO FIG. 4.6.5G FOR A TYPICAL EXAMPLE OF THE USE OF THESE GRAPHS

FIG. 4.6.5D. Thickness of insulation for curved surfaces – single material.

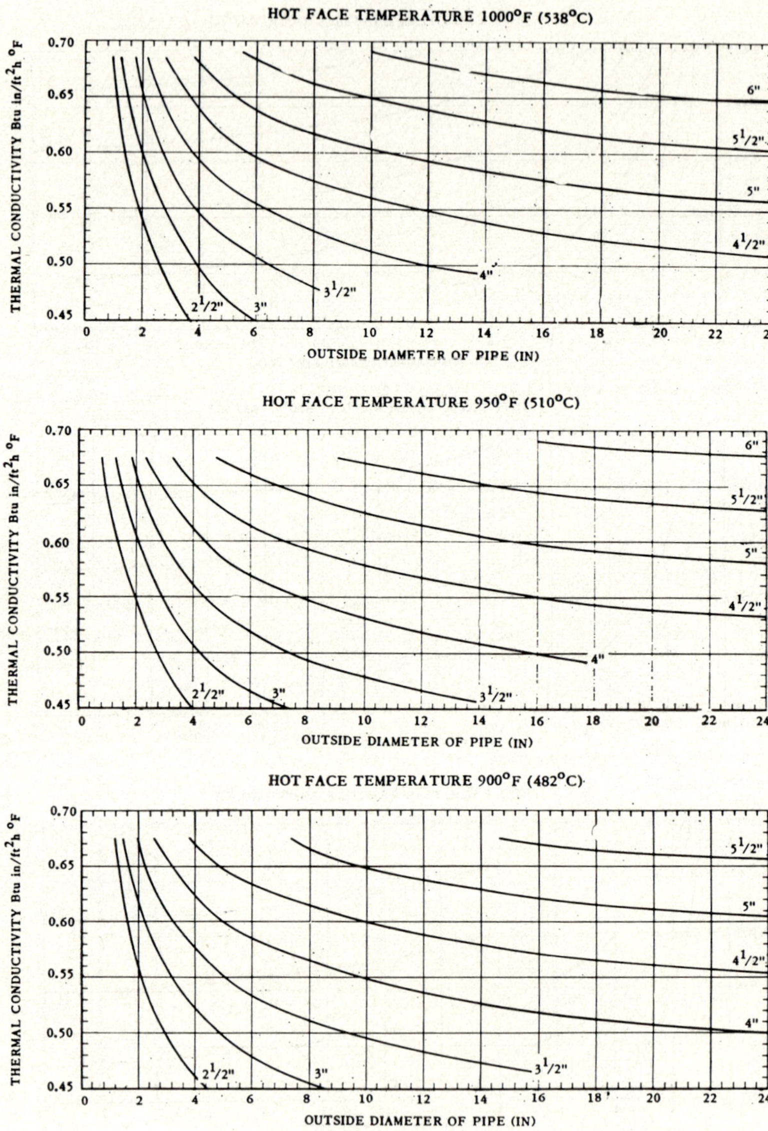

FIG. 4.6.5E. Thickness of insulation for curved surfaces — single material.

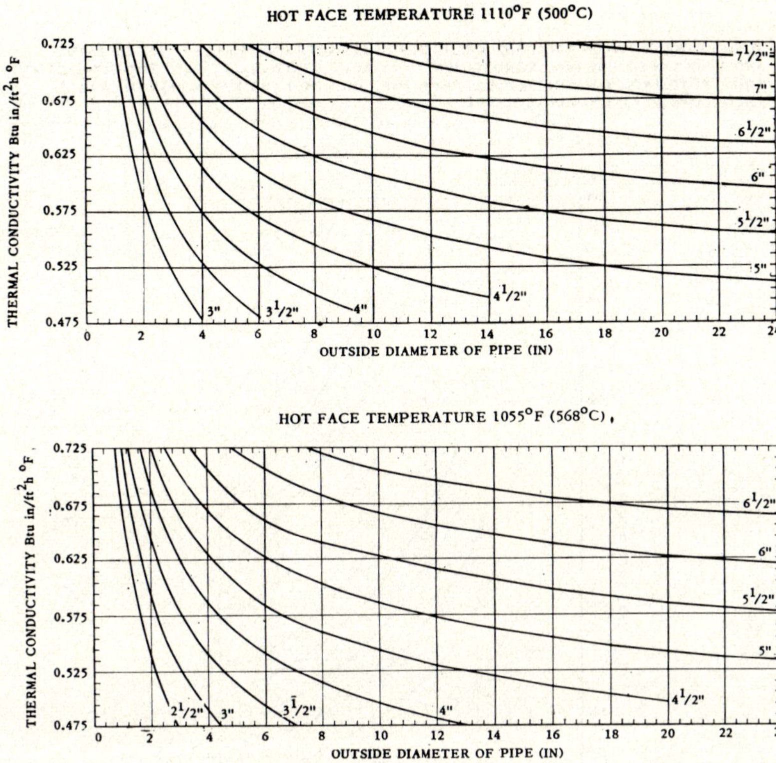

REFER TO FIG. 4.6.5G FOR A TYPICAL EXAMPLE OF THE USE OF THESE GRAPHS

FIG. 4.6.5F. Thickness of insulation for curved surfaces – single material.

4.8. WATER STORAGE TANKS

These represent a big space requirement, for instance at least 1,300,000 gallons of town water should be stored to cover station requirements for 24 h on 4×500 MW units. Reserve feed water should be available for at least 1 h MCR supply for each boiler. This vast amount of water is not now stored at high level in the boiler or turbine house, since it is uneconomic in terms of steelwork to support this load, and a disaster would occur if for instance a tank weld failed.

Space can normally be found for unit feed water storage at boiler basement level, and town water storage is best placed as an adjunct to the water treatment plant.

The fire service pumps on the station should normally draw from town water storage and it is necessary to ensure that there is a reserve of town water always available in one tank.

Vertical cylindrical tanks are the most economic. The height (and consequently the ground load in terms of weight per ft²) can be varied with the base diameter and selection of size can be a compromise between the ground area available, its capacity for load bearing and the height available. Tanks 30 to 40 ft diameter and 30 ft high are now quite common on advanced units.

(a) IS AN EXAMPLE TAKEN FROM FIG. 4.6.5A. FOR A HOT SURFACE TEMPERATURE OF 650°F AND A DECLARED THERMAL CONDUCTIVITY VALUE OF 0.5 BTU IN/FT² H °F THE COINCIDENT POINT READ FROM THE GRAPH IS BETWEEN THE LINES OF 3 AND 3½ INCHES THICKNESS THEREFORE 3½ INCHES IS REQUIRED

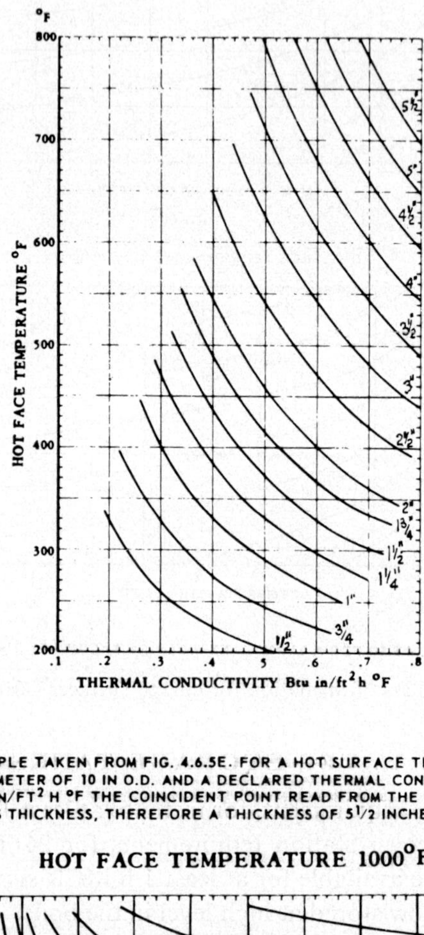

(a)

(b) IS AN EXAMPLE TAKEN FROM FIG. 4.6.5E. FOR A HOT SURFACE TEMPERATURE OF 1000°F, A PIPE DIAMETER OF 10 IN O.D. AND A DECLARED THERMAL CONDUCTIVITY VALUE OF 0.625 BTU/IN/FT² H °F THE COINCIDENT POINT READ FROM THE GRAPH IS BETWEEN 5 AND 5½ INCHES THICKNESS, THEREFORE A THICKNESS OF 5½ INCHES IS REQUIRED.

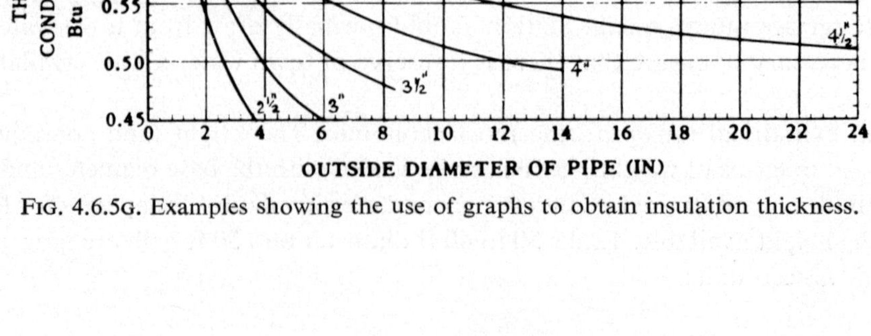

FIG. 4.6.5G. Examples showing the use of graphs to obtain insulation thickness.

4.9. WATER TREATMENT PLANT

The unit arrangement of the treatment plant is best left in expert hands, but the layout engineer must provide the necessary space in the right place relative to the main plant. The most favourable position is at the centre of the station, but good road access is necessary for chemical deliveries and it is normally best to keep sulphuric and hydrochloric acid storage tanks outside the building proper, with a bund wall around them enclosing sufficient volume to retain a full tank of acid in the event of a tank failure.

A central position should, however, be the first consideration since pumping of evaporated water from the unit evaporators will be considerably simplified and more economic. Similarly, distribution of treated water back to each unit is much less involved.

4.10. CHLORINATION PLANT

Chlorine is a poisonous gas and stringent safety measures are always adopted by the specialist contractors engaged in this work.

It is now becoming the practice to have an outdoor bulk tank installation which is refilled from road or rail tankers. This installation is built very securely with strict inspection of all work and can be remote from main areas of general access. Special procedures are laid down for operators concerned with the plant.

The chlorine is led in the liquid phase through a pipeline with all welded joints (radiographed) to electric evaporators and immediately into the water chlorinating equipment. The evaporators and chlorinators should be housed in a weatherproof building with clear access in and out for personnel. The building must be provided with forced ventilation via a plenum duct at floor level.

The chlorine dosing solution is piped in rubber lined piping to the injection points at the condensers. The length of this pipeline is relatively unimportant compared with the need to have the plant well away from areas of general access, and the pipeline need only be sized to carry the dose required to the farthest injection point, since each injection point dose is programmed through a master clock and usually only one dose is injected at any given time.

With some British waters chlorine injection is needed at the CW intake to prevent mussel or organic growth on the CW culvert walls. Depending on the distances involved it may be economic in some cases to provide a separate chlorinating plant at the CW intake position.

4.11. CW PUMPHOUSE

The position of this plant relative to the main station is dictated usually by site and civil engineering considerations, but the number of pumps provided, the type of culvert, and the standby capacity has varied widely in the past. The decisions made affect the intrinsic layout of the pumphouse plant and the condenser connections. The trend appears to be towards 100% capacity unit pumps with one standby 100% pump for a four-unit station. The five pumps discharge into a common bus chamber with four culverts, one running to each unit. Flow velocities in the culverts will be high, in the order of 10–12 ft/sec to scrub

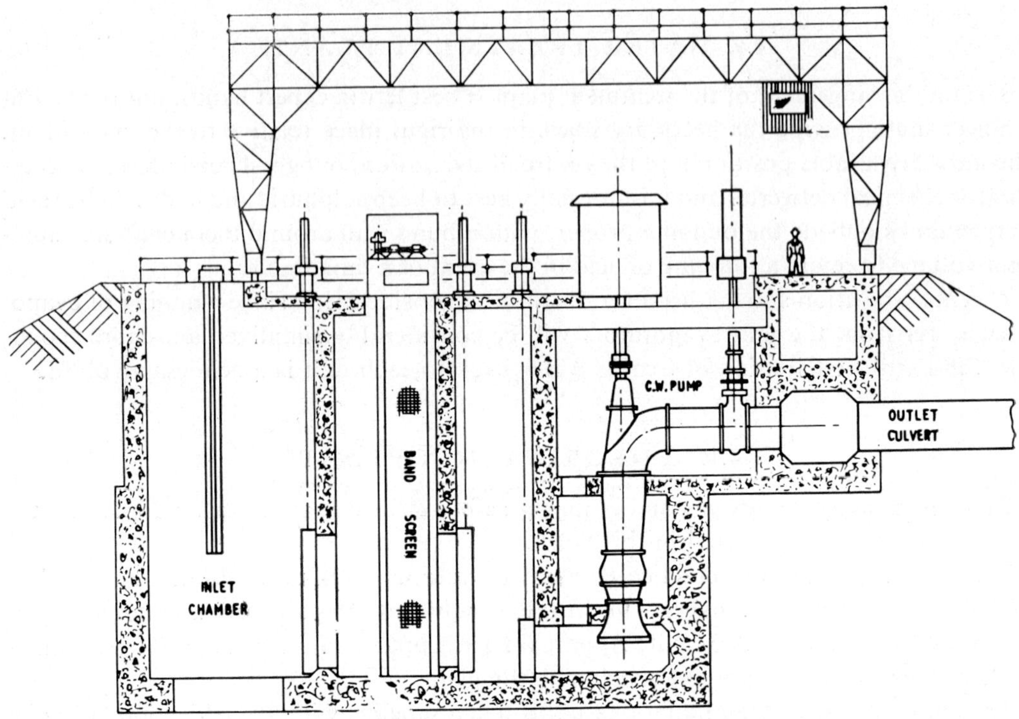

FIG. 4.11A. Cross-section through CW pumphouse for 200 MW unit.

the inside surfaces of the culverts. Because of this the culverts will be circular since this shape presents the smallest wetted perimeter ratio to flow area and pressure drop is minimised.

Screening plant is dealt with in detail together with types of CW pumps in Volume 3, Chapter 4. This plant has to be laid out economically, with due consideration of the expensive civil engineering work involved.

The minimum possible number of penstocks should be included in the layout. These are expensive and infrequently operated. However, facilities must be provided to isolate each screen and pump for maintenance. If each pump and screen has a unit inflow channel, then one penstock is needed on the inlet to the screen. The CW pump can be used for dewatering down to minimum suction levels and both items are then available for maintenance.

Because of the great mass of water in the discharge culverts, modern pumps are fitted with controlled discharge isolating valves of the butterfly type which close automatically if pressure head is lost from the pump. The valve closing time is variable and set by practical experiment at site to avoid any possibility of heavy back pressure surge from the discharge culvert. Also when the pumps discharge into a common chamber there will be a maximum time limit on reverse rotation of the idle pump due to delivery pressure from the working pumps; this is largely set by pump bearing design, and varies between 10 and 20 min for pumps of 100,000 gal/h capacity.

Where there is a high tidal range (on the Severn for instance there is a difference of 40 ft between low and high tide) the suction pressure at each pump will vary and in some stations it is economic to provide each pump with speed variation to maintain constant pump output head without throttling valves. Tidal conditions vary from station to station, and an economic appraisal must be carried out for each layout to decide whether variable speed pumps are justified.

The pumphouse must be equipped with an overhead crane over the pumps, screens and valves and a loading bay with free access for vehicles is also essential.

Figure 4.11A is a cross-sectional elevation showing the installation of a 100,000 gal/min pump for a 200 MW unit. The pumps are in the open and served by a Goliath type crane;

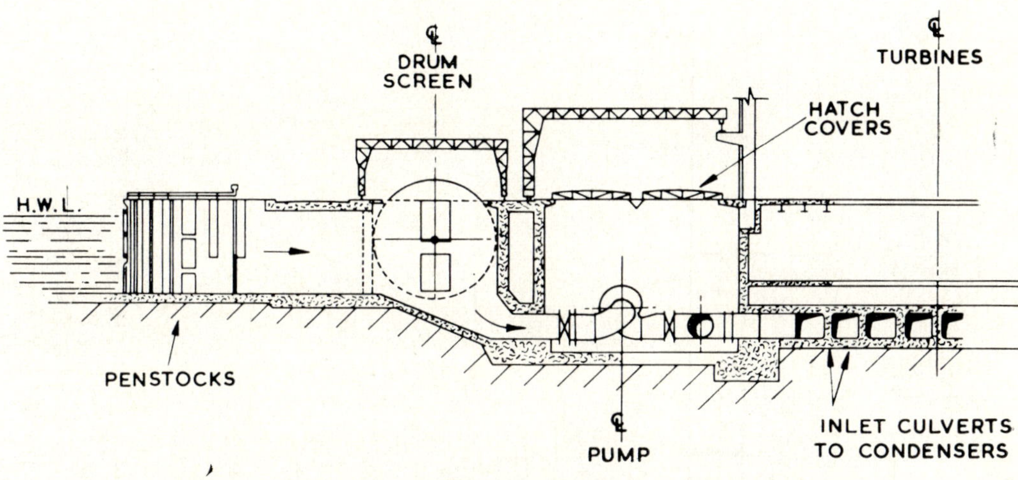

FIG. 4.11B. Sectional elevation through Fawley CW pumphouse.

the arrangement works well, but maintenance in wet cold weather is uncomfortable and a light curtain walled building would not be expensive. Figure 4.11B shows the arrangement adopted for Fawley power station. In this case the pumphouse is at the end of the turbine house and is served by the turbine house crane. Another interesting feature is that the CW intake forms a canal which can accommodate coastal shipping and the berth for these steamers is below by the turbine house crane rails.

4.12. PEAK LOAD GAS TURBINES

All modern stations are now provided with unit "House Sets" which are gas turbo-alternators of about 25 MW output, although a number of 17·5 MW capacity have been installed. The gas generators are intended primarily to provide standby supplies to unit auxiliaries, particularly feed pumps, but they can also be used for peak loading.

The relative position of the gas turbines to the remainder of the plant is less important than the convenient siting of chimneys for the exhaust gases and the coldest possible air suction for combustion, since there is a significant effect on the efficiency of the plant if these two items are badly arranged. Figures 4.12A to F show various chimney

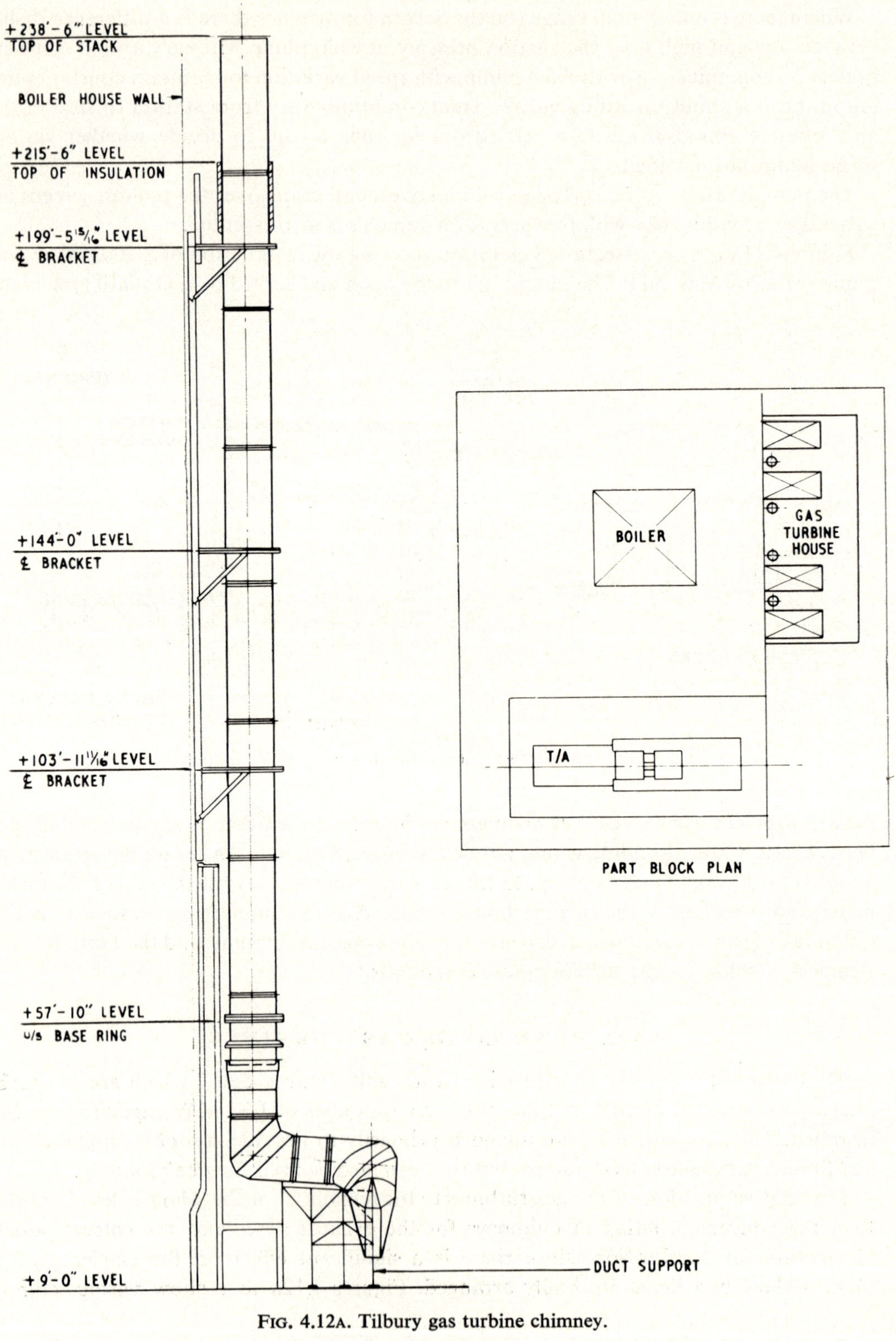

FIG. 4.12A. Tilbury gas turbine chimney.

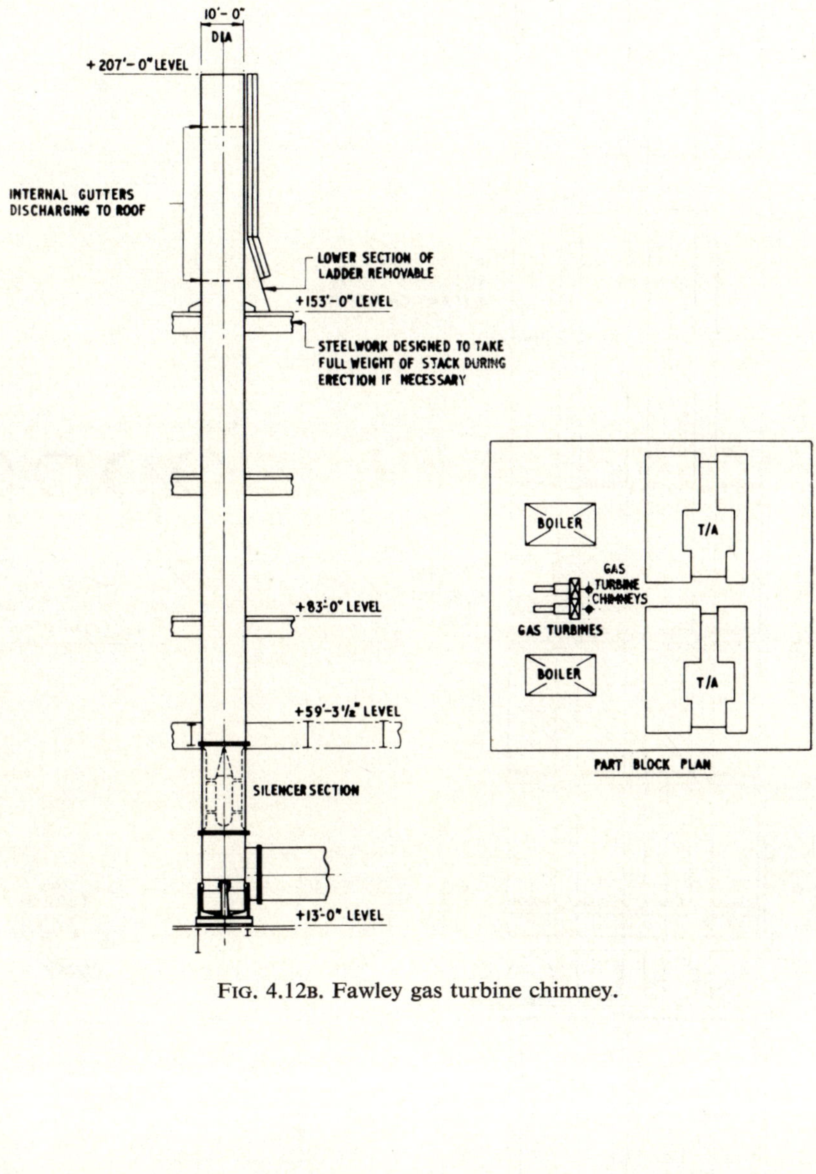

FIG. 4.12B. Fawley gas turbine chimney.

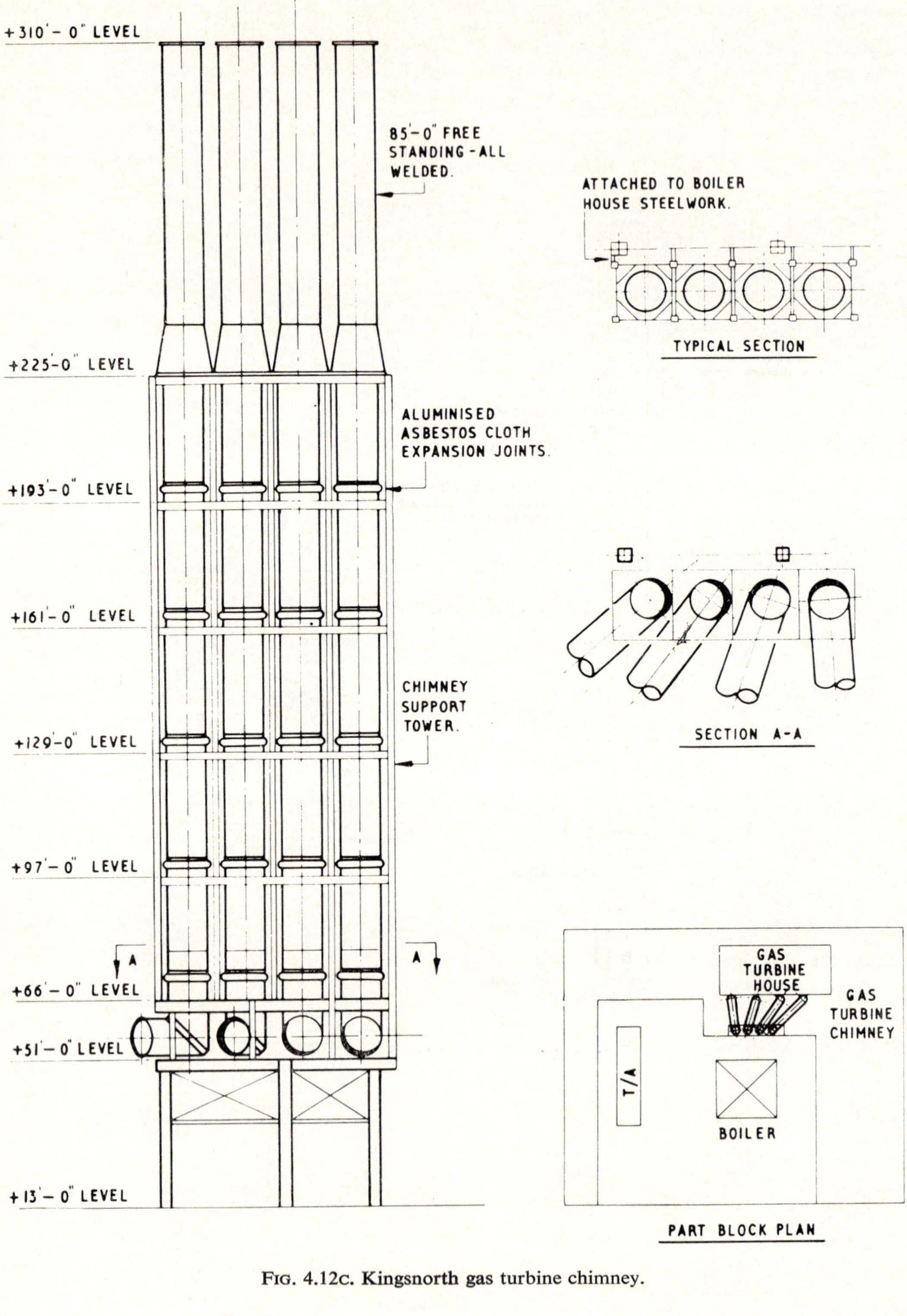

FIG. 4.12C. Kingsnorth gas turbine chimney.

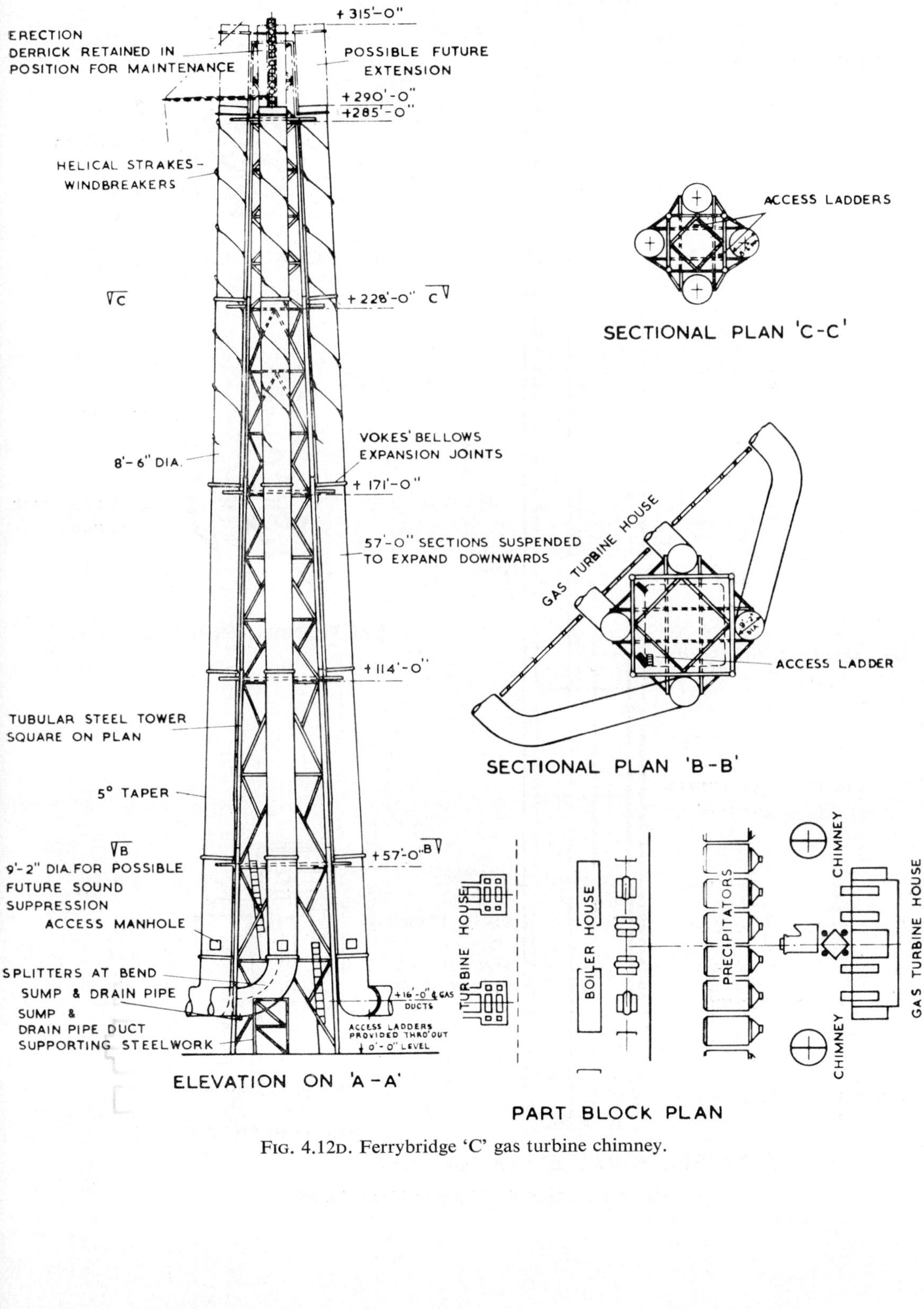

FIG. 4.12D. Ferrybridge 'C' gas turbine chimney.

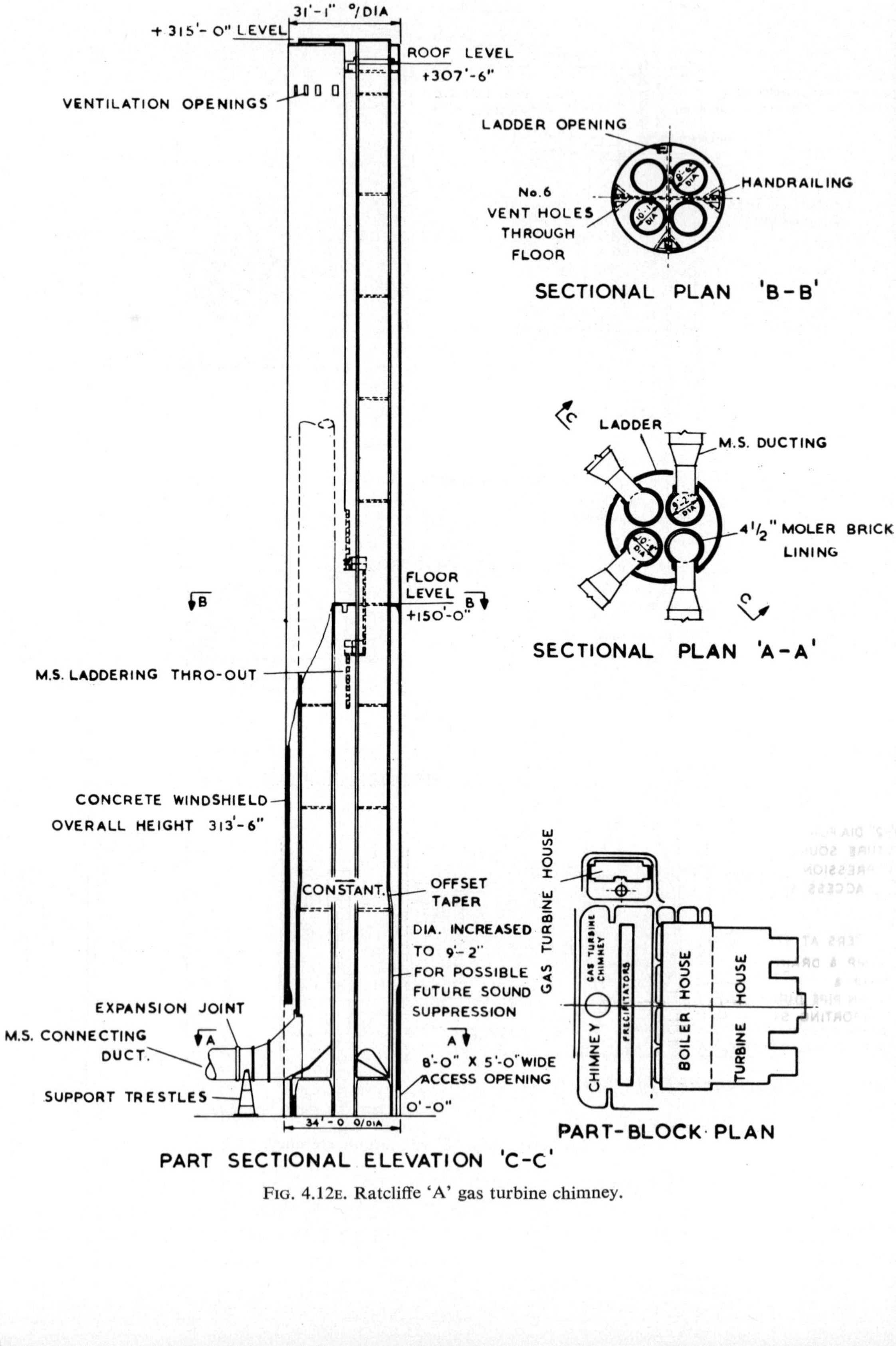

FIG. 4.12E. Ratcliffe 'A' gas turbine chimney.

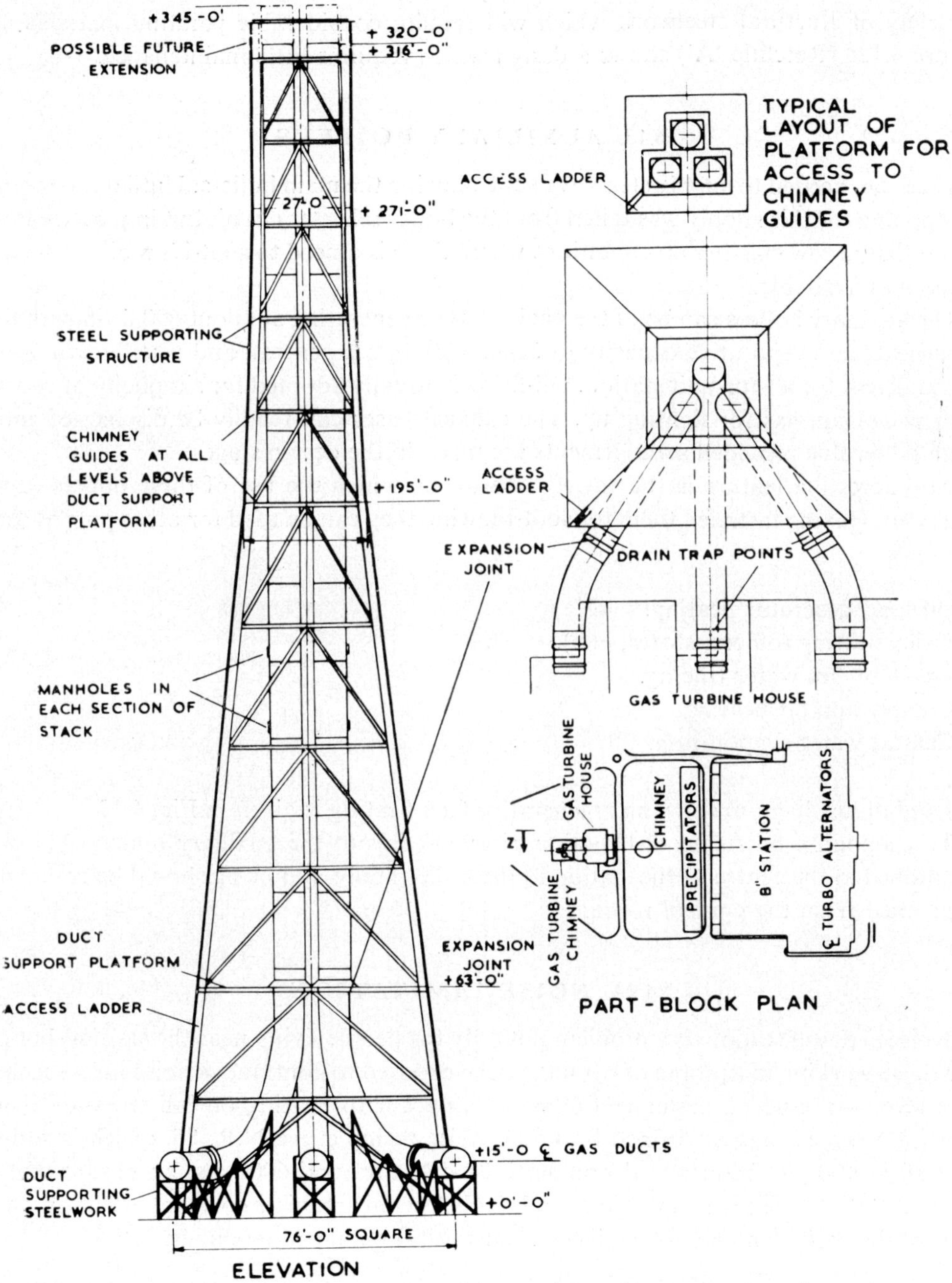

FIG. 4.12F. Aberthaw 'B' gas turbine chimney.

arrangements adopted at a number of stations and also give block plans showing the relative positions of the gas turbines. Many of these have steel chimneys supported by a variety of stuctural steelwork which will require considerable painting maintenance. Figure 4.12E (Ratcliffe 'A') shows a design which requires little maintenance.

4.13. AUXILIARY BOILERS

These are needed to supply steam for soot-blowing the main boilers. Until quite recently the tapping for this supply was taken from the boiler reheater circuit but in practice it was found that at low starting up conditions where there is a need to soot-blow air heaters, no steam was available.

The auxiliary boilers can be of the package type generating at about 600 lb/in with little superheat; an evaporative capacity of 30–45,000 lb/h is general, and usually two boilers are sufficient for a four-unit station. Oil-firing is usually adopted for simplicity of control, with petroleum gas for lighting up. The exhaust gases can readily be discharged into a main boiler flue provided arrangements are made in the design stages.

An interesting feature is the variety of ways in which the use of these boilers can be extended. Having installed them for soot-blowing they can be used for other purposes, for example:

Off-load deaerator heating.
Space heating (offices, stores, etc.).
Steam or hot water tracing.
Combustion air heating.
Central water evaporation.

An ambitious diagram showing regenerative feed heating is given in Fig. 4.13.

To economise on steam distribution pipework layout the auxiliary boilers should be positioned at the centre of the station in the boiler house and firing should be completely controlled from the control room.

4.14. NOISE LIMITATION

Noise in power stations is a problem not only for people living near the station, but also for those working in it, some of whom may be exposed to continuous noise for a full shift. The level above which permanent damage can occur to hearing on full exposure is now generally agreed as that defined by I.S.O. noise rating curve N.R. 85, which is roughly 100 dB at 100 c/s frequency dropping to 80 dB at around 6000 c/s. To ensure that the general station noise is maintained at a level below this requires constant attention on the part of the design and layout engineers. The critical plant can readily be listed.

Unit generator transformers.
The turbo-generator set.
Gas turbines.
Compressors.

Draught plant.
Milling plant.
Steam exhausts.

A great deal can be achieved by lining buildings with a good thickness of mineral wool faced with perforated board or sheeting, and in most modern turbine houses a belt of perforated absorbent tiles is usually installed on the walls at operating floor level.

Continuing efforts are necessary to encourage plant designers to adopt means of limiting noise at source particularly with transformers and rotating plant. The latest 500 and 660 MW turbines appear to have a general noise output above N.R. 85, and the problem of reducing this level is currently under investigation.

There exists a good deal of information on limiting noise, but the first need is to institute preventative methods at the design stages and then to judge results by the actual noise spectrum existing in the station environment.

Simple attenuation systems for gas turbine exhausts are shown in Figs. 4.14A and B and the principle is to present a maximum noise absorbent surface (formed by acoustic panels), to the gas flow without significantly impeding the flow. Similar techniques are used for the air supply to the gas turbine.

Steam exhausts can also be dealt with quite simply and the following notes give useful information to enable this to be achieved. Calculation of fundamental frequency:

For jet noise:

$$f = 0{\cdot}25\,\frac{v}{d}\,\frac{C_0}{C_j}$$

where v = velocity of flow (ft/sec),
d = jet diameter, in feet,
C_0 = ambient speed of sound (ft/sec),
C_j = local speed of sound in jet,

and in the case of a sonic jet:

For reverberating noise:

$f = \dfrac{C}{2L}$ for an "open" pipe, with overtone frequencies proportional to 1, 2, 3, 4, 5, etc.,

$f = \dfrac{C}{4L}$ for a "closed" pipe, with overtone frequencies proportional to 1, 3, 5, 7, 9, etc.,

where C = velocity of sound in the steam (ft/sec),
$= 68{\cdot}3\,(GkpV)^{1/2}$,
L = length of pipe (ft),
Gk = adiabatic index,
p = average steam pressure in pipe (lb/in² g),
V = average specific volume of steam in pipe (ft³/lb).

A pipe with a cowl at the outlet would be classed as a "closed" pipe, a pipe completely straight and open at the end would be classed as an "open" pipe.

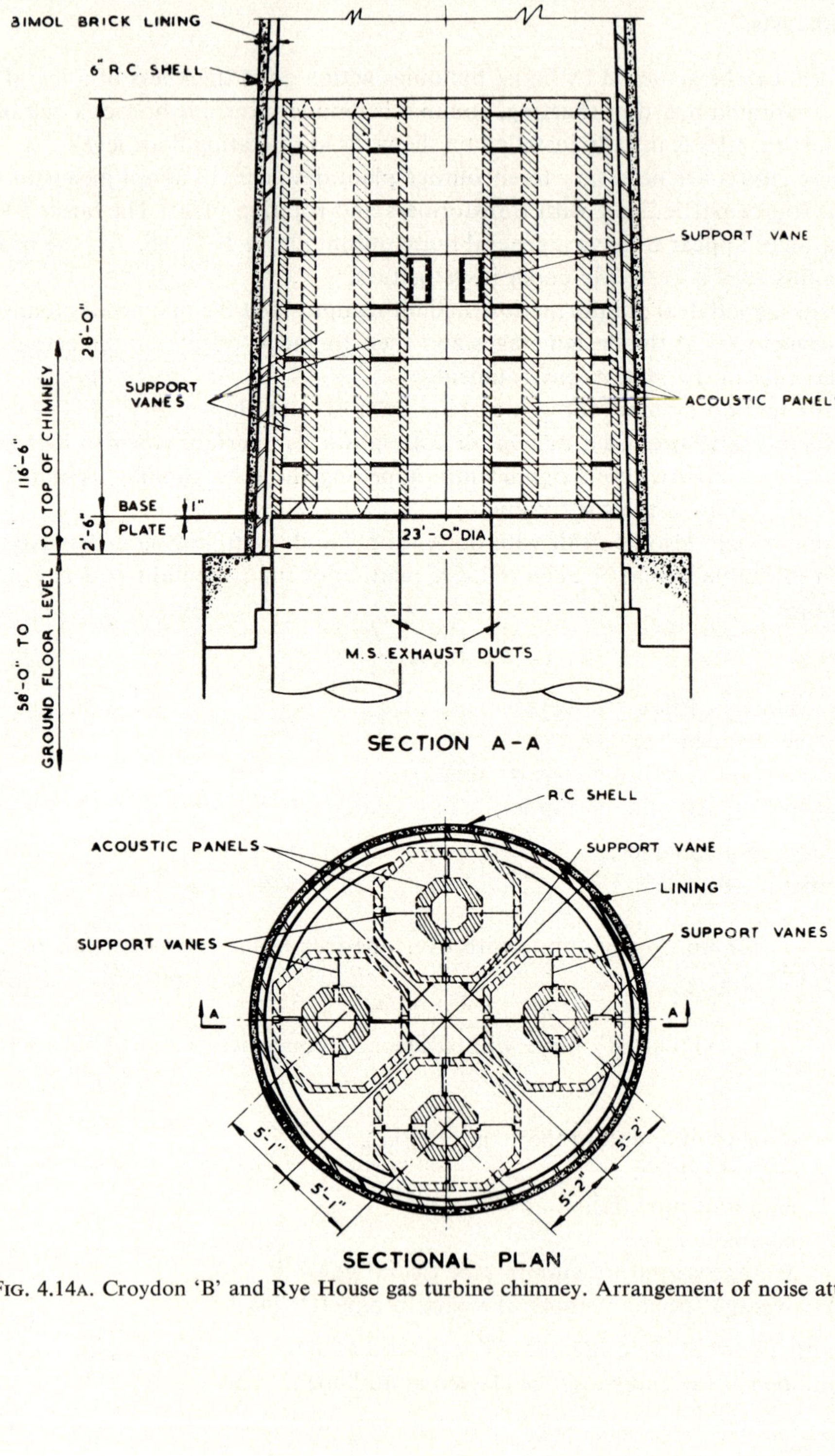

FIG. 4.14A. Croydon 'B' and Rye House gas turbine chimney. Arrangement of noise attenuator.

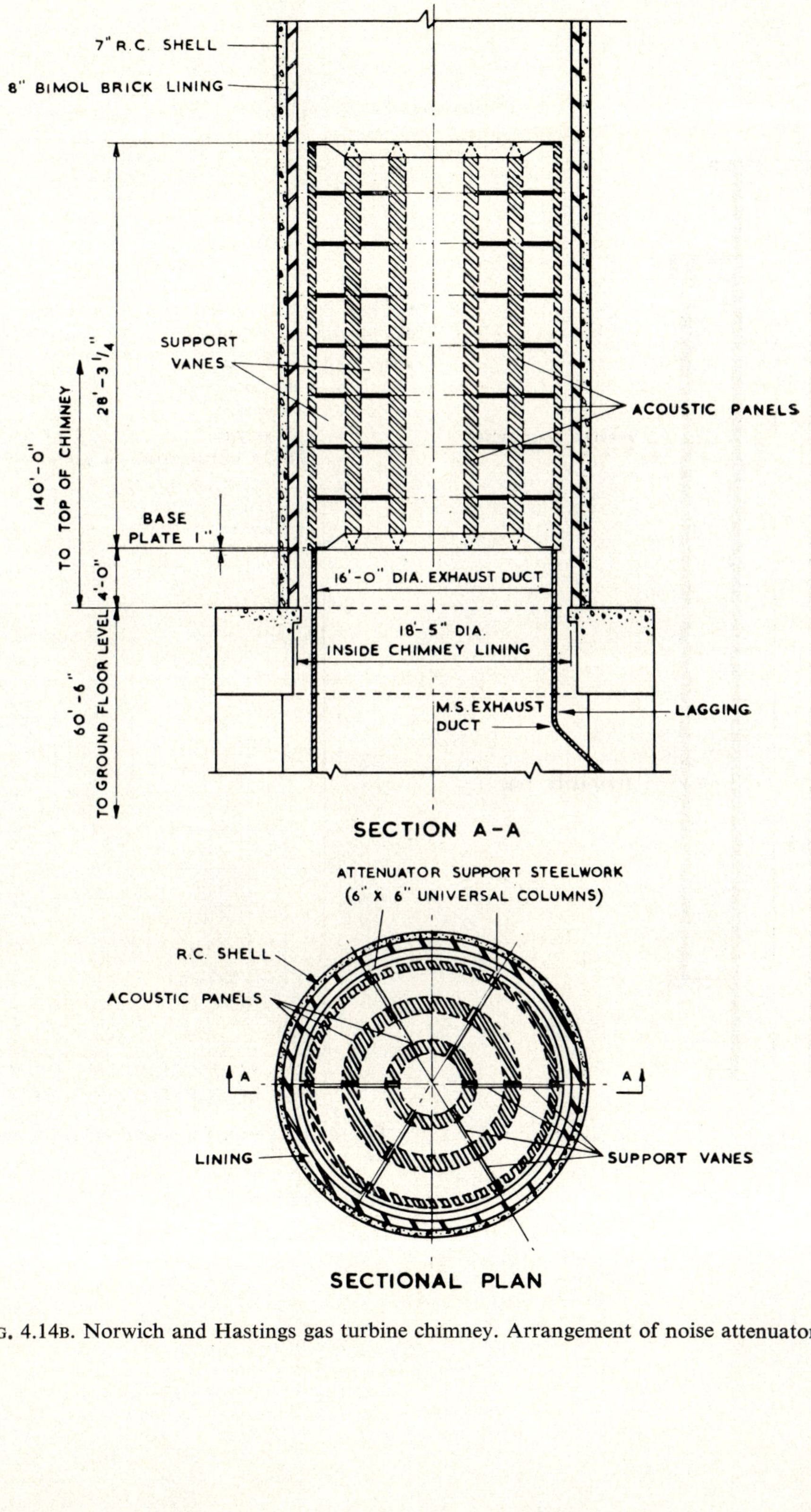

FIG. 4.14B. Norwich and Hastings gas turbine chimney. Arrangement of noise attenuator.

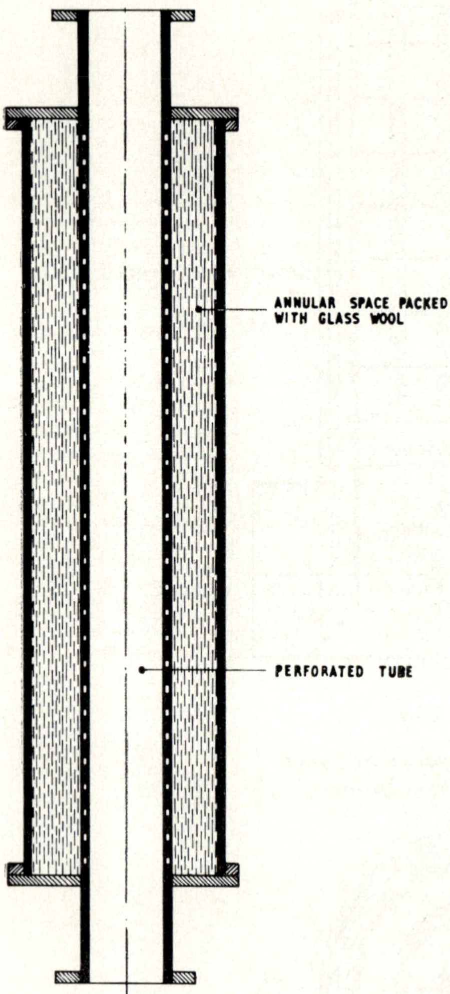

FIG. 4.14C. Typical absorption silencer.

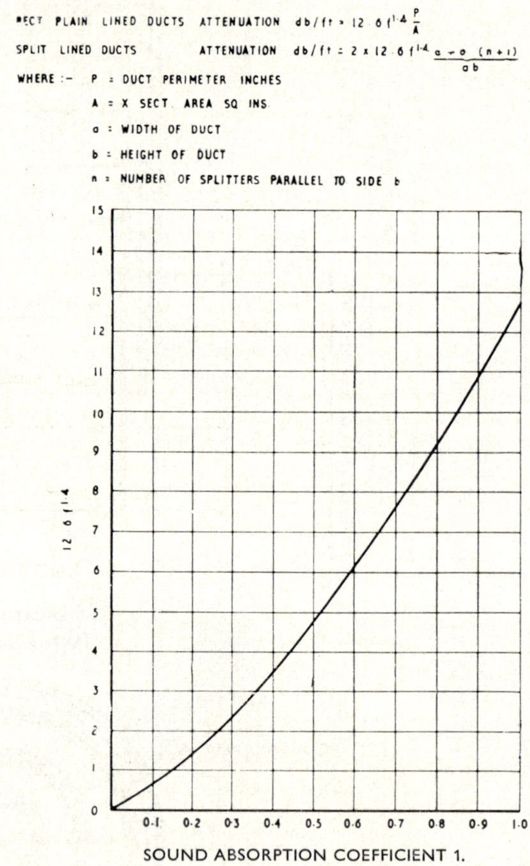

FIG. 4.14D. Curve for sound attenuation in ducts.

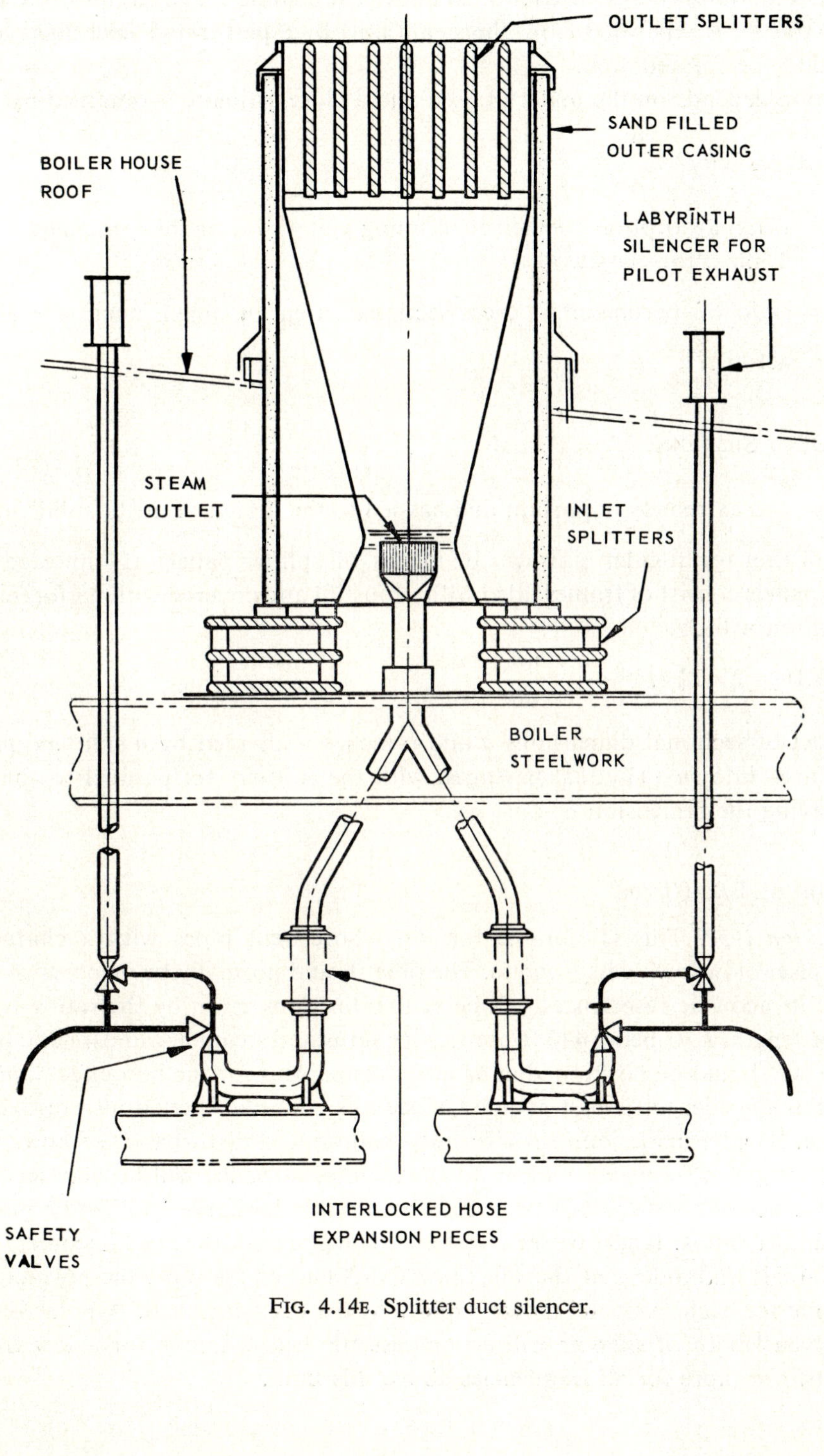

FIG. 4.14E. Splitter duct silencer.

Absorption Type Silencer

A typical silencer of this type is shown in Fig. 4.14c. The silencing effect depends on reduction of the sound energy by friction in fibrous or cellular materials termed the acoustifil. The acoustifil is separated from the steam flow by a perforated steel liner, having a minimum 30% perforated area.

Attenuation depends on the material used and a close estimate is obtained by:

$$\text{Loss, dB/ft} = 12{\cdot}6f^{1{\cdot}4}\frac{P}{A}$$

where f = mean absorption coefficient of lining material over the frequency band (see Table 5 for glass wool),

$\frac{P}{A}$ = ratio of perimeter to cross-sectional area of lined duct (see also Fig. 4.14d).

Splitter Duct Silencing

The ratio $\frac{P}{A}$ is extremely important and has led to the design of ducts subdivided into a number of thin rectangular passages by means of splitter panels as indicated in Fig. 4.14e. The panels consist of frames filled with acoustifil and covered with perforated sheet; the attenuation will be:

$$\text{Loss, dB/ft} = 2\times 12{\cdot}6f^{1{\cdot}4}\times\frac{a+b(n+1)}{ab}$$

where a duct of sectional dimensions a and b inches is divided by n splitters pl aced to divide the area into $(n+1)$ equal passages, with the splitters set parallel to the side of the duct having the dimension b.

Application of Each Type

Absorption type. This is suitable for large bore vent pipes with a characteristic resonant noise (blowdown vessel vents). The pipe should normally be "closed" as defined earlier and its acoustic resonances will be at the positions given by the ratios 1, 3, 5, 7, etc., to pipe length, L, C being 440 ft/sec for dry saturated steam at atmospheric pressure.

The silencer should be positioned at acoustic resonances for the best effect, and a long tail pipe after the silencer is best avoided, since this could be excited into resonance by, for example, a badly made joint or deformity leading to a disturbance of flow.

Since the range of frequencies likely to cause noise nuisance will lie between 100 and 6000 c/s, the positions of the last resonance before the discharge end can be plotted for the odd number ratios. It will be seen that commencing with the fundamental frequency where the single antinode is at the source or vessel end of the pipe, the positions of the last antinode for each overtone moves rapidly to the discharge end. A point is reached where a given length of silencer will encompass the last antinode for a low frequency limit and two or more for all frequencies above this limit.

For design purposes the lower limit should be based on approximately 100 c/s, but if on practical use an unpleasant low-frequency noise persists, the vessel end of the vent pipe will require attention. In this case a spray as described in the following section will be more economic and practical. Where an "open" pipe is being considered it is observed that an antinode for each frequency throughout the range will exist at each "corrected" end of the vent pipe.

The acoustifil can be leached out from this type of silencer and may be a source of silica contamination in recovered drains water, consequently use should be restricted to situations where all drainage is either retreated or discharged to waste.

Splitter type. These are suitable for steam jets and common use could be made for superheater safety valve escapes and superheater circulating vents at roof level.

To date the Board have no experience with this form of silencing except on the air intakes and discharges of peak load gas turbine generators. An application to particular steam exhausts will, therefore, require detailed fundamental design work which should call upon the experience and knowledge built up by specialist contractors in this field.

Spray Water Silencing

A properly sprayed steam atmosphere produces a sound carrying medium analogous to gas/solid medium such as glass wool where attenuation is due to friction at the boundaries of the mixed mediums and also to reflection of the sound with reversed phase.

One of the graphs plotted in Fig. 4.14F for dry saturated steam at atmospheric pressure shows the major decrease in acoustic velocity with introduction of spray, the

Table 5. Sound Absorption Coefficients for Glass Wool

Treatment	Thickness in	62	Absorption coefficients at frequencies of (c/s)								Mean coefficient over range 200–2000 c/s
			125	250	500	1000	2000	4000	6000	8000	
Unfaced											
Bitumen bonded	1		0·10	0·35	0·50	0·55	0·70	0·70	0·65	0·75	0·55
Resin bonded											
Type 300	1		0·10	0·35	0·55	0·65	0·75	0·80	0·75	0·75	0·60
	2		0·20	0·50	0·70	0·80	0·75	0·80	0·80	0·80	0·70
Type 600	1		0·10	0·25	0·55	0·70	0·80	0·85	0·90	0·95	0·60
Type 900	1		0·10	0·25	0·60	0·75	0·85	0·90	0·90	0·90	0·60
Tissue faced											
Fine fibre	1		0·10	0·25	0·45	0·60	0·70	0·70			
	2		0·20	0·45	0·65	0·75	0·80	0·80			
	4		0·45	0·75	0·80	0·85	0·90	0·85			

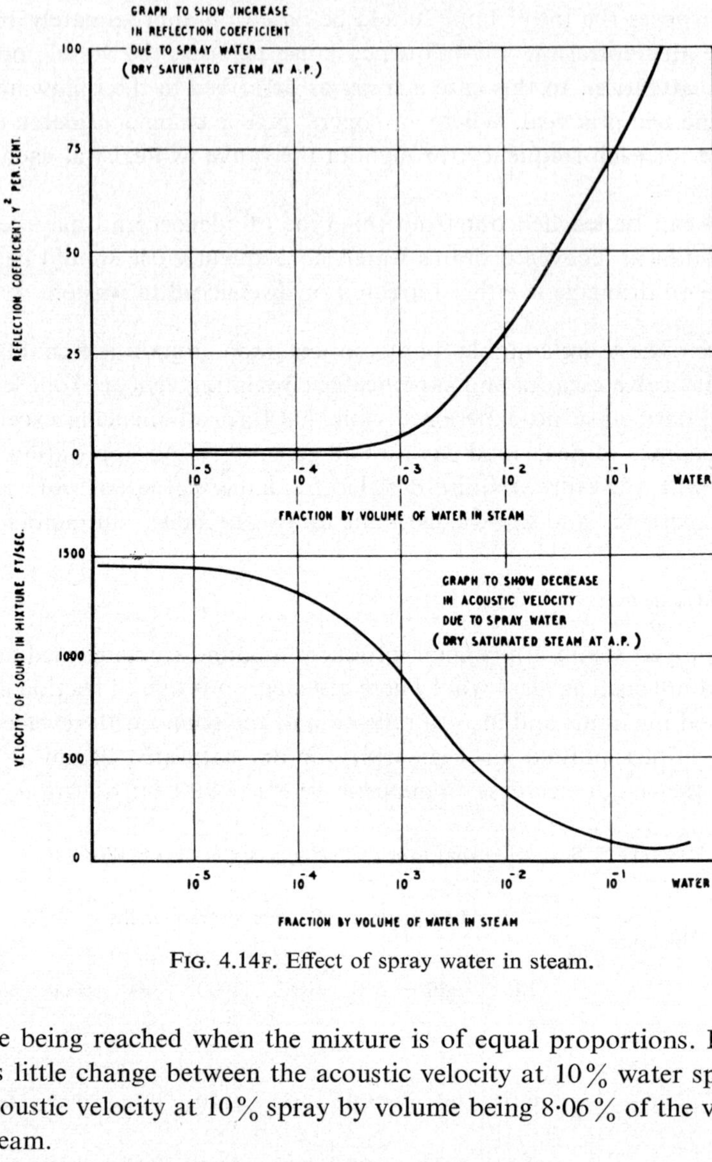

FIG. 4.14F. Effect of spray water in steam.

lowest figure being reached when the mixture is of equal proportions. It is also shown that there is little change between the acoustic velocity at 10% water spray and that at 50%, the acoustic velocity at 10% spray by volume being 8·06% of the value for the dry saturated steam.

The increase in the reflection coefficient r^2 with increasing water spray is also shown, in Fig. 4.14F, to reach a maximum of nearly 100% when the mixture is of equal proportions. The major increase is again, however, between nil and 10% of spray water. Spray silencing can be used for all blowdown vessel vent applications, provided the vessel vent is handling only flash steam from boiler blowdown water at atmospheric pressure (*superheated steam should not be discharged via this vent except under emergency conditions*).

The practical maximum spray quantity will be 10% by volume of steam flow and this will attenuate to the order of 30 dB, but it is worth noting that attenuation of 3 dB *halves* the noise level (10 log 2).

Experience so far with spray silencing leads to the following conclusions:

(a) Most of the spray introduced emerges as spray from the vent pipe outlet and special measures are necessary to prevent the water becoming a nuisance; an exhaust head of the type shown in Fig. 4.14G has proved very successful in this respect.
(b) It is essential for the success of spray silencing that the water droplets are dispersed evenly over the volume of the vent pipe; sharp bends in the pipe can be expected to cause some separation. The spray nozzle should give a cone discharge such that the minimum of water impinges on and runs down the wall of the pipe.
(c) The spray control should be automatic, the control valve being opened on a temperature rise signal from the drain vessel. A vee port or needle valve should also be installed, set by practical experiment to pass the minimum quantity for silencing maximum flow of steam.
(d) Where a pumped drains recovery system is in use, the spray water will be additional and pump sizing should budget for the extra quantity.

4.15. USE OF SCALE MODELS

Industrial experience in this country and America of the use of scale models as an aid to the design and layout of plant and buildings, shows that a more efficient and technically improved layout of plant is achieved. The advantage lies in having immediately available in the design office a three-dimensional aspect of the project, growing and being modified as design proceeds, so that comprehension and resolution of any problem does not depend on an appraisal of many different drawings. The model also encourages quick decisions and reduces very considerably the time spent on producing drawings. Estimates of the actual total time saved vary and depend upon whether the model is used to the fullest extent possible. A model must be generously planned; the main reason why some modelling attempts have been a failure is that the effort has been half-hearted. The model must be a strong and professionally made job, well detailed and dimensionally accurate, and should be readily available to the engineers and draughtsmen concerned with the project.

The basic model, e.g. base, floors and plant items, requires careful planning and construction by expert craftsmen using proper tools and equipment. Piping, cable runs and lighting services are best placed on the model by draughtsmen who are normally found to require little training since the manipulation of standard parts needs only elementary practical skill.

4.15.1. Scale, Extent and Materials

The complete boiler/turbo-generator unit should be modelled to a minimum scale of $\frac{3}{8}$ in = 1 ft. Where more floor area is available it is always advantageous to increase the scale ($\frac{1}{2}$ in, $\frac{3}{4}$ in and 1 in to the foot scales are commonly used), particularly where piping services are complex; but since a 500 MW unit at $\frac{3}{4}$ in scale will occupy an area of about 8 ft$\times$20 ft, it is unlikely that larger scales will be used, except when it is desirable to model

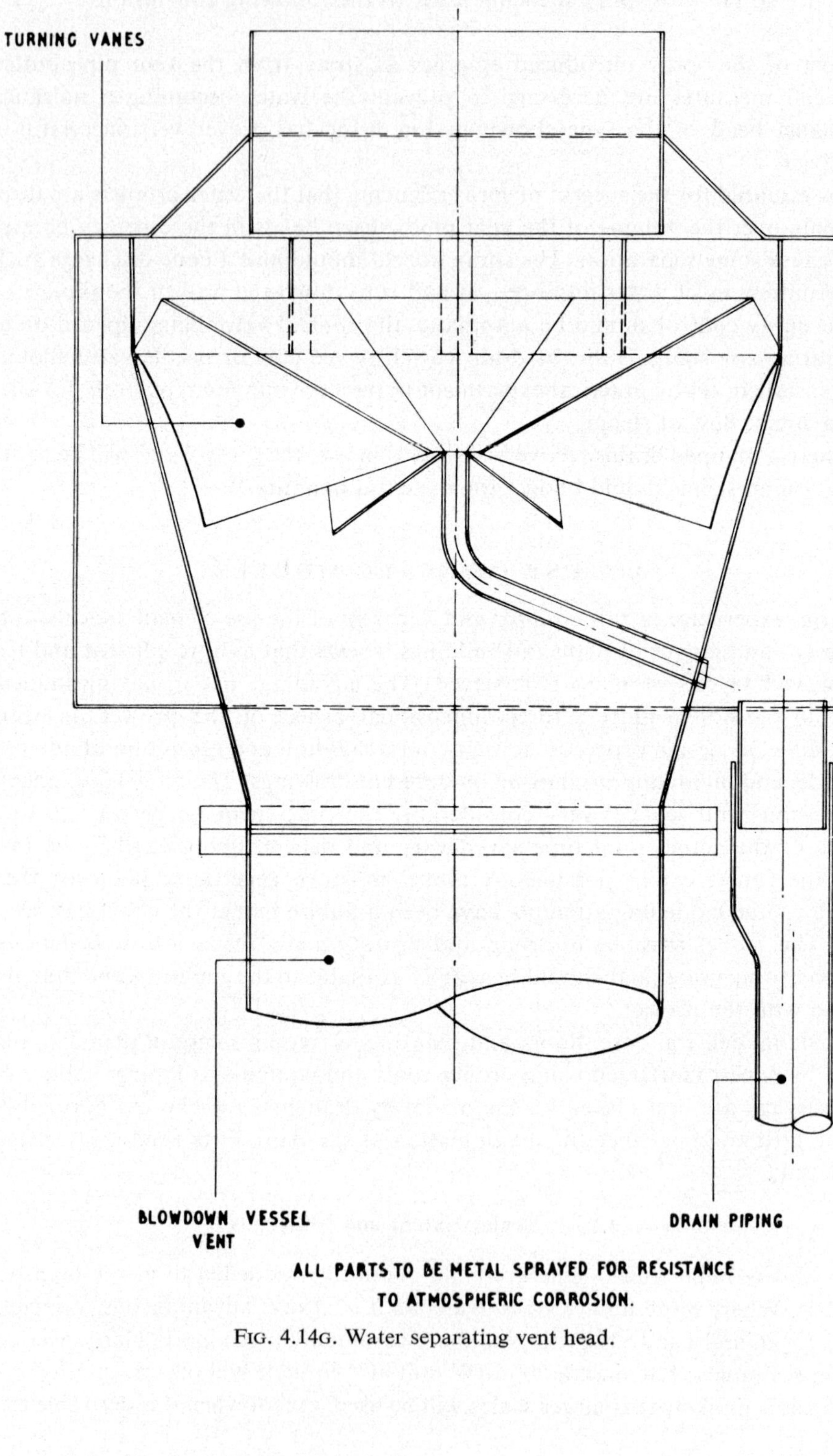

FIG. 4.14G. Water separating vent head.

a small portion for verification of, for example, a boiler burner gallery layout or a particularly complex local piping arrangement.

The model must be designed to split down into suitable portions, not only for the model to be transported, but also to facilitate access to the model during construction of piping runs and galleries, etc.

In the very early stages of design a small scale block model about $\frac{1}{8}$ in $=$ 1 ft is useful to determine the most economic space relationship of plant items, access, cranage and pipe route reservations.

For plant items, wood or plastic material is used, the shape being symbolic rather than an accurate copy, provided all overall dimensions and piping terminal points are exactly true to scale. The main base board is usually of timber and represents raft or pile cap level; culverts and drains below this level are painted or drawn on the base in plan form. Floor and gallery levels above base level are best formed in clear perspex sheet; access stairways are shown in true shape and to scale.

Structural steelwork is modelled in timber of rectangular section to the scale dimensions of R.S.J. channels, etc.; the true section is not copied since at the small scale, strength is insufficient. The rectangular sections are, however, true in width and depth and soffit faces (the underside of beams) are always correctly positioned in space.

Piping runs are shown in two forms—wire and disc or true to scale plastic piping. The wire and disc system represents pipe centre lines with the disc representing the maximum finished outside diameter. Tee joints are soldered, valves are represented by a short length of wire to the limit of a rising spindle with a disc for the handwheel. The wire and disc system is structurally robust and easily modified: it does not obscure other model piping below or around it from access by the modeller.

The commercially available scale plastic model piping is only true at $\frac{3}{8}$ in scale; but it can be used in other scales with some loss of accuracy of bend radius and other standard dimensions. All the commercially available model pipe fittings are scaled to American piping and valve standards, but British Standard models may be available as the market demand grows. Plastic piping on a model gives an immediate pictorial idea of the actual installed piping and, provided it can be installed on the model in a rigid robust manner, it is more truly indicative of clearances, etc., especially when modelling pipework above 12 in (real) outside diameter.

4.15.2. **Application of the Model**

Uses to which a model can be put are:

(a) DESIGN OFFICE

(i) Determination in initial stages of the project of the most economic relationship in space of main and auxiliary plant items (small-scale model).

(ii) Reservation of maintenance access, cranage, service pipework routes and cable runs (small-scale model).

(iii) Physical layout of plant items, structural steel, floor levels, galleries, pipework, cable runs and permanent lighting (minimum scale $\frac{3}{8}$ in $=$ 1 ft).

(iv) Verification of contractors' proposals involving complete and early implication of contractors' engineers in the model technique.

(b) Site Office

The model, or a duplicate, should be available on site as early as possible in the construction period, so that the model can be used in the following ways:

(i) Economic co-ordination of contractors' activities by areas, e.g. sequences of erection, piping assembly, lagging, cable laying, etc., arrangements to avoid congestion.

(ii) Assessment of strength of construction teams required.

(iii) Planning of major lifting requirements.

(iv) Planning of scaffolding and temporary access on a station wide basis to ensure maximum use by all contractors.

(c) Station Operating Staff

The site model should be available for use by early appointed operating staff, but in due course should become the property of the station for use in training of staff, work-study and maintenance planning.

4.15.3. Savings Achieved

Experience indicates four ways in which great savings can be made by the use of these models.

A. Reduction in Total Project Design Time

The models should function as a rapid communication of ideas and requirements without resort to production of layout drawings and cross-checking of contractors' drawings. It is difficult to give an objective estimate of engineering savings but consider the time ratio between the following:

(i) Preparation of drawings compared with physical layout on a model.

(ii) Close checking of contractors' drawings compared with physical checking on a model.

B. Site Modifications due to Fouls and Poor Access are Eliminated

It can be shown that the use of a model in the design of two 200 MW units currently being erected would have eliminated modification work due to fouls and poor access, costing £12,000.

C. Tenders for Piping Services Particularly are Invariably Lower in Price when the Tenderer has Access to a Model

Because of the "final" detailed nature of piping services shown on a model, tenderers can bid very accurately. If the contractor is given reasonable access to the model he does not need to include in his estimate cover charges for variations in final pipe routings and the special difficulties that have always been implicit in pipework enquiries.

D. The Plant Commences Production Earlier

All time saved on project engineering works towards an earlier commissioning date for the units and since one 500 MW unit represents revenue to the Board of at least £10,000 per 24 h period, it is clear that any small gain in meeting commissioning dates is of considerable value.

4.15.4. **Examples of Models**

The illustrations shown should be referred to in detail.

Figure 4.15.4A shows a trial small-scale turbine layout for Littlebrook 'D'. This may not be exactly the final arrangement; but it was constructed to demonstrate the feasibility of transverse arrangement of the unit mentioned in an earlier section.

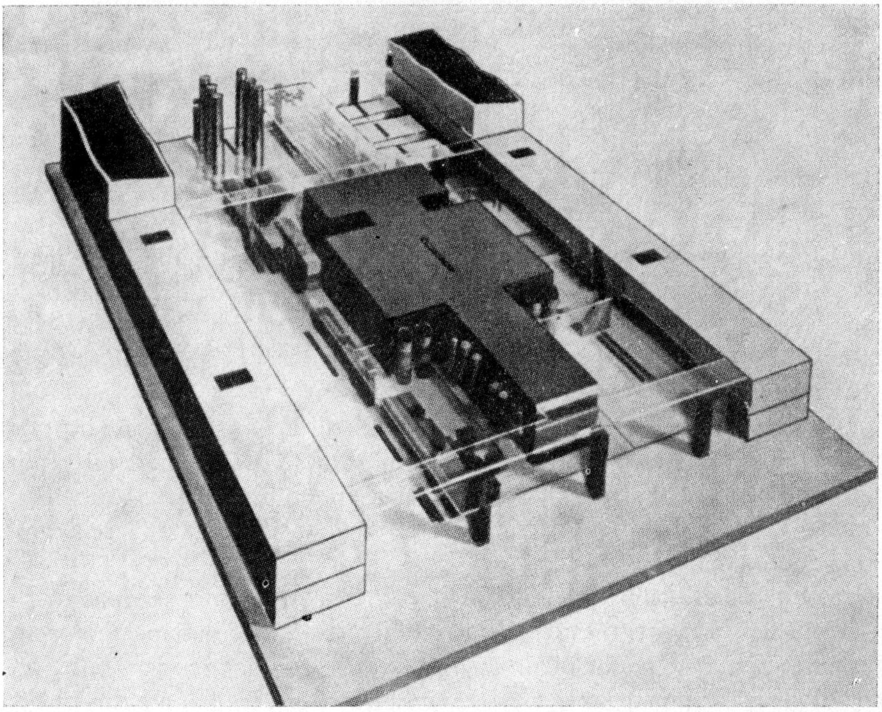

FIG. 4.15.4A. Model–Littlebrook 'D' turbine area.

FIG. 4.15.4B. Model–Fawley deaerator bay.

Figure 4.15.4B shows a typical deaerator bay with all piping installed. The method used is wire and disc, and the black discs can be seen extremely well. Note also the access provisions, i.e. galleries, stairways, landings, etc. Comments on the provision of lifts in a previous section of this lesson are high-lighted when the height of the main stairway in this bay is seen.

Figure 4.15.4C has been included to give an idea of the size and complexity of a $\frac{3}{8}$ in scale model. This is of Fawley power station and shows how the model is broken down into smaller sections so that people can work closely on piping detail, etc. The portion of the model in the foreground carries a quarter of the turbine and generator. Space required is approximately 12 ft×25 ft.

Figures 4.15.4D and E are models of Fawley turbine and boiler sections to $\frac{3}{8}$ in scale. Figure 4.15.4D shows clearly the arrangement of the suspended basement floor mentioned in an earlier section; the major plant such as pumps, auxiliary coolers, etc., are on the suspended floor level; piping services, etc., run below this, so that a clean turbine house is seen from the operating floor.

FIG. 4.15.4C. Model–Fawley layout.

FIG. 4.15.4D. Model–Fawley turbine area.

FIG. 4.15.4F. Model–Fawley steelwork arrangement.

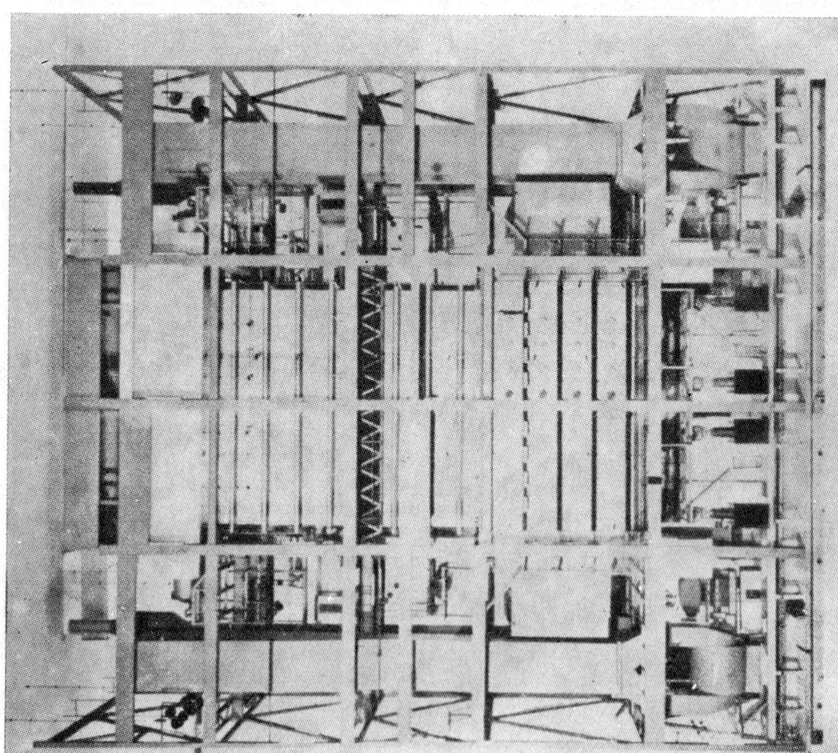

FIG. 4.15.4E. Model–Fawley boiler area.

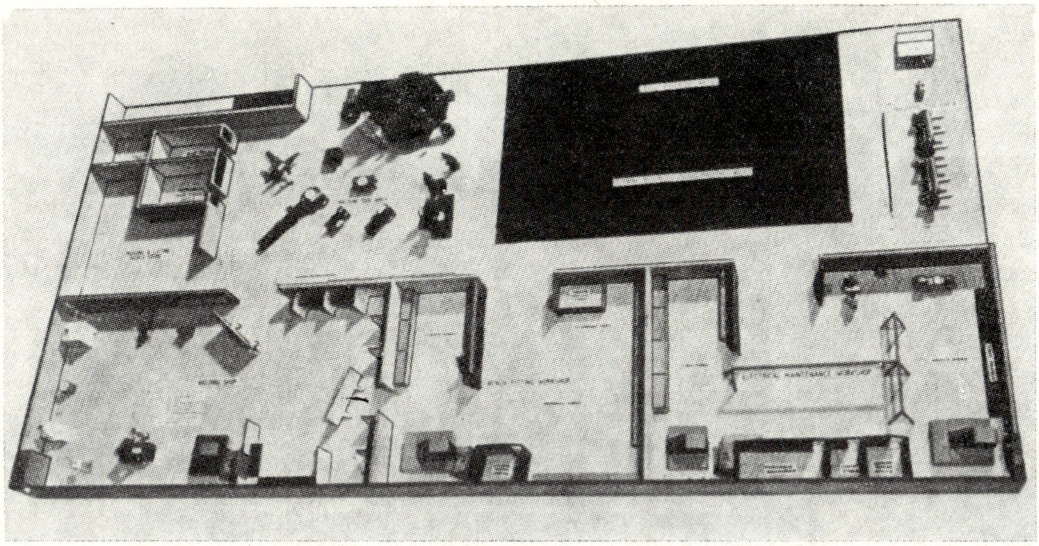

FIG. 4.15.4G. Model — workshop layout.

Figure 4.15.4F is of a section of steelwork modelled to $\frac{1}{4}$ in scale entirely to assist in the erection of steelwork at site; in modern power station construction the steelwork is extremely complex and a very large number of drawings are required to cover a small section such as in shown in the illustrations. If a model is available, people can proceed with erection and sorting of steelwork as it arrives, in an economic manner.

Figure 4.15.4G shows a $\frac{1}{8}$ in scale layout of the workshop area at Fawley power station; by the use of small-scale die castings of particular machine tools the maintenance superintendent was able to agree with the project engineers the exact disposition of all the machines in a single afternoon.

Figure 4.15.4H shows a very small-scale model of the instrument workshop and maintenance centre. Notice the air lock to the physical instrument workshop where the atmosphere is maintained under slight pressure to prevent ingress of dust.

EXERCISES

1. Figures E.1, E.2 and E.3 show three different approaches to modern plant layout. Make an appraisal of each layout (under the main headings of the chapter) relative to the other two and conclude with a reasoned argument on the layout you consider to be most favourable to the Board from the standpoints of construction, operation, maintenance and welfare.

2. It has been customary to provide two turbine house cranes for maintenance on a four-unit station. These are capable of acting together to lift one generator stator. Discuss:

(a) Circumstances which may justify provision of more than two cranes.

(b) How Goliath cranes could be used instead of beam cranes.

(c) How would two Goliath cranes be linked to act together for the stator lift?

(d) At what level would the Goliath travel tracks be placed, what would the effect be on the turbine house structure and routing of copper connections to the generator transformers?

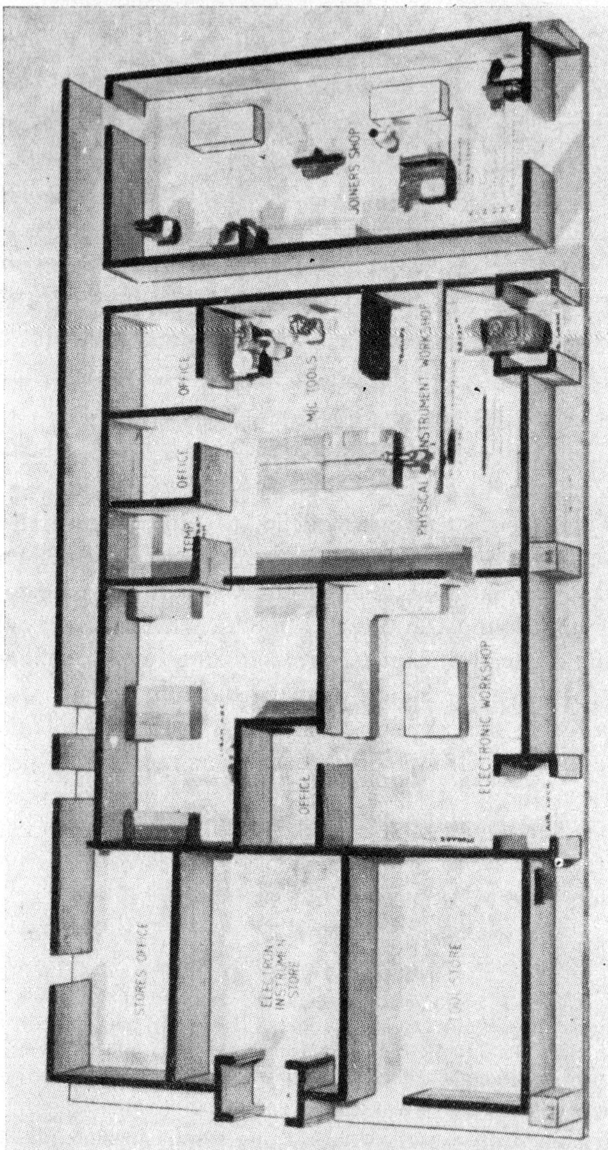

FIG. 4.15.4H. Model — instrument workshop layout.

3. Detail the information required and process followed, to establish the economic distance between centres of two adjacent coal-fired boilers having first postulated the relative positions of mills, bunkers and the turbo-generator. Use freehand sketches to illustrate particular points where possible.

Assuming a single multi-flue chimney, what are the factors which determine the position of the chimney relative to the boilers?

4. Draw, freehand if necessary, a cross-sectional elevation of a boiler/turbo-generator unit, derived from one of the figures in the lesson and detail all the considerations necessary to arrive at fixed economic levels for each plant item shown, including the deaerator, main turbine house cranes and if you decide to have it, the mechanical annexe crane.

5. In Section 4.5.3 the process used to establish the feed pump suction pipe diameter is given in detail. Assume that the flow given of $3{\cdot}52 \times 10^6$ lb/h is the turbine MCR flow rate, and assess the difference in margins available when the flow increases to a rate 10% in excess of this.

Show on a line diagram from condenser to deaerator, where the elements of static and frictional pressures given in the system characteristics for extraction pump (Section 4.5.3) are situated.

The extraction pumps must clearly operate at a controlled duty point where they will normally pump the water flow required by the plant and yet be capable of transient increase or decrease in demand as load changes or other variations to normal running are imposed. Discuss this requirement with particular reference to the form of flow control needed on the extraction pumps, particularly, comment upon the ways available to derive a signal to the flow control and maintenance of condensate levels in the condenser and deaerator.

How would you propose to protect the generator coolers from over-pressure on the water side when the arrangement shown in Fig. 4.5.3B is used?

6. Sketch a typical reheat steam service from the h.p. cylinder exhaust on the set to the boiler and back to the i.p. cylinder inlet steam chests, include a supply of low-temperature reheat steam to a 100% steam turbine driven feed pump and comment upon the following:

(a) The effect on reheat steam circuit design by the introduction of a turbine-driven feed pump.
(b) What are the elements of pressure loss in the reheat circuit? List these and state how you would go about the job of assessing total pressure drop through the system.
(c) What provisions would you make for periodic hydraulic testing of the reheat section of the boiler?
(d) What major differences would you expect to see between the low- and high-temperature reheat steam pipes?
(e) What drainage provisions should be made for cold and hot starting?
(f) What process occurs in the reheat steam circuit when the unit is shut down for a long outage period?

7. Discuss in detail the application of the model technique to design and layout, assuming that you were charged with starting a special section to develop the use of models. Detail your management plan and the overall policy you would adopt.

8. The furnace bottom ash "make" of a station is 250 tons per day, all of which is discharged to a settling pond in the day shift. The ash is grabbed from the pond and drained on a concrete apron, then grab loaded on 5-yd^3 capacity road vehicles; 5-ton capacity grabs are to be used and the ash has a specific density of 35 ft^3/ton (wet), 27 ft^3/ton (drained).

Determine a layout of drainage aprons, cranes and vehicle access which will enable the total make to be removed during the 8 h day.

Assume:

1. Draining ash should be limited to 3 ft high on the aprons, drainage time to 24 h.
2. An average crane cycle time of 5 min (unless your layout affects this considerably).
3. An average grab load efficiency of 75%.

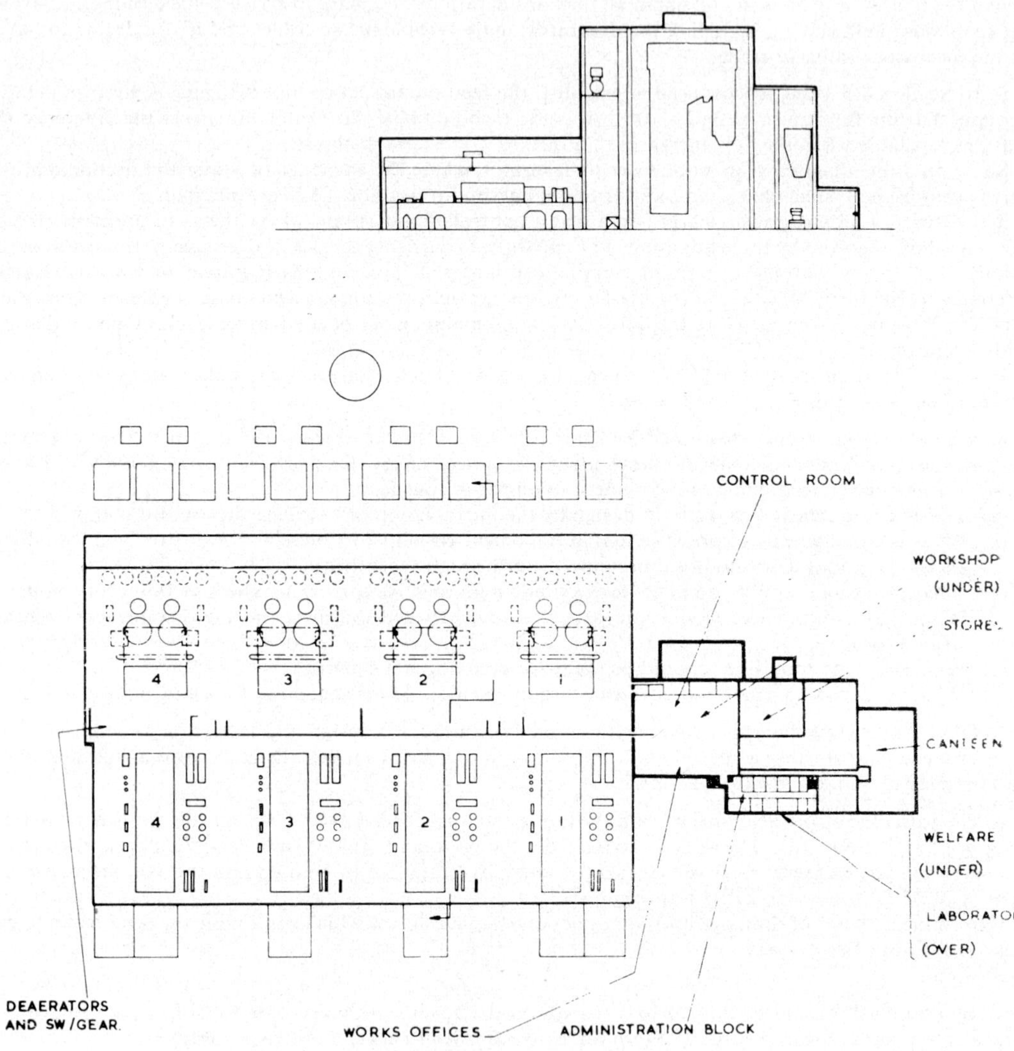

FIG. E.1. Modern plant layout.

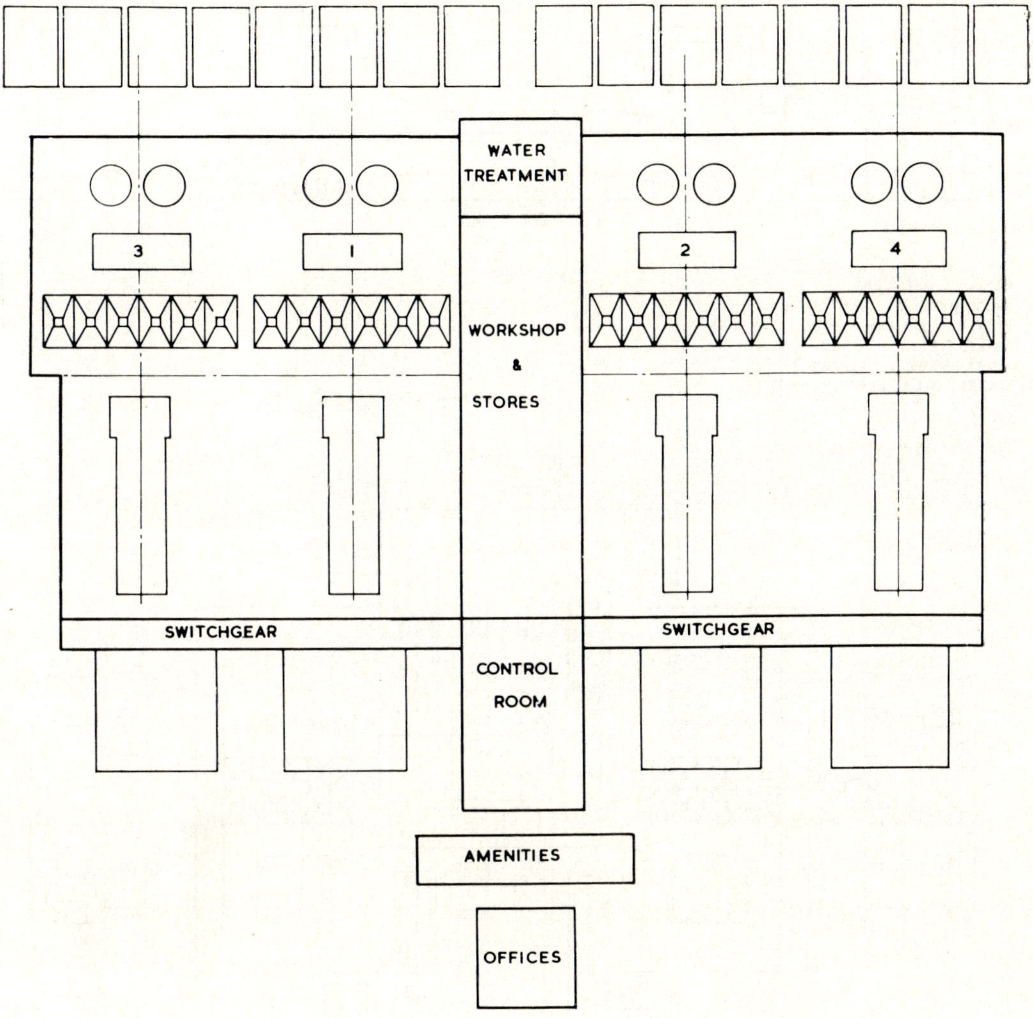

FIG. E.2. Modern plant layout.

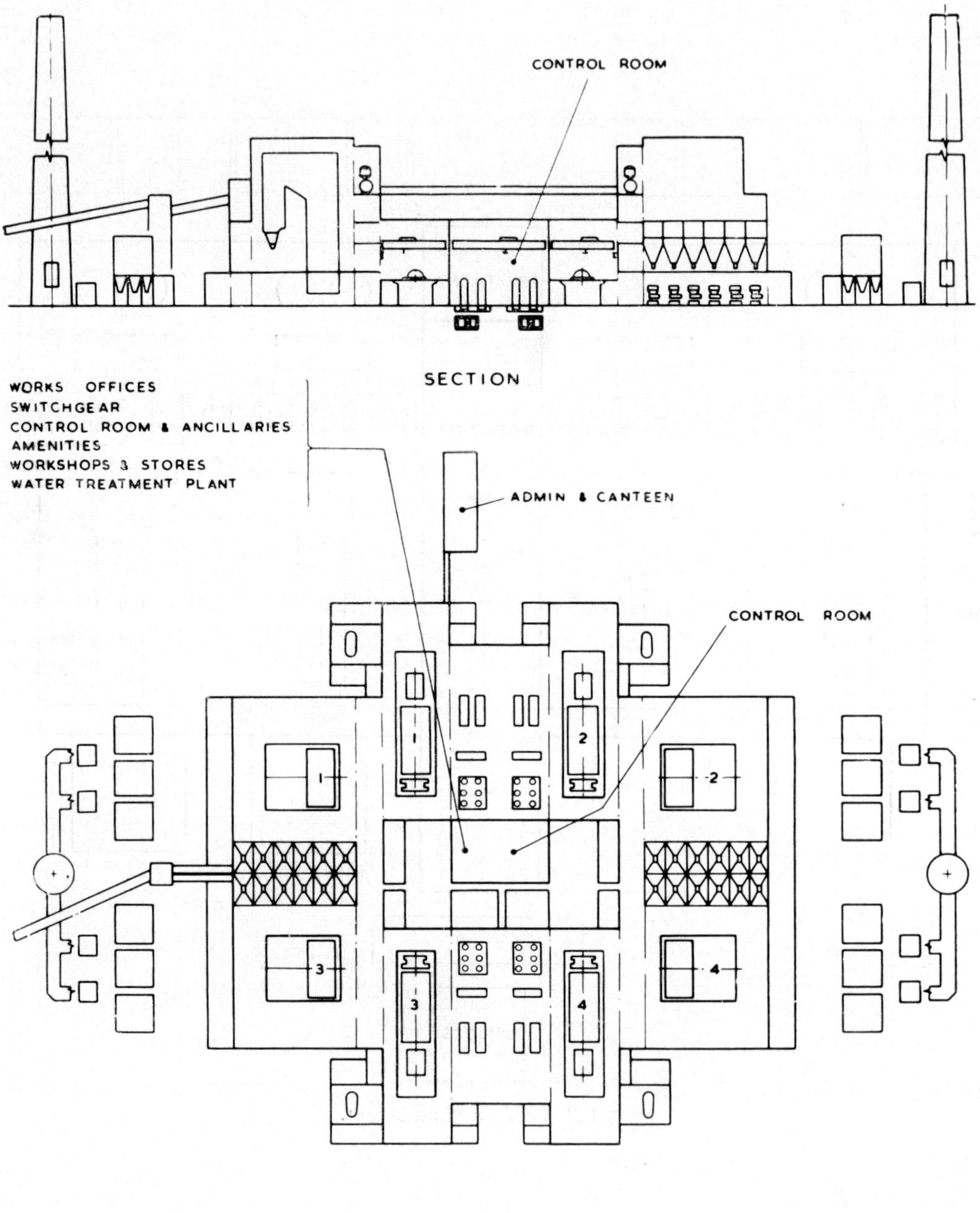

FIG. E.3. Modern plant layout.

CHAPTER 5

ELECTRICAL PLANT LAYOUT

5.1. INTRODUCTION

This chapter deals with the facets associated with the selection, location and accommodation of power station electrical plant. No reference is made to motors as their type and location is governed by economic considerations as well as by the plant they drive.

Since it is impossible within the compass of a relatively short space to cover all the variations in size and number of units, station location, main plant layout, etc., the examples given are related to a typical station of 4×500 MW units.

Where relevant, the requirements of larger units and the trend of development are discussed.

5.2. MAIN SWITCHGEAR (132 kV, 275 kV and 400 kV)

5.2.1. General

Main switchgear layout is very fully discussed in Volume 4, Chapter 2, and this section, therefore, deals very briefly with the broad principles as they affect its relation to the power station proper.

The exact location of the switching station relative to the power station will depend on a number of factors:

(a) Space availability on the power station site. The physical sizes of some grid substations are now bigger than the power station itself. This is especially applicable to some 400 kV substations of the outdoor type.

(b) Physical considerations and the difficulty in negotiating wayleaves for overhead lines may necessitate their termination some distance away from the substation involving the extensive use of cables.

(c) The h.v. connections between the generator transformers and their associated switchgear; these may be made by overhead lines or underground cables.

(d) Whether an outdoor or indoor form of construction will be used for the substation.

(e) The location of the station transformers; these may be housed either in the power station or the substation.

It is therefore obvious that a decision on the location of the substation cannot be made unless due consideration is taken of factors (a) to (e) above.

An outdoor type of substation is usually the most economic form of construction, but where there is a danger of salt storms (as on a coastal site), moderately heavy industrial pollution, or where an alternative site in close proximity to the power station cannot be found, and the site is restricted, the gear will be housed in a building.

A building will, of course, always be required for the auxiliary plant, such as the batteries, relays, air compressing and oil treatment plant, etc. Since the cost of h.v. cables is great, an economic balance will have to be struck regarding the cost of h.v. feeder cables as in (b) against (c).

Regarding (e) the choice will depend on whether it is cheaper in first cost to transmit the power at the main system voltage, or at the voltage of the power station auxiliary system, usually 11 kV.

For 132 kV and 275 kV the circuit-breakers may be of the air-blast or oil-break type, both designs being available in the very high breaking-capacity ratings now required. However, for 400 kV only the air-blast type is available at present for 35,000 MVA.

The interrupting capacity of the circuit-breakers will be chosen to suit the fault level of the system at the point at which they are installed. The usual ratings are 3500 MVA for 132 kV, 15,000 MVA for 275 kV and 35,000 MVA for 400 kV.

It is sometimes necessary to introduce reactors between sections of the busbars in order to keep the fault level down to these figures.

A key diagram of a 400 kV and 132 kV switching station, controlling four 500 MW generators for outgoing circuits at 400 kV and 132 kV, is shown in Fig. 5.2.1.

It will be seen that busbar selector switches, which also function as isolating switches, allow each circuit to be connected to either the main or reserve busbar. These selector switches are always interlocked so that it is impossible to couple the busbars by closing both at once, unless the bars have already been coupled by closing the appropriate busbar-coupler circuit-breaker.

Isolating switches and earthing switches are arranged so that any part of the equipment may be disconnected and earthed, in order to allow safe access for inspection and maintenance.

A comprehensive system of interlocks is incorporated to ensure that current cannot be made or broken by the isolating or selector switches, or the earthing switches closed on to a live circuit. The interlocking is achieved by means of "Castell" figure locks and by electrically operated bolts. Interlocking is also needed to ensure that any doors or gates giving access to normally live equipment cannot be opened until the appropriate isolating and earthing switches have been operated. In the case of circuits connected to the power station, i.e. the generators and station transformers, the interlocking scheme must embrace the equipment at that end also, to ensure that the circuits cannot be energised from the power station when the corresponding h.v. equipment is to be isolated and earthed.

Purely electrical interlocking is used only for operational interlocks—that is to say, to prevent mal-operation of the plant. Interlocking upon which the personal safety of the operator depends is always by means of "Castell" locks.

The switchgear will usually be controlled from the power station—an exception would be the case where the switching station is also an important transmission centre, having

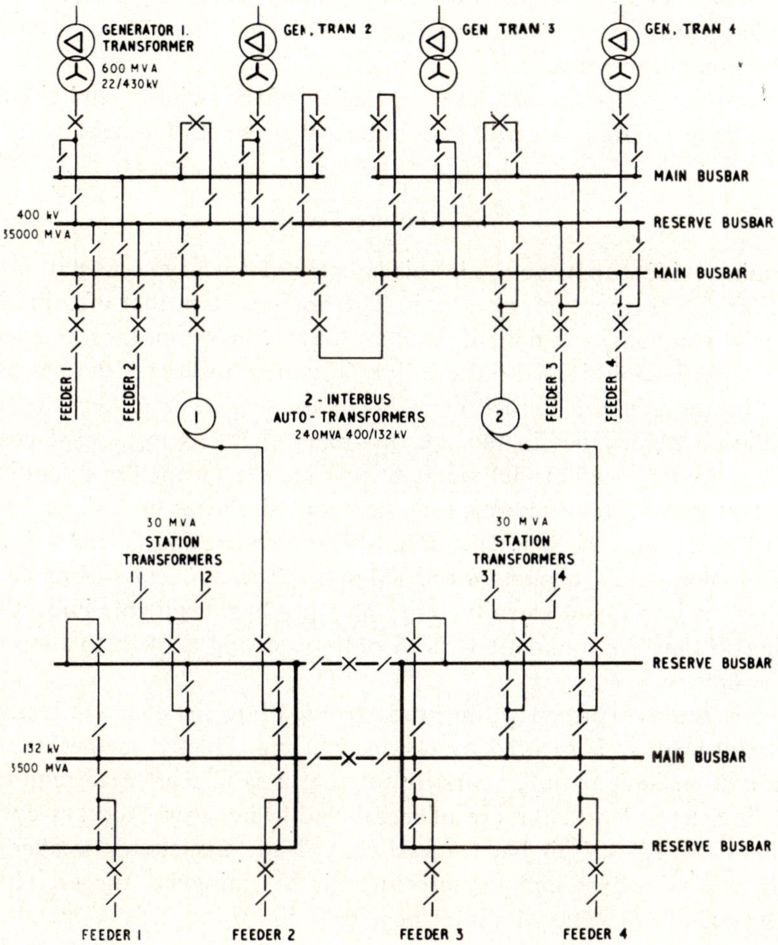

FIG. 5.2.1. Key diagram of 132 kV and 400 kV switchgear for generating station of 4×500 MW units.

its own control room—but the generator and station transformer circuit-breakers must obviously be controlled from the power station in all cases.

It has now become standard practice to employ a 50 V "interposing" system for control and indication. Where this system is adopted, the small size of the control switches and similar apparatus makes possible a considerable reduction in the size of the control panel, often a great advantage.

The 50 V system necessitates the use of interposing relays to operate the switchgear, and interposing transformers with transducers to "scale down" the normal output of current and voltage transformers.

Indications of the "open" and "closed" position of the switches and circuit-breakers, and all the necessary alarms, are also transmitted over the 50 V system.

5.2.2. **Future Trends**

It is becoming evident that the use in this country of open-type substations at 132 kV, 275 kV and 400 kV is approaching a limit. The obvious step, that of using indoor substations, is helpful but only to a limited extent so long as open-type air insulated gear alone is available. A partial solution is to use solidly insulated busbars with open-type apparatus. By this means a useful reduction of occupied volume may be achieved, but the complete solution plainly lies in the use of totally enclosed metal-clad gear. For this purpose SF_6 (sulphur hexafluoride) seems to provide the means for a solution by using it as an insulant and a circuit-interrupting medium. Work has been done on designs for 132 kV and 275 kV and the first SF_6 circuit-breaker operating at 132 kV, 3500 MVA, has now been successfully commissioned (March 1966). SF_6 circuit-breakers are quiet in operation, thus improving amenity. They also produce negligible impact reaction on the mounting so that the foundations need only be strong enough to support the dead weight of the unit.

A simple concrete raft incorporating studs grouted into the concrete is needed and to these the circuit-breaker is secured by clamps bearing directly on the base channel. It would appear from design studies carried out that a saving in space required of up to 85% might be achieved with the use of metal-clad switchgear. Another current line of development is the 275 kV air blast, metal-clad, dead tank circuit-breaker using high-pressure air as an insulant and low-pressure air to extinguish the arc. Figure 5.2.2B shows the layout of a 3×500 MW unit power station using 275 kV metal-clad circuit-breakers.

The layout for one unit is shown in greater detail on Figs. 5.2.2G and H. Of necessity all connections to the switchgear will be carried out in 275 kV cables, since physical centres for overhead line connections would not line up with those of the circuit-breaker. The alternative and present arrangement for an indoor 275 kV substation with the same number of circuit-breakers is shown on Fig. 5.2.2A, and in greater detail on Figs. 5.2.2E and F. An additional two figures, 5.2.2C and D, are included in order to illustrate the differences between the two types of arrangement in a schematic manner. It must not be forgotten that there are a number of proposed sites for power stations, especially in the London area on the River Thames, where land is at a premium, and any saving in space

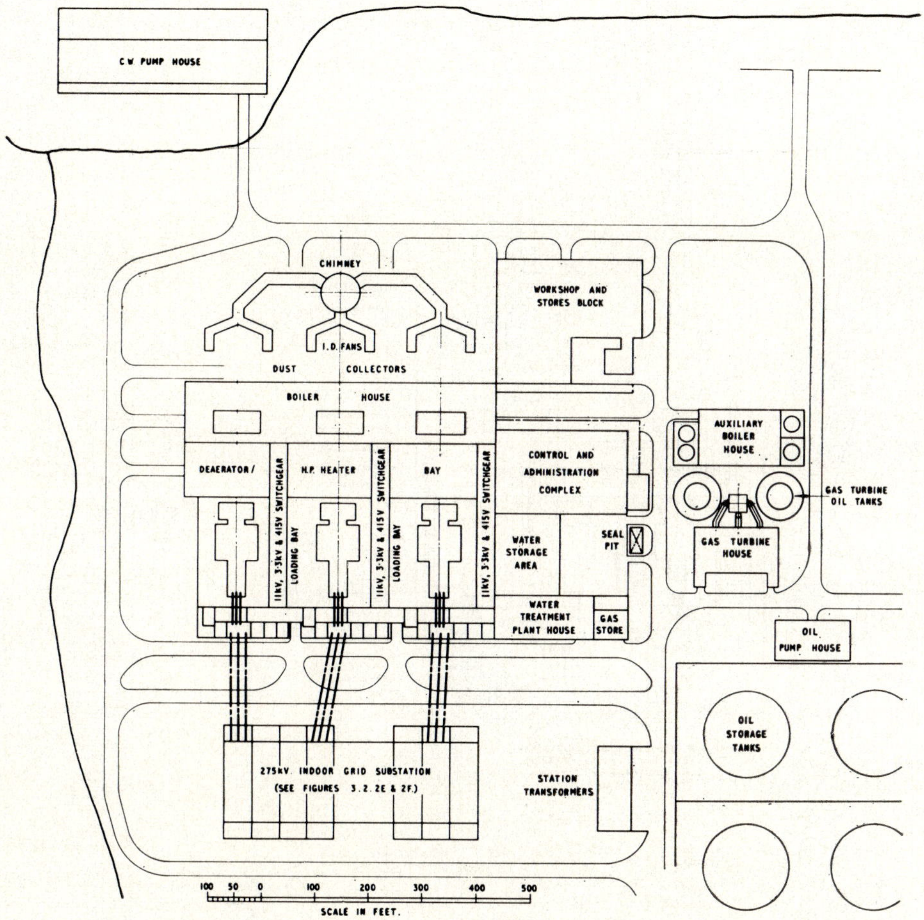

FIG. 5.2.2A. Layout of a 3×500 MW unit power station using conventional 275 kV air blast circuit-breakers.

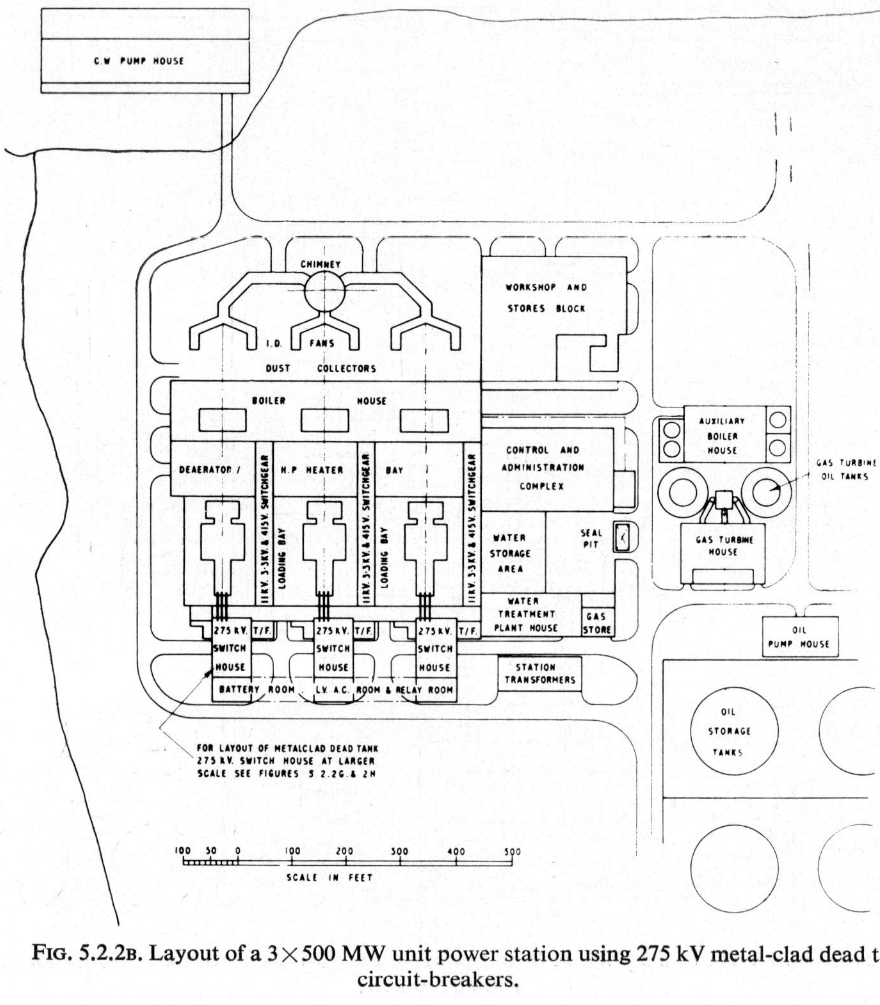

FIG. 5.2.2B. Layout of a 3×500 MW unit power station using 275 kV metal-clad dead tank circuit-breakers.

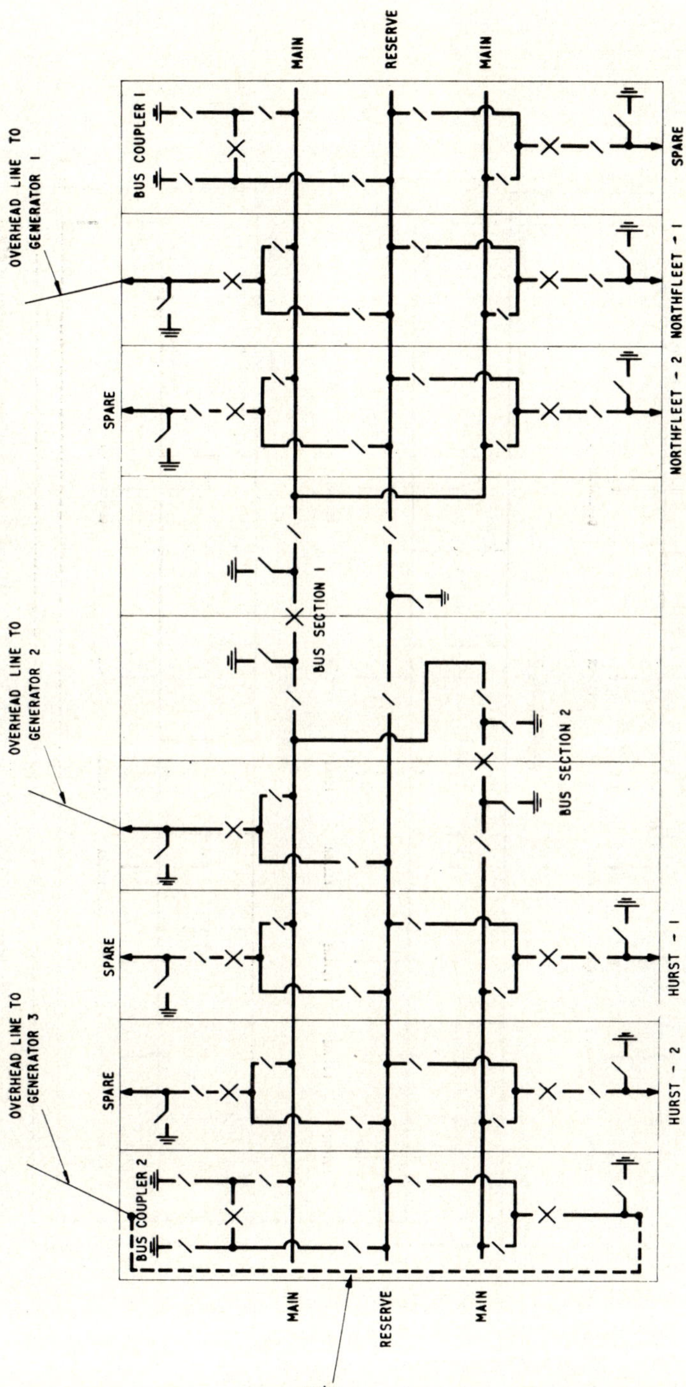

FIG. 5.2.2C. Schematic diagram of 275 kV system using conventional air-blast circuit-breakers for a power station of 3×500 MW units.

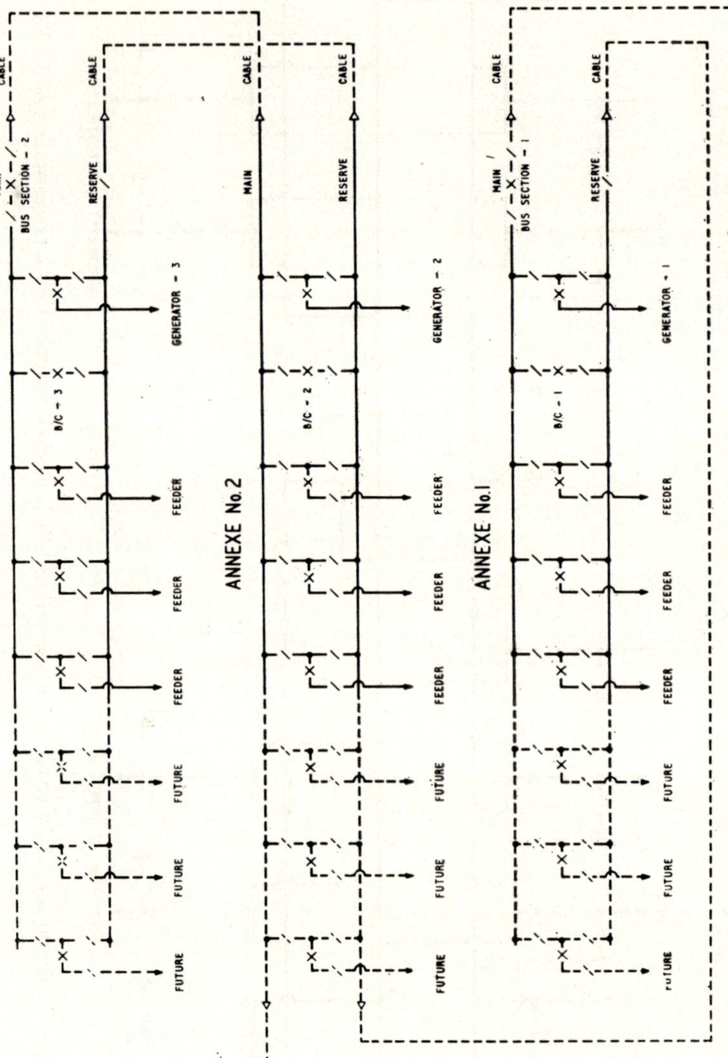

FIG. 5.2.2D. Schematic diagram of 275 kV system for a 3×500 MW unit power station employing metal-clad dead tank circuit-breakers.

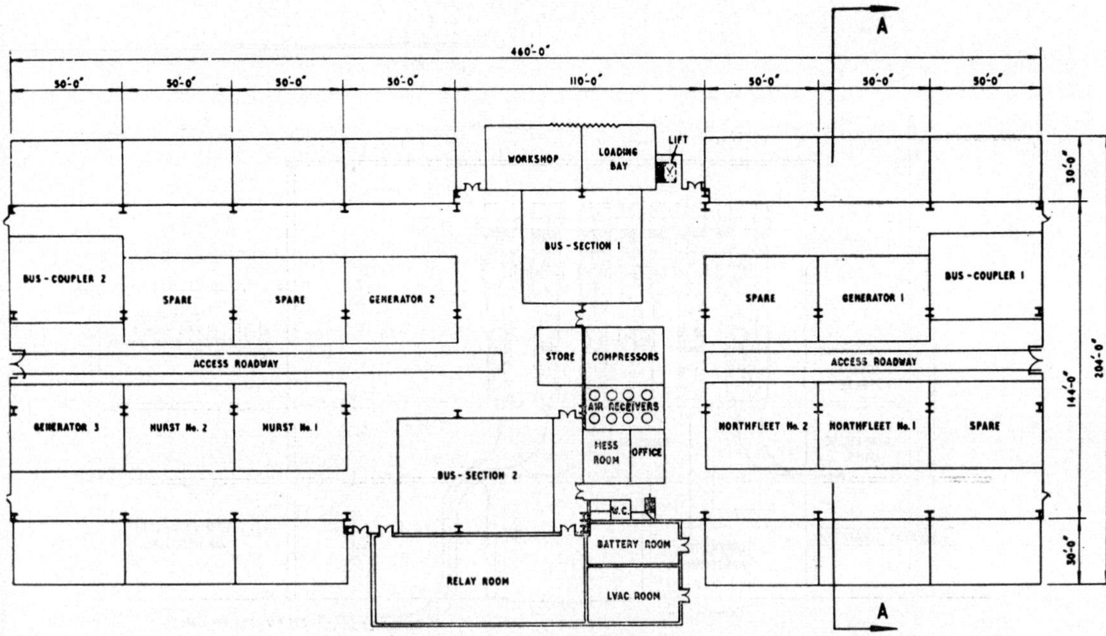

FIG. 5.2.2E. Plan of a conventional 275 kV indoor grid substation for a power station of 3×500 MW units.

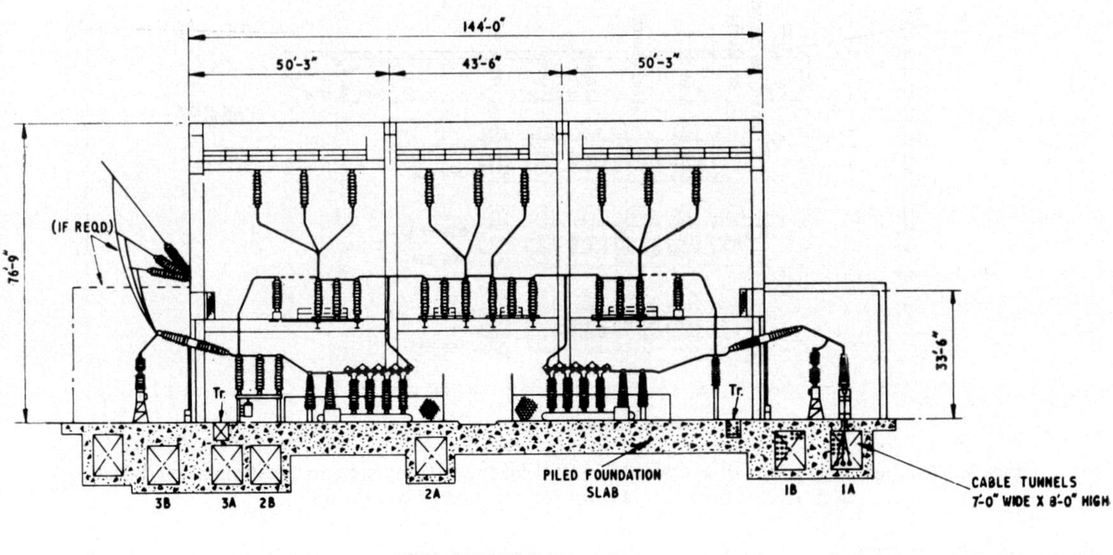

FIG. 5.2.2F. Cross-section of a conventional 275 kV indoor grid substation for a power station of 3×500 MW units.

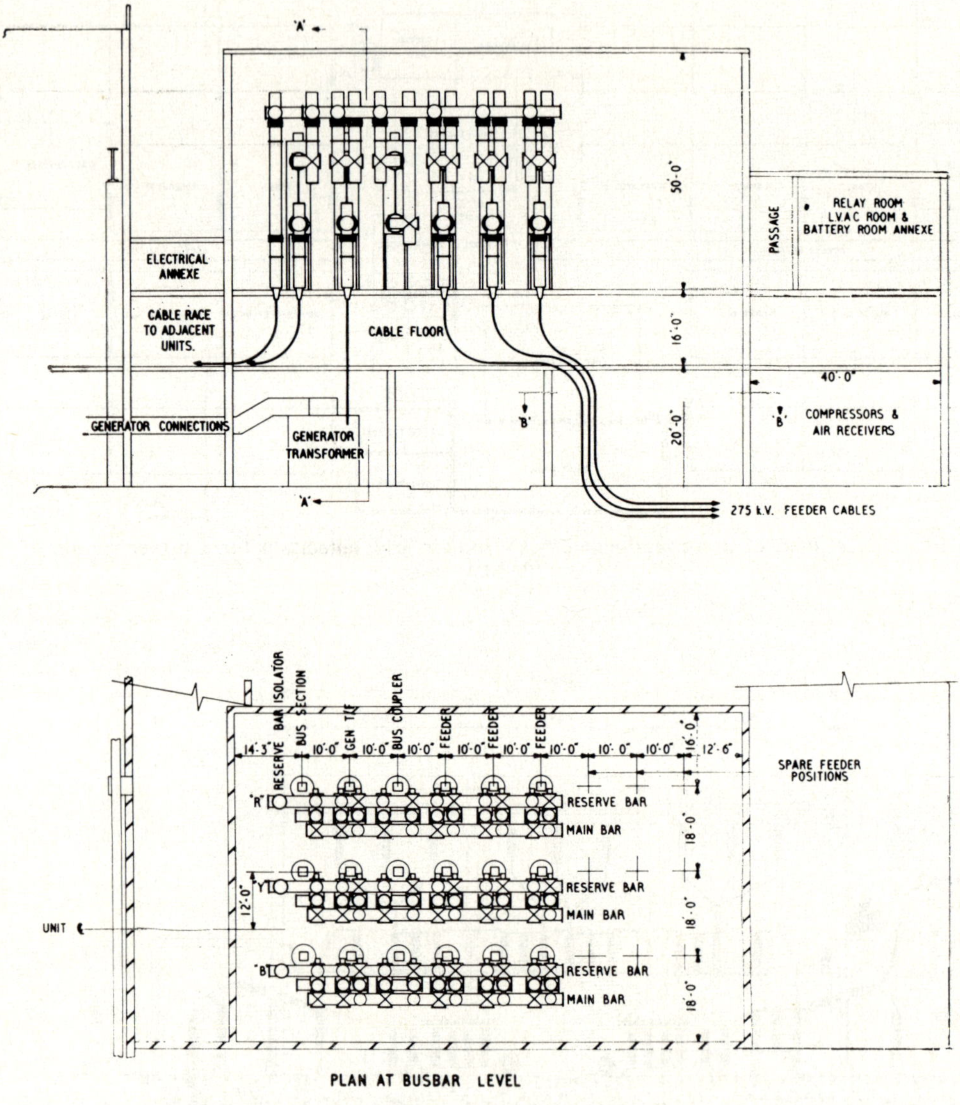

FIG. 5.2.2G. Plan and section of proposed 275 kV grid substation using air blast metal-clad dead tank circuit-breakers (shown for one 500 MW unit only).

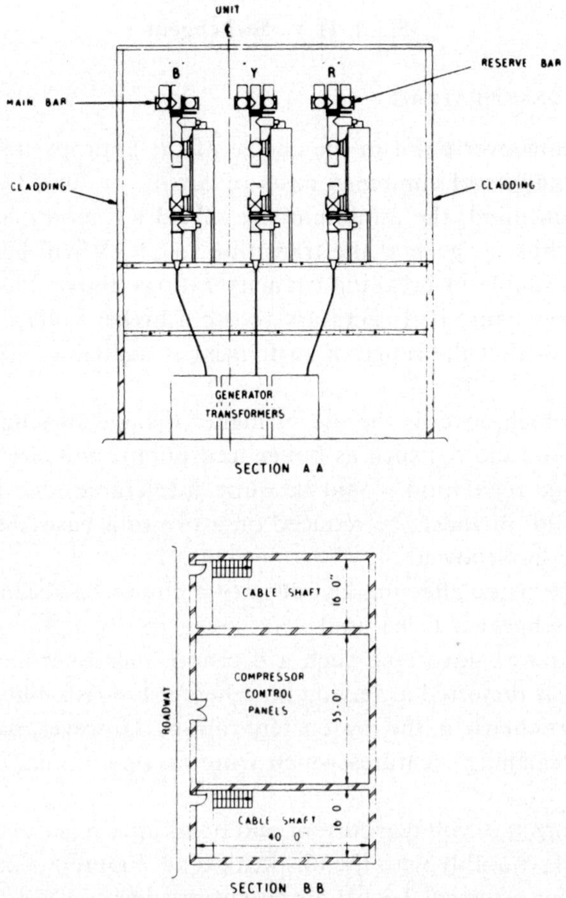

FIG. 5.2.2H. Sections of proposed 275 kV substation using air blast metal-clad dead tank circuit-breakers (shown for one 500 MW unit only).

which could be achieved with an arrangement such as that depicted on Figs 5.2.2B, D, G and H, would be a very attractive proposition.

Finally one further development which offers some prospect for the future is a range of high-voltage circuit-breakers using vacuum interruptors.

5.3. AUXILIARY SWITCHGEAR

5.3.1. General

Two voltage levels are used in generating stations for units up to 200 MW output and three voltage levels above this rating.

The high-voltage system may be at 3·3 kV, 6·6 kV or 11 kV, the low-voltage system invariably being at 415 V.

5.3.2. H.V. Switchgear

5.3.2.1. General Considerations

The economic changeover point in the choice of the appropriate voltage for a motor is governed by an analysis of combined costs of cables, motors, transformer losses and capacity. With this in mind, the minimum size of 3·3 kV motor has been fixed by the Board at 150 h.p. whilst in general the transition to 11 kV will be at 3000 h.p. 3·3 kV switchgear is not available in breaking capacity ratings above 250 MVA, so when the fault level exceeds this figure it is necessary to use a higher voltage. In this case, 11 kV has proved suitable in that the range of switchgear is available, viz. 500 MVA and 750 MVA.

Another feature which governs the use of higher voltage switchgear such as 11 kV is the starting currents on motors such as boiler feed pumps and circulating water pumps. At 3·3 kV the voltage regulation would be quite intolerable, due to the large currents involved, which would obviously be reduced on a pro rata basis the higher the working voltage for the same horsepower.

No decision can be made affecting layout until a choice has been made as to whether air- or oil-break switchgear is to be used.

There are many factors governing such a decision, mostly economical, but generally air-break switchgear is preferred as having no inherent fire risk, but it occupies considerably more room, particularly in the low current ratings. However, some air-break switchgear have built in "earthing" features which from an operational aspect are highly desirable.

The range of ratings (i.e. voltage, current and breaking capacity) available in air-break gear now compares favourably with the oil-break type. Figure 5.3.2.1A shows the spring-closing mechanism for a typical 3·3 kV air circuit-breaker.

A recent development of interest has been the use of a "fused" circuit-breaker for the control of 3·3 kV motors which considerably reduces the explosion severity if a fault occurs in a motor terminal box (see Fig. 5.3.2.1B).

5.3.2.2. Installation and Layout

For illustration purposes let it be assumed that either Scheme I (Fig. 5.3.2.2.A) or Scheme II (Fig. 5.3.2.2B) is adopted. In this case 3·3 kV 150 MVA, and 11 kV 500 MVA and 750 MVA gear will be installed in the station under consideration.

5.3.2.3. 11 kV and 3·3 kV Station Boards

With the number of switch units and their approximate size settled, a tentative layout of the switchroom can be made and its location considered. While some degree of compromise will undoubtedly have to be accepted, the following requirements must be borne in mind.

FIG. 5.3.2.1A. Typical 3·3 kV air circuit-breaker showing spring closing mechanism. (Reproduced by kind permission of The English Electric Co. Ltd.)

FIG. 5.3.2.1B. Typical 3·3 kV air circuit-breakers showing back-up H.B.C. fuses. (Reproduced by kind permission of The English Electric Co. Ltd.)

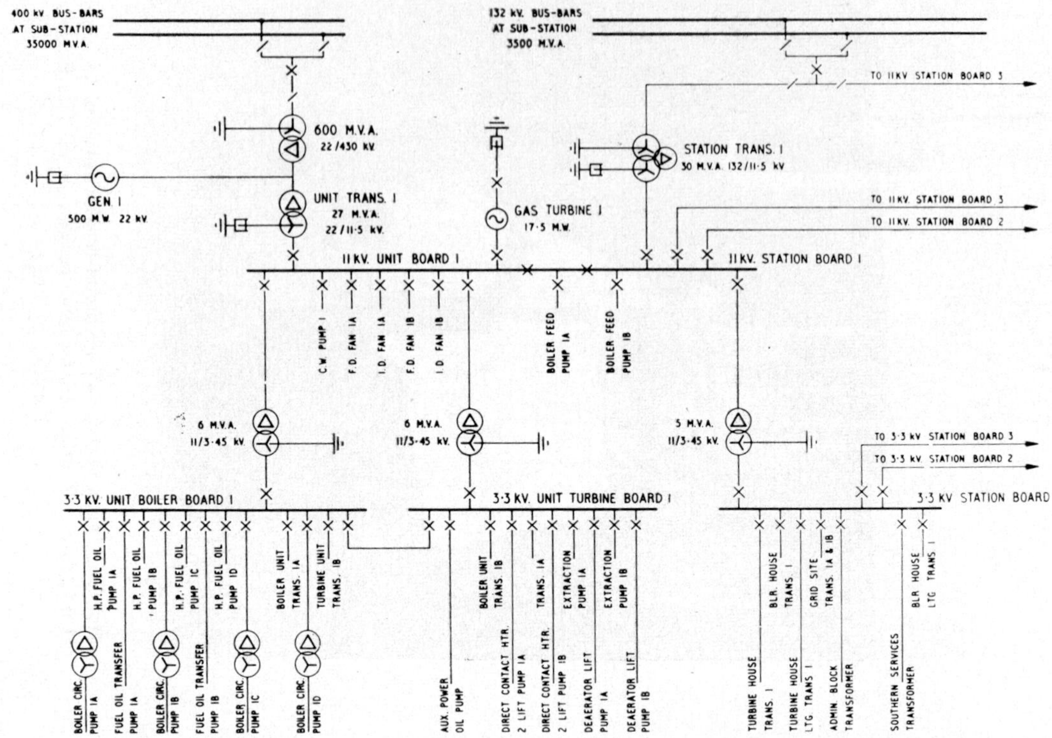

FIG. 5.3.2.2A. Key diagram of auxiliary supplies for 4×500 MW units, Scheme I (oil-fired).

(a) (i) Regarding Scheme I, since there is a station board for each unit, it is necessary to site this board as near its own unit as possible.

(ii) In the case of Scheme II (i.e. one station transformer feeding one station board) it is preferable from a cabling aspect that these switchboards be situated centrally in the power station.

(b) There must be a practicable access route, not only for the initial installation of the switchgear, but for possible removal, overhaul or replacement. Advantage will be taken of any facilities provided for other purposes, such as the turbine house crane, or it may be necessary to provide a hatchway in the floor or roof of the switch-house with a lifting beam over.

(c) There must be convenient operational access.

(d) The location must be considered in conjunction with possible cable routes. Switchgear should have immediately beneath it a cable chamber, or trench from which the power and control cables can be deployed, each cable passing up to its associated cable box on the switchgear. In some installations a marshalling kiosk is built into the switchboard for multicore control cable terminations.

(e) It is imperative that any water tanks or pipes are not positioned directly above switchgear, otherwise leaks percolating on to the switchgear could have disastrous consequences.

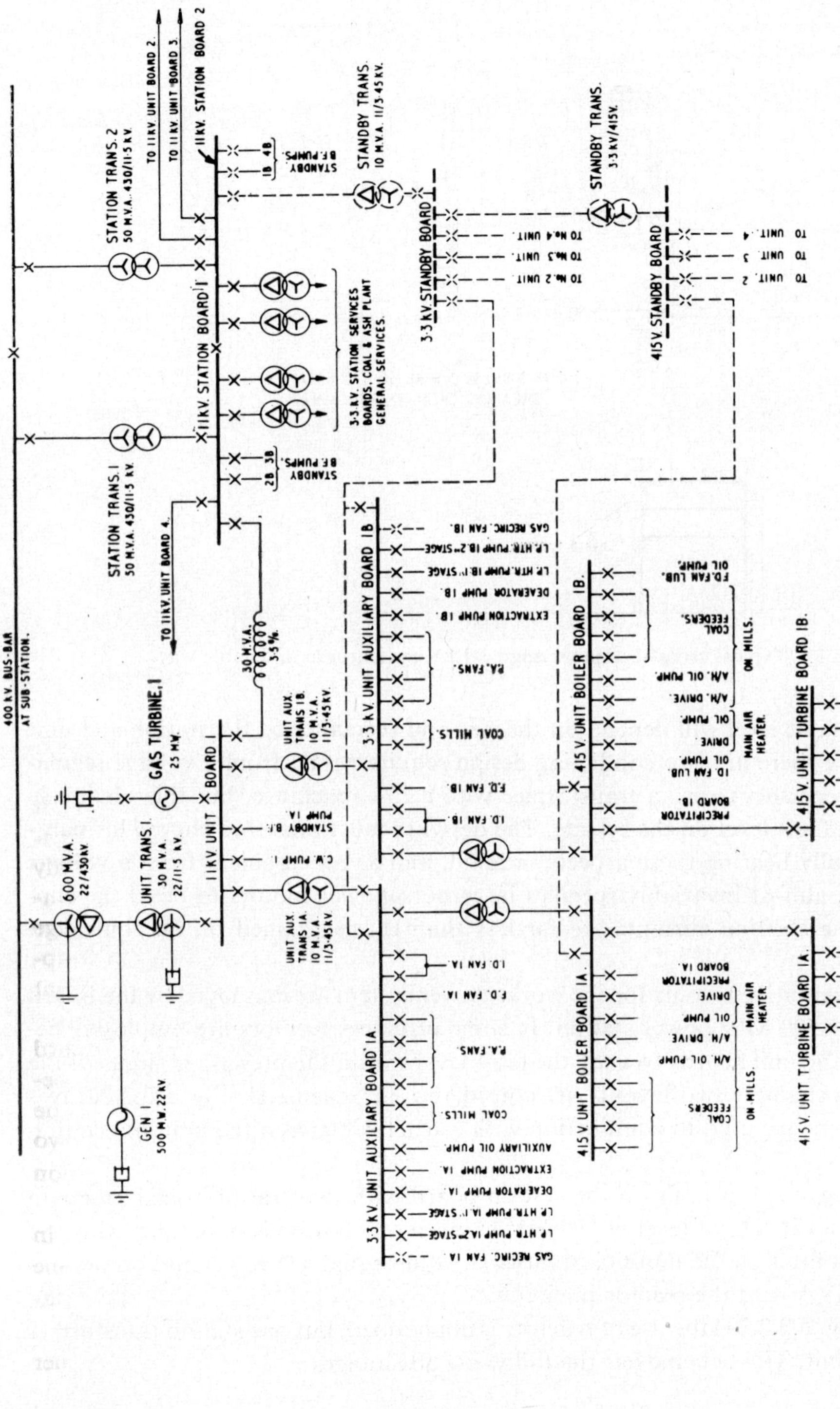

FIG. 5.3.2.2B. Key diagram of auxiliary supplies for 4×500 MW units, Scheme II (coal-fired).

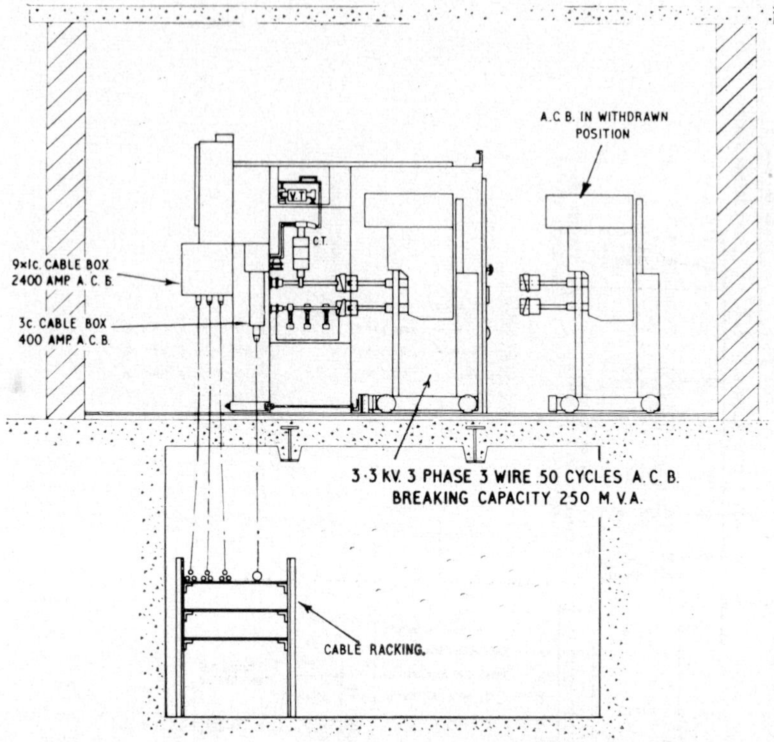

FIG. 5.3.2.3A. Typical cross-section through 3·3 kV switchroom and cable way.

The maximum fault level will depend on the size and reactance of the station and unit transformers. Here there are two conflicting design requirements; from a voltage regulation aspect it is necessary to use a transformer with a low reactance, but this obviously results in a bigger fault level on the system. The designer must therefore choose his parameters very carefully bearing these aspects in mind, and as an easement from a voltage regulation aspect, almost invariably specifies large horsepower motors to be of the slip-ring type since the starting currents are far less than those attained on an equivalent squirrel-cage type.

A number of alternative layouts for the works power system are employed by the Board in the case of a 4×500 MW power station. In some instances reactors are employed between the station and unit boards to keep the fault level within the prescribed limits of the switchgear. As an example two layouts are considered. In Scheme II (Fig. 5.3.2.2B) two station transformers are used in conjunction with a reactor between the unit and station board.

With this arrangement, a fault on the station board with two station transformers in parallel would result in a fault level of 690 MVA; when one station is in parallel with one unit transformer a fault on the unit board busbars would equal 475 MVA and on the station board 525 MVA with the reactor in circuit.

In Scheme I (Fig. 5.3.2.2A) the use of reactors is abandoned, but one station transformer is employed per unit. This scheme has the following advantages:

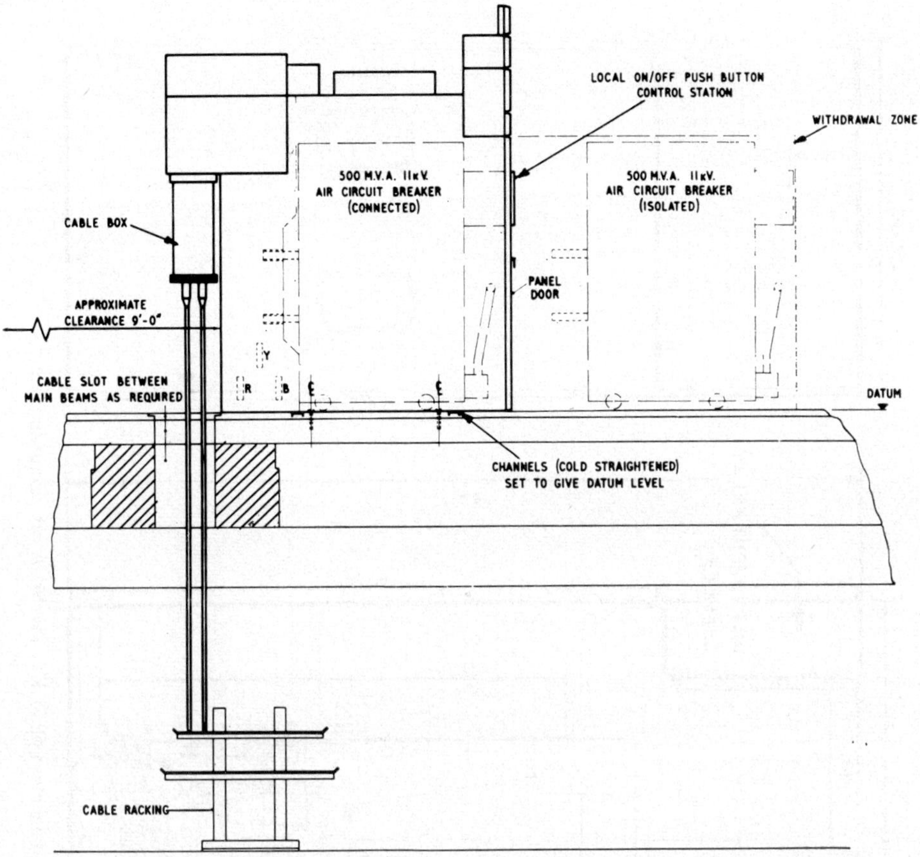

FIG. 5.3.2.3B. Typical section through 11 kV switchboard switchroom and cable floor.

(a) Lower short-circuit fault levels are encountered on the 11 kV switchgear.
(b) Station boards can be "unitised".
(c) Less restriction on the number of units that can be started simultaneously if one station transformer is out of service.

5.3.2.4. Layout of Switchroom

The switchgear is laid out so as to allow sufficient space at the front of the gear to withdraw and manoeuvre the switch trucks, and at the rear to allow access for cable jointing, maintenance and inspection of protective relays if positioned there.

The design of the floor must take account of any fixing arrangements and floor runners for switch trucks, and it is essential to ensure that holes or slots for cables are clear of floor steelwork or reinforcement.

Switchrooms for air-break switchgear should be provided with heating sufficient to prevent condensation notwithstanding the fact that heaters may be built into the switch units. Conversely switchgear located in a high ambient temperature, say on the boiler side of the building, should be adequately ventilated in order to avoid excessive temperature.

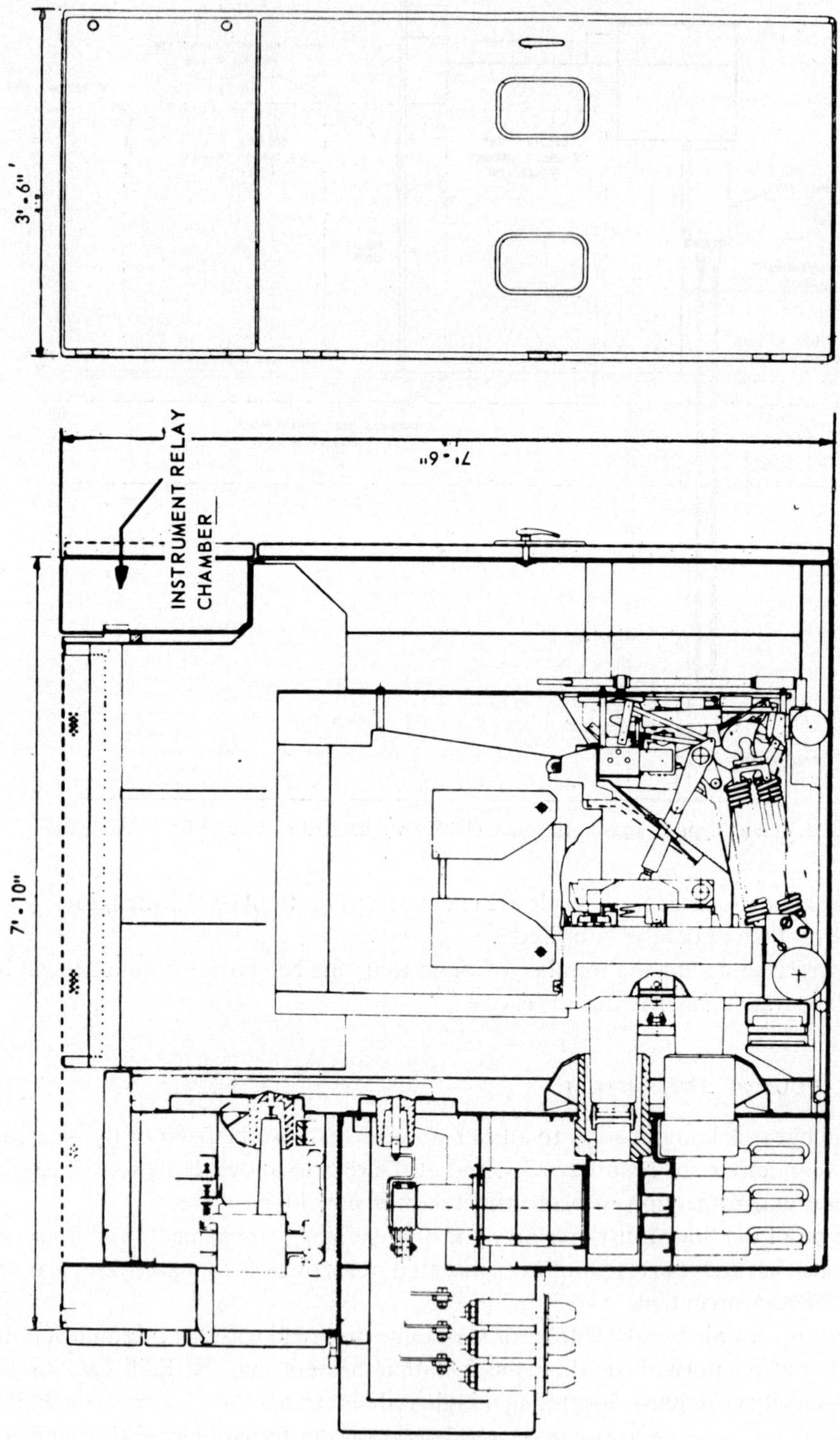

FIG. 5.3.2.3C. Typical general arrangement of 11 kV, 750 MVA air circuit-breaker.

For safety reasons, there must be at least one alternative means of exit. Doors should open outwards and the "escape" door should lead on to a safe landing, and not, for example, on to the top of a ladder or stairway.

For a current 500 MW unit the turbo-generator length is greater than the boiler width, whereas its width including auxiliaries, but excluding h.p. and l.p. heaters and feed pumps, is approximately 65 ft less. If a transverse layout is adopted, this extra space may be divided between a loading bay and a switchgear annexe across the turbine house parallel with the centre line of the machine (see Fig. 5.2.2A). The annexe would have a cable floor at basement level with switchgear mounted on the floor 12 ft above. The roof level of the annexe would be coincident with the operating floor level and be used as a secondary loading area. A bay for the heaters, feedpumps and deaerator would be provided between the turbine and boiler and the switchgear annexe could be extended into this bay. However, instead of just one level of switchgear, as in the turbine house, the switchgear in the heater bay could be tiered, cable floor, switchgear floor, cable floor, etc., with the levels coinciding where possible with boiler gallery levels, and the individual switchgear located at the best level for the particular duty. A cable slot would be provided either side of the heater bay switchgear annexe behind the switchgear to enable free access for cables between all floors, for elevation through switchgear annexe (see Fig. 5.3.2.4A).

5.3.2.5. 11 kV and 3·3 kV Unit Boards

The principles already discussed will apply also to the unit boards.

In fixing the location, it is evident that the switchboard should be positioned near to the "centre of gravity" of its load, namely between the boiler and the turbine. This, unfortunately, is not always possible, as this area may be congested with pipes, gas ducts, etc., which would necessitate a compromise being made.

Difficulty in finding accommodation may sometimes be resolved by splitting the unit board into two sections, one supplying the turbine auxiliaries and the other the boiler auxiliaries. Some economy in cabling may also be achieved by adopting this means. It is also extremely important from an operational aspect and possibility of human error failure to ensure that the relative position of switchgear controlling a unit and switchgear of other units be separated in a similar manner.

5.3.2.6. Trend of Development

The examples given in this section relate to a station of 4×500 MW units. It has been established that with the 500 MW class of unit, a three-voltage auxiliary system, i.e. 415 V, 3·3 kV, 11 kV, is essential, from a flexibility and economical aspect. Even with this system it is observed that there are minor variations in the works power system which are dictated mainly by economic considerations.

Stations now under construction for the Board with one or two exceptions incorporate 500 MW single-shaft units. Currently being considered are designs incorporating super critical boilers with unit size of 660 MW and it may transpire that the age of the 750 MW

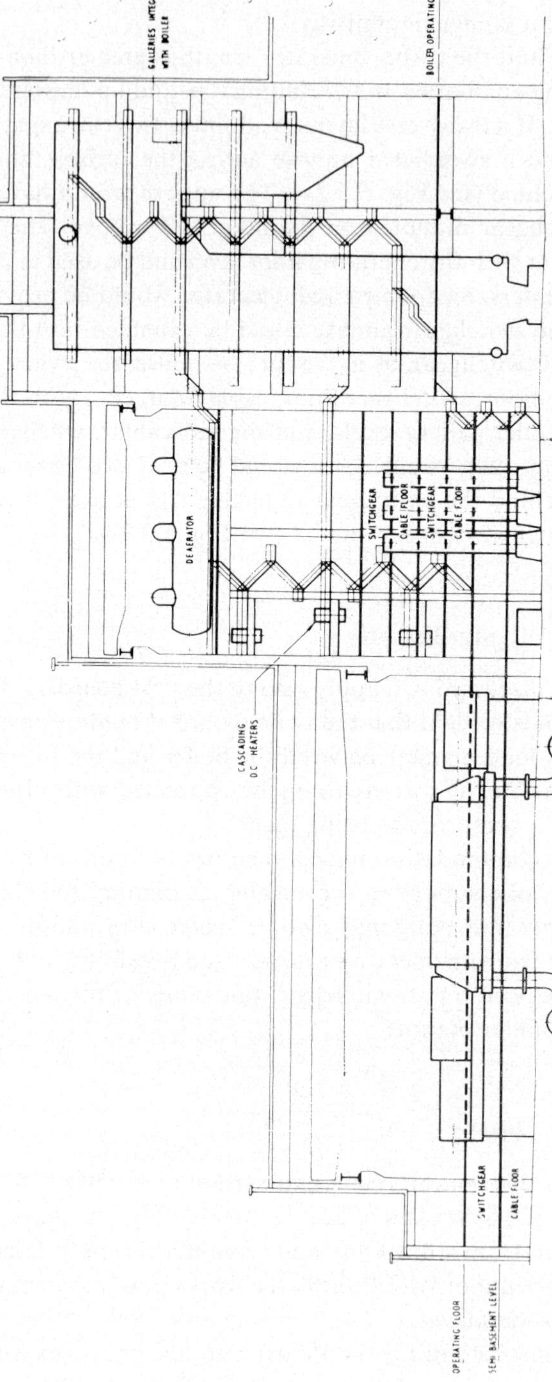

FIG. 5.3.2.4A. Transverse arrangement of 500 MW unit showing location of switchgear.

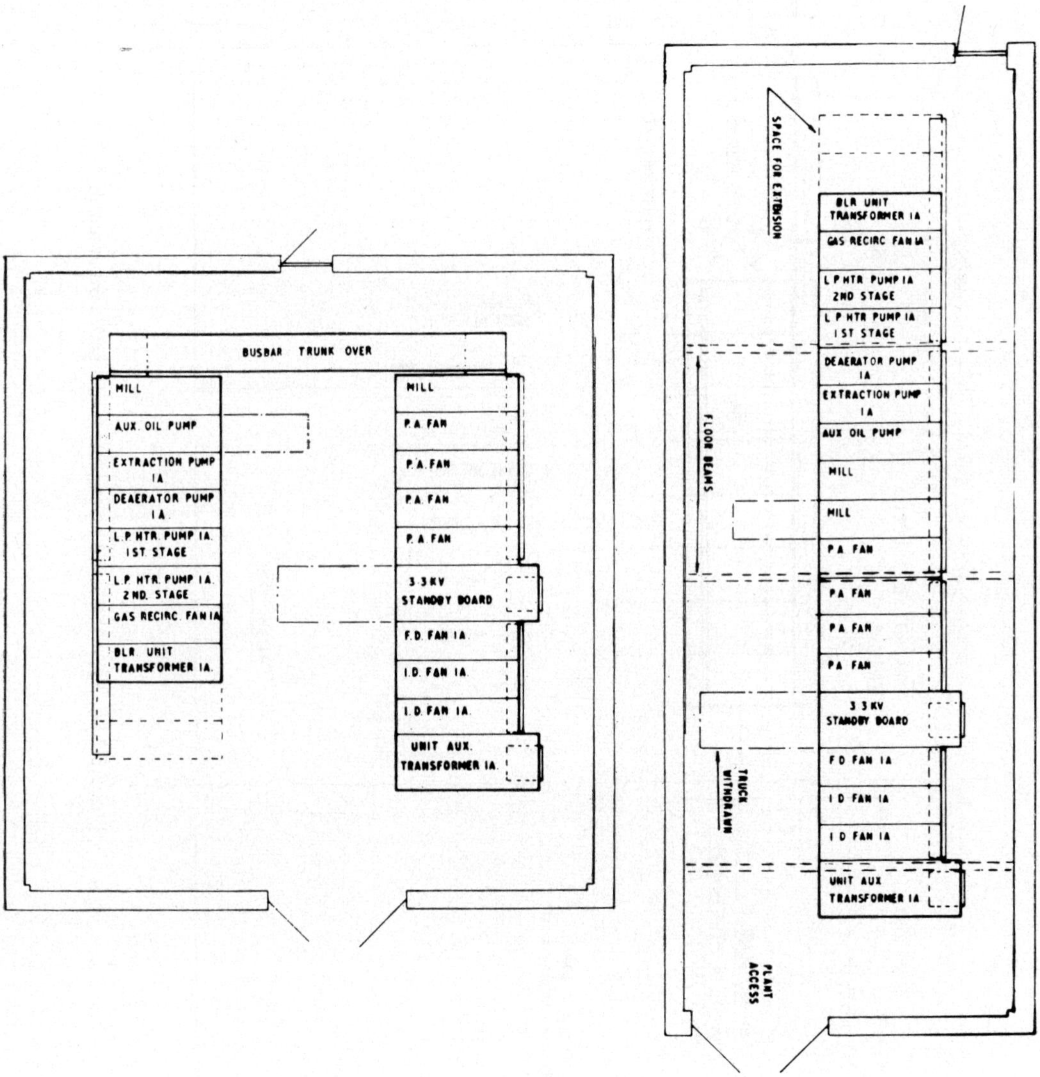

FIG. 5.3.2.4B. Typical general arrangements of 3·3 kV unit switchboard.

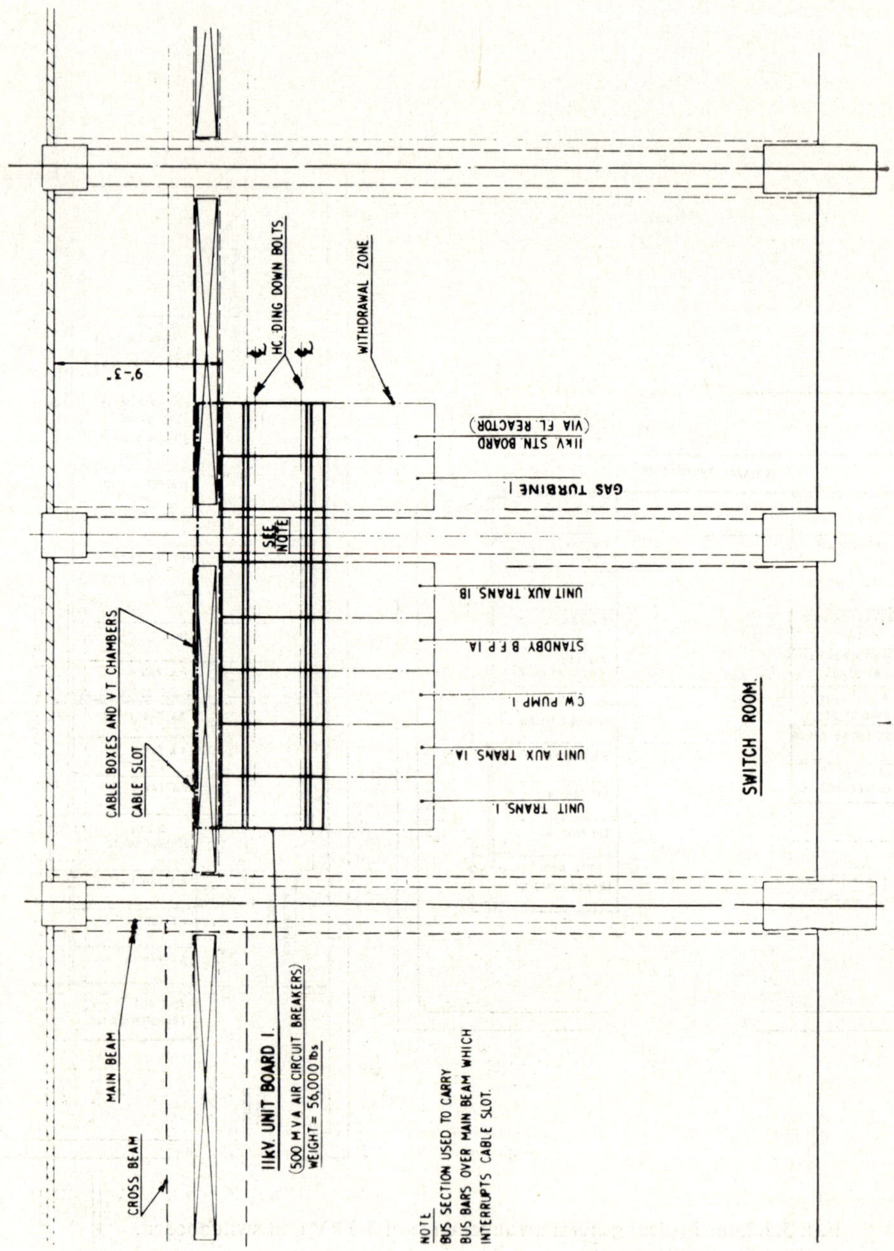

FIG. 5.3.2.4C. Typical general arrangement of 11 kV switchboard.

class of machine is not too far distant. The ramifications of a bigger unit size from an electrical aspect would be:

(a) Assuming electric standby boiler feed pumps would be employed, a radical change to a higher auxiliary voltage, i.e. from 11 kV to 13 or 15 kV, still retaining a 3·3 kV and 415 V system.
(b) Station and unit transformers will be proportionately larger. At present the limits of standard 11 kV switchgear from a current rating aspect are 3000 A at 750 MVA, which, of course, would be exceeded if a station transformer larger than 57 MVA was used. This limit would be exceeded if a unit of 700 MW is envisaged, giving a further argument for the use of higher voltage switchgear with a corresponding reduction in the current to within the manufacturing capacity and rating of switchgear.

5.3.3. **415 V Switchgear**

5.3.3.1. General Scheme

Again it may be advisable to incorporate a marshalling kiosk in the switchboard for reception of multi-core cables. Advantage can also be taken of this design to incorporate the interposing relays or solid state switching devices in the marshalling kiosk, the latter being essential for a remote centralised or computer control scheme.

Fig. 5.3.3.2a. Typical 415 V auxiliary switchboard, door open showing air circuit-breaker. (Reproduced by kind permission of The English Electric Co. Ltd.)

FIG. 5.3.3.2B. Typical 415 V auxiliary switchboard, front view showing insulated busbars. (Reproduced by kind permission of South Wales Switchgear Ltd.)

Depending on economic considerations 3·3 kV/415 V class C type transformers can be incorporated at the end of the switchboard thereby obviating the necessity of running large low-voltage cables usually from outside the power station where the transformer is positioned to the respective switchboard.

5.3.3.2. LOCATION

As before, the general services board should have a central location, and the unit auxiliary boards should be adjacent to their respective units. These boards should, if possible, be mounted at the operating floor level, as ready access is required in the normal course of operation. For typical views of 415 V switchgear see Figs. 5.3.3.2A and B.

The switchboards outside the main building, such as the jetty, coal and ash plant boards, need the protection of switch-houses, and where the position is exposed it will be advisable to provide heating in order to avoid condensation. Switchgear should not be located where there is a risk of corrosion, e.g. in the same room as chlorination plant or batteries.

The sub-distribution switchboards, which will be quite numerous, will be located conveniently to the plant which they serve, and as far as possible in positions not exposed to falling water, moisture and dirt. In extreme cases, for example in some parts of the coal or ash plant, or in the boiler-house basement, it may be advisable to provide a canopy or some similar protection.

5.4. TURBO-GENERATOR AUXILIARIES

5.4.1. General

Considering first the turbo-generator itself. In, for example, West Burton, a modern 4×500 MW station (Fig. 5.11.1A), the 500 MW single-shaft machines are arranged lengthwise along the turbine hall which is 850 ft long by 130 ft wide by 85 ft 6 in high. The overall length of each machine is 162 ft 6 in. The generators are rated at 500 MW, 22 kV, at 0·85 power factor lag, maximum current rating 15,400 A.

To operate a set of this size will require a substantial amount of auxiliary equipment and the turbine maker will usually have set ideas as to its location. Most of the items have natural positions in relation to the main plant and present a simple problem of co-ordination and efficient use of space.

The following is a list of auxiliaries that need to be considered from an electrical layout point of view.

Excitation equipment.
Turbovisory panel.
Busbar cooling fans.
Auxiliary oil pump motor.
A.c. and d.c. seal oil pump motors.
A.c. and d.c. flushing oil pump motors.
Oil tank vent blower motor.
Turbine oil purifier centrifuge motor.
Turbine oil purifier water pump motor.
Turbine oil purifier heaters.
Jacking oil pump motor.
Barring gear motor.
Stator winding coolant pump motor.
Distilled water pump motors.
Hydrogen equipment for rotor cooling.
Quick start exhausters.
Rotary air pump motors.
Condensate extraction pumps.
Direct contact heater extraction pumps.
Cathodic protection.
Auxiliary CW strainer motor.
"Taprogge" equipment for condenser tube cleaning.

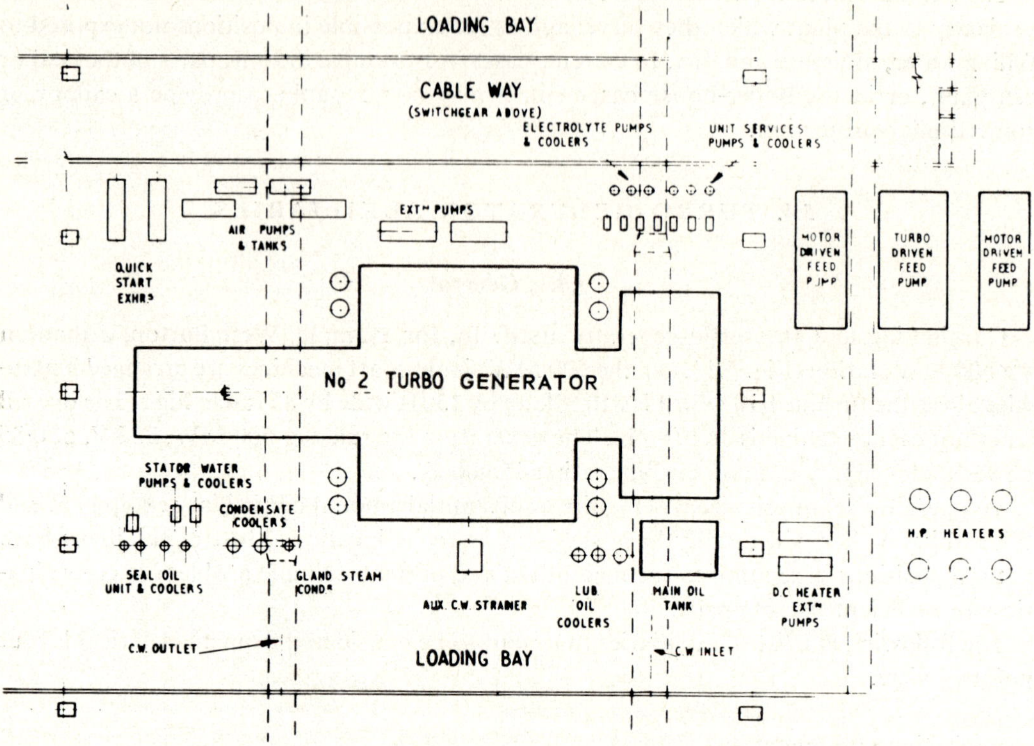

FIG. 5.4.1. Typical arrangement of 500 MW turbo-generator auxiliaries.

Figure 5.4.1 shows a typical arrangement of some of the above-mentioned auxiliaries in relation to the main set. All the auxiliaries mentioned, with the exception of the excitation equipment and the turbovisory panel will—since their prime purpose is mechanical in nature—be laid out by the mechanical engineer in conjunction with the turbine manufacturer. However, the electrical layout engineer must be sure that he knows of all such equipment as may require electrical services, in order that such civil work as may be required to cater for cable routes can be assessed and conveyed to the civil designer as soon as possible. The electrical engineer will have to specify the type of cable box and type of gland. These factors in turn will be dependant on the size of the motor involved and the type of cable used. Generally speaking when any equipment is to be mounted on the operating floor care must be taken not to encroach on the generator rotor withdrawal space, or obstruct the handling of heavy turbine parts during overhaul.

5.4.2. **Excitation Equipment**

The present tendency for units now under construction or in the design stage is to provide an a.c. exciter and rectify the output from the exciter by means of silicon-diode cells. This does eliminate certain problems, for example commutation problems, but of course does not entirely eliminate brush gear, because the pilot and main exciters and the main alternator field require slip rings. Figure 5.4.2A shows a typical a.c. excitation scheme for a 500 MW unit.

The complete equipment will usually comprise a "suite" of panels containing rectifiers, a.c. and d.c. switchgear and field suppression equipment. Since large currents are involved entailing heavy connections, the equipment should be positioned so that the runs are kept as short as possible and also access does not constitute a problem. The connections are of

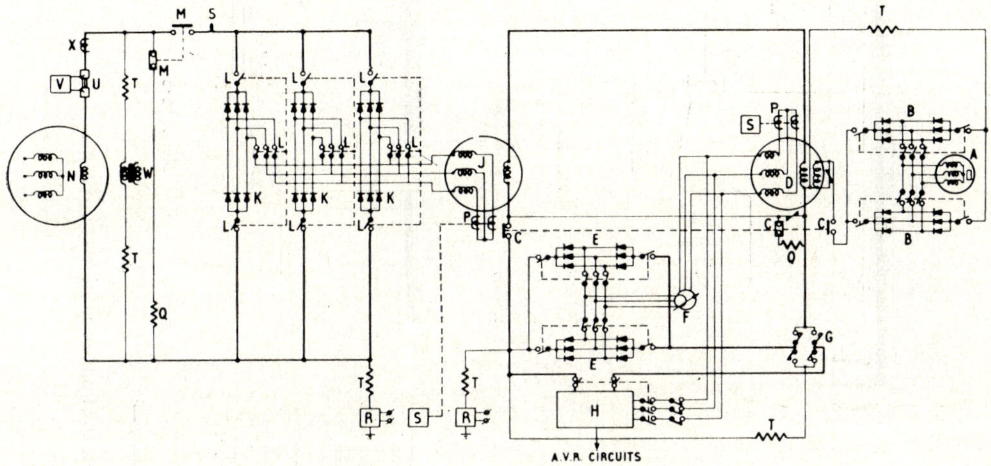

A 3 PHASE PERMANENT MAGNET BIAS GENERATOR.
B PILOT EXCITER FIELD RECTIFIER (IN A.V.R SUITE)
C MAIN / PILOT EXCITER FIELD CIRCUIT BREAKER
D 3 PHASE PILOT EXCITER
E MAIN EXCITER FIELD RECTIFIERS (IN A V R SUITE)
F MANUAL CONTROL INDUCTION REGULATOR. (MOTOR OPERATED).
G AUTO - MANUAL CHANGE OVER SWITCH
H 3 PHASE FLUX RESET MAGNETIC AMPLIFIER.
J 3 PHASE MAIN EXCITER
K GENERATOR FIELD RECTIFIER CUBICLE
L GENERATOR FIELD RECTIFIER A C / D C ISOLATOR.
M GENERATOR FIELD CIRCUIT BREAKER 4550A. D.C.
N TURBO - GENERATOR 500 MW 22 K V PF 0 85.

P CURRENT TRANSFORMER.
Q DISCHARGE RESISTANCE.
R EARTH FAULT.
S OVER CURRENT.
T RESISTANCE.
U SHUNT.
V UNDER CURRENT.
W VOLTAGE TRANSFORMER
X CURRENT TRANSDUCTOR

FIG. 5.4.2A. A.C. excitation scheme.

copper section, suitably enclosed in trunking with removable covers to facilitate cleaning and mounted on post insulators as shown on Fig. 5.4.2B. Although not visible in the figure, the copper connections should have either phase barriers or be taped to make them as fault-free as possible, i.e. to avoid flash-over, particularly near the terminals.

Typical C.M.R. is 4550 A d.c. at 494 V, main excitation alternator C.M.R. 3710 A a.c. at 390 V 100 c/s. The higher frequency giving better response times and better utilisation of materials. Excitation equipment physical sizes and weight are considerably reduced with this method compared with those of an equivalent d.c. system.

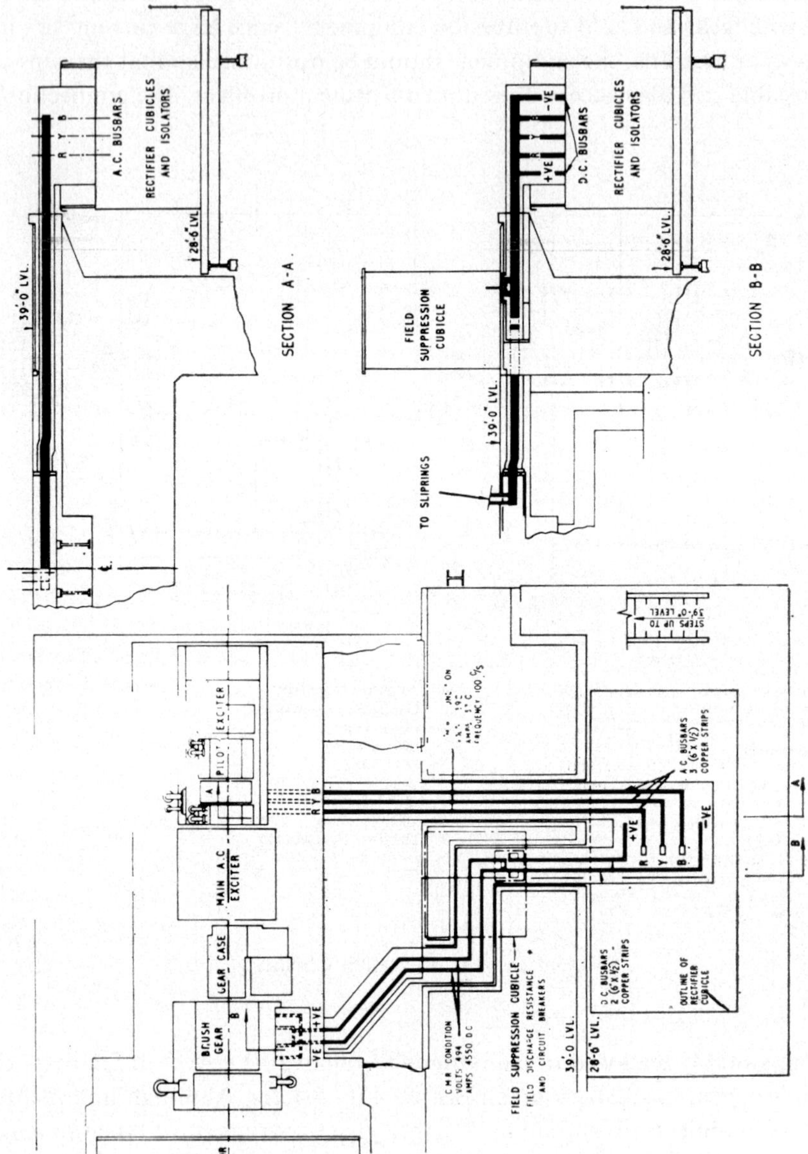

FIG. 5.4.2B. General arrangement of a.c. excitation equipment.

5.4.3. Hydrogen Equipment

The hydrogen panel, seal oil pumps and the hydrogen and CO_2 cylinders are mounted at basement level close to the foundation block of the generator, care being taken to ensure that equipment is located to prevent accumulation of pockets of gas. Clear access to the cylinder banks must be allowed, as routine exchange of full for empty cylinders has to be made.

5.4.4. Distilled Water Pumps

These are also placed at basement level close to the foundation block.

5.4.5. Turbovisory Equipment

With the advent of remote centralised control philosophy, it is becoming increasingly common to incorporate such instruments which record and indicate shaft eccentricity, differential expansion, pedestal vibration, etc., in a panel situated in the control room.

In general, only the minimum of indication and control features are retained local to the turbo-generator.

5.4.6. Stator Water-coolant Panel

This panel is generally positioned at basement level, near or even adjacent to the hydrogen panel in order that associated cables can be run with advantage along a common route.

It should be noted that all panels associated with the turbo-generator should be so positioned that in the event of a steam or water leak damage to the panels is obviated or kept to a minimum.

5.4.7. Neutral Earthing Resistor

It is usual to earth the neutral or star point of the main alternator, unit transformer secondary and the gas turbine as shown in Figs. 5.3.2.2A and B. Various methods of achieving such earthing exist, and modern power station practice is to use the liquid type resistor connected between the system neutral point and earth. The advantages to be gained by the use of liquid earthing resistor are many, some of the most important being:

(1) The electrolyte possesses a negative temperature coefficient, valuable for protective purposes.
(2) The earth—fault current can be limited to a predetermined value.
(3) A faulty section can be isolated by the application of suitable earth fault protection.

The electrolyte used may be turbine condensate or other suitable water with a high specific resistance, so that with the addition of a quantity of sodium carbonate the desired value of resistance for particular circuit conditions may be obtained. It is obviously necessary to have the liquid resistor tank effectively earthed in order to complete the circuit; also if the tank is situated outside it will be necessary to provide a thermostatically controlled immersion heater to prevent the liquid freezing in winter.

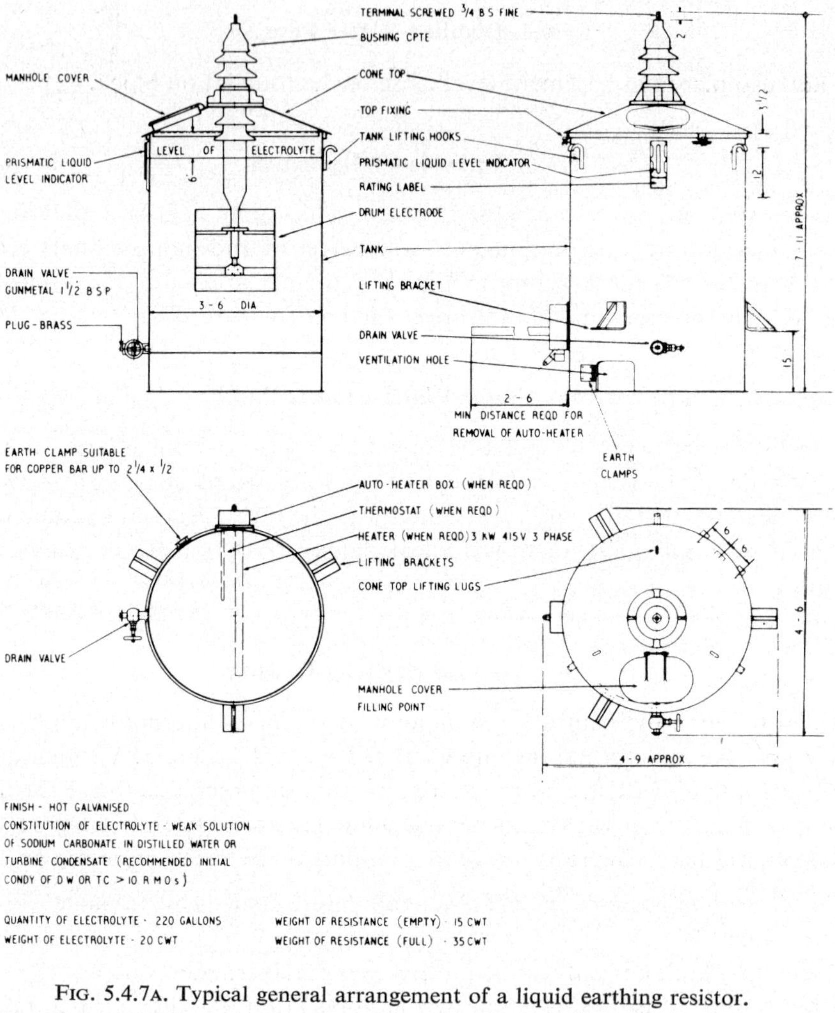

FIG. 5.4.7A. Typical general arrangement of a liquid earthing resistor.

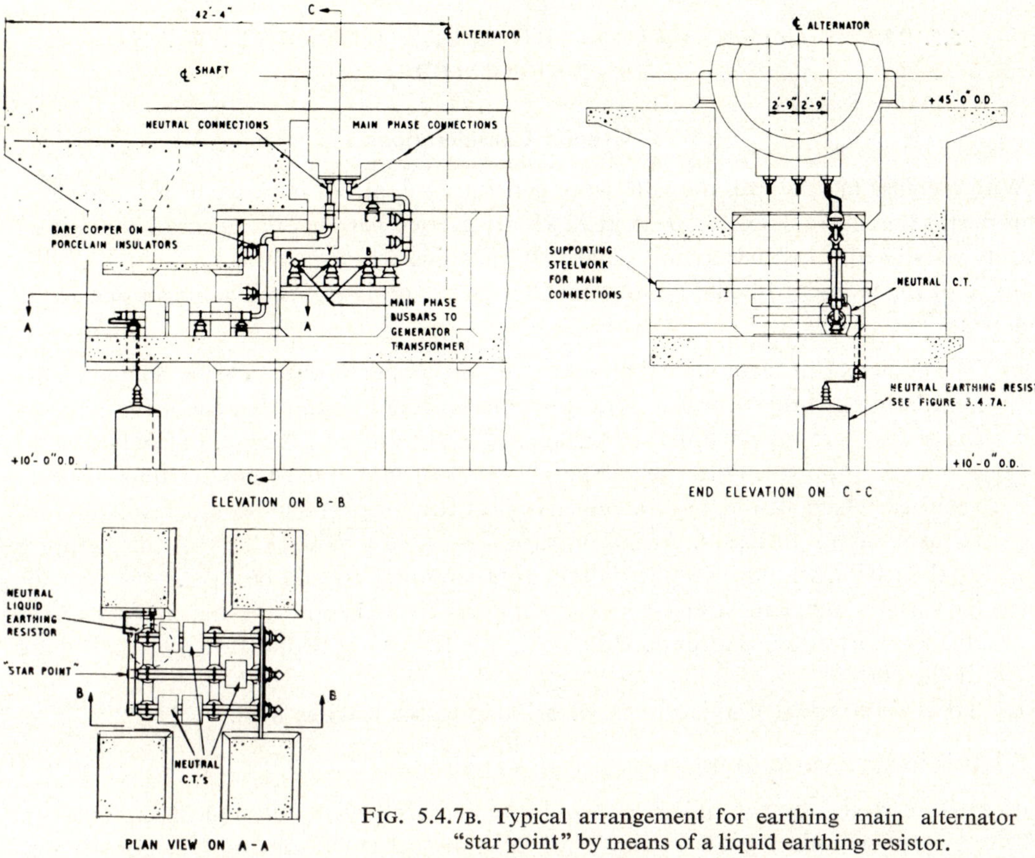

FIG. 5.4.7B. Typical arrangement for earthing main alternator "star point" by means of a liquid earthing resistor.

For the main alternator, the liquid resistor is located as near as possible to the alternator "star point", but preferably outside the building. For the transformer earthing it is usually more convenient to position the resistor in a brick cell adjacent to the transformer concerned. It is preferable that the connections to the resistor be of bare copper on porcelain insulators, and rated so as to carry the maximum earth fault current that would result in the event of an internal flashover occurring in the resistor, thereby "shorting" itself out. This would only occur when an earth fault is present.

The tank should be positioned so that:

(1) The immersion heater may be easily withdrawn and cabled.
(2) The electrolyte visual level gauge may be clearly visible from outside the compound.
(3) The electrolyte drainage valve and refilling point are well placed for maintenance.
(4) Restricted earth fault current transformers mounted in copper work are readily accessible.

Figure 5.4.7A shows the typical general arrangement of a liquid earthing resistor and Fig. 5.4.7B a typical arrangement for earthing the main alternator "star point" by means of a liquid earthing resistor.

5.5. MAIN CONNECTIONS BETWEEN GENERATOR AND TRANSFORMER

5.5.1. General Considerations

With increase in unit size, the voltage of generation does not increase proportionately. This means that currents of 15,500 A at 22 kV are carried between the generator terminals and its associated transformer, for a 500 MW machine.

Some of the factors to be taken into account on the design of these connections are as follows:

(a) The choice of an economical cross-section of copper busbar. This is related to the ventilation or other working arrangements bearing in mind that the losses appearing as heat are proportional to the square of the current. It may, of course, be an economic proposition to use a bigger cross-sectional area of copper than actually required which would have a higher initial capital cost but would result in subsequent saving in losses. An amortisation exercise would be carried out before a final decision would be made on the cross-sectional area to be used.

(b) To ensure sufficient strength in the supports to withstand the mechanical forces due to short-circuit currents. Here again the forces are proportional to the square of the current.

(c) To avoid heating of adjacent steelwork due to the leakage field.

It is impracticable to use cables for this application due to the following reasons:

(i) A large number of cables would be required with consequent difficulty in accommodating the sealing ends both at the generator and transformer terminals.

(ii) It would be almost impossible to ensure equal current sharing between the cables in such a short run.

(iii) It is exceedingly doubtful whether any cable would carry the short-circuit currents encountered for the time required.

Conductors are usually made of high conductivity copper since the relative costs of copper and aluminium have hitherto not been such as to outweigh the advantages of the former's high conductivity. The double-angle arrangement of conductors shown in Fig. 5.5.2D is being used for the higher current ratings as it combines a low skin-effect ratio, large surfaces for cooling by radiation and natural convection between the "channels", ease of manufacture and good mechanical strength to withstand the electromagnetic forces which can occur during short-circuits. For large angles, however, the maximum lengths available are limited to about 12 ft so that, in order to make up complete runs which may be 100 to 180 ft long, several lengths must be joined end to end. These joints are usually made to absorb thermal expansion over the solid lengths by bolting on flexible connectors made of laminated copper strips. The flat horizontal arrangement shown on Fig. 5.5.2A is generally the most convenient way of arranging the three phases. The conductors are supported on post-type porcelain insulators which must be exceptionally robust and sufficiently closely spaced to withstand the short-circuit forces which in the case of a 500 MW unit at the unit "tee-off" would amount to 3500 lb/ft run during the first half cycle of the fault. The insulators used are of the outdoor type in order that a higher factor of safety from a creepage and clearance aspect is maintained. The arrangement adopted should be such that the porcelain is stressed in compression, as it is well known that porcelain is weak in tension.

The individual 12 ft lengths of conductor are clamped to the insulators so that they are held rigidly at one end but are free to slide at the other in order to absorb thermal expansion while still being fully restrained against sideway movement.

The connections must be totally enclosed for obvious safety reasons, also to exclude dust and vermin.

Experience has shown that in general for large units it is necessary to cater for forced air cooling of the main connections chamber. This will usually take the form of exhaust fans suitably positioned in the chamber and alternator pit. On Figs. 5.5.2A and D the ventilating equipment installed at West Burton power station can be seen. At Ferrybridge power station, however, the main connections have been enclosed within an open mesh construction for about two-thirds the length of the run and natural ventilation is expected to suffice. The star point chamber at Ferrybridge is more enclosed, but has been provided with sufficient air outlet points (filtered) to provide adequate temperature control. No cables or pipes should be run inside the conductor chamber as the risk of induced currents from the connections is very great. It will of course be necessary to run wiring from the "star point" current transformers, but this should be routed so as to have an absolute minimum length actually within the chamber, and should not be enclosed in a ferrous covering or armouring in the vicinity of the chamber because of the possibility of induced heavy currents which would cause heating. Any access doors to the main connections must be interlocked so that it is impossible to enter the chamber unless the unit is off-load. The enclosures are constructed of concrete (which must not, however, contain steel reinforcement), brick, or of sheet aluminium. Care must be taken that in the latter case, any fixing bolts used for securing the sheet aluminium to the building should not be in contact with any steel reinforcement. This is particularly important in the case of the steel reinforcement in the alternator block. All access doors should preferably be also of wood construction or other non-magnetic material.

For a 500 MW generator the current is approximately 15,500 A. In order to carry such a current with the naturally air-cooled arrangement as previously described, a conductor size of at least 2 in $\times$ 12 in $\times \frac{3}{8}$ in would be needed in order to keep the average temperature rise of the conductors above the ambient outside the enclosure, within the 50 degC maximum permitted by B.S. 159 which is usually specified. Although such a conductor arrangement as previously described may well be acceptable in the relatively unrestricted spaces between the foundation block and the power station wall and out to the generator transformer, its large size has presented problems inside the confines of the foundation block, especially where the bar primary current transformers have to be accommodated over the neutral conductors. It has been found that with the axial and integral designs of condensers now being used the space available inside the foundation block for generator connections is so restricted that it is difficult to accommodate such large conductors

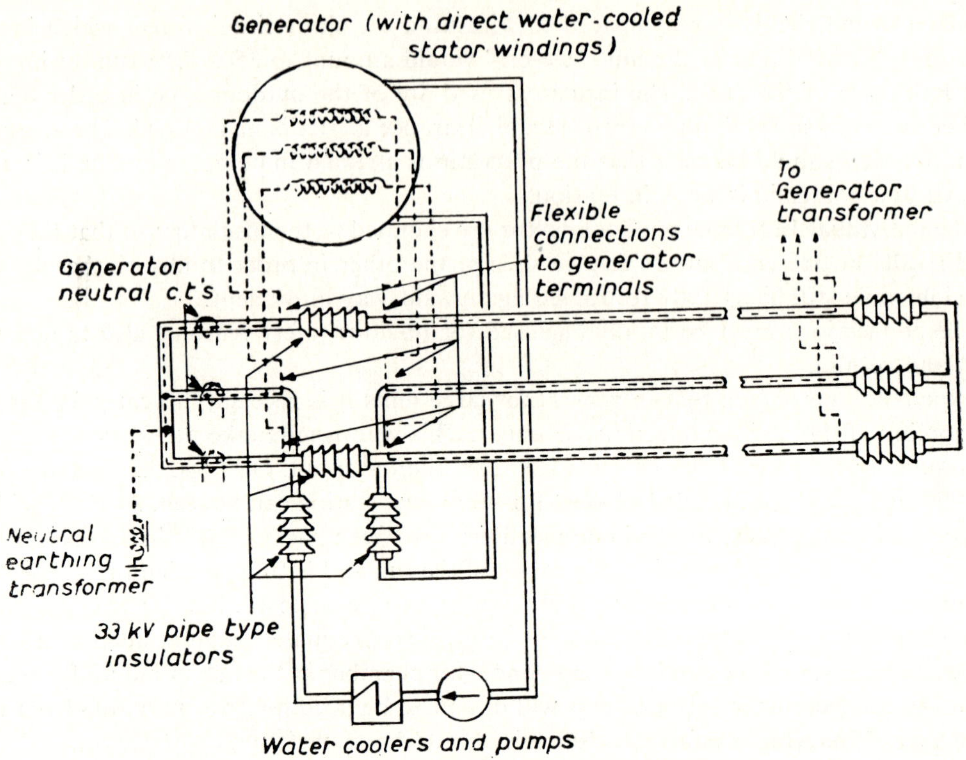

FIG. 5.5.1A. Schematic arrangement of water-cooled busbars (Aberthaw 'B'). (Reproduced by kind permission of The English Electric Co. Ltd.)

with the specified electrical clearances, and still provide adequate access for maintenance and inspection. Coupled with the space problem there is perhaps an even more pressing one of heat dissipation, the most critical area being within the foundation block. In addition the size of conductors required for a naturally air-cooled installation and the need to clamp them securely have all led to very high initial costs.

For Fawley and Pembroke power stations the turbine operating levels coincide with the existing ground level, hence it was necessary in both cases to bring the main connections away from the top of the alternator. This arrangement gives a bonus since it avoids the critical area under the block altogether (see Figs. 5.11.1J and L). Ample room is then available for mounting the current transformers and running naturally air-cooled busbars each totally enclosed in a separate sheet metal tube horizontally out through the station wall.

With separate enclosures, phase-to-phase faults are impossible, and being freely surrounded by air, sufficient heat should be dissipated to prevent high temperature rises in conductors and current transformers. Although this arrangement solves some problems, it does, as is quite common with power station layout, give rise to others. For example. much space which would otherwise be available for turbine hall crane access is so no longer.

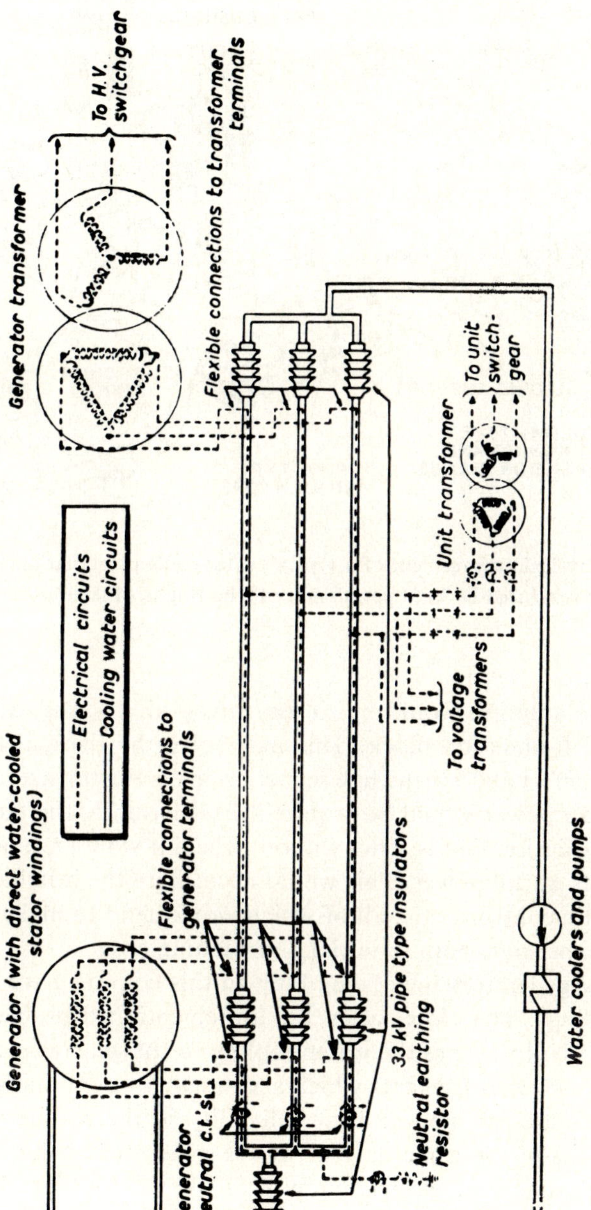

FIG. 5.5.1B. Schematic arrangement of water-cooled busbars (Eggborough). (Reproduced by kind permission of The English Electric Co. Ltd.)

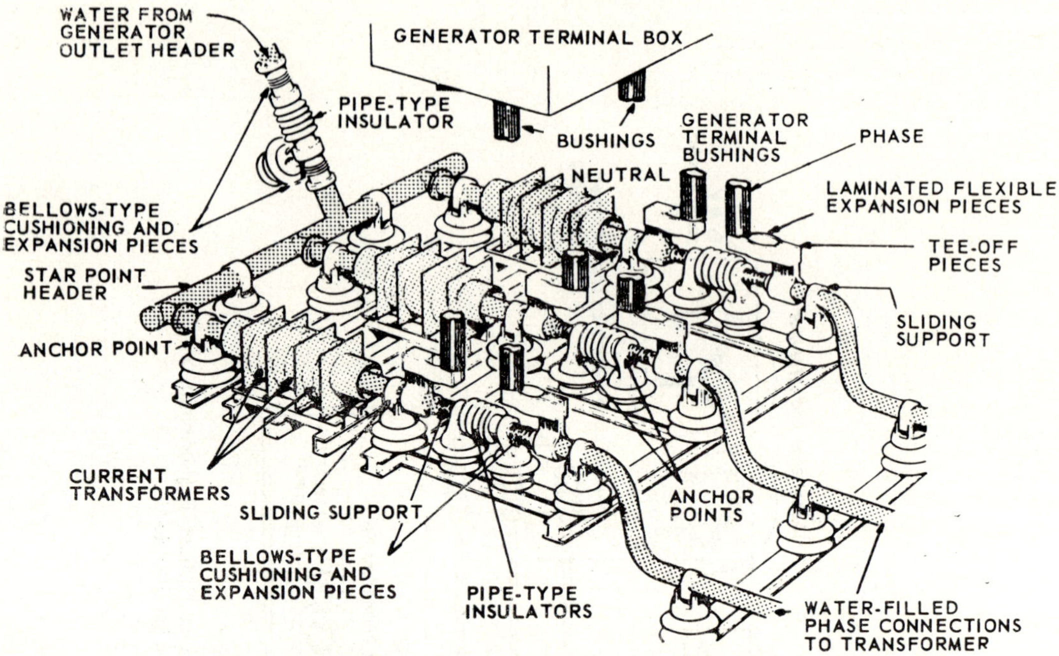

FIG. 5.5.1C. Physical arrangement of a typical water-cooled installation at generator end. (Reproduced by kind permission of The English Electric Co. Ltd.)

Another possible solution would be to use forced air cooling, at least in the critical regions within the foundation block. This may solve the thermal troubles but not the space problem. To effect any worthwhile reduction in conductor size, a quite unreasonably large amount of fan power would be required. Furthermore, the fans would have to be classed as vital auxiliaries, that is, ones without which it would not be possible to operate the generating units at full power. This would necessitate the installation of standby fans with fully automatic controls, etc., all of which would tend to make such an installation complicated and expensive, both in initial and running costs.

All the 500 MW generators now being built in this country have direct water cooling of the stator windings. Therefore, when the aforementioned problems are encountered one solution might be to extend the water cooling to the generator connections. Figures 5.5.1A and B show two arrangements which are in use. By combining the cooling circuits no extra auxiliaries are required and the reliability of the cooling water system of the connections is virtually the same as that of the generator. It has been found that by divorcing the hydraulic and electric circuits as shown in Figs. 5.5.1A and B a much better balance in flow, cooling, strength and other requirements can be obtained.

The diameter of the high conductivity copper tube chosen for the phase and neutral conductors for 500 MW water-cooled generator connections is 6 in o.d. 3 S.W.G. representing a copper cross-sectional area of 4·55 in^2 whereas the double 12 in$\times$12 in$\times\frac{3}{8}$ in angles have a cross-sectional area of 18 in^2, i.e. nearly four times as much.

Thus the use of water-cooled tubes offers a substantial economy in copper and space. The tubes are obtainable in lengths of up to 30 ft so that only a few longitudinal joints are needed in the run.

Figure 5.5.1C shows the physical arrangement of a typical water-cooled installation at the generator end.

5.5.2. **Layout**

The run of main connections will vary according to whether the turbo-generators are arranged in a longitudinal, transverse, or diagonal manner. It must be borne in mind, however, that the height of the turbo-block and the generator transformer will depend on the individual manufacturers, therefore every busbar installation must be tailor made and may only be identical in the case of repeat power stations.

Longitudinal Arrangement

With this arrangement shown in Figs. 5.5.2A and B the terminals are shown in the pit below, but they could be above the alternator if required (as Figs. 5.11.1J and L), the choice being dictated by the height of the low-voltage generator transformer bushings relative to the block height, in order that a horizontal run of busbars can be achieved with a minimum of angles or bends.

Diagonal Arrangement

The above remarks also apply, but this arrangement has the distinct advantage of a shorter run of busbars with consequent savings in capital cost and losses (see Fig. 5.5.2B).

Transverse Arrangement

To avoid right angles and bends, it is obvious that the main connections must emerge underneath at the exciter end of the block as shown in Fig. 5.5.2C. This could introduce an additional hazard in that the copperwork associated with the excitation system must be avoided. However, with an arrangement of exciter copperwork (as shown in Fig. 5.4.2B) there should be no problem. With a transverse arrangement the length of connections will be governed by the space requirements for rotor withdrawal, which in turn fix the distance of the block from the turbine house wall.

Returning now to Fig. 5.5.2A, this depicts the layout of the 15,500 A connections for a 500 MW generator.

The connections accommodated in the non-ferrous enclosure (see Fig. 5.5.2D for cross-section) terminate on through wall bushings, the outside enclosure will be fixed to the wall and also to the transformer top plate which will be provided with suitable flanges. This portion of the enclosure, being out-of-door, must be weatherproof. Fig. 5.5.2E shows the arrangement at the transformer terminals. At the alternator end the star point is completed by connecting together the three neutral terminals. These connections are made in the pit beneath the machine and incorporate the bar primaries of the generator

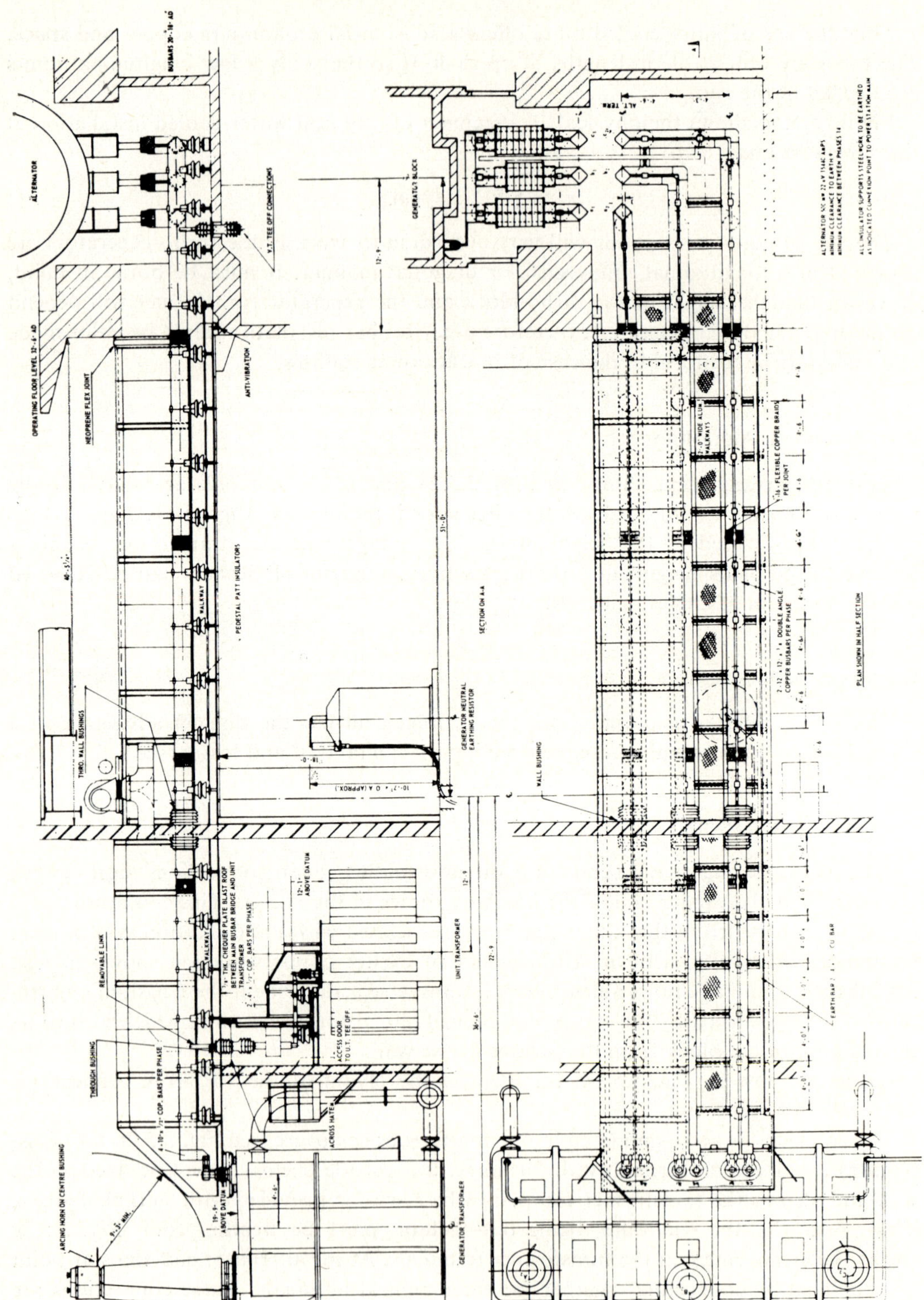

FIG. 5.5.2A. Main connections for a 500 MW generator.

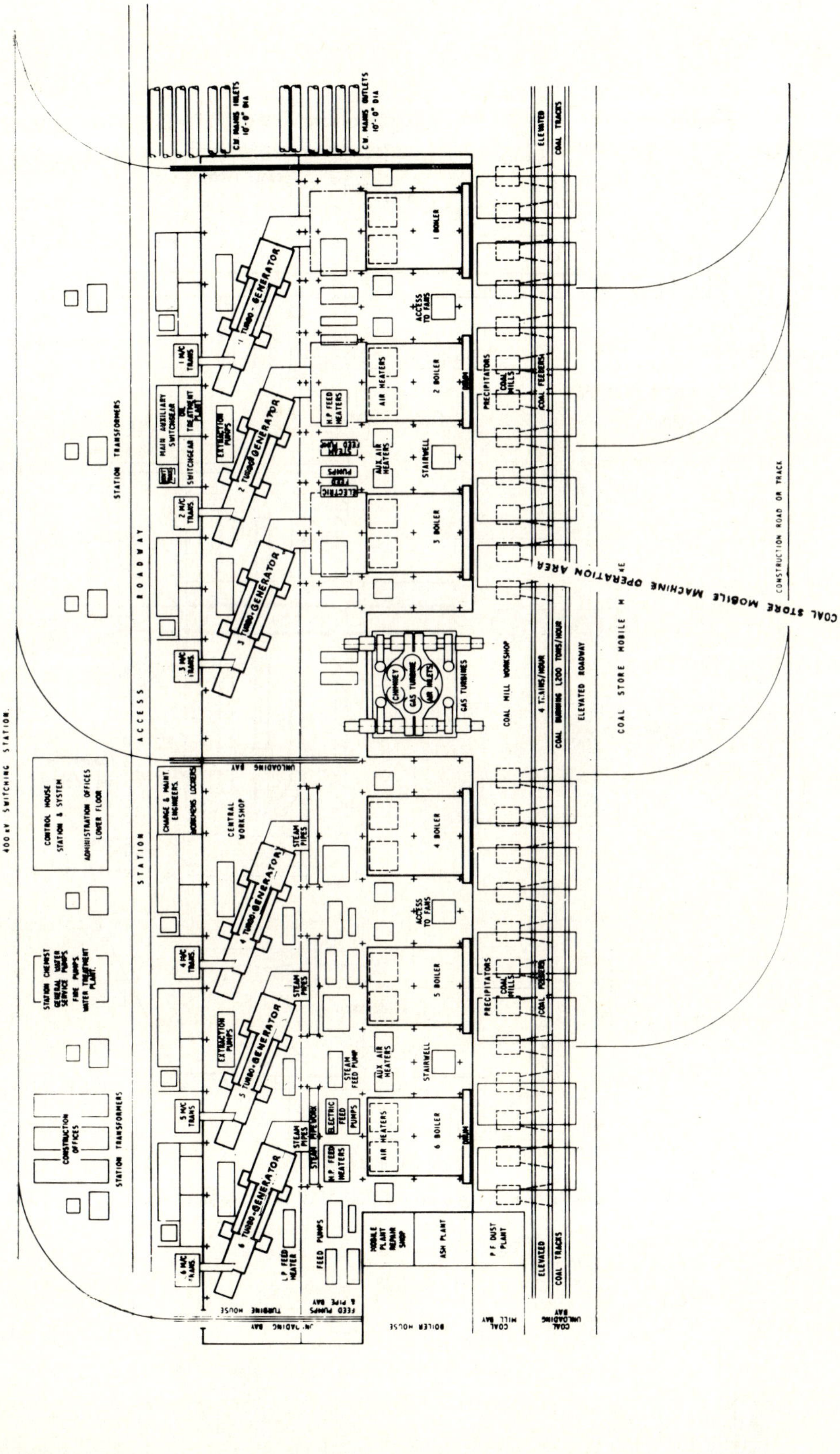

FIG. 5.5.2B. Diagonal turbo-generator arrangement as proposed for Drax power station.

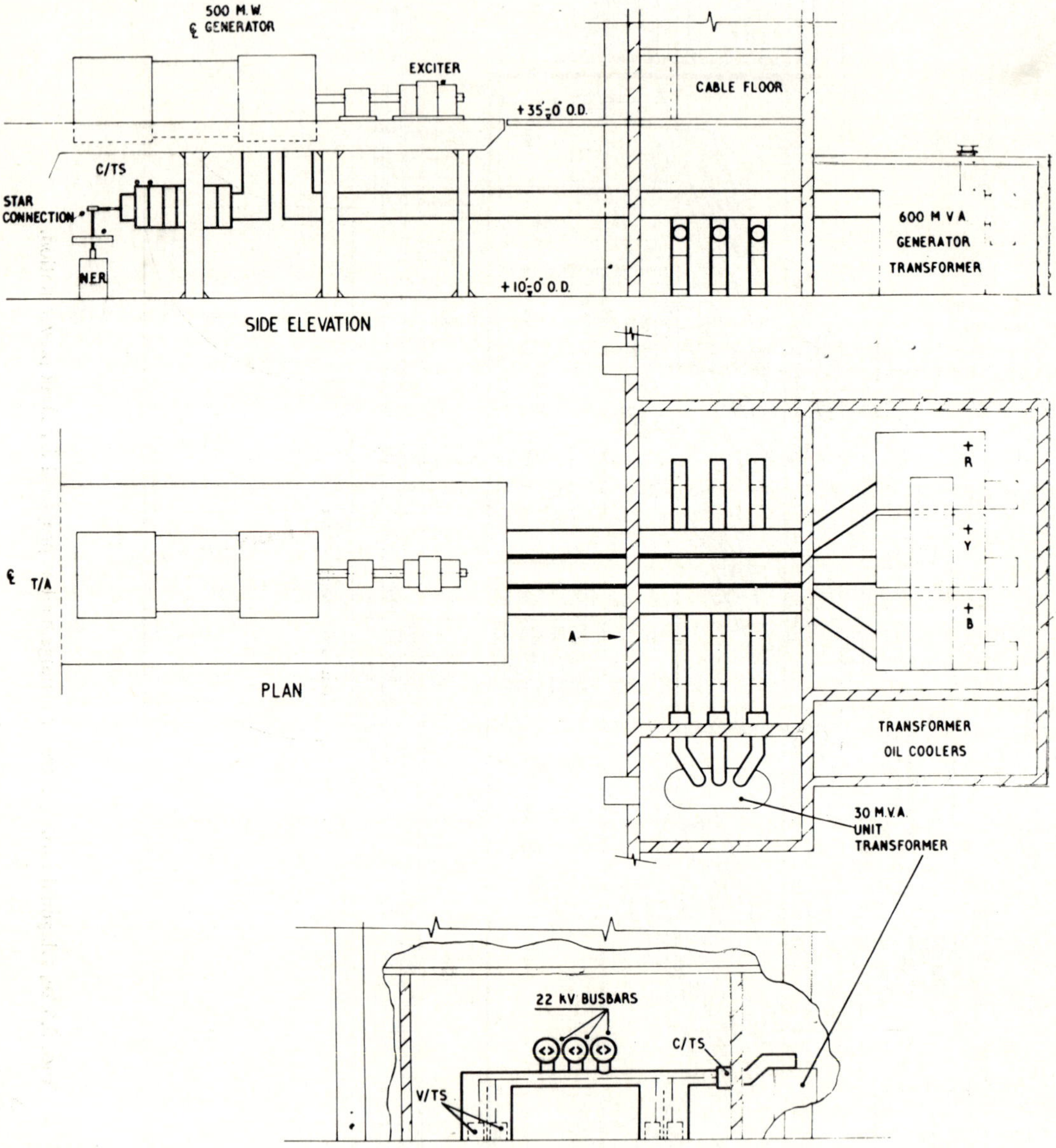

FIG. 5.5.2C. Transverse arrangement of 500 MW turbo-generator.

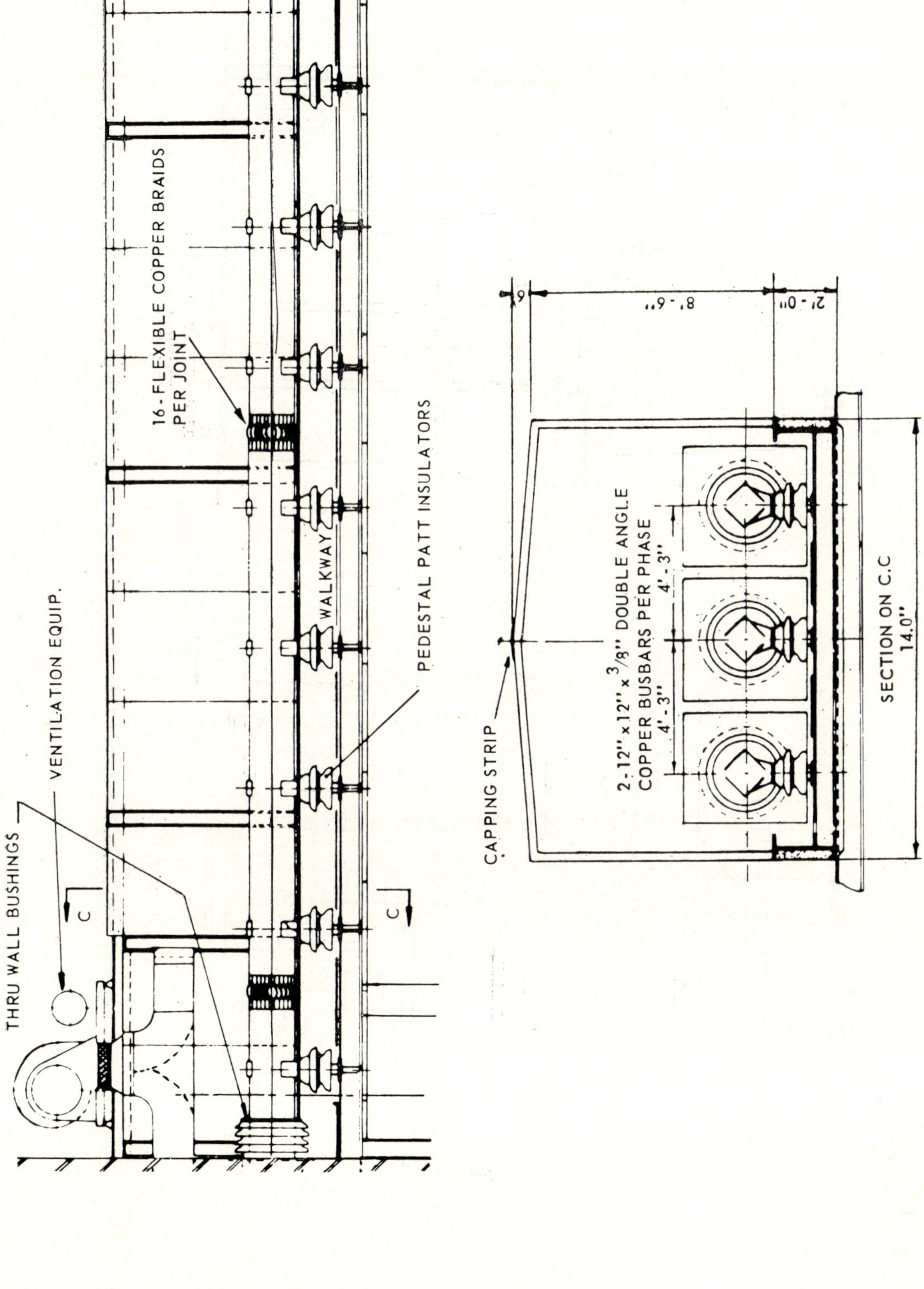

FIG. 5.5.2D. Sections of main connections for 500 MW generator.

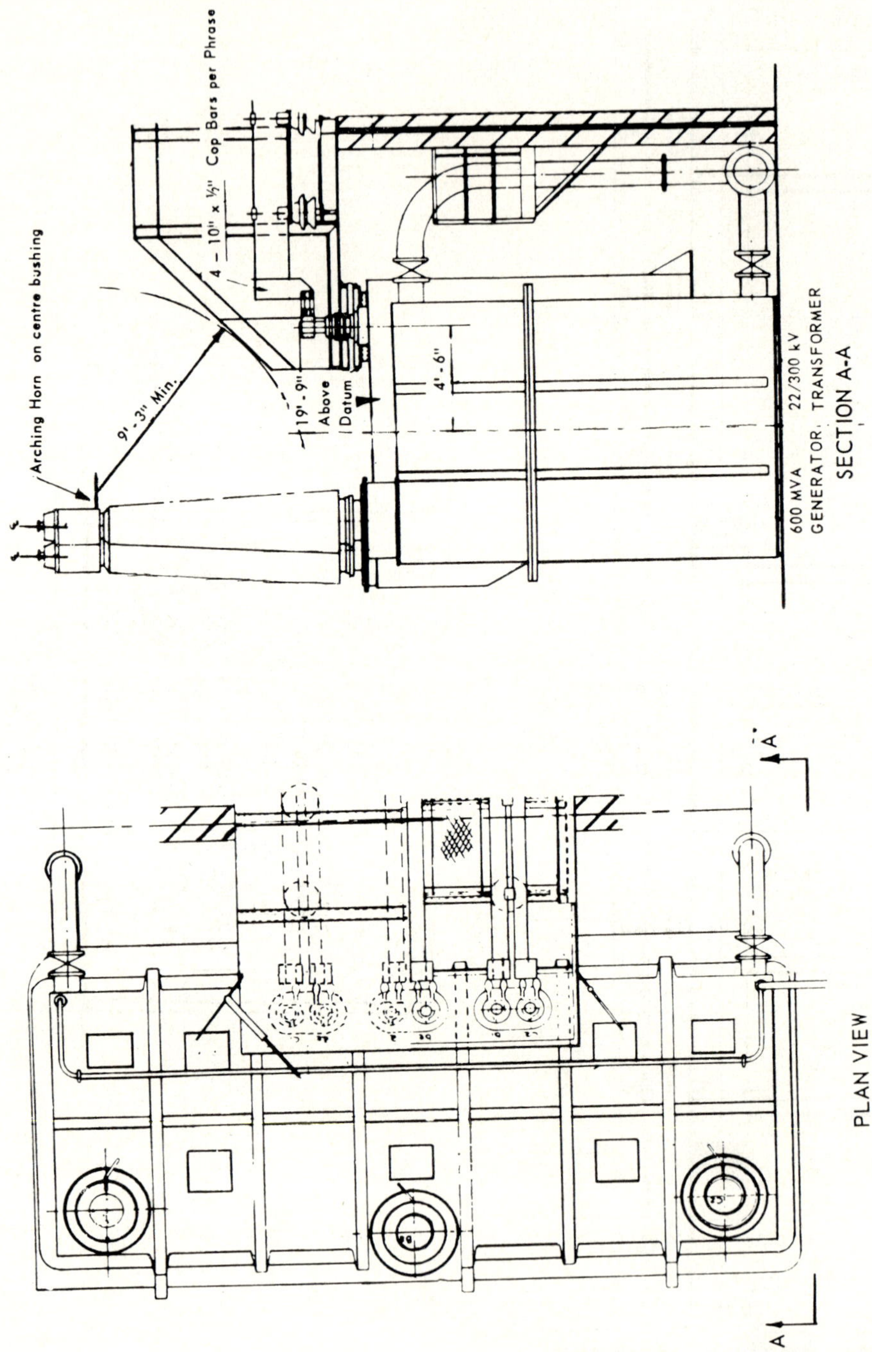

Fig. 5.5.2e. Layout of 600 MVA transformer showing connections to terminals.

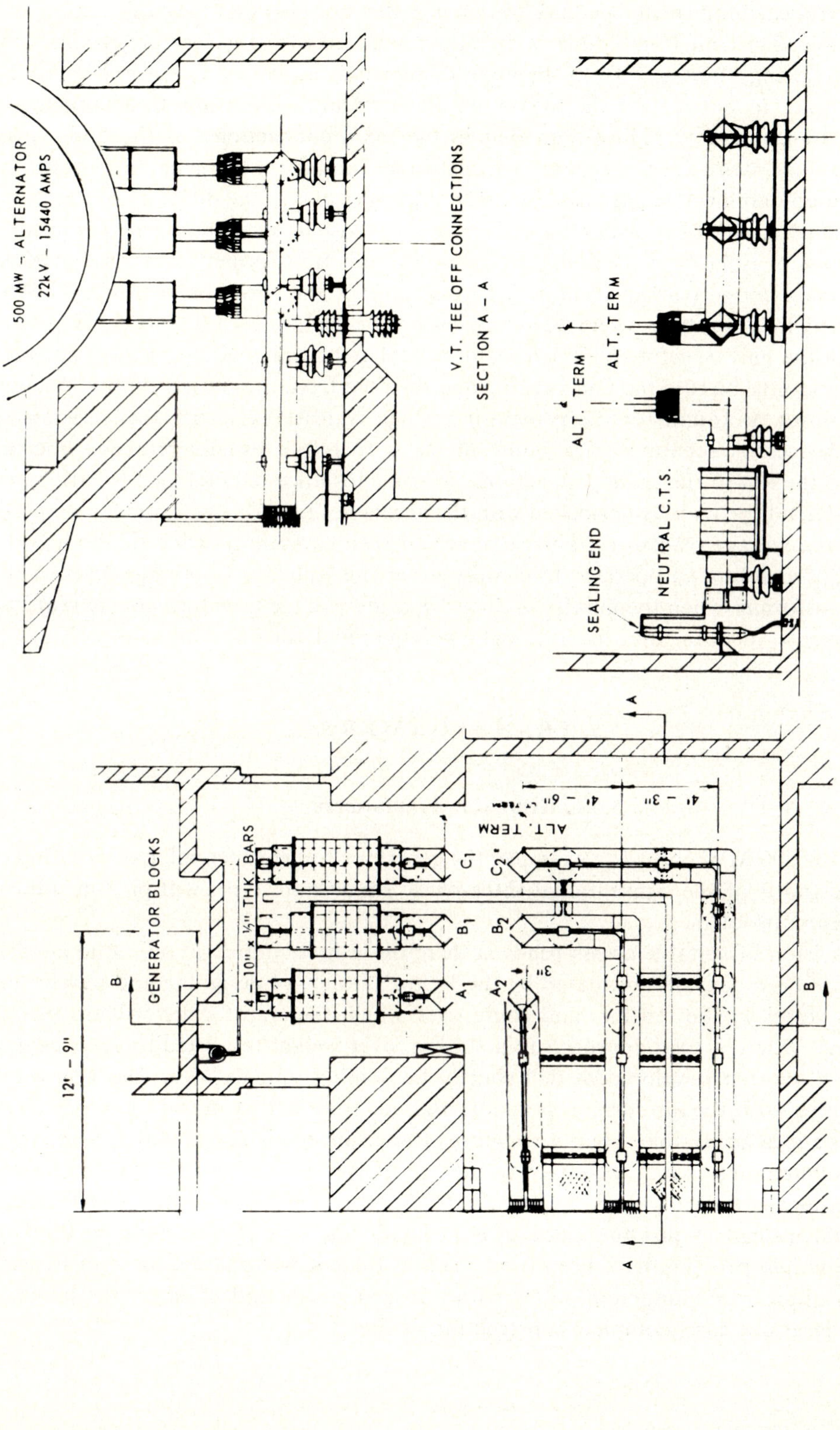

FIG. 5.5.2F. Main connections for 500 MW unit showing alternator connections.

neutral current transformers clearly visible in Fig. 5.5.2F. The star point is connected to the neutral earthing resistance tank by cable in this example and the cable sealing end is also visible. The unit transformer is directly connected to the generator terminals, its connections being "teed-off" from the main connections above as shown in Fig. 5.5.2A. It is important to notice that the maximum short-circuit current due to a fault in the "tee-off" is considerably higher than that in the main connections, as the fault is fed both from the system, via the generator transformer, and from the generator itself.

The "tee-off" connections are usually made in copper due to the difficulty of obtaining cable which is capable of withstanding the temperature rise under short-circuit conditions.

Some form of isolation is desirable to allow the unit transformer to be disconnected from the main copperwork run, and this usually takes the form of removable bolted links located in a position where access is easily obtainable. By this means, if a fault develops in the unit transformer it can easily be isolated thereby allowing the associated turbo-generator to be run, the unit board being then fed from the station supply Accommodation must be found for the generator voltage transformers with their associated isolating devices. A separate brick built cubicle is therefore usually provided, located adjacent to the run of the main connections. In some instances a combined isolating and earthing switch is provided interlocked with the access door. In this way, the high voltage supply to the voltage transformers is isolated and then earthed allowing the door to be opened, thus enabling an operator to change either the high or low voltage fuses of any voltage transformer when the unit is running. Separate voltage transformers are required for instruments, metering and the automatic voltage regulator.

5.6. TRANSFORMERS

5.6.1. Generator Tranformers

Taking the 500 MW unit as an example, the generator transformer will have a rating of 600 MVA and if of the conventional three-phase single tank type, will present quite a transportation problem.

The provision of suitable access roads is therefore a first consideration, although this problem has been considerably eased by the Board's policy of designing and purchasing a suitable vessel to shorten the land route over which the transformer will travel (see Fig. 5.6.1A). Where a road approach to a station, over weight-restricted bridges, cannot be avoided, another development will shortly be coming into its own. The Board are experimenting with the air cushion principle and adapting it to a heavy trailer (see Fig. 5.6.1B). The idea being that when a weight-restricted bridge is encountered, a separate vehicle-mounted blower unit is coupled to the transporter, which has been suitably modified, the trailer is then "floated" across the bridge the load being spread over a much bigger area, instead of just the axles as is normally the case. Other possible ways of easing the weight problem have been tried, such as three single-phase units, transformer tank fabrication from aluminium, filling with nitrogen gas instead of oil for the journey, and, in at least one case, complete construction at site.

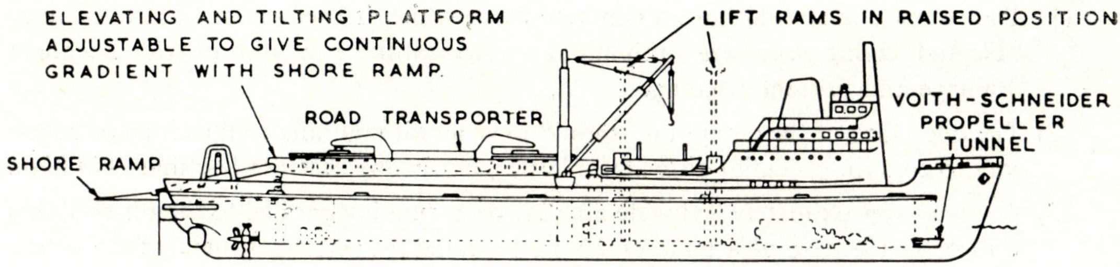

FIG. 5.6.1A. Heavy lift vessel.

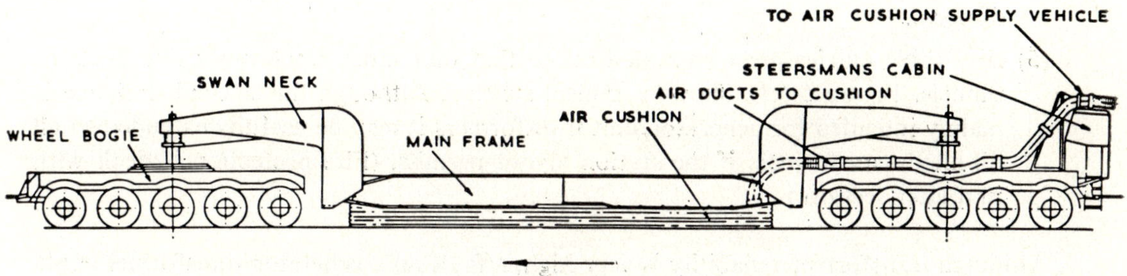

FIG. 5.6.1B. A typical heavy-load transporter trailer fitted with an air cushion.

In general the increase in generator transformer capacities has led to an increase in the time required for site erection of same. One manufacturer has developed a PVC tent, pressurised 1 or 2 lb above atmospheric pressure and sufficiently large to contain a generator transformer. Suitable orifices for accepting the h.v. bushings are provided in the roof, and an air lock large enough to contain a small lorry is provided. A small blower is necessary to pressurise the tent and the air is heated, thus "factory conditions" are reproduced at site and any delays in site erection due to bad weather are eliminated.

In order to keep the heavy and expensive connections between the transformer and its associated generator as short as possible, of necessity this means the positioning of the transformer immediately outside the turbine house wall opposite the generator, in an open top cell with high flanking walls to restrict possible fire damage.

The flat base of the transformer may well rest on resilient pads to minimise structure borne noise. The floor of the cell must be designed to carry not only the transformer as installed, but the concentrated load of jacks which may be used to place it in position. The provision of at least two securely anchored haulage eyes, into which tackles can be hooked, will assist the latter operation.

The total quantity of oil in the transformer will be approximately 30,000 gal. The fire protection system if operated would add another 4500 gal/min for say 5 min. Thus in the event of a burst tank or pipe coupled with a fire approximately 52,000 gal of oil/water mixture would have to be disposed of from a single generator transformer.

It is obvious that with quantities of this magnitude special consideration must be given to compound drainage and disposal of oil/water mixture at the layout stage, and the following are briefly the main requirements:

(1) The compound should have a concrete floor, which is level under the transformer base and having elsewhere a minimum gradient of fall 1 in 60 to provide adequate drainage and prevent ponding.

(2) The open face of the compound should have an oil retaining wall as high as necessary to contain potential oil/water mixture, but in any case not less than 9 in high.

(3) A trench approximately 2 ft wide and 2 ft deep running parallel to the oil-retaining wall or in the case of a large compound around the whole perimeter, fall again 1 in 60.

(4) Pipes of a suitable size to convey the oil/water mixture quickly from the compound trench to an oil catchment pit where the mixture is allowed to settle out before disposal.

(5) Any cable entries must be sealed off so that oil cannot reach any cable ducts or tunnels. Figure 5.6.1C shows a typical scheme. Although this special drainage is mainly intended for generator/unit transformers it may be usefully extended to all oil-filled transformers if the station layout permits. (Fire protection is dealt with in a later section.)

Although transformer reliability is very high, a fault on a generator transformer could at worst impair the security of the grid system and at best incur financial loss to the Board because of the idle plant. To cover emergency conditions temporary replacement units could be used.

The Board are now proposing to purchase a limited number of "national spare" generator transformers. The proposal at present is to have three nominal types though each type will allow for some variation. They are as follows:

425 MVA	18·5/300 kV
600 MVA	22/300 kV
600 MVA	22/430 kV

It is not clear at present whether three separate single-phase double wound transformers or the conventional three-phase transformer will be adopted.

A survey of recent installations has been carried out in order that the proposed design of the national spare may be integrated in as many stations as possible, should the need arise.

All future stations will of course have their generator transformer compounds designed so as to reduce to a minimum the time required for erection and installation on site, and at the same time provide facilities for ease of transportation with a minimum of parts to be removed.

Figure 5.6.1D shows an arrangement utilising overhead lines or aerials for connecting the high-voltage terminals of the generator transformer to its associated switchgear.

It is admirably suitable for use when the switching station is positioned so that the anchor points for the lines can be accommodated on the power station and substation building structures.

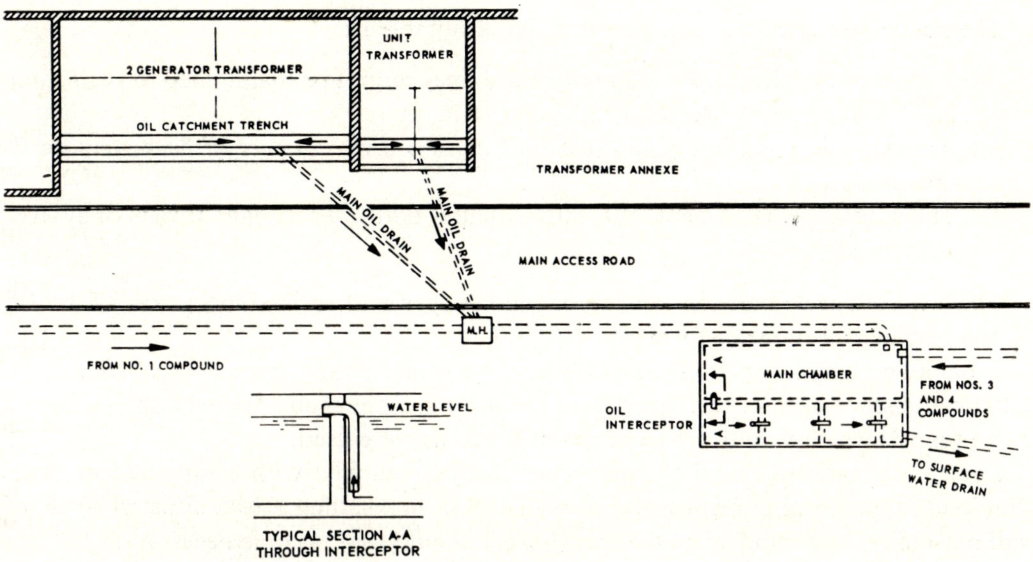

FIG. 5.6.1C. Suggested method of oil/water drainage from generator/unit transformer compounds.

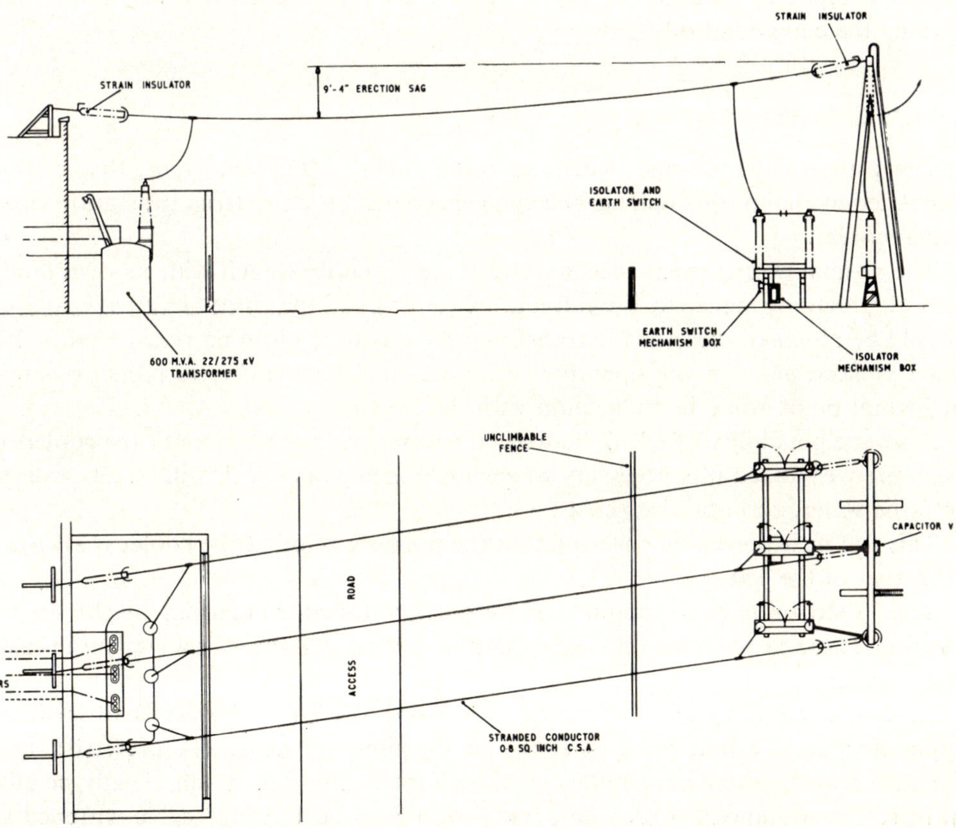

FIG. 5.6.1D. Layot MW transformer showing 275 kV aerial connections.

The above arrangement may, however, be unsuitable if:

(a) The necessary clearances of the overhead lines cannot be maintained to such equipment as the conservator and explosion vent.
(b) Atmospheric pollution is so great that there is a real danger of flashovers on the line insulators.
(c) The substation is so geographically situated that intermediate towers may have to be used.

The low-voltage connections will be of solid copper as described in Section 5.5.1 taken to the bushings of the transformer.

From an economic aspect overhead lines are cheaper than using e.h.v. cables with the latter's attendant costs of sealing ends and jointing requirements. There is also the cost of cable tunnels which would not be required if aerials were used.

For voltage adjustment, the transformer will be equipped with a tap changer of the "on-load" type, local control being provided in a marshalling kiosk, situated so that it will not impede the removal of the transformer should that become necessary.

A suitable route should be provided for the instrument and control cables from the marshalling kiosk to the control room. It is preferable that any cables leaving the transformer compound do so at about 1 ft above ground level as this will greatly assist in making the compound oil-tight.

5.6.1.1. Cooling

The generator transformer will have cooling of the "O.F.W." type, that is, the oil is circulated by pumps through heat exchangers cooled by water from the station circulating water system.

The normal arrangement incorporates three oil coolers each with its own pump, any two of which are sufficient for full-load requirements, the third acting as a spare. They should be mounted close to the transformer, so as to avoid long runs of oil piping, and must be accessible for the operation of valves and observation of pressure gauges. An important point arises in connection with the coolers.

Since the possibility of a leak between the water and the oil sides of the coolers cannot be entirely excluded it is necessary to ensure that any such leak will result in oil passing into the water, and not vice versa.

This can be achieved by ensuring that the pressure of oil at the cooler is always higher than that of the water.

This could be done by mounting the conservator at a sufficient height to give the required static head to the oil and by putting the cooling water on a syphon to reduce the water pressure.

However, the flow of water required for a 600 MVA transformer is so great that the syphonic system would prove uneconomic on pumping costs. The preferred method is the use of water/water intercoolers each with its own pump. It will usually be advisable in these circumstances to provide some form of "trace heating" cable wrapped around the body of the coolers to prevent freezing under extreme climatic conditions.

5.6.1.2. Transformer Noise Enclosures

Some power stations are situated in close proximity to residential areas where the noise emission from large transformers would be a nuisance. In these circumstances brick or concrete and/or steel enclosures would be built around the transformers to keep the noise level down to an acceptable minimum. Figure 5.6.1.2 shows a simple form of enclosure which should provide adequate sound insulation for the majority of cases and all generator transformer compounds should be arranged so that they may be erected later if required.

5.6.2. Station Transformers

For a station of 4×500 MW units, either two or four station transformers can be used (see Figs. 5.3.2.2A and B).

Since they will usually be mounted some distance away from the main station they will have "O.F.B." cooling, that is, oil pumped through separately attached radiators and cooled by means of forced air provided by fans. This system of cooling is adopted since the capital cost of providing a water supply from the power station is usually too expensive and cannot be justified economically, as in the case of the generator transformers.

On-load tap-changing equipment is provided complete with the necessary ancillary equipment housed in a marshalling kiosk. It is the usual practice for station transformers to be fitted with on-load tap-changing equipment in order that the voltage on the station busbars may be kept sensibly constant under all conditions, bearing in mind the voltage variations encountered in the substation, and the fluctuations in load on the station transformers themselves.

5.6.3. Unit Transformers

Each unit transformer for a 500 MW generator will have a rating of 30 MVA. Since the h.v. connections are connected in parallel with the generator transformer, to keep them as short as possible, the transformer should be installed as near as possible to it.

As previously mentioned in Section 5.5.2, the high-voltage connections will be in bare copper, but on the low-voltage side the transformer will be provided with integral cable boxes for terminating the associated 11 kV cables. A suitable route must therefore be provided into the turbine house for these cables as well as the associated multi-core cables.

In this instance the transformer will probably have the "ON" type of cooling, that is, natural oil circulation through integral radiators so that no external oil or water connections are required.

An "off-load" tap changing mechanism is fitted which makes a marshalling kiosk unnecessary.

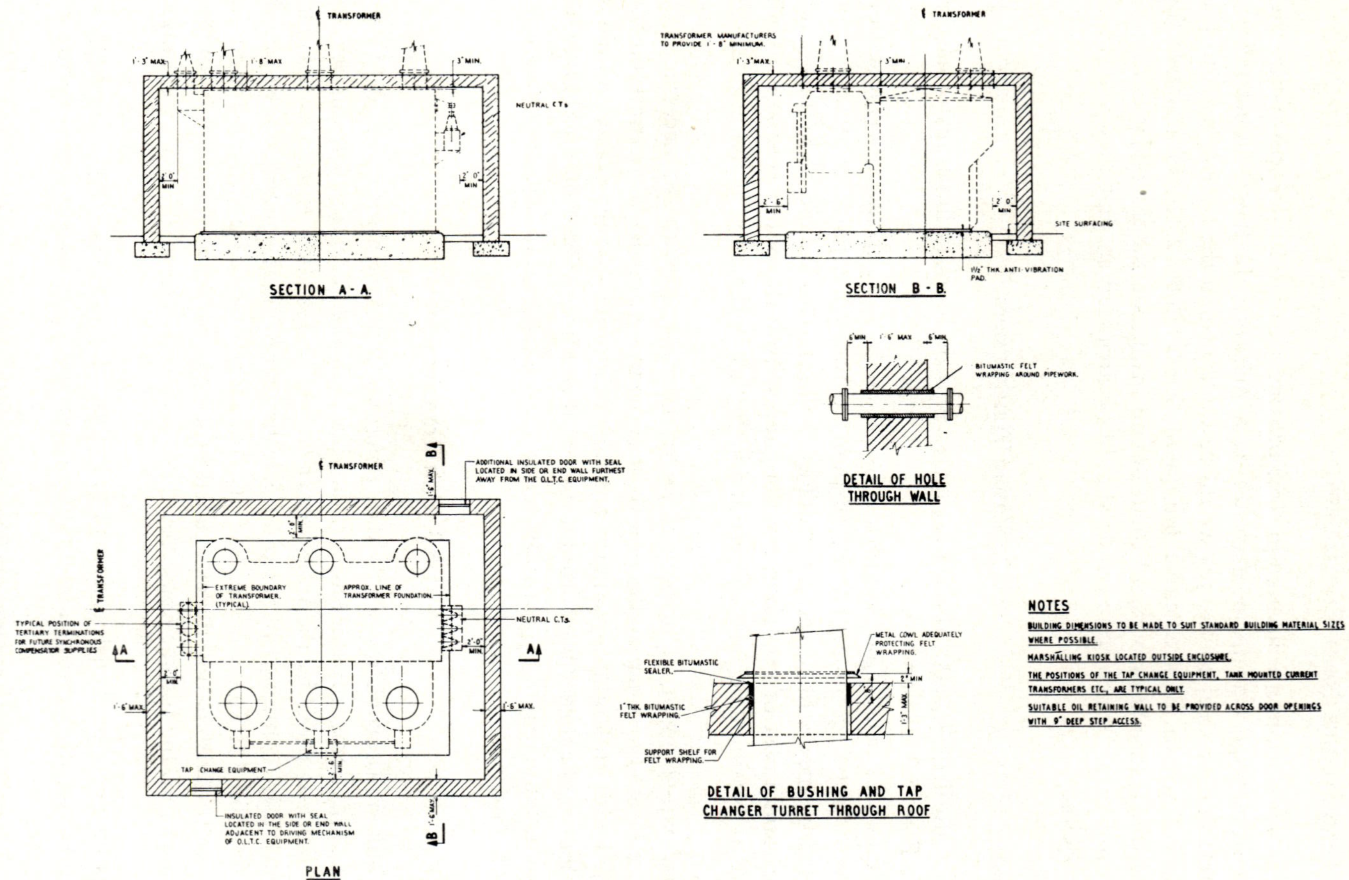

FIG. 5.6.1.2. Typical 275 kV transformer noise enclosure.

5.6.4. Auxiliary Transformers

Three ranges of auxiliary transformers are in general use, namely:

(a) 11/3·3 kV, with ratings up to 20 MVA.
(b) 3·3/415 kV, having ratings varying from 2500 kVA to 150 kVA.
(c) 415/240 V or 415/110 V, having ratings up to 50 kVA. These transformers in the latter case may either be three or single phase on the l.v. side.

Figure 5.6.4C shows the vector relationship of auxiliary transformers employed on a typical power station.

In the case of (a) these transformers are invariably mounted outdoors and will be of the "ON" type (see Fig. 5.6.4B).

Regarding (b) and (c) these transformers will either be of the "ON" cooled or the air-cooled (class "C") type (see Fig. 5.6.4A).

In the case of the larger rating transformers the 415 V cable connections will be quite heavy, i.e. a 2500 kVA transformer having a secondary current of 3500 A involves $4 \times 1{\cdot}0$ in^2 cables per phase if these are laid in air. From an economic aspect, therefore, the transformer should be sited as near as possible to its associated l.v. switchboard. There are a number of alternatives available to meet this situation:

(i) The use of a class "C" dry-type transformer in which case no special fire precautions are necessary as there is no oil present. This transformer can with advantage be built into the switchgear, thereby abandoning the use of low-voltage cables in favour of small lengths of copper busbars from the transformer terminals to the switchgear.

(ii) Substituting for oil as an insulating medium a non-flammable organic liquid such as "pyrochlor" in a conventional transformer. In this case again no special fire precautions are needed but the foundations of the transformer have to be more robust as the weight per kVA is much greater. The transformer could be positioned in an indoor situation provided adequate arrangements are made for draining away any leakage that might occur.

(iii) The employment of a conventional "ON" type transformer preferably positioned outdoors, but to diminish the route length of the l.v. cables it may be necessary to provide an indoor situation, in which case the fire risk must be met by enclosing the transformer in a substantial cell.

Such cells need to have open or grilled fronts in order to allow for the necessary air circulation and must embody arrangements for containing (or safely draining away) any oil which might be released as a result of a transformer failure (see Section 5.6.1).

It is obvious that the ultimate choice of the above arrangements is governed largely by economic considerations.

With the increasing use of small transformers built into switchboards such as (c), it cannot be over-emphasised that these should be of the "dry" type.

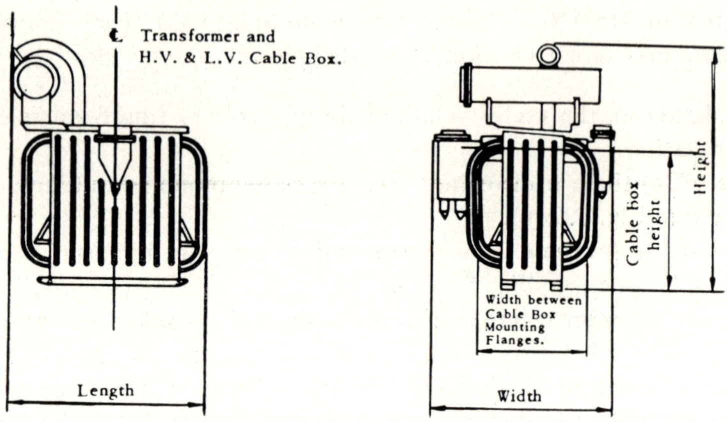

All dimensions in inches								
Code No.	Capacity kVA	Envelope dimensions			Gross weight (including oil) lb	Approx. oil quantity gal	H.V. & L.V. Cable box mountings	
		Length	Width	Height			Width between flanges	Height
35/22/010	500	75	56	102	6,200	120	27	56
35/22/011	750	82	65	110	8,200	180	29	65
35/22/012	1000	90	67	116	10,000	220	31	68
35/22/013	1500	100	88	130	14,000	300	48	78
35/22/014	2000	105	90	142	19,000	400	56	85
35/22/015	2500	114	100	150	23,000	480	60	90

FIG. 5.6.4A. Auxiliary transformers "ON".

11,000/3,450 Volts. 2.5 to 15 MVA

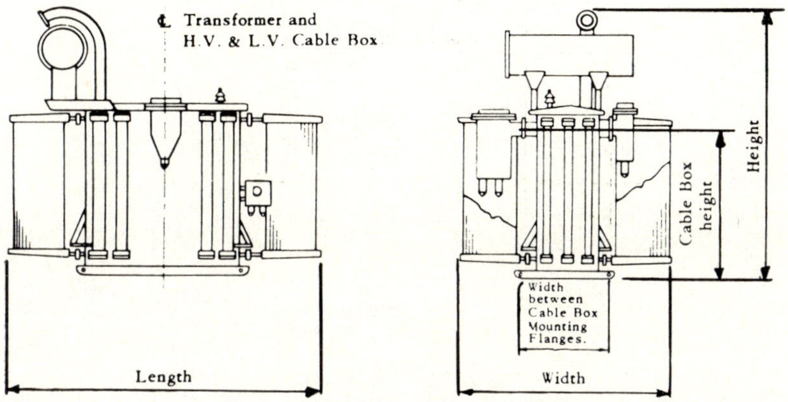

All dimensions in inches								
Code No.	Capacity MVA	Envelope dimensions			Gross weight (including oil) lb	Approx. oil quantity gal	H.V. & L.V. Cable box mountings	
		Length	Width	Height			Width between flanges	Height
35/22/515	2.5	110	82	146	20,000	450	56	86
35/22/518	5.0	135	126	161	34,000	750	56	103
35/22/519	7.5	175	142	173	45,000	1,100	56	109
35/22/520	10.0	205	150	189	58,000	1,300	60	113
35/22/521	12.5	213	154	195	69,000	1,600	60	114
35/22/522	15.0	238	162	200	84,000	2,000	64	118

FIG. 5.6.4B. Auxiliary transformers "ON".

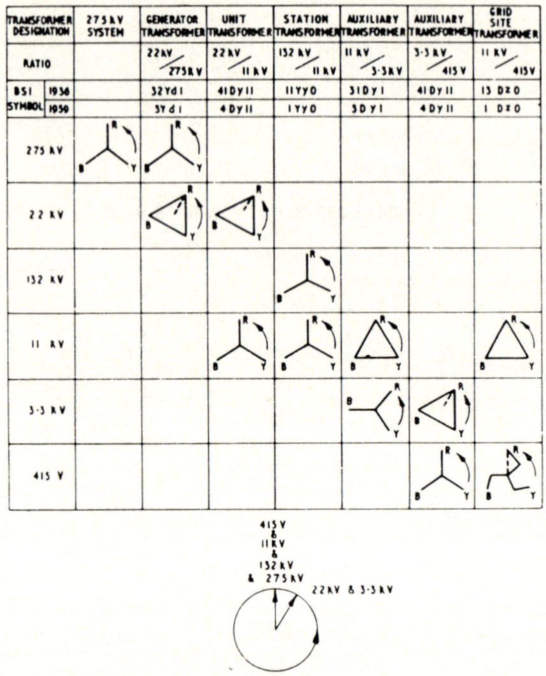

FIG. 5.6.4C. Vector relationship of typical power station auxiliary transformers.

5.6.5. Oil-handling Plant

5.6.5.1. Fixed Installation

It may be valuable, especially in the case of the large transformers, to provide a fixed oil filtration plant. This comprises the filtration unit itself, clean and dirty oil storage tanks, and ancillary pipework with the necessary standpipe connections in each transformer cell. For maintenance of the tap change equipment or examination of the top yoke of the transformer it is necessary to drain away and filter large quantities of oil which can asily be acco mplished with a plant of this description.

5.6.5.2. Portable Filtration Unit

Where transformers are positioned away from the precincts of the power station, it is far too expensive to run costly pipework from the filtration plant. In this case, it is prudent to employ a portable unit which can easily be manoeuvred to accomplish this task.

5.7. CABLES

5.7.1. Cables (Layout)

The general scheme of cable layout has been dealt with in Chapter 4, but it is worth emphasising that the following parameters should be closely followed.

(a) Cables should be segregated and run on a unit principle so that a fire or explosion in a cable tunnel or chamber should not affect the running of other units.

(b) Main and standby cables associated with a particular unit should be run over separate routes.

(c) Power cables should be run separate from control and instrumentation.

(d) If possible, controls should be separated from instrumentation but can follow power cable routes.

(e) Instrument cables liable to induced voltage troubles should have large clearance from all other cables and in particular current generating devices, e.g. 275 kV cables, high current a.c. and d.c. circuits, earth straps, etc.

(f) Special precautions in the form of duplicate supplies run over different routes should be adopted for cables associated with such common priority services as fire alarms, external G.P.O. communications, etc.

5.7.1.1. Cable Tunnels and Sub-basements

The importance of providing adequate accommodation for cables cannot be overemphasised. In many power stations inadequate facilities have, in the past, been provided with subsequent overloading of cable traywork, and segregation of cables is now an impossibility.

If cable tunnels are used, advantage can be taken during their construction, of building in the "inserts" for supporting the racking which will at a later period be erected. Adequate access facilities will be provided at certain locations along the route to enable the cables to be drawn in, also to serve as suitable points for "fire fighting" purposes.

The use of sub-basements is also an alternative to cable tunnels, segregation of unit cables being accomplished by installing fire-proof barriers of asbestos after the cables are installed.

It is necessary, however, that suitable provision is made for ventilation and drainage requirements in both sub-basements and cable tunnels.

5.7.2. **11 kV Power Cables**

The 11 kV cables can either be of the following types:

(a) Butyl rubber-insulated, polychloroprene-sheathed or polythene-insulated PVC-sheathed and in the case of three phase cables steel wire or tape armoured. Butyl rubber is preferred to PVC as it is a better insulant which results in a smaller cable, and also has a higher temperature withstand.

(b) Paper-insulated lead-sheathed, non-draining type steel tape or wire armouring can again be used in the case of three-phase cables. Although not essential in some applications, a serving over the armour is very desirable to protect it from corrosive agents. A serving of PVC can be used which is relatively cheap, does not support combustion, is impervious to most agencies, and has a smooth finish which facilitates handling and drawing into ducts. This is generally specified when cables are buried in the ground.

5.7.3. 3·3 kV Power Cables

The 3·3 kV power cables can either be as type (b) above, or of the PVC-insulated PVC-sheathed type.

These PVC cables do not require sealing against moisture as is the case if type (b) cables are used, but are terminated by using compression type glands. PILC cables, however, require the use of compound-filled sealing boxes and plumbed glands, the jointing of which requires a high degree of skill.

5.7.4. 415 V Power Cables

For this application these cables can either be of the PILC or the PVC type and with the further choice in the case of small power circuits of mineral insulated copper-sheathed (MICS) cable. This MICS cable has many advantages, it is small, resists high temperature and will withstand a great deal of deformation without breakdown. They are admirably suitable for supplies to sootblowers, oil lighting-up equipment and other auxiliaries around the boiler and turbine where ambient temperature is high and the risk of mechanical damage great. For use in corrosive conditions, the copper sheath may be served with PVC at the expense of some reduction in the maximum allowable temperature. Having a low-impulse strength, however, they are not suitable for use on inductive d.c. circuits or where surge voltages may appear.

It is sometimes necessary to provide electrical trace heating on pipework either to avoid freezing of the liquid, or to keep it at a certain temperature as on oil fuel lines to maintain the correct viscosity.

The installation on the pipework usually comprises MICS cable wrapped in spiral formation and afterwards covered with lagging.

5.7.5. Sizes of 11 kV, 3·3 kV and 415 V Cables

In determining the cross-sectional area of cables to be used, due regard must be paid to the following:

(i) Full load current of the circuit.
(ii) Voltage drop.
(iii) Magnitude and duration of the short-circuit.

The cross-section of the conductor must therefore be such that the copper temperature during short-circuit conditions does not exceed the value which will cause deterioration of the insulation. The evaluation of this temperature is quite involved, and depends on a number of assumptions. The essential data, however, are published in a convenient form by the cable manufacturers. PVC cables have a short-circuit rating somewhat lower than paper cables owing to the lower permissible temperature of the insulant; this means that a larger cable would be required than if paper insulation were used and a cost comparison of the installation using both types must be made before a decision is made. Nevertheless, PVC cables are now being used extensively at 3·3 kV.

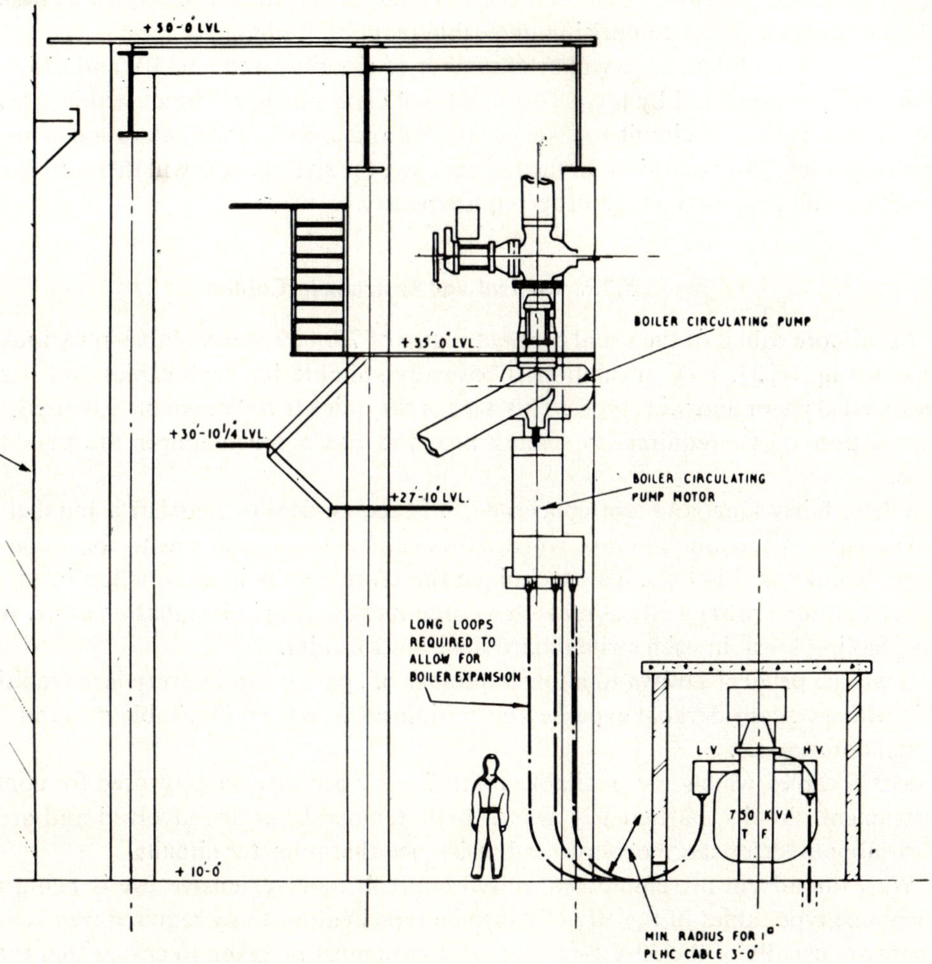

FIG. 5.7.5. Cabling to boiler circulating pumps.

Three-phase cables will be generally used in sizes up to 0·5 in^2, but at this size the minimum bending radius of about 6 ft must be borne in mind, particularly in laying out ducts.

For example, take the case of a 30 MVA unit transformer, with a full load current of 1570 A at 11 kV. If the cables are run on racks in free air 2–0·75 in^2 or 3–0·5 in^2 single-core cables per phase would be used and these would be run in trefoil groups as described in Chapter 4, each group comprising one cable from each phase.

It will be found that there will be a number of circuits on the 3·3 kV and 415 V systems which will be protected by h.v.c. fuses and will clear on short-circuit within a few cycles. The remainder will be circuit-breaker controlled and therefore will take a much longer time to clear under fault condition. In the former case, the size of cable will therefore be dictated by either full load or voltage drop requirements.

5.7.6. **Control and Instrument Cables**

Multi-core cables of the usual conductor size of 7/0·029 are available in various number of cores up to 37. PVC insulation is basically suitable for such cables and has rapidly superseded paper and v.i.r. types. This type is far quicker to "make off", thereby reducing installation costs, requires no sealing box and has a much longer life than the v.i.r. type.

Where fairly long runs are concerned, it is economical to marshal a number of individual circuits at a junction box, whence they can be so connected as to form one or more large "trunk" multi-cores. This technique therefore results in a reduction in the number of runs of numerous small cables with a consequent saving in installation costs. A built-in marshalling kiosk in each switchboard achieves this aim.

It will be prudent always to allow a number of spare cores in every large multi-core as this will save considerable expense and untidiness when the inevitable demand for additional cores arises.

MICS cables which are available up to 7 cores are also largely used for control and instrument circuits, particularly where high temperatures are involved and are almost invariably used for thermocouple and resistance thermometer circuits.

With the advent of remote centralised control, more extensive use is being made of telephone type cables of the 10 or 20 lb/mile type in almost any required number of pairs. These are usually of the 50 V type and great care must be taken to ensure that their use is restricted to this voltage.

5.7.7. **Cleating and Supports**

Large power cables will be supported on galvanised steel racks at intervals of 2 ft 6 in to 3 ft and securely cleated. Where extensive runs of cable are to be installed the cables are laid on ladder racking, which easily allows the changes in the vertical runs of cable without resorting to the cutting of trays if these were used.

Smaller power cables and control and instrument cables will usually be run on perforated steel trays of suitable widths. Provided they are adequately "cleated" a neat run can be made of a number of small cables on the trays.

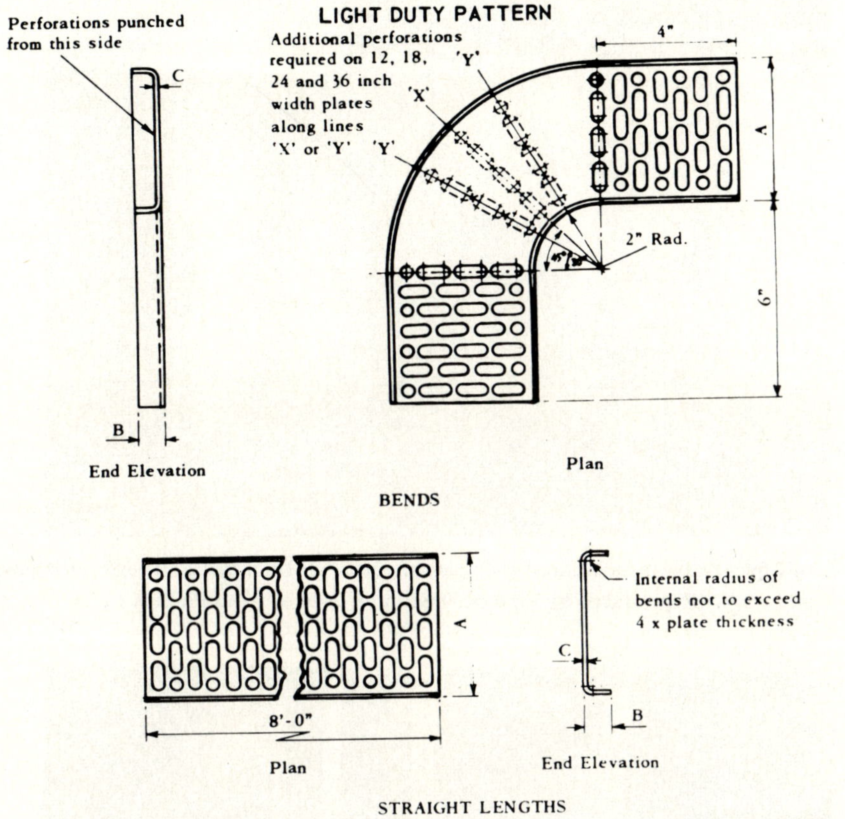

C.E.G.B. Code No.		Dimensions - Inches		
Bends	Straight Lengths	Width A	Depth B	(S.W.G.) C
12/17/252	12/17/262	4	0.5	20
12/17/253	12/17/263	6		
12/17/255	12/17/265	12	0.75	16
12/17/257	12/17/267	18		
12/17/259	12/17/269	24		14
12/17/261	12/17/271	36		

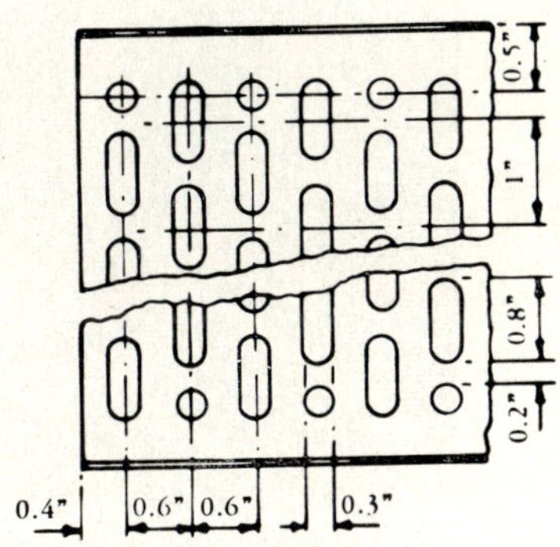

DETAILS OF PERFORATION

FIG. 5.7.7A. Cable trays.

FIG. 5.7.7B. Typical cable installation showing multi-core and MICS cables on perforated galvanised trays. (Reproduced by kind permission of Johnson and Phillips Ltd.)

FIG. 5.7.7C. Typical power cabling installation showing cleating arrangement. (Reproduced by kind permission of Johnson and Phillips Ltd.)

The principal runs of trays and racking needs very careful planning, otherwise the installation becomes very untidy especially in areas underneath the control room where severe congestion is likely to occur (see Fig. 5.7.7A for standard cable tray and Figs. 5.7.7B, C, and D for typical cable installations).

FIG. 5.7.7D. Typical installation of PVC cabling showing method of support. (Reproduced by kind permission of Johnson and Phillips Ltd.)

5.7.8. **Future Trends**

When suitable 11 kV fuses are available, it may be possible to fit these to motor control circuits in the same way as 3·3 kV and 415 V fuses have been applied at these voltages with a subsequent saving in the size of cable used.

With the increasingly large number of cable ends associated with both power and control cables to be terminated, involving the employment of a large number of skilled personnel, crimped terminations have obvious advantages. These are now available in numerous ranges and are suitable for use with 1·0 in^2 cables. The crimped termination has the recognised advantage of withstanding short-circuit conditions much better than a brazed or soldered connection, and the possibility of corrosion is greatly reduced.

From price considerations alone, it may be that in the future the aluminium conductor may eventually supersede the copper conductor, much, of course, depending on the supply and comparative costs of these two.

5.8. STATION EARTHING

5.8.1. General Considerations

In order to ensure personal safety, and to provide a predetermined path for earth-fault currents, all metal parts of electrical apparatus, other than the conductors, are bonded together electrically and connected to a low-resistance earth electrode.

The current paths must have a current-carrying capacity sufficient to deal with the maximum fault currents, and a resistance low enough to prevent a dangerous voltage appearing between any points which a man could reach simultaneously.

TABLE 1. TYPICAL SITE SURVEY CARRIED OUT AT EGGBOROUGH POWER STATION FOR EARTHING

Soil The tests were carried out in wet, red sand which covers the site to a depth of 30 ft at least, except at position 14 where the soil was a mixture of fine brown clay and red sand.

Depth	4 ft	8 ft	12 ft	16 ft	20 ft	24 ft	Site Posn. No.		
Soil resistivity (ohm/cm)	2460	2150	2305	3990	5750	7370	1	West side	400 kV switch house
	6912	5535	6450	6450	6525	6912	2		
	6912	5535	4610	4300	5000	5530	3		
	23,000	18,430	16,130	15,370	13,450	11,050	4	East side	
	4610	5840	6910	5530	8450	6450	5		
	9980	13,830	16,130	14,760	16,900	16,130	6		
	13,450	18,430	20,750	18,430	18,430	15,650	7	Turbine house	
	7860	9220	9220	9830	9610	9680	8		
	6140	9980	11,040	13,200	13,450	14,300	9		
	11,520	16,900	20,750	18,430	21,100	23,050	10	Boiler house	
	11,520	11,520	11,070	9530	11,140	9670	11		
	6140	9980	10,600	13,200	13,050	13,800	12	Cooling towers and chimney area	
	8450	8450	9680	10,140	8830	10,150	13		
	1075	1537	1842	2150	1920	2300	14		

The earthing arrangements should also be such as to ensure that, under fault conditions, the lowest practicable voltage appears between earthed points of the equipment and the main body of the earth, so that insulation breakdown or burning does not occur on equipment which is earthed outside the station, e.g. apparatus connected to pilot or telephone cables, cable sheaths, etc.

It is important for a site survey to be carried out to ascertain the value of soil resistivity and the depths at which this can be used to maximum advantage. The main station power earthing system should be entirely separate from the lightning and static protection earthing system, and is usually installed by the main cable contractor.

A test rod driven to 4 ft, 8 ft and 12 ft returned resistance values of 120 ohms, 60 ohms, and 44 ohms respectively.

With a fault level of 35,000 MVA at 400 kV it is estimated that the current will be 50,000 A and the short circuit period be taken as 5 c/s (i.e. 0·10 sec) at least.

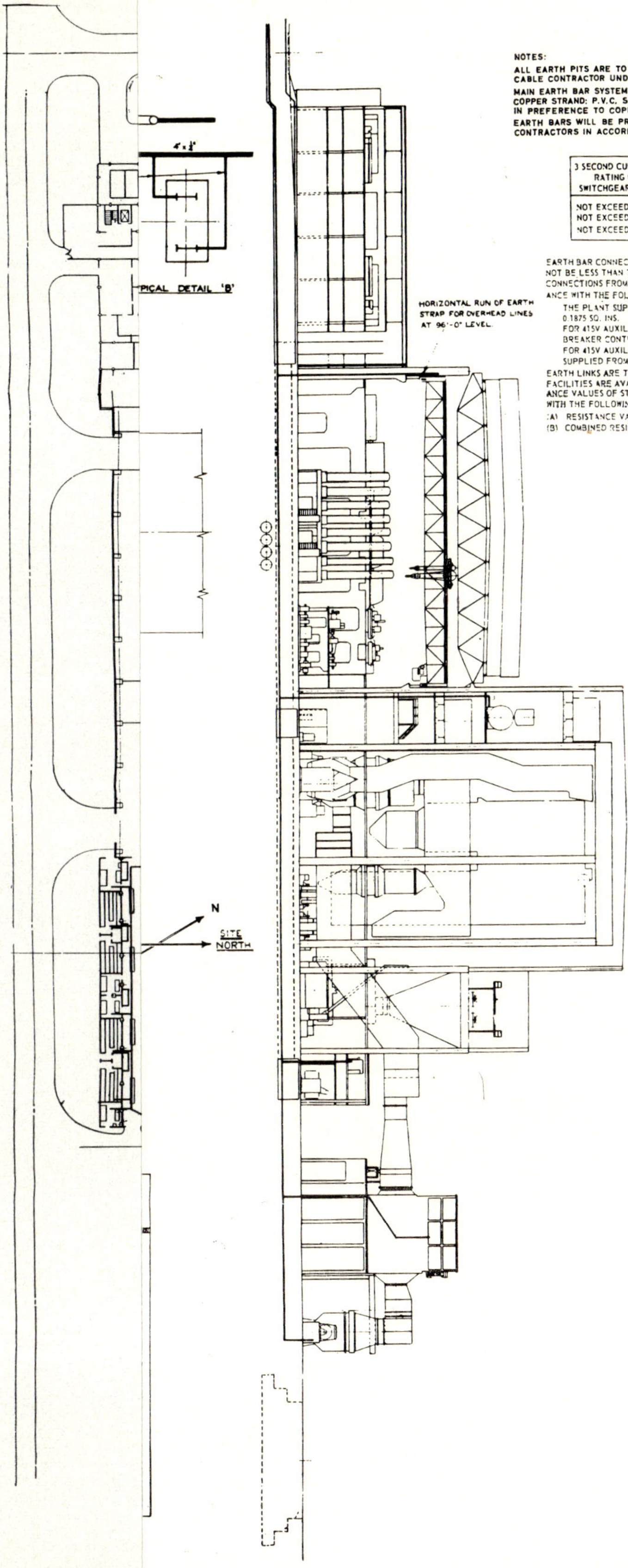

NOTES:

ALL EARTH PITS ARE TO BE ESTABLISHED BY MAIN AUXILIARY AND MULTICORE CABLE CONTRACTOR UNDER CONTRACT E.G. 66 67 68 E 8A

MAIN EARTH BAR SYSTEM TO COMPRISE 0.5 SQ. IN. COPPER STRIP S.P.V.C. SHEATHED COPPER STRAND: P.V.C. SHEATHED COPPER STRAND TO BE USED WHERE PRACTICABLE IN PREFERENCE TO COPPER STRIP

EARTH BARS WILL BE PROVIDED ON THE SWITCH BOARDS BY THE SWITCHGEAR CONTRACTORS IN ACCORDANCE WITH THE FOLLOWING:-

3 SECOND CURRENT RATING OF SWITCHGEAR (KA)	SIZE OF COPPER EARTH CONDUCTOR	
	CROSS SECTION IN SQ. INS.	DIMENSIONS INS
NOT EXCEEDING 22	0.1875	1" x 3/16"
NOT EXCEEDING 30	0.281	1½" x 3/16
NOT EXCEEDING 40	0.500	2" x 1/4

EARTH BAR CONNECTIONS FROM SWITCHGEAR EARTH BARS TO MAIN EARTH BAR SHALL NOT BE LESS THAN THE SIZE AS STATED IN ABOVE TABLE BRANCH EARTHING COPPER CONNECTIONS FROM THE MAIN EARTH BAR SHALL BE OF SECTIONAL AREA IN ACCORDANCE WITH THE FOLLOWING:-

THE PLANT SUPPLIED FROM THE 11 kV AND 33 kV SYSTEMS – NOT LESS THAN 0.1875 SQ. INS.

FOR 415V AUXILIARY SWITCHBOARD AND PLANT SUPPLIED FROM CIRCUIT BREAKER CONTROLLED CIRCUITS – NOT LESS THAN 0.281 SQ. INS.

FOR 415V AUXILIARY SWITCHBOARDS, DISTRIBUTORS BOARDS AND PLANT SUPPLIED FROM THE FREE CONTROLLED CIRCUITS – NOT LESS THAN 0.125 SQ. INS.

EARTH LINKS ARE TO BE LOCATED ON SITE SUCH THAT REASONABLE GOOD ACCESS - FACILITIES ARE AVAILABLE TO THE LINKS FOR TESTING AND RECORDING OF RESISTANCE VALUES OF STATION EARTHING SYSTEM TO BE CARRIED OUT IN ACCORDANCE WITH THE FOLLOWING:-

(A) RESISTANCE VALUES OF EACH INDIVIDUAL EARTH PIT (TO)

(B) COMBINED RESISTANCE VALUES OF PAIRS OF EARTH PITS

Fig. 3.8.14. Earthing layout for typical 4 x 500 MW unit station.

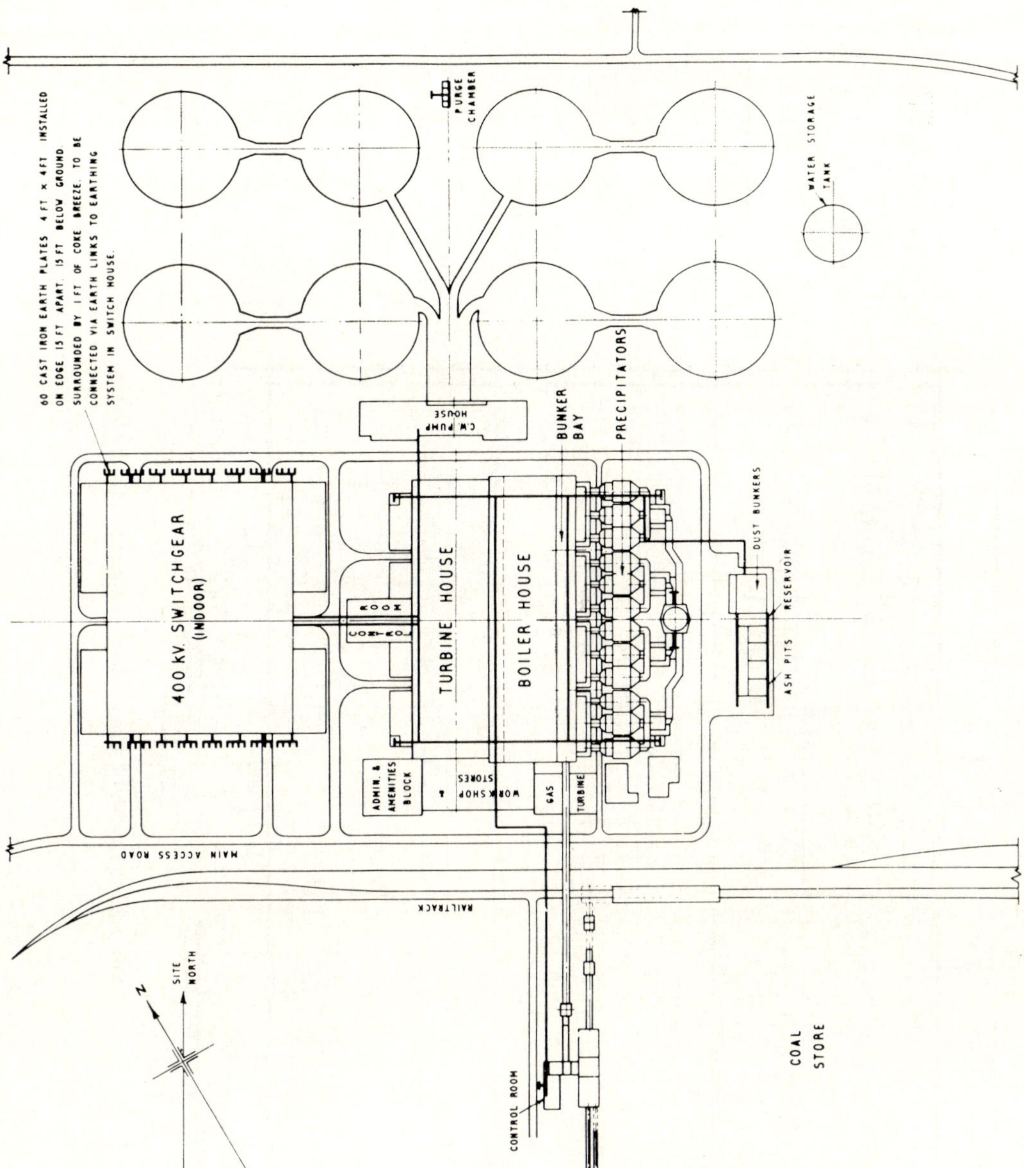

FIG. 5.8.1B. General **site plan** showing earthing points for a typical 4×500 MW unit station.

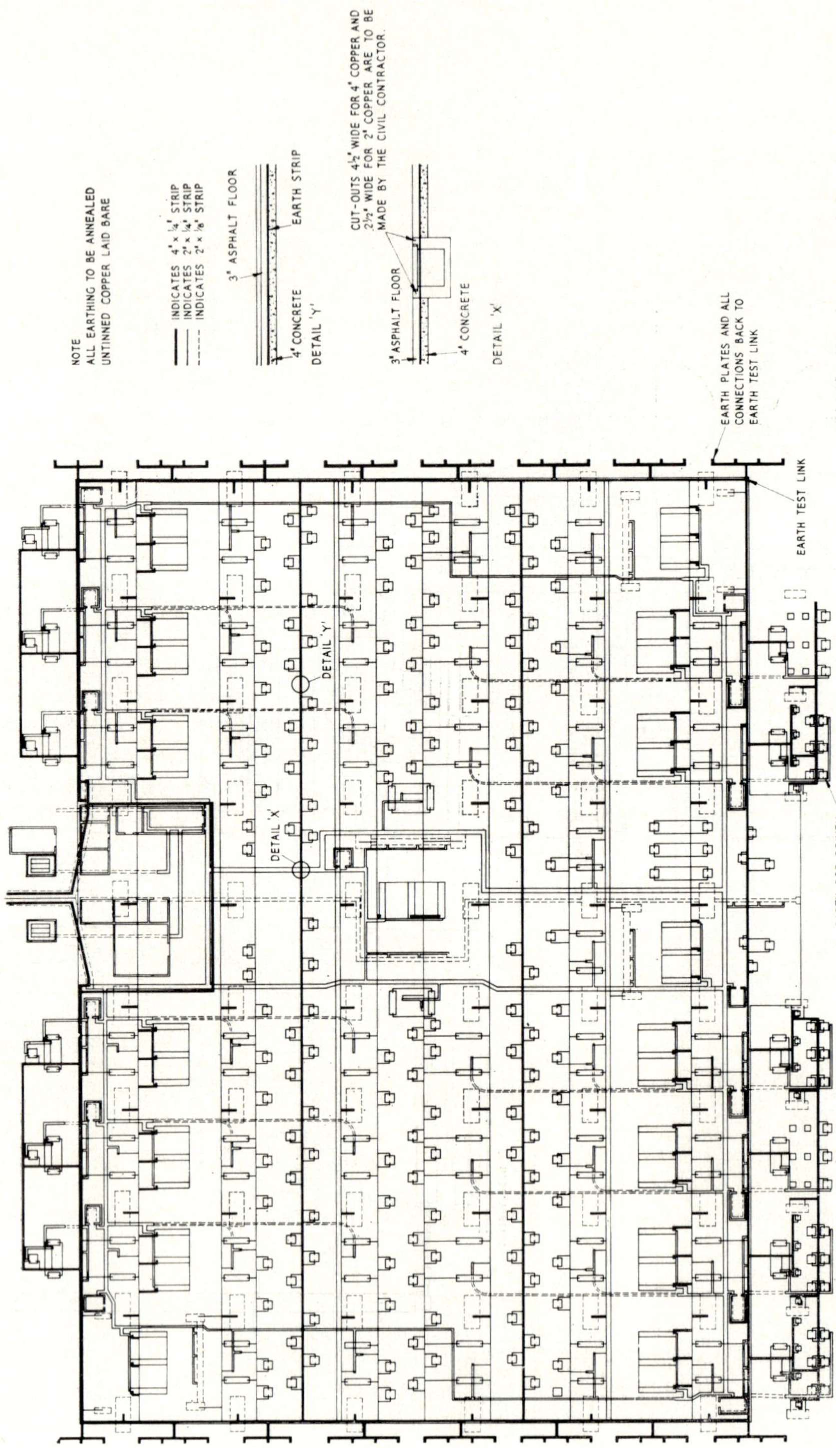

FIG. 5.8.1C. Layout of ground floor earthing for 400 kV switch house.

Taking the soil resistivity readings at all depths an average of 10,000 ohms/cm is obtained, and using sites 1, 2 and 3 as being the nearest to the selected earth park, the average figure to a depth of 12 ft is 5000 ohms/cm.

From these results either of the following schemes can be employed:

(a) 60—4 ft×4 ft cast iron earth plates, installed on edge, 5 ft below ground level, spaced 15 ft apart and surrounded by 1 ft of coke breeze. This scheme should give an estimated earth resistance of 0·50 ohm.

(b) 30—4 ft×4 ft cast iron earth plates, installed on edge, 5 ft below ground level, spaced 15 ft apart and surrounded by 1 ft of coke breeze. This scheme should give an estimated earth resistance of 1·00 ohm.

The bonding for the above schemes to be with 0·5 in^2 stranded copper earth cable with a PVC serving throughout for both interconnecting the plates and connections to the earth links on the power station site.

As a possible alternative to the above schemes the following suggestion of an "earth mat" could be employed:

(c) 650 yd of 2 in×$\frac{1}{4}$ in copper strip to carry a short-circuit current of 60,000 A for 0·10 sec.

(d) 600 yd of 2 in×$\frac{1}{4}$ in copper strip to carry a short-circuit current of 50,000 A for 0·10 sec.

The above strip should give an earth resistance estimated at 0·50 ohm.

Table 2 covers the quantities of 4 ft × 4 ft plates and 2 in×$\frac{1}{4}$ in strip required for various conditions.

Due to the effect of sheath corrosion, cables not suitably protected either with QR 2QH2 or PVC serving should be installed away from coke assisted electrodes.

If two or more stations are adjacent on what may be considered to be one site, the earthing systems and the stations are interconnected to form a single earthing system. Where the stations actually adjoin, the extremities of their earthing system are connected together so that the whole area is enclosed by the earthing systems. Where the separation is too large to treat as adjoining stations, an interconnecting cable (s) of substantial cross-section is run to ensure that, as far as practicable, fault currents are diverted from cable sheaths and armour. This is of particular importance where fault current flowing in one station is provided from the adjoining station, e.g. where a switching station adjoins a power station site so that an earth fault in the switchgear causes current flow between the two sites in order to reach the system neutral at the generators or transformers, such interconnection between sites should include links suitably disposed to assist in testing. Figures 5.8.1A and B show the earthing arrangements adopted for a 4×500 MW station, and Fig. 5.8.1C shows the earthing layout for the adjoining 400 kV indoor substation.

5.8.2. **Earth Electrodes**

There are many forms in which electrodes can be provided:

(a) Rod-type, consisting of copper rods as shown in Fig. 5.8.2A which can be driven into the ground to varying depths.

TABLE 2. EGGBOROUGH POWER STATION—EARTHING ANALYSIS OF PROPOSED REQUIREMENTS FOR VARIOUS CONDITIONS

Soil resistivity (ohms/cm)	10,000	10,000	10,000	10,000	5000	5000	5000	5000
Maximum S.C. period (sec)	0·20	0·10	0·20	0·10	0·20	0·10	0·20	0·10
Estimated E.R. (ohms)	1·00	1·00	0·50	0·50	1·00	1·00	0·50	0·50
	Short-circuit 60,000A							
(a) No. of 4 ft×4 ft plates required, S.C.	15	10	15	10	10	8	10	8
Ditto, E.R.	30	30	60	60	15	15	30	30
(b) Length of 2 in×$\frac{1}{4}$ in strip required, S.C. (yd)	1000	650	1000	650	650	500	650	500
Ditto, E.R. (yd)	300	300	600	600	150	150	250	250
	Short-circuit 50,500A							
(a) No. of 4 ft×4 ft plates required, S.C.	12	8	12	8	8	6	8	6
Ditto, E.R.	30	30	60	60	15	15	30	30
(b) Length of 2 in×$\frac{1}{4}$ in strip required, S.C. (yd)	800	600	800	600	600	450	600	450
Ditto, E.R. (yd)	300	300	600	600	150	150	250	250

S.C., short-circuit.
E.R., earth resistance.

(a) 4 ft×4 ft cast iron earth plates.
(b) 2 in×$\frac{1}{4}$ in copper earth strip.

(b) Abandoned sheet steel piling as shown on Fig. 5.8.2B, also reinforcing bars in concrete foundation piles suitably welded together.

(c) Plate-type, comprising a galvanised plate buried into the ground and surrounded by coke which tends to keep the area permanently moist.

Using a combination of the above methods, a low resistance to earth is invariably obtained with ample current carry capacity. Figure 5.8.2csh ows a typical earthing arrange

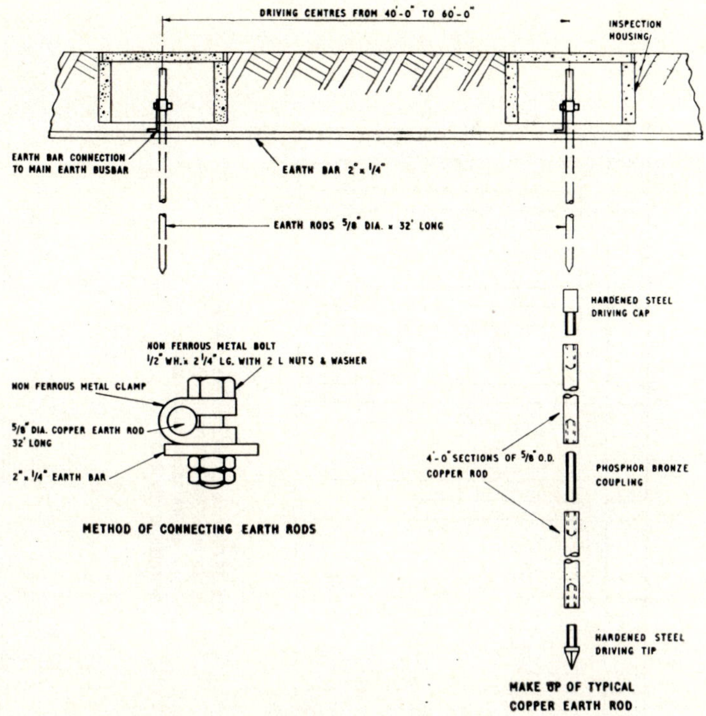

FIG. 5.8.2A. Method of earthing utilising copper earth rods.

ment for a five-unit coal-fired station. Removable links should be provided at the connection to the electrodes whatever form they may take so that individual periodic measurements of the resistance can be made.

5.8.3. Earth Bars

Earth bars are of copper and range in size from $4 \times \frac{1}{2}$ in for the main bars and risers down to 1 in $\times \frac{1}{8}$ in tape for the earthing of small distribution boards and the like.

The main elements of the system are interconnected to form a mesh so that accidental disconnection of one bar will not interrupt the path to earth. Major plant and the earth bars of switchboards should be connected to the station system at not fewer than two points for the same reason.

The cross-sectional area of the earth bar is proportioned to the fault current it may have to carry, so that the maximum allowable copper temperature of 300°C will not be exceeded with a fault duration of 3 sec.

Under normal conditions a single phase to earth fault is limited in the case of the generator, 11 kV and 3·3 kV systems by the neutral earthing resistor. For example, assuming two phases are earthed on the 11 kV 750 MVA switchboard, then the earth connections would have to carry 30,000 A and a total copper section of 0·45 in^2 would be required.

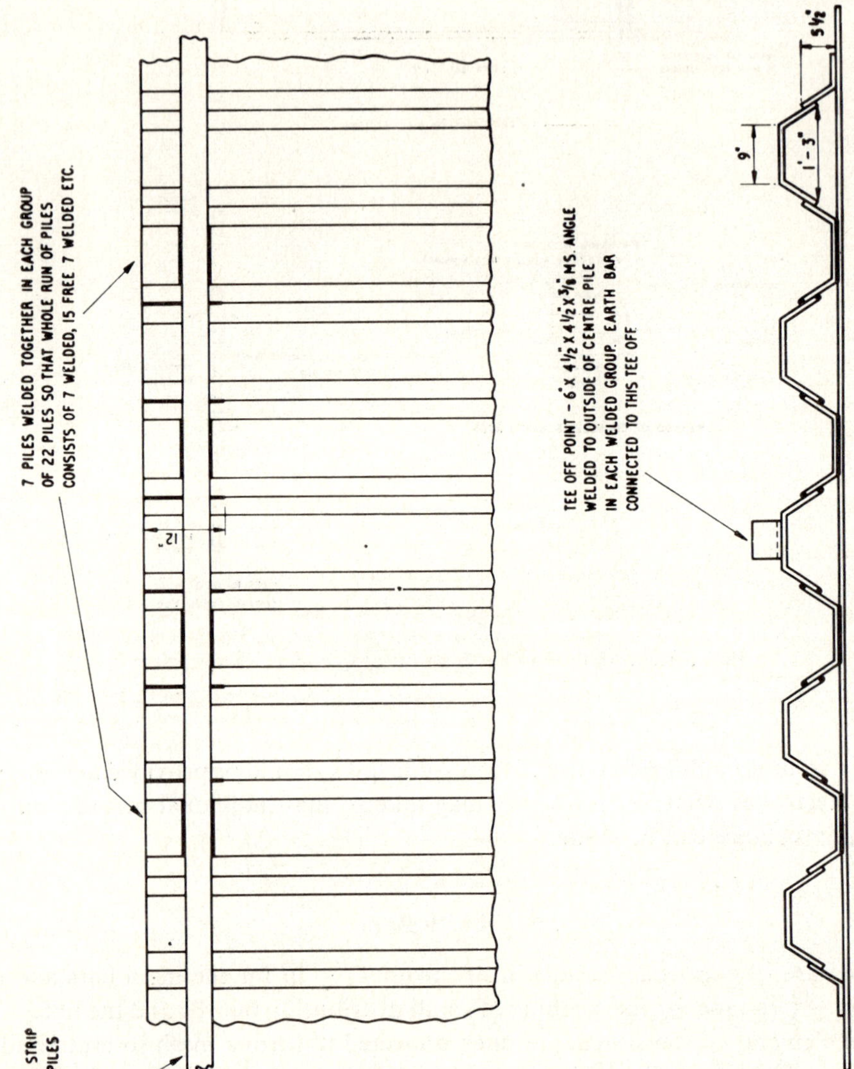

FIG. 5.8.2B. Method of earthing by pile bonding.

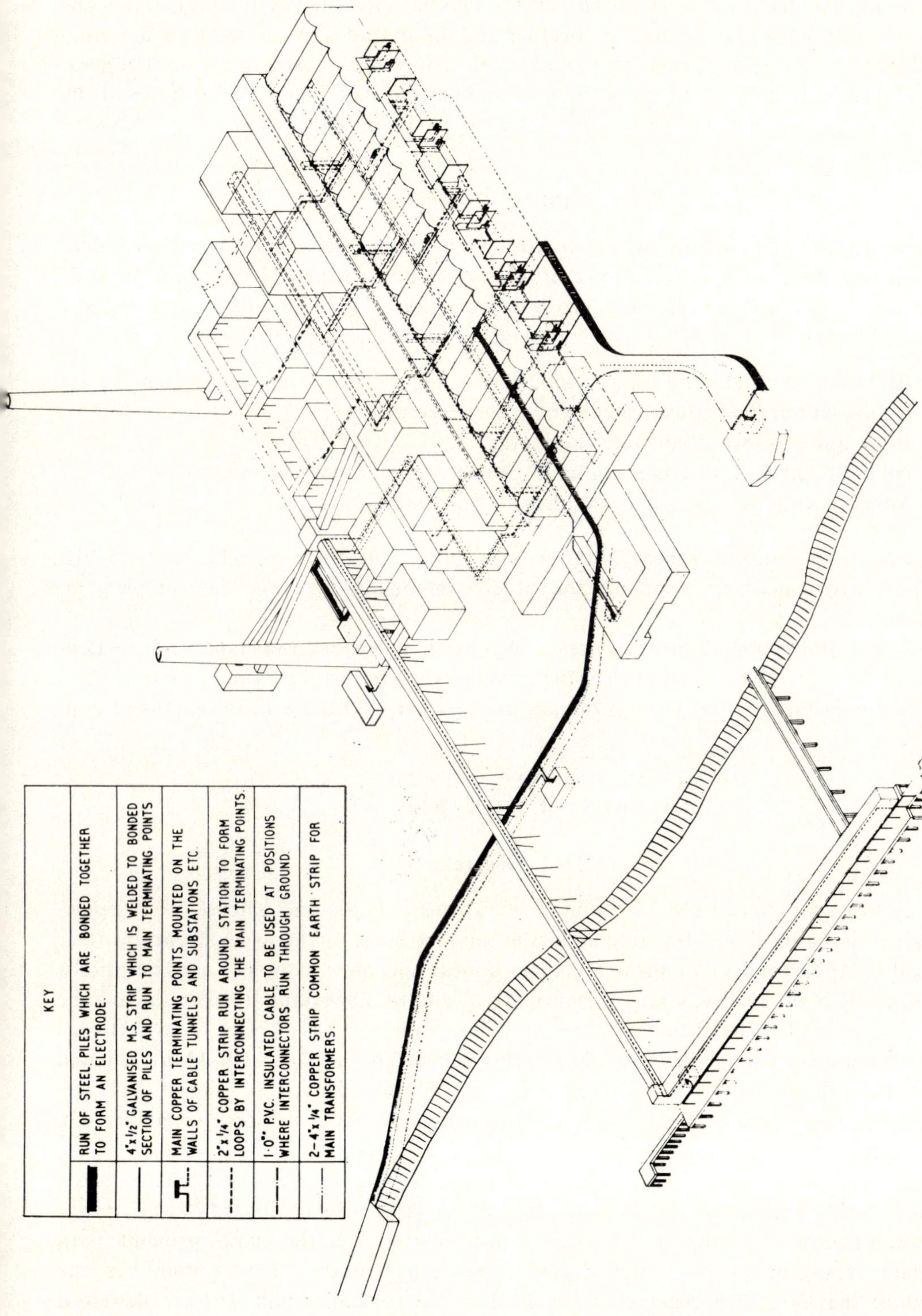

FIG. 5.8.2C. Typical arrangement of main earthing scheme.

There are two methods in common use for making joints between conductors. The first is by riveting the two conductors together and the second is by low temperature brazing. Solder cannot be used since this would melt before the maximum copper temperature is reached. It cannot, therefore, be over-emphasised that joints should be carefully made to minimise resistance and prevent failure at the joints when heavy current flows.

5.8.4. Lightning Conductors

The protection of structures against lightning and static discharge is covered by a British Standard code of practice. However, this does not adequately cover the Board's requirements, and a C.E.G.B. Standard interprets the basic requirements and application as follows:

(a) All buildings (including transmission) exceeding 75 ft in height and which are not wholly within the protective zone of other structures.
(b) Brick and reinforced concrete chimneys.
(c) All buildings and structures containing explosives or flammable material.
(d) Metal fencing for site boundaries, transformer compounds, etc.

Figure 5.8.4A shows a typical lighting and static protection installation, and Fig. 5.8.4B arrangement of air terminals and other components necessary for complete installation.

Lightning conductors should have their own earth electrodes and should not be connected in any way to the station earthing system as, owing to the extremely steep wave front of the discharges, dangerous voltages may appear over quite short lengths of conductor.

5.9. BATTERY ROOMS, BATTERIES AND CHARGING EQUIPMENT

5.9.1. General

Many storage batteries will be required of various voltages and ampere-hour capacities. There is no hard and fast rule as to the number used, as this will depend on their physical location relative to the load, their duties, and also economic considerations. For instance, if the load is a long distance away, and of a substantial nature, it may be cheaper to provide a local battery than provide a supply cable of large cross-sectional area to counteract voltage drop problems. This particularly applies to buildings situated remote from the power station such as coal and ash plant, jetty, CW pump house, etc. The advent of the gas turbine which provides, amongst other services, a standby a.c. source can be used with effect to reduce battery ampere-hour capacity. For example, the charging equipment can be connected to this source thus enabling the battery to be charged during emergency conditions. Due to the large size of 500 MW unit power stations, it is now becoming the tendency to instal batteries on the "unit" principle with adequate arrangements for interconnection between batteries. This system has the advantage that faults are necessarily confined to the particular unit and are therefore

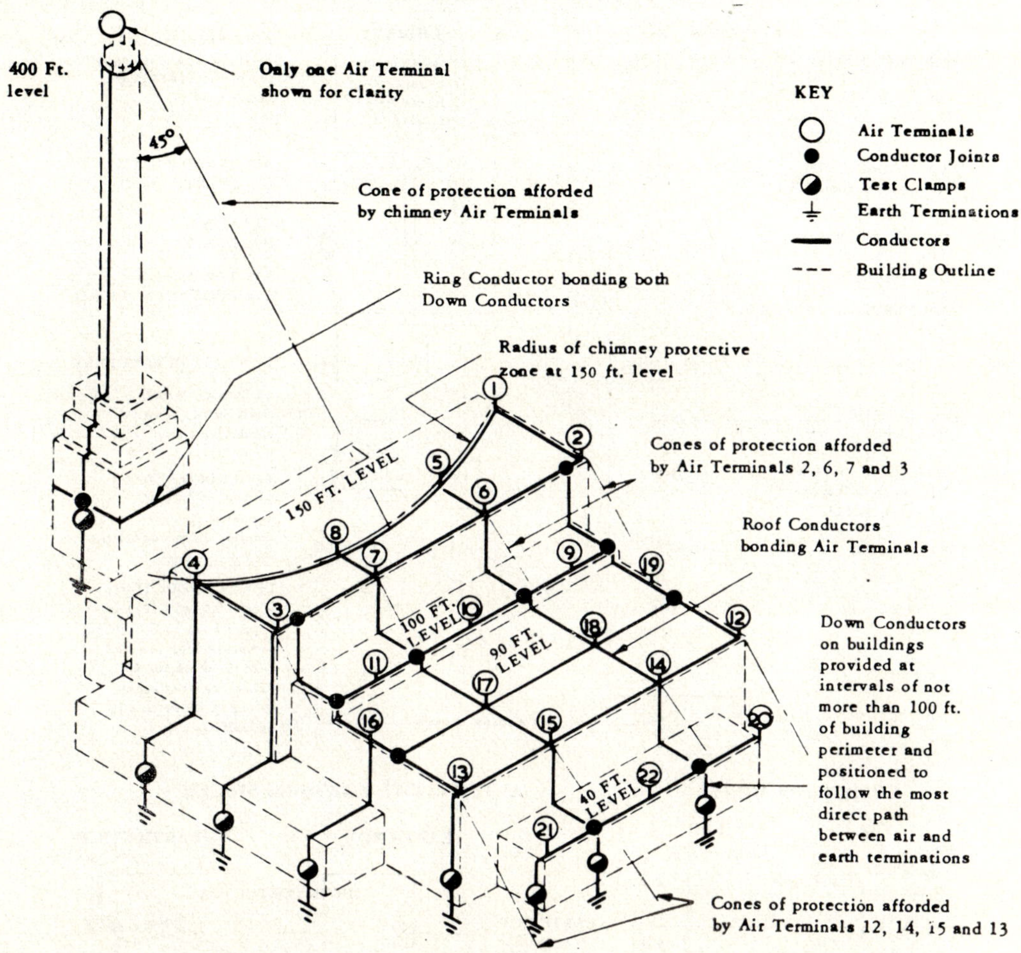

150 FT. LEVEL	Air Terminals 1, 2, 3 and 4 placed at salient points outside chimney protection zone. 5, 6, 7 and 8 placed to limit spacing between air terminals to 100 ft. maximum.
100 FT. LEVEL	Air Terminals 9, 10 and 11 placed at locations unprotected by 2, 6, 7 and 3.
90 FT. LEVEL .	Air Terminals 12 and 13 placed at unprotected salient points. 14, 15, 16, 17, 18 and 19 placed to limit spacing between air terminals to 100 ft. maximum.
40 FT. LEVEL	Air Terminals 20 and 21 placed at unprotected salient points. 22 placed to limit spacing between air terminals to 100 ft. maximum.

FIG. 5.8.4A. Lightning and static protection.

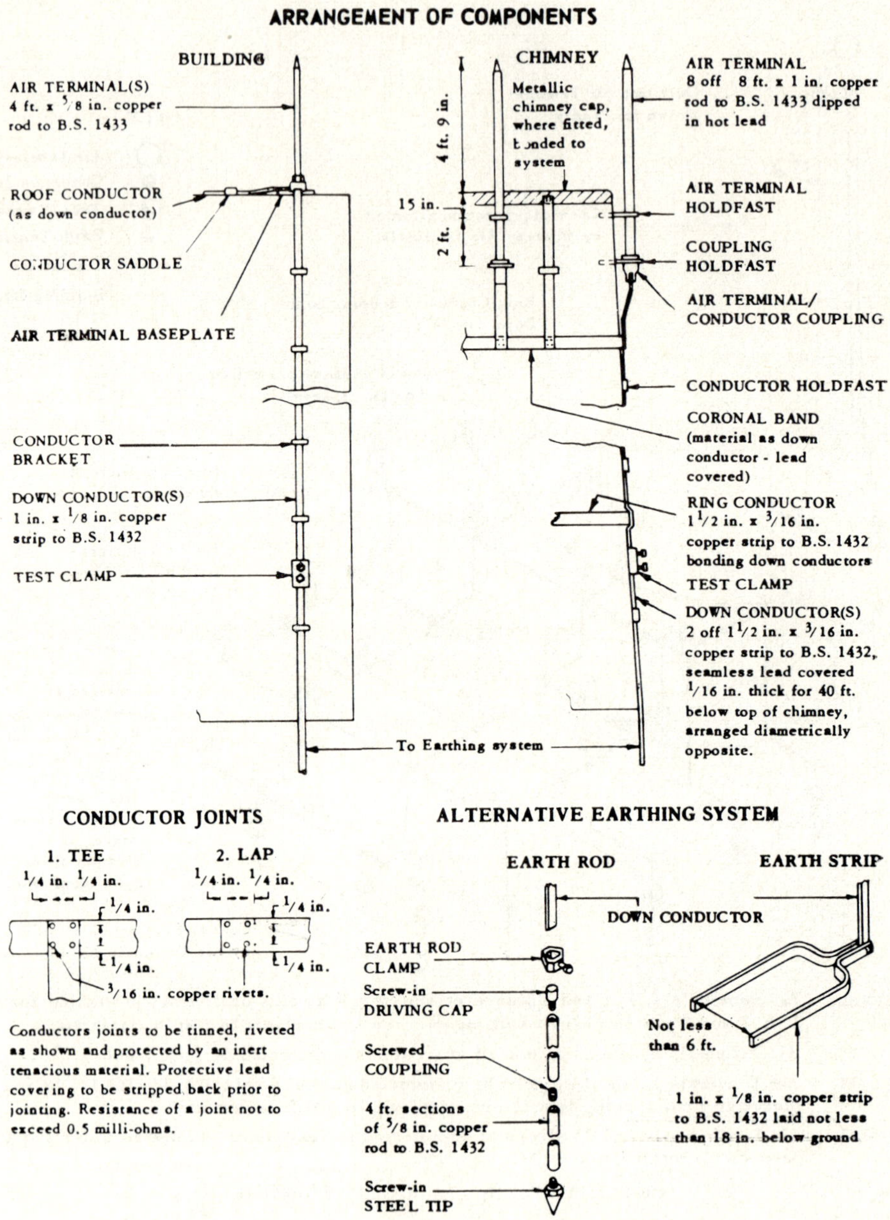

FIG. 5.8.4B. Lightning and static protection.

more readily rectified, particularly in the case of a 50 V battery with a large number of control circuits. Since the "no break" supplies for instrumentation are derived from an invertor or motor generator set driven from a battery, they would be necessarily unitised from the source (see Section 5.11.4). The present trend is to reduce the size of batteries for the automatic start of a d.c. oil-driven generator in duplicate when required.

It is preferable that the telecommunications battery be a separate unit, as requirements dictate that the positive pole be solidly earthed. This raises certain complications, particularly regarding control circuits which should derive their supplies from an unearthed battery, otherwise earth faults might give rise to spurious indication or operation. Another reason is, of course, that telephone circuits are not subjected to a fluctuating voltage.

Battery duties and voltages are listed below:

At 240 Emergency lighting.
Emergency oil pumps for turbo-generator shaft lubrication.
Emergency seal oil pumps for generator hydrogen system.
Instrument "no-break" supplies.

At 110 Auxiliary switchgear, closing, tripping and indication.

At 50 Closing tripping, indication and supervisory control.

At 50 Telecommunications only.

5.9.2. **Battery Rooms**

Batteries are of the lead-acid type usually in sealed glass boxes mounted on wooden stands which may be single- or double-tiered according to the space available. Each cell must be readily accessible for maintenance and inspection so that ample passageway between trunks is necessary. The layout of a typical 1000 Ah battery is shown in Fig. 5.9.3A.

The battery room should be centrally located, as voltage drop may necessitate uneconomically large cables if long runs are involved. The room must not be exposed to high ambient temperatures. For the latter reason, a situation on the side of the station remote from the boilers is preferred. Good natural ventilation is necessary and the floors and walls are tiled or otherwise treated to resist the action of any acid which may be spilt. A sink and water supply are also required.

The electrical connections are made with bare copper rod supported on porcelain insulators, and are led out of the room through wall bushings.

5.9.3. **Switchgear and Charging Plant**

It is usual for each battery to have two sets of charging equipment invariably of the metal rectifier type. These are so designed that under normal conditions the whole of the load is supplied by the chargers therefore leaving the battery "floating" in a fully charged state.

Manual control arrangements are provided on the charger for "normal" or "boost" charging to be carried out, the latter being required when the battery is discharged after an emergency operation. It should be noted that "boost charging" cannot be carried out unless all the load is disconnected from the busbars as the battery voltage under these conditions is too high.

The charging equipment, together with the d.c. switchboards, should be situated as closely as possible to the battery room. The switchboards usually of the sheet steel cubicle type comprise fused switches for the outgoing circuits and the necessary incoming switches for the battery and rectifier supplies.

The distribution system associated with 50 V batteries, since the advent of remote centralised and computer control techniques, have now become quite enormous in their own right. It is appreciated that with a 50 V system of interposing control requiring a large number of relays and contacts for control and indication, these in turn must derive their supplies from a suitable source fused where necessary. For a typical arrangement see Figs. 5.9.3A and B.

A convenient position for installing such equipment would be as near the control room as possible. The exact composition of the d.c. distribution boards would vary according to requirements but would usually comprise fuse monitoring and d.c. change-over equipment, etc.

5.10. FIRE PROTECTION

5.10.1. General

Protection of transformers and switchgear of the oil-break type is achieved by fixed automatically operated water-jet equipment usually referred to under the trade name of "Mulsifyre".

5.10.2. Mulsifyre Equipment

The apparatus to be protected is surrounded by a number of sprinklers or projectors from which water spray is discharged at high velocity. The impact of the water particles on the oil forms an emulsion which will not burn. Normally water reserve tanks of sufficient capacity for the maximum fire risk are needed as the town main will rarely sustain the high rate of flow involved. These tanks are located where convenient, but usually at ground level on a large station. The required pressure of 50 to 150 lb/in^2 is supplied by diesel- or motor-driven pumps. It is essential that diesel engines start in sequence, and that in the case of motor-driven pumps these are provided with alternative supplies with automatic changeover and alarm in case of failure of one of them.

The pressure tank, like the reserve water tanks, should be mounted in a central position.

For plant situated indoors a "wet system" may be used since freezing is not a problem. With this system the water is run to a fusible control detector, mounted above the protected plant, which acts as a shut-off valve. When the temperature exceeds 65°C the fusible link breaks admitting water to adjacent pipework and hence to projectors.

For outside installations where there is a risk of freezing, as in the case of outdoor transformers, the pipes are not normally charged with water; instead, a second system of small-bore pipework charged with air and fitted with detectors is run around the transformer and then back to a special valve suitably positioned indoors (see Figs. 5.10.2A, B and C).

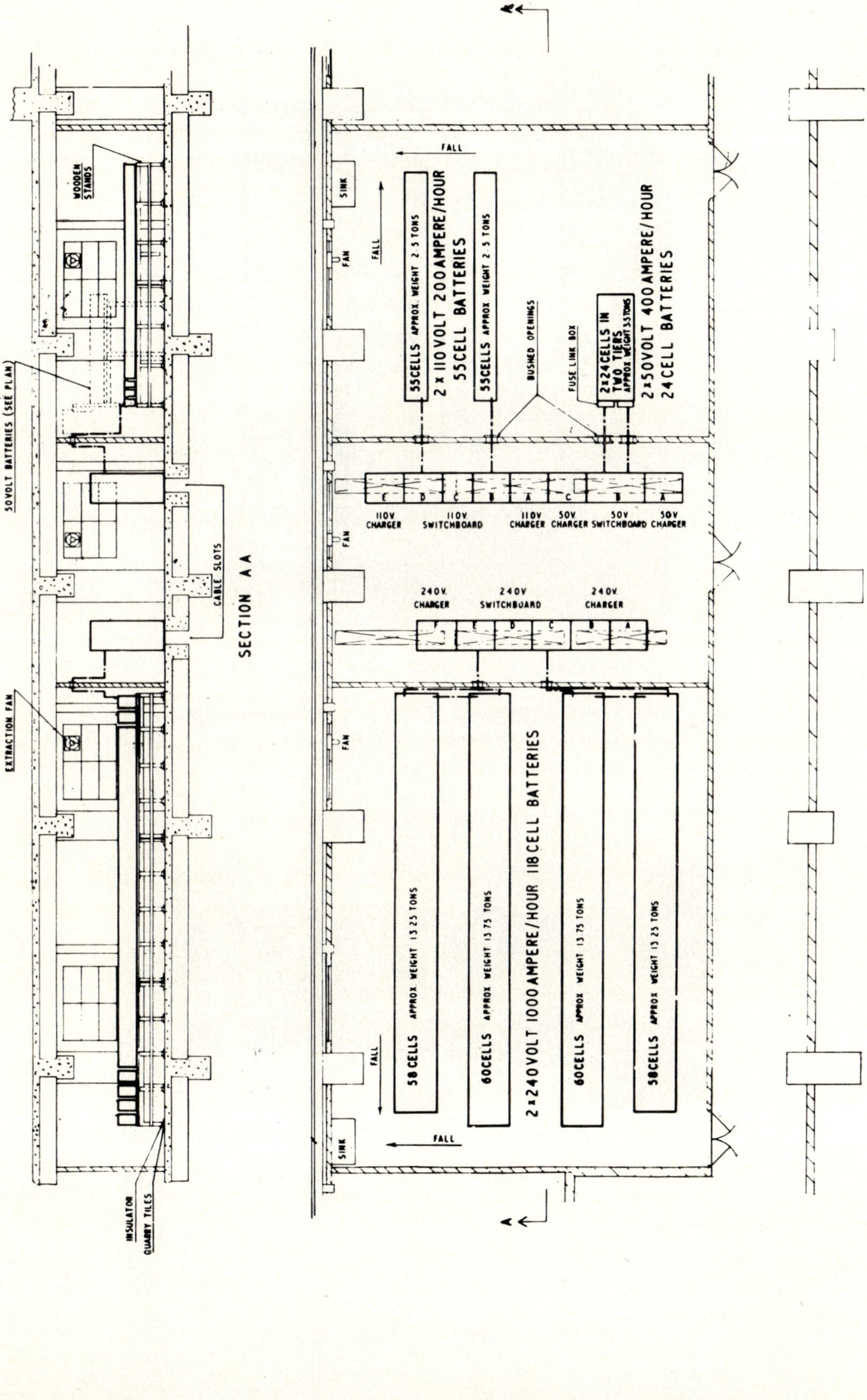

FIG. 5.9.3A. Typical layout of battery room.

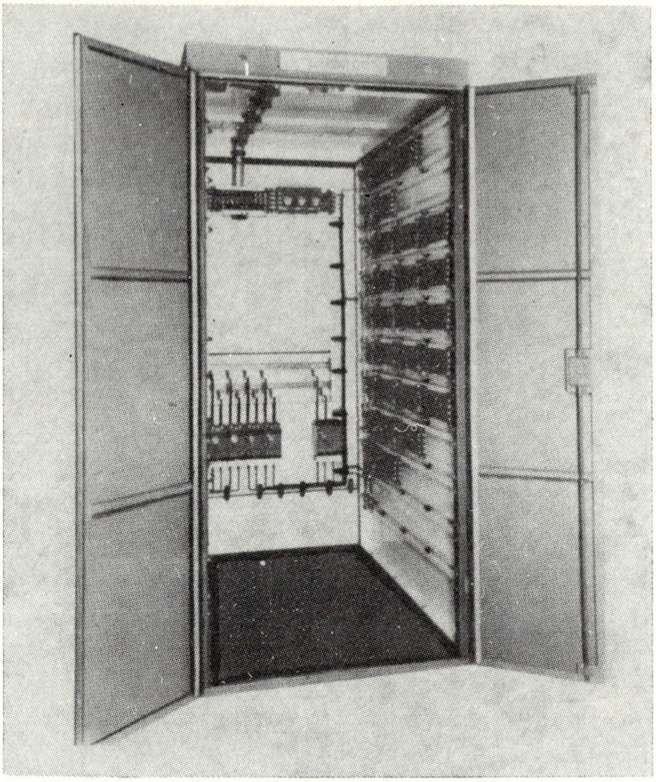

FIG. 5.9.3B. Typical arrangement of 50 V d.c. distribution board.
(Reproduced by kind permission of Aish and Co. Ltd.)

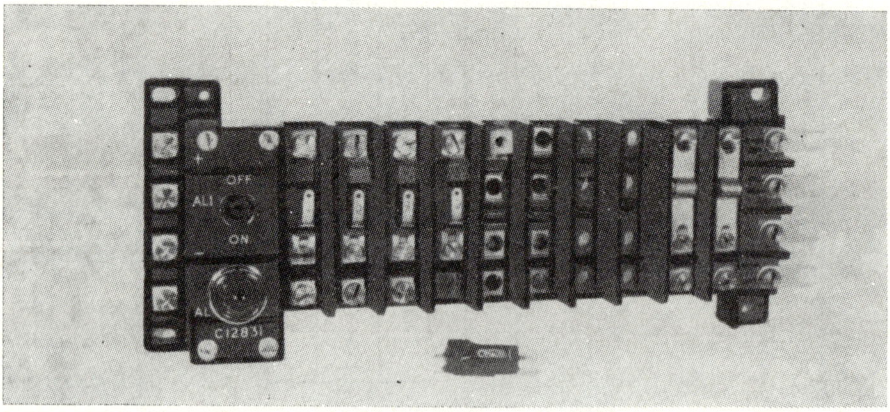

FIG. 5.9.3C. Enlarged view of one fuse centre.
(Reproduced by kind permission of Aish and Co. Ltd.)

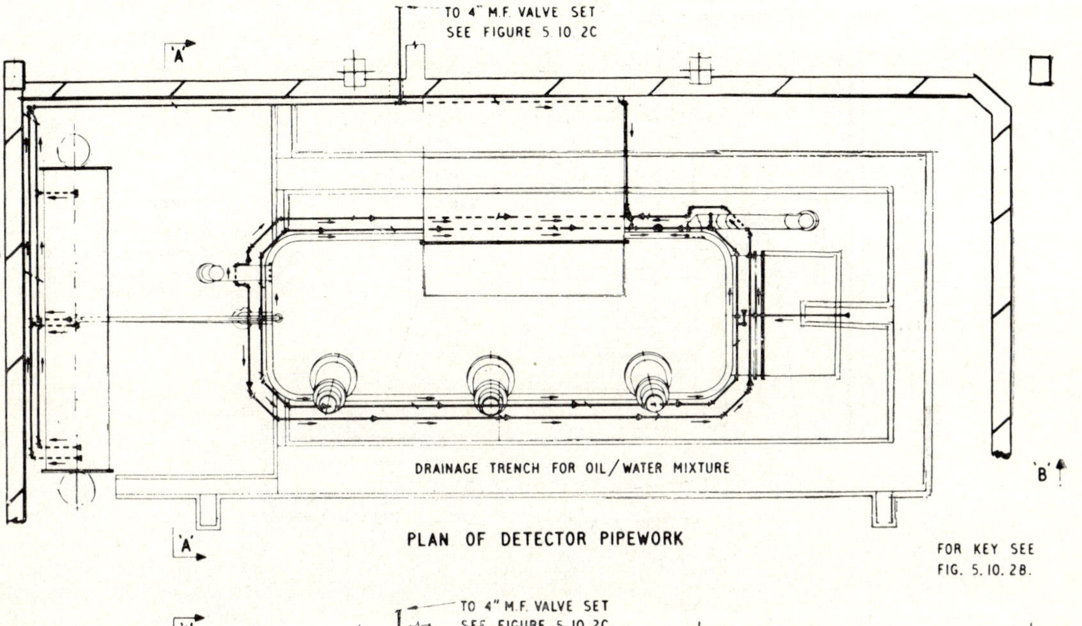

FIG. 5.10.2A. Plan views of typical "Mulsifyre" fire protection system for a generator transformer.

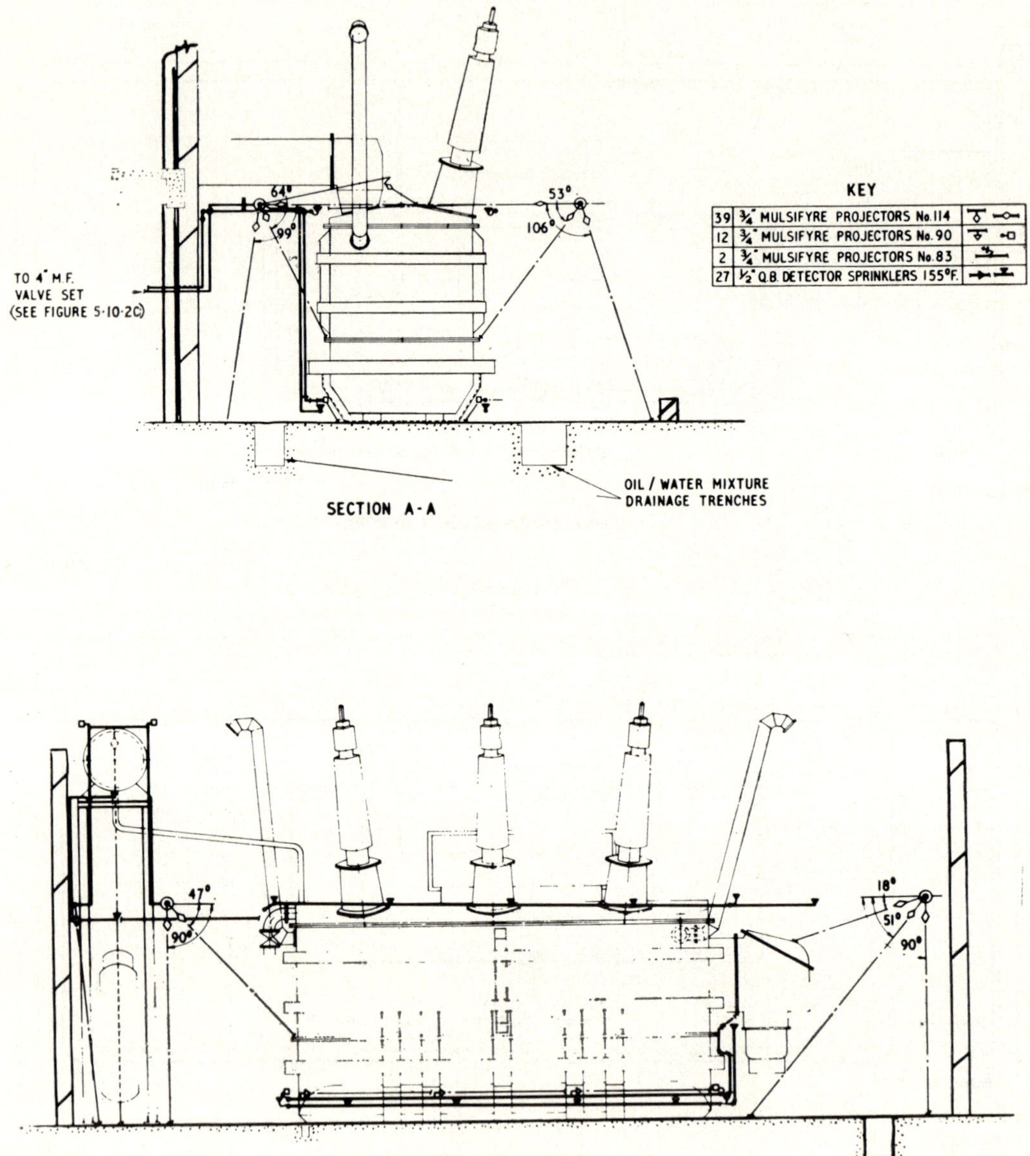

FIG. 5.10.2B. Sections of typical "Mulsifyre" fire protection system for a generator transformer.

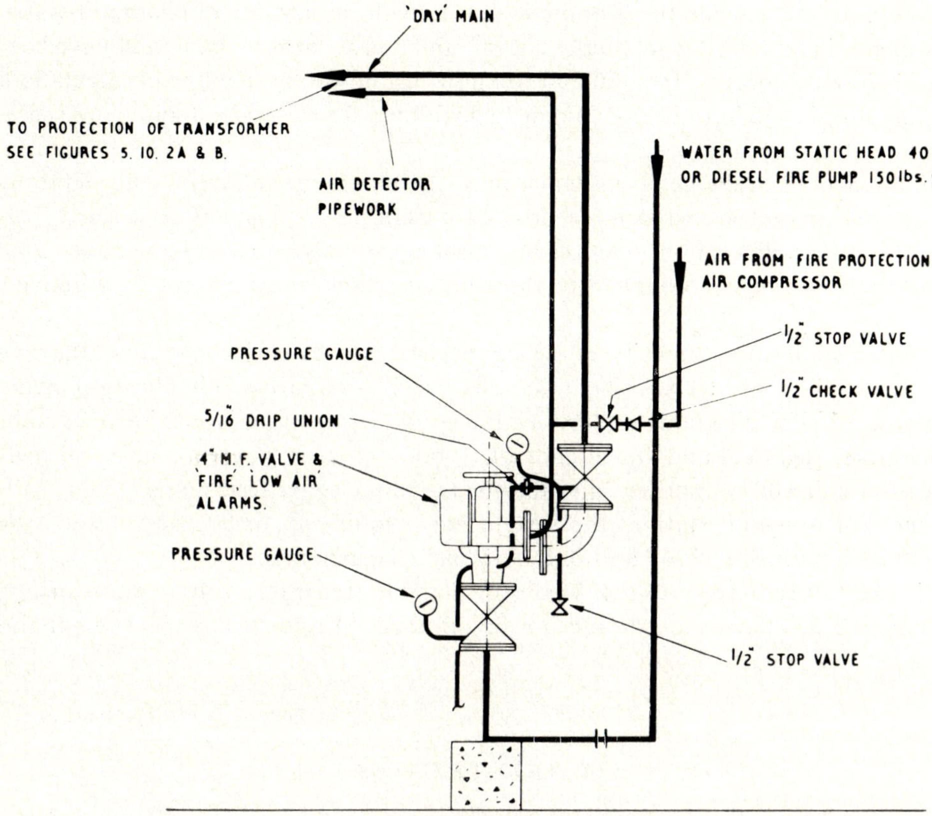

FIG. 5.10.2C. Typical installation of "Mulsifyre" "dry" control valve.

The valve incorporates a diaphragm with water one side and air the other. Under normal conditions the pressures are approximately equal, the air pressure being supplied by a fire system air compressor and the water pressure by a static head.

If the temperature in the transformer compound rises above 150° F, the fusible link breaks, and the air side of the diaphragm is reduced to atmospheric pressure. Simultaneously water is admitted to the fire protection pipework, the diesel fire pump is automatically started in order to boost the water system to approximately 150 lb/in^2 and an alarm is initiated. A low air pressure alarm may also be fitted in order that a possible leak in the detector system is brought to the operator's notice before the system operates for what is a false alarm.

The fire protection of plant is a skilled job and as such is usually designed and installed by specialist contractors. It is quite usual for the civil work to be completed, i.e. drainage, etc., and possibly the plant installed before the fire protection proposals are available and for this reason it will be necessary to estimate roughly the potential amount of water to be disposed of in the event of the system operating. Suggested figures are 1·2 gal/min per square foot of surface area of plant to be protected, and an additional 0·5 gal/min per square ft of floor area in the immediate vicinity of plant, say 5 ft to 6 ft all round.

It is to be noted that while the automatic feature will initiate the discharge of water, the flow will continue until it is manually cut off, and a great deal of water will have been discharged before this occurs. It is difficult to know just how long to allow in calculations but since any fire should have been extinguished within a few seconds, 5 min would seem to be a reasonable time to allow.

The containment or drainage of this water must therefore receive careful consideration, and may indeed present a considerable problem where indoor plant is concerned. It is necessary to have switchgear protection arranged as selectively as possible, such that when it operates its range is confined to the affected switch unit adjacent to it on each side.

On a modern station the "mulsifyre" system may be very extensive and cover the protection of such plant as: transformers standby diesel alternators, oil filtration plant, hydrogen store, boiler oil mains and boiler burner galleries, transformer/rectifier units for precipitators, electrical and steam ends of turbo-alternator, governor gear, oil relay piping, seal oil unit and oil purifier, and the steam-driven boiler feed pump.

For a modern four-unit station, the system may require up to 80 control and stop valves, etc., each with one "fire" and one "low air" alarm contact.

In the interest of reducing fire risk wherever possible, generator voltage transformers for instruments and relays, etc., should not be oil-filled. SF_6 gas insulation is a suitable alternative for this particular application.

5.11. CONTROL

5.11.1. General

The siting of the control block is very important from a construction, cabling and operational aspect. It is desirable that this building is completed well before the commissioning of the first unit and there is merit, therefore, in separating it from the rest of the power station. On the other hand, it should preferably be near the rest of the station to provide easy access for personnel and to keep the many cable runs which terminate in the control room as short as possible.

A compromise is, therefore, made between these conflicting requirements.

It is now usual practice to control the turbo-generator/boiler unit with its ancillary plant from the control room.

A number of alternative layouts for accommodating the necessary control panels or desks can be adopted:

(a) A separate plant control room per unit.
(b) A combined control room containing all the plant unit control panels.

In addition to the above, accommodation is required for the necessary panels or desks associated with the control of the 400 kV, 275 kV or 132 kV switchgear, generators, 11 kV, 3·3 kV and 415 V switchgear. These may be located in a separate room from the plant control panels or may be amalgamated.

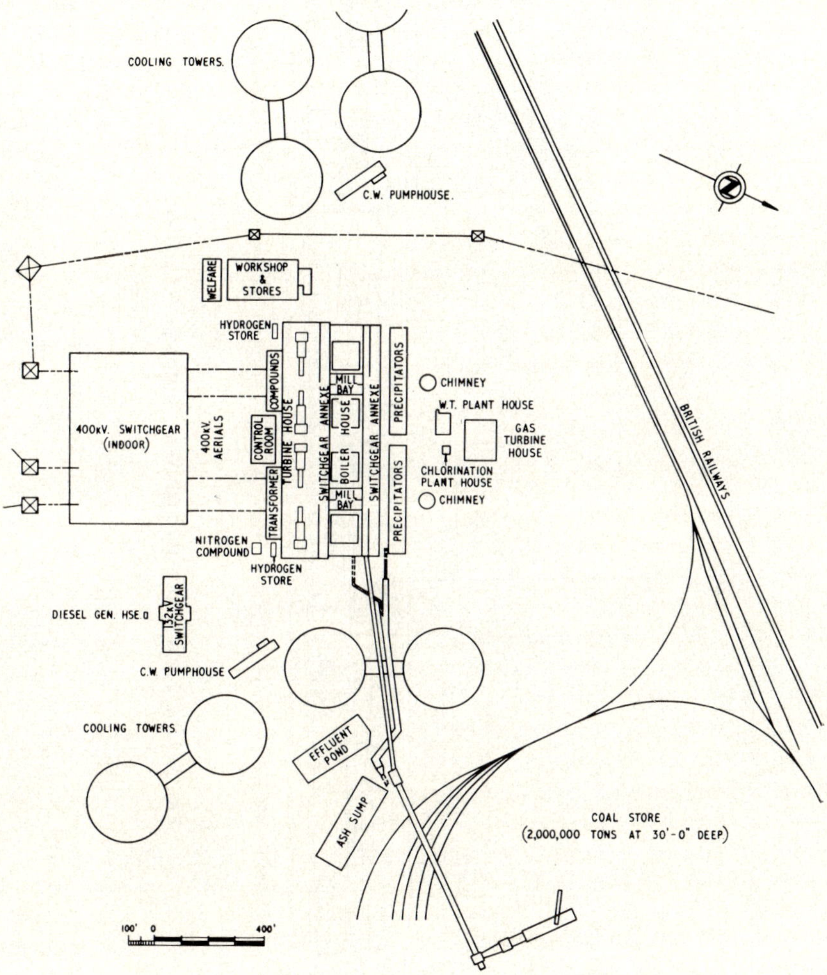

FIG. 5.11.1A. West Burton power station, 4×500 MW, coal.

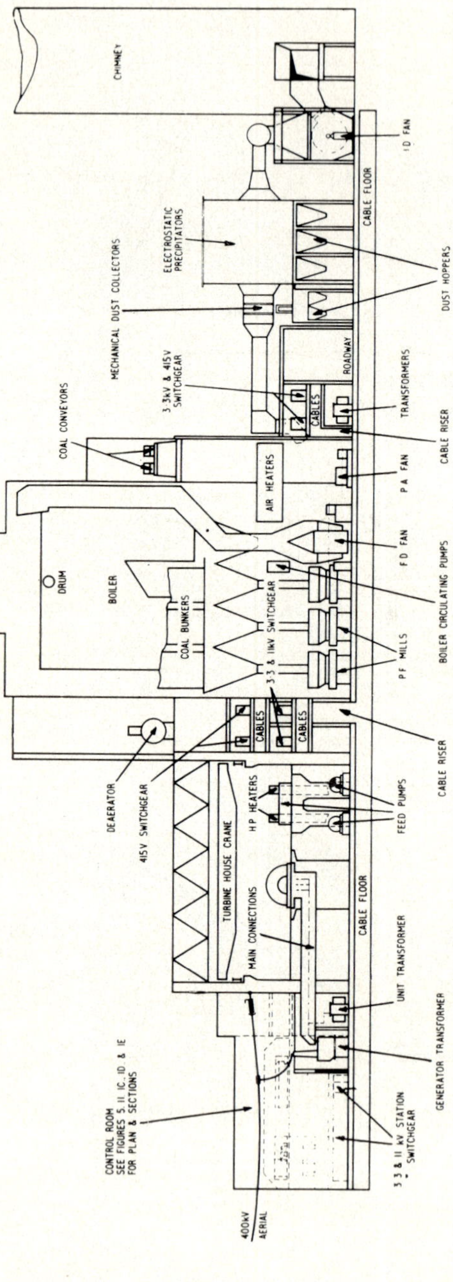

FIG. 5.11.1B. Cross-section through West Burton power station.

There are merits to using each of the alternatives suggested above which cannot be examined in detail in this exercise. Sufficient is it to say that a technical analysis will be made of the factors involved, coupled with an examination of the manning schedule of the power station (see Figs. 5.11.1A to L inclusive). A 10 ft cable room should be allowed for under the control room (see Figs. 5.11.1D and E).

5.11.2. **Plant Unit Control Panel**

All plant control panels are housed in a room which should, if possible, be situated in a locality not subject to high ambient temperatures. The humidity and temperature should be carefully controlled by a filtered air conditioning system to exclude dirt and dust from the instruments.

The architect will obviously pay great attention in the layout to such aspects as the comfort of the operators and also the amenities.

The control room should have immediately beneath it a chamber of the same size to accommodate the large number of control and other cables to be marshalled and deployed.

There is in addition a considerable amount of ancillary plant. For example, alarm and other relay panels, automatic "start-up" equipment, electronic equipment associated with boiler control, etc., which can best be accommodated in a separate room either adjacent to or underneath the control room, or as near as possible to the centre of gravity of the unit (see Figs. 5.11.2A and B for desk detail).

Control rooms should also have a filtered air system adequate for the number of operators plus hot water wall radiators fed by l.p. steamheat or electrode boiler, and use being made of combined fan and radiator units.

The local control and instrumentation of coal mills and feeders is considered essential, and it may be thought desirable to house this panel in a field equipment room along with other transmitters and thermocouple cold junction boxes, etc.

5.11.2.1. Unit Control Panel Layout

The increasing complexity of modern plant involves a corresponding increase in the number of instruments and controls. In designing the panels, therefore, much thought is given to ensuring that all essential items can be suitably mounted on a panel of reasonable size. Apart from the problem of accommodation, every increase in the size of the panel tends to defeat the object of centralised control. Economy of space can be achieved by:

(a) The relegation of items which need not be under constant observation to "inferior" positions, e.g. at the rear of the panels. Some recording instruments come in this category and in some cases a separate recording panel immediately behind the unit control panel is deployed for this purpose.
(b) Adopting desk-type panels which afford a larger available area for a given floor space.
(c) The use of small instruments. When the instrument can be mounted within a foot or two of the observer's eye a 2 in dial instrument is generally large enough.

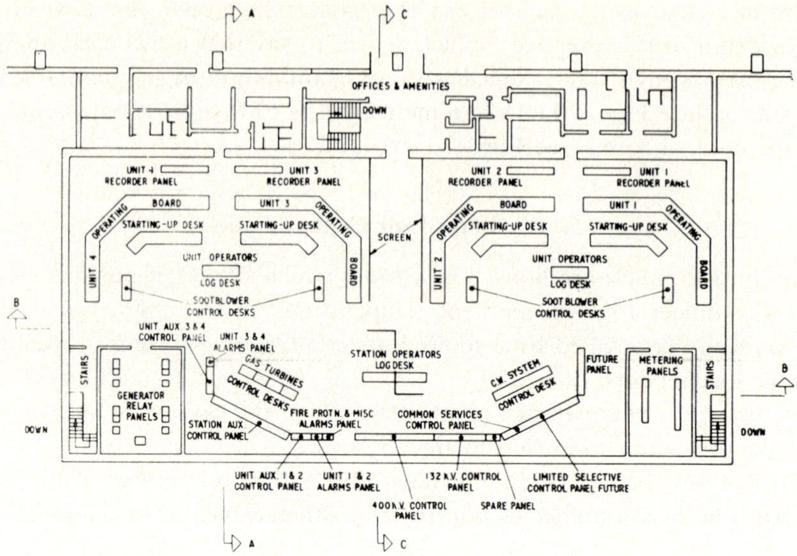

FIG. 5.11.1C. Plan on operating floor at +32 ft 6 in level, West Burton control room.

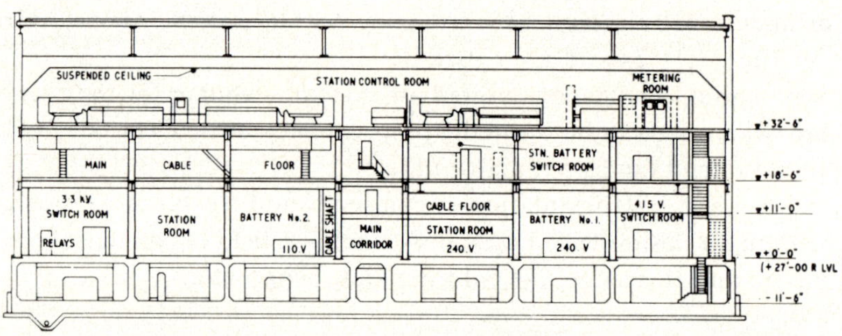

FIG. 5.11.1D. Longitudinal section B–B, West Burton including control room.

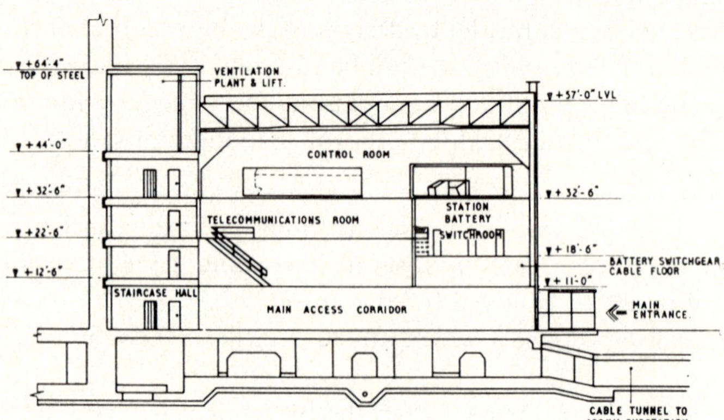

FIG. 5.11.1E. Section C–C, West Burton including control room.

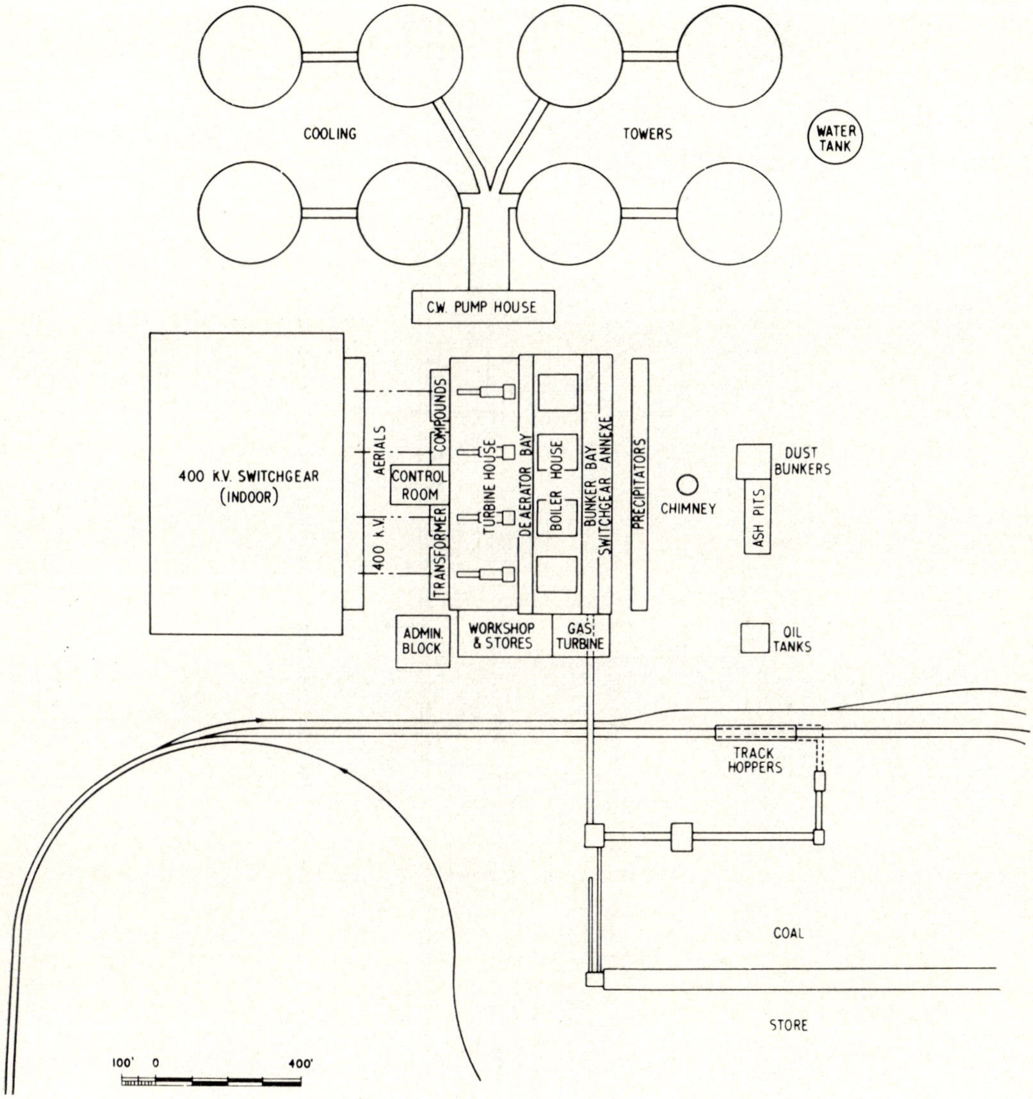

FIG. 5.11.1F. Eggborough power station, 4×500 MW, coal.

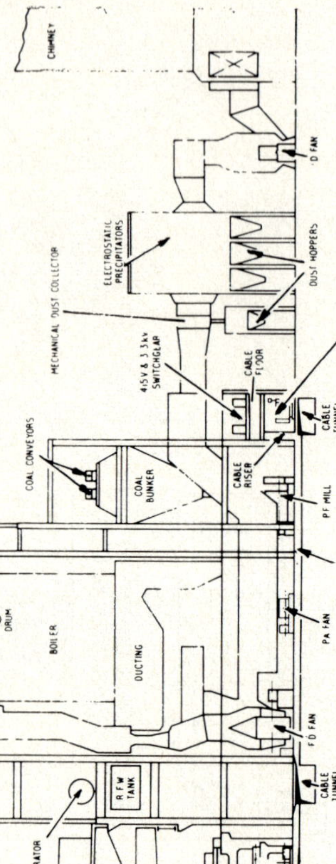

FIG. 5.11.1G. Cross-section through Eggborough power station.

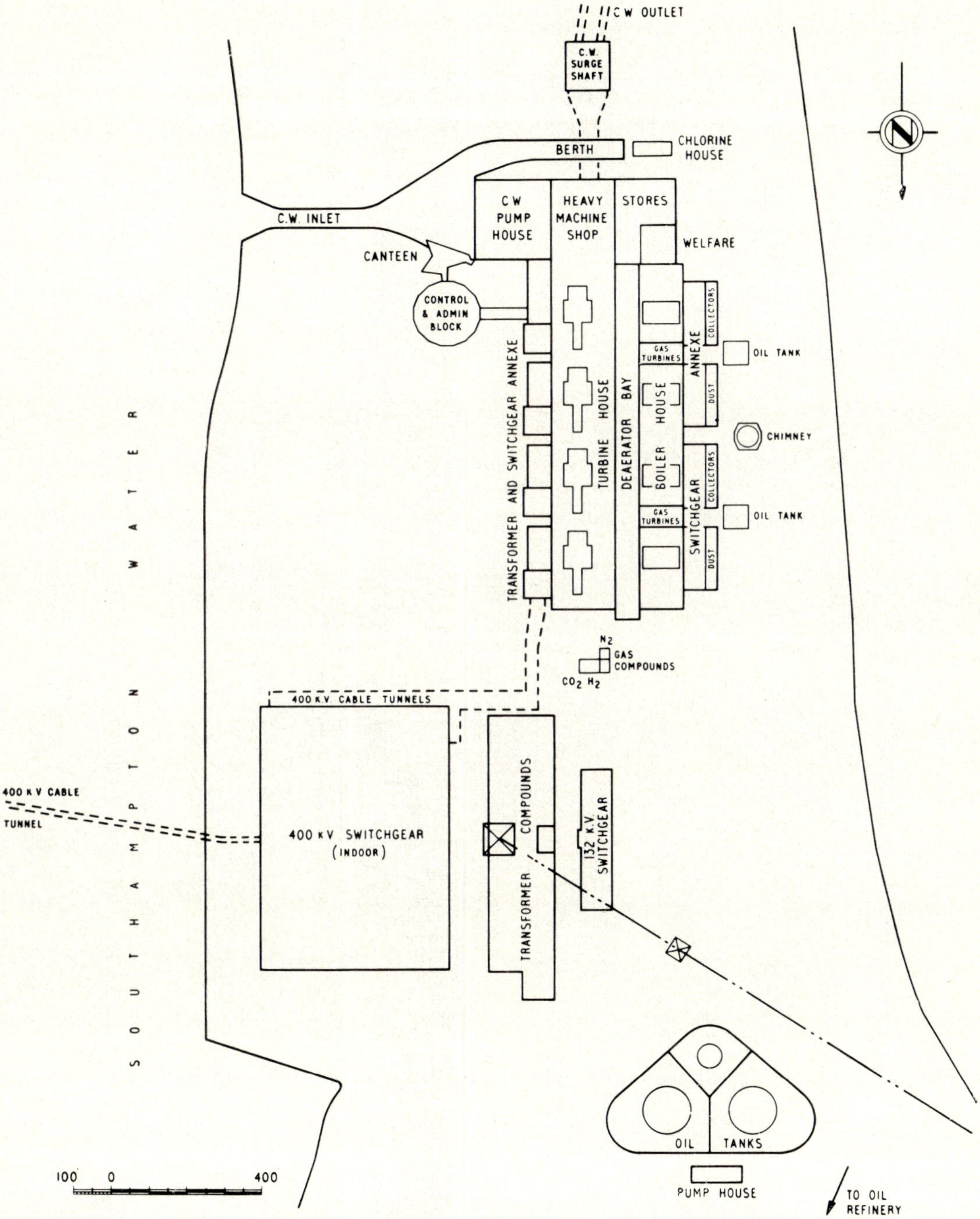

FIG. 5.11.1H. Fawley power station, 4×500 MW, oil.

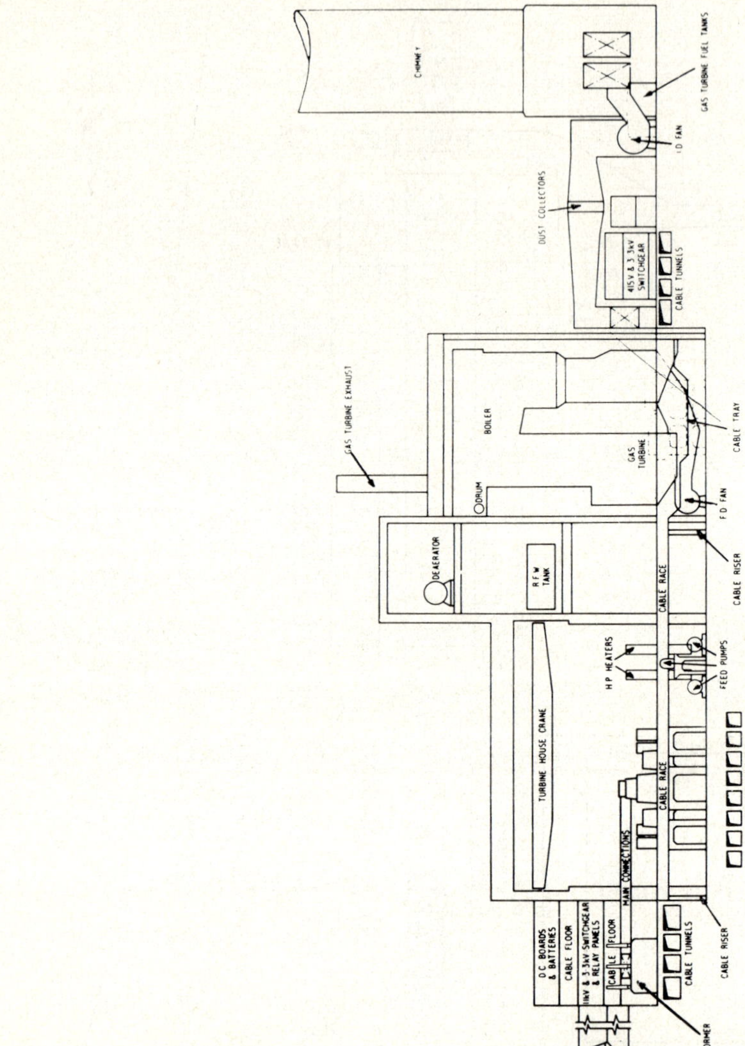

FIG. 5.11.1J. Cross-section through Fawley power station.

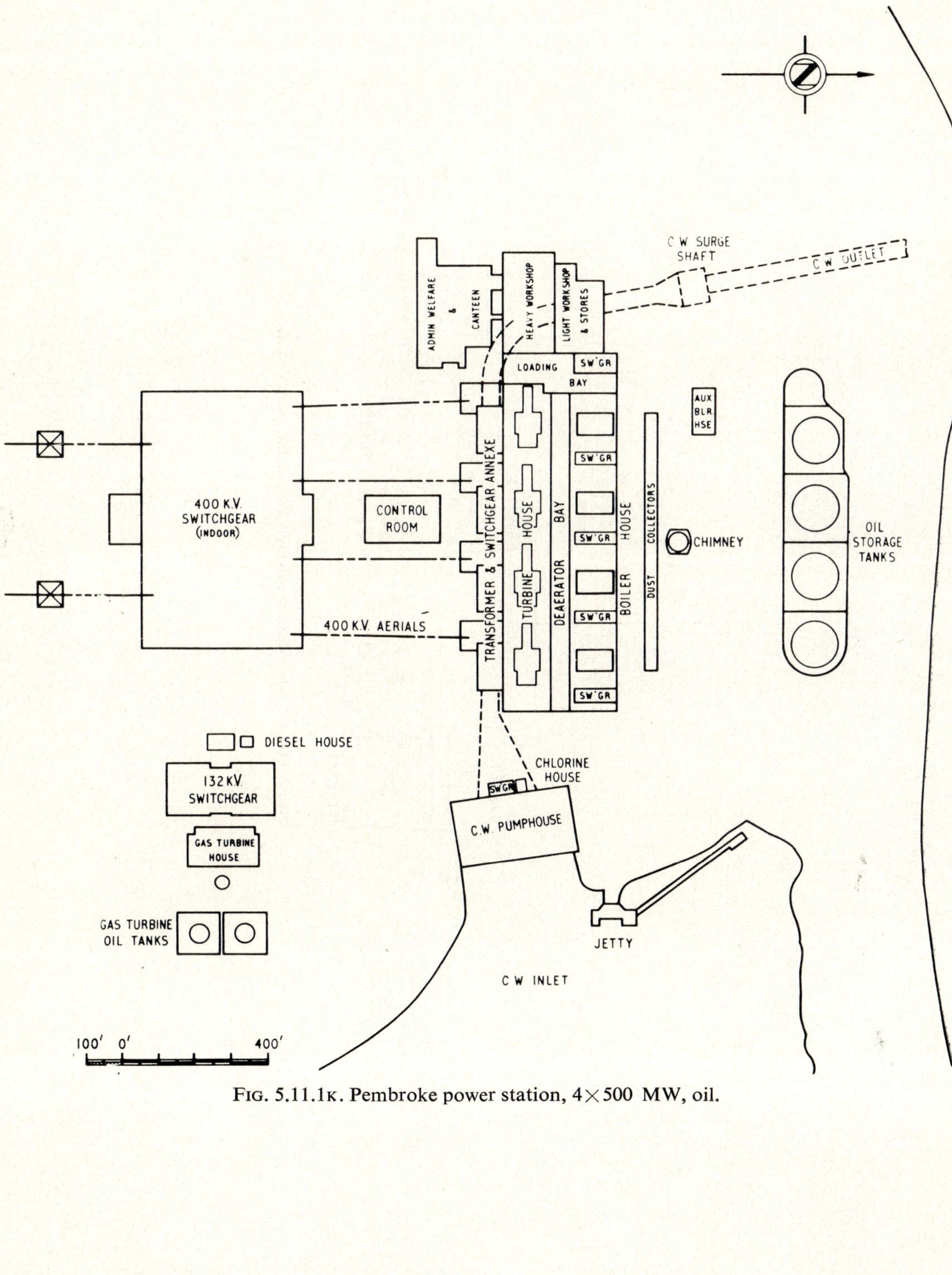

FIG. 5.11.1K. Pembroke power station, 4×500 MW, oil.

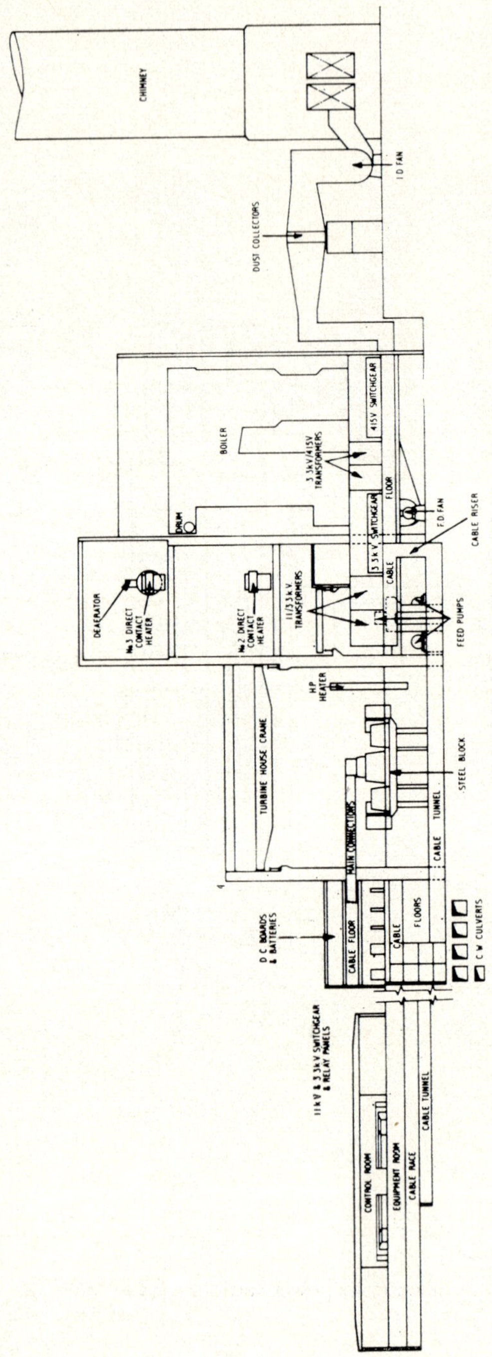

FIG. 5.11.1L. Cross-section through Pembroke power station.

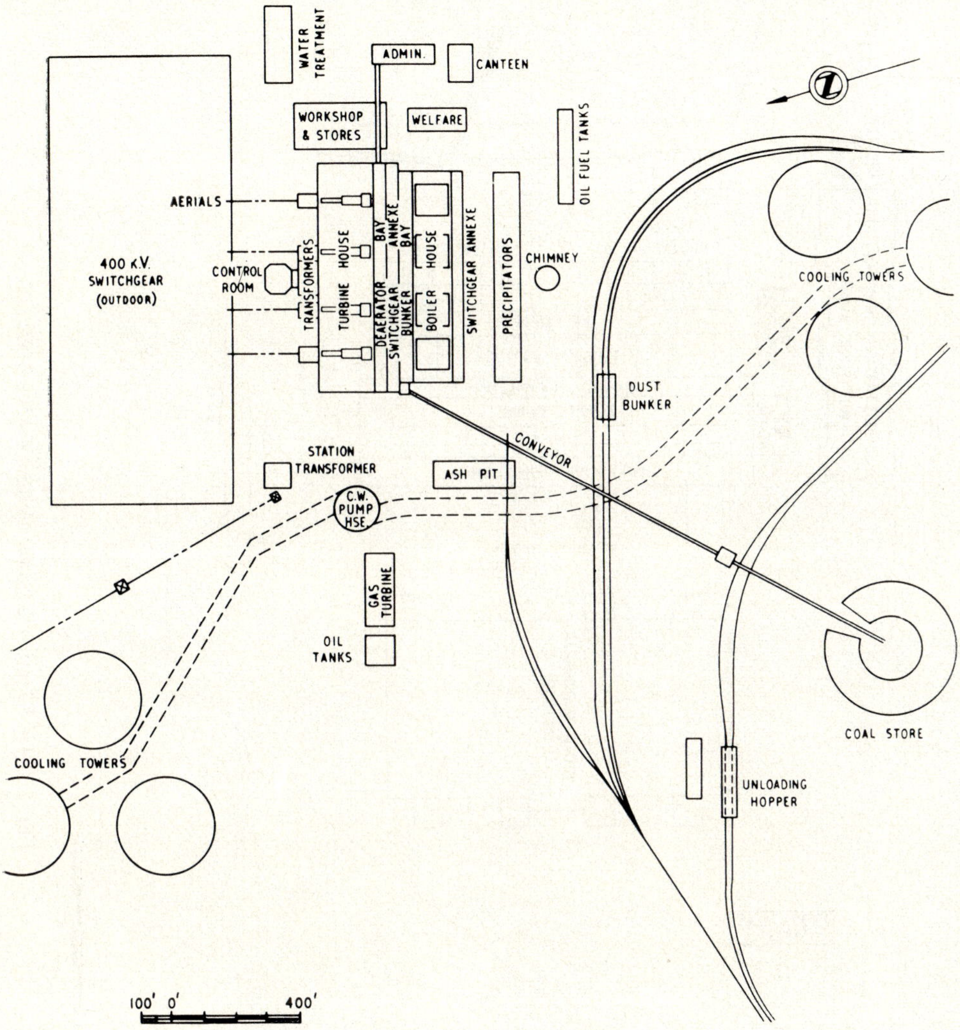

FIG. 5.11.1M. Didcot power station, 4×500 MW, coal.

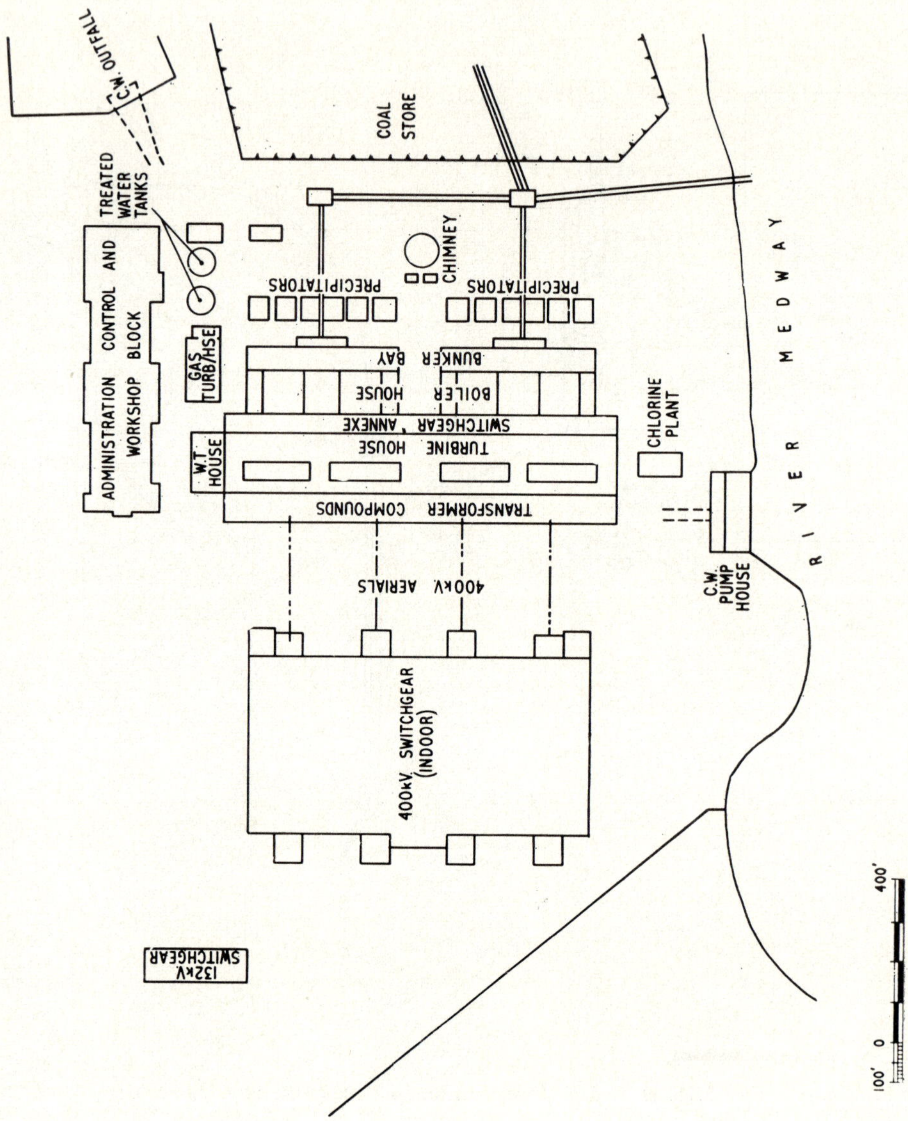

FIG. 5.11.1N. Kingsnorth power station, 4×500 MW, oil/coal.

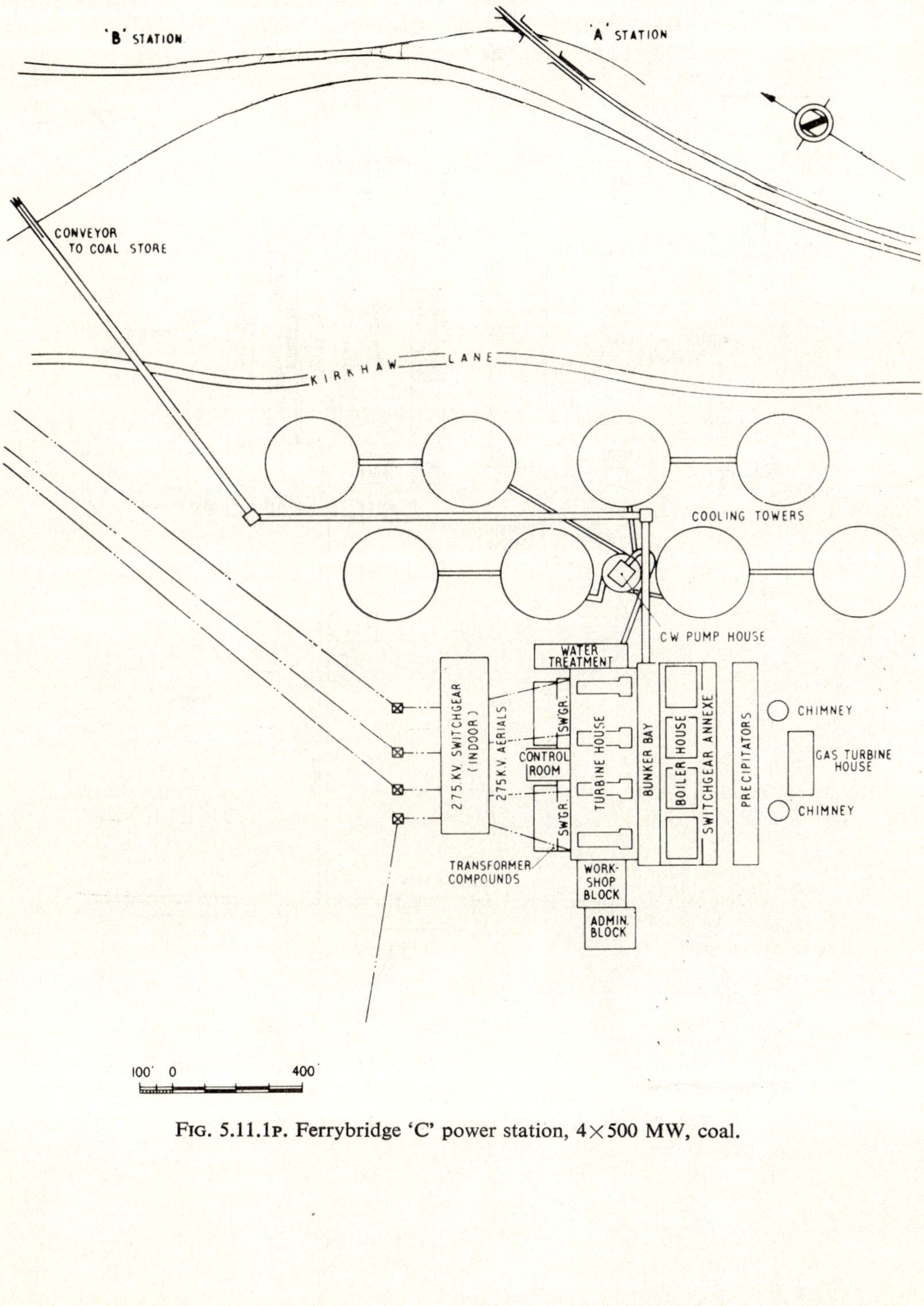

FIG. 5.11.1P. Ferrybridge 'C' power station, 4×500 MW, coal.

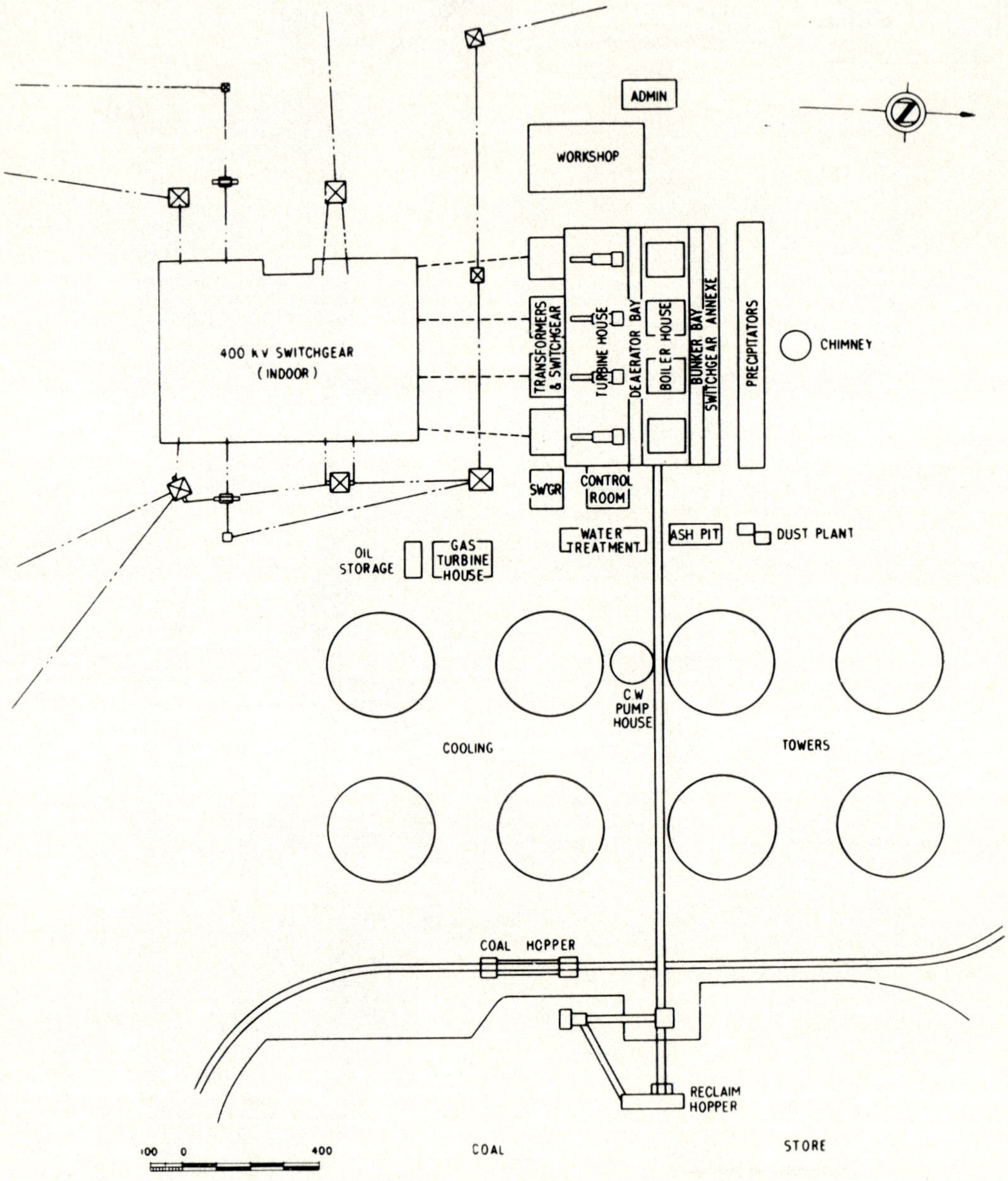

FIG. 5.11.1R. Cottam power station, 4×500 MW, coal.

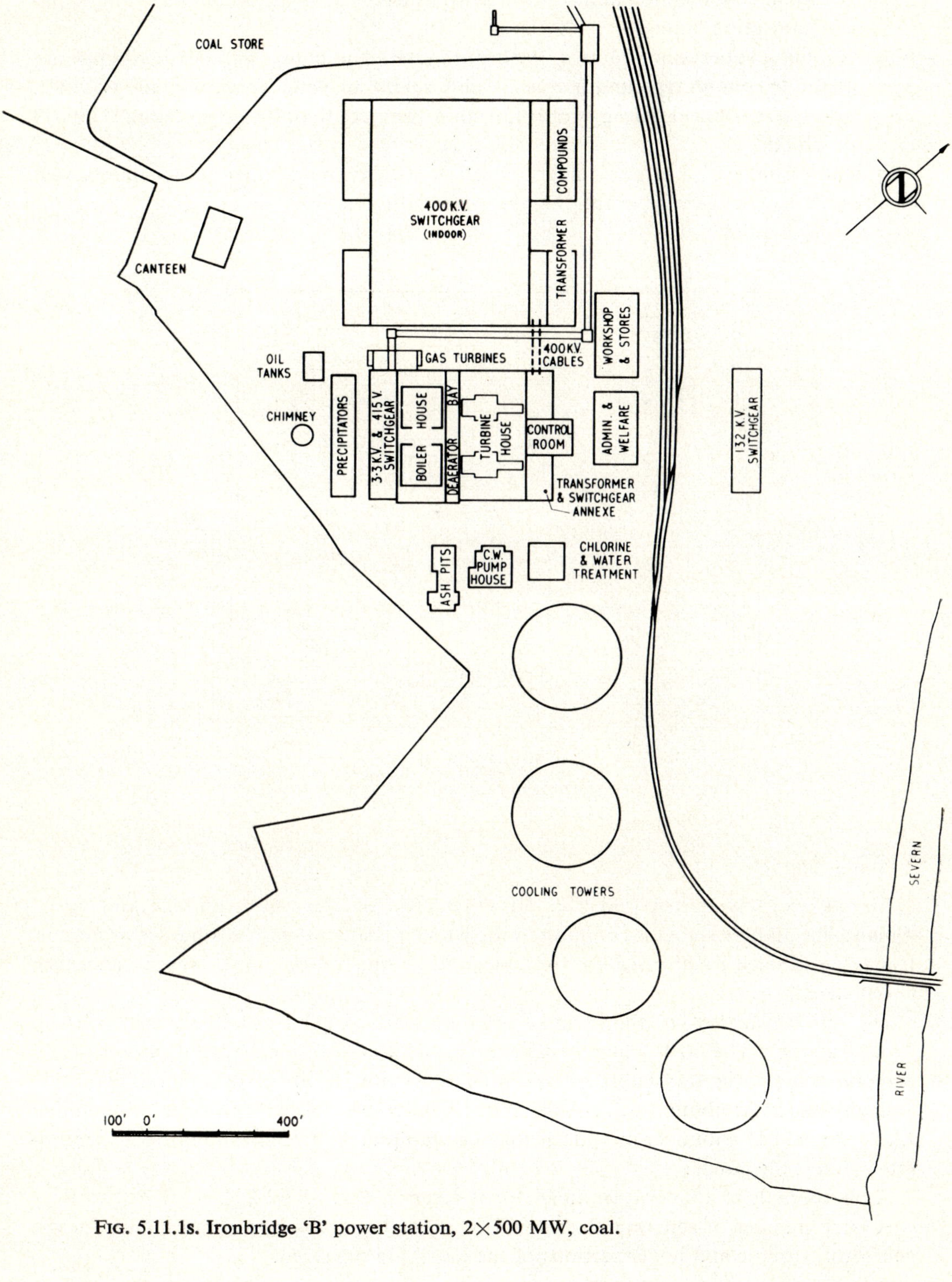

FIG. 5.11.1s. Ironbridge 'B' power station, 2×500 MW, coal.

(d) A combined indication and control centre, thereby saving space in providing separate indicating lamps.

(e) Adopting supervisory control. With this system the controls indicating lamps are of the telephone type and a considerable saving of panel space is made possible. The internal panel wiring is also far more compact than the conventional 7/0·029 in wiring.

System (e) involves a large number of relays which would either be accommodated locally at the switchgear or in the chamber referred to earlier.

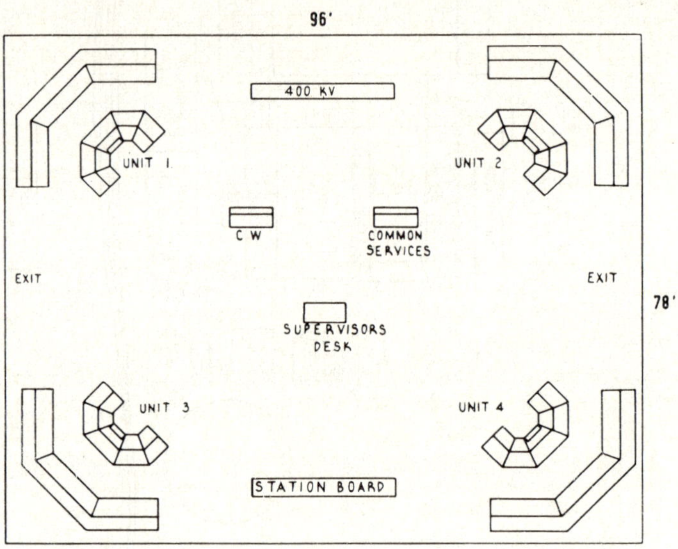

FIG. 5.11.2A. Typical layout of control room for four units.

5.11.2.2. ALARMS

Alarms are now transmitted over 50 V circuits, and give both audible and visual signals. The audible signal is common to the whole panel and may be a buzzer or hooter. It may be silenced by pressing an "Accept" button which does not, however, cancel the visual signal.

The visual signal takes the form of an illuminated legend usually described as an "Annunciator". The 50 V alarm system can be of a conventional type using telephone equipment, or of the transistorised type which does not require associated relays.

Each window is about $1\frac{1}{2}$ in square and the units are formed into panels containing 12, 24, 36 or such a number of units as may be required. A "test" push button is incorporated to test the lamps.

A photograph of a six-way annunciator is given in Fig. 5.11.2.2.

A large number of separate alarms will be needed and separate annunciators for the generator, turbine and boiler sections of the panel are provided.

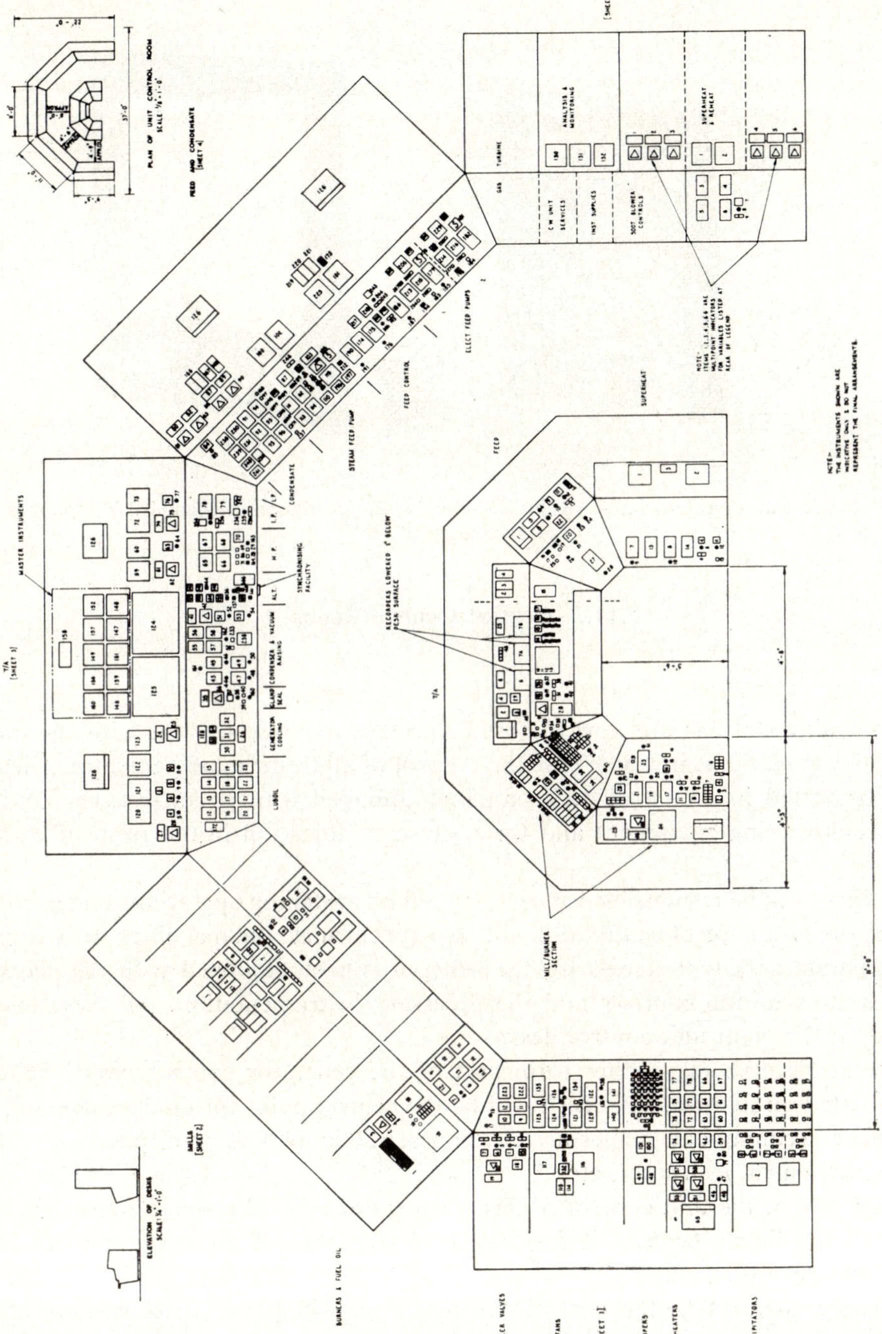

FIG. 5.11.2B. Typical layout of control room desk and panel.

FIG. 5.11.2.2. Alarm annunciator. (Reproduced by kind permission of Standard Telephones and Cables Ltd.)

5.11.3. **Central Control Room**

5.11.3.1. LAYOUT

New switching stations are now almost invariably required to deal with the output from new power stations, in which case the control of all feeders and generators, circuits, etc., will be vested in a control desk or panel equipped with circuit-breaker controls, manual synchronising equipment and the necessary indication in the form of a mimic diagram.

The operator will be responsible for carrying out all switching operations under instructions from the grid control centre and will also synchronise the machines. It is usual to provide separate desk-type panels for the generators each equipped with the necessary governor and excitation controls and the necessary instrumentation. All these may be duplicated on the plant unit control desks.

The operator's desk may either form part of the generator control desks' suite, or may be situated on its own. This desk will be the central point for communication, and will therefore accommodate numerous telephones, public address transmitter, fire alarm initiator, grid control instructor, etc.

As in the case of the unit control rooms, the central control room, if separate, needs a chamber immediately beneath it for orderly deployment of the many control cables and ancillary apparatus, etc.

Immediately adjacent to the central control room will be the telecommunications, relay and metering rooms each provided with a filtered air conditioning system.

5.11.4. **Instrumentation "Safe Supply" Equipment**

Provision of a "safe supply" is now specified in C.E.G.B. contracts for power station instrumentation. Following the major failure of the electricity supply system in 1963, when freezing conditions caused a succession of faults resulting in a break-up of the 275 kV system into isolated sections, it was found that under these conditions certain essential a.c. operated instruments would need a stabilised voltage/frequency supply for satisfactory operation.

The usual methods of providing a "safe supply" source, at either 240 V or 110 V a.c. is by means of static invertors or motor generator sets, these in turn being fed from trickle-charged batteries. There is merit in unitising this equipment and locating it as near the centre of gravity of the load as possible, especially when the supply voltage is 110 V, in order to overcome voltage drop problems which could be quite large on a 500 MW unit with an estimated load of 25 kW. If motor generator sets are used possible disturbance to sensitive relays, etc., should be considered when deciding the layout. The use of anti-vibration mountings may help in achieving a really integrated layout.

Similar equipment to that mentioned above has also been specified for at least one 4×500 MW power station now under construction, in order to afford a standby 415 V three-phase 50 c/s a.c. supply to certain essential valve actuators, though in this case there is only one 52 kW, 65 kVA, 0·8 P.F. equipment per unit. This standby supply equipment may become accepted practice for all stations and space should be allowed for installing it in all future station layouts.

5.11.5. **Future Trends**

There is an unmistakable move for establishing automatic starting facilities on boiler turbine units of 500 MW output and above. This can be achieved by the provision of a central control desk in addition to the conventional unit control console.

From this position the operator will be able to control a unit hot start from burner ignition to full load operation by the control of an independent sequence system from a number of push buttons.

With the ever-increasing size of turbo-alternators has come an increase in the complexity of auxiliary equipment and the need for high speed decisions on the part of the unit operator in order to safeguard valuable plant. To assist the operator what is really required is a high-speed method of centralised data presentation, which is compact, versatile and easy to read. At least one manufacturer is now offering a suitable system which is becoming known as "Instantaneous visual data presentation for supervision and control".

The system uses a 21 in cathode ray tube for the display of data, which may be: (1) Alarms, up to 80 or more being accommodated in four columns according to the brevity of the description. (2) Bar graph displays of up to 32 functions. Provision may also be made for remote repeat displays and microfilm data logging, etc.

For the alarms the "visual equipment" may be used in conjunction with a digital computer and an electronic scanner. This would analyse a fault pattern, deduce the primary cause, and then present a statement of the action required.

For the bar graph display the format is such that identification legends or codes appear at the left-hand side of the tube in column form, each followed by the associated instantaneous numerical value. Then, occupying most of the tube face, there is a series of bars whose lengths are analogues of the instantaneous numerical values. Such a display produces a characteristic pattern on the tube face. The operator will, in general, come to recognise this pattern and will be able to detect divergencies from it without having to read the numerical values. A departure from the pattern thus produced is very readily seen by the operator who can then read the numerical values and take any appropriate actions.

The immense value of the immediate display lies in the fact that the operating staff are presented only with essential information. Furthermore, in the stress of emergency conditions they are relieved of the onus of making correct deductions, if these had been foreseen and programmed.

5.12. LIGHTING AND SOCKET OUTLET SYSTEMS

5.12.1. General

The lighting scheme for the power station does not differ in any essential respect from that for ordinary industrial applications, but one or two areas require special consideration.

In the turbine hall the lighting must be mounted to avoid interference with the operation of the crane, and positioned such that normal maintenance can be carried out from the crane. An intensity of about 15 lm/ft^2 at the operating floor is usual and this can be provided either by a relatively small number of sources of high light output or by a much larger number of smaller units. Suitable concentrated sources such as the mercury vapour/fluorescent lamp are available, but modern practice favours the latter, using fluorescent tubes available in lengths of up to 8 ft. The cold cathode type having a life of up to 25,000 h is also particularly suitable in this situation. Control room lighting also requires special treatment. An intensity of about 18/25 lm/ft^2 is needed on the face of the panels. Shadows and specular reflection must be avoided and the lighting problem is considerably eased by the use of non-reflecting glass on the instruments and by employing a "matt" finish on all the panels. Some form of diffused lighting is therefore necessary and is usually provided by a "luminated ceiling" or similar proprietary device with fluorescent tubes as the light source.

In the interests of safety, efficiency and amenity there is a lot to be said for providing a lighting tube maintenance floor above the control room, and the grouping of all the fluorescent light starting equipment in a choke room. Outside lighting follows accepted road lighting practice, using mercury vapour, sodium, or fluorescent lamps, on brackets using a column of the steel or concrete pattern. Automatic control by time switch or photo-electric cell can be applied with advantage. Specialised treatment is also required for lighting large areas such as coal stores and sidings and this generally takes the form of floodlights situated in strategic positions. The lighting of the roads and jetty may have to be agreed with the appropriate harbour or river authority, which may have statutory powers in this respect. Screening of lights from certain angles of approach may be called for, and the usual navigation obstruction lights provided. Aircraft warning lights are required on chimneys, cooling towers and high transmission towers. It will usually be necessary to provide a multi-lamp fitting at each of the four points of the compass and repeat this at up to three levels. Only one lamp in a given fitting would be on at once, the idea being that if a lamp fails, its failure would initiate an alarm and the next lamp in the particular fitting would be selected quite readily from a remote point. Thus the unplanned attendance of a steeplejack is hardly necessary. Figures 5.12.1A and B give some idea of typical installations and access to same. Special multi-lamp fittings have been designed to overcome the problem of access, corrosive atmosphere, etc., associated with chimney lighting.

Chimneys for future stations will be 1000 ft tall or even higher. In the United States of America stacks with heights of up to 1200 ft are under construction. Chimneys of this height could be partly in cloud, and present forms of lighting would not give stray aircraft adequate warning. This problem is under active consideration at the moment and a rotating flashing beacon as used by aircraft themselves for anti-collision warning may be the answer. Another solution might be the use of an ultrasonic device, though any warning system which required the carrying of a special detector unit within the aircraft would not be very acceptable to aircraft users, particularly light aircraft users who would probably be the ones at most risk from high chimneys, etc. An auxiliary plant room for chimney lighting should be provided either adjacent to, or in the base of the chimney in all future stations. Elsewhere in the station fluorescent tubes can be economically used to illuminate fairly large unobstructed areas. Boiler galleries, tunnels, and situations congested with plant and pipework call for a large number of small robust units. Tungsten lamps are most suitable here, usually with vitreous enamelled reflectors or in bulkhead fittings. Special attention is required for illuminating areas where there is a risk of explosive gases such as hydrogen, also acid-resistant fittings will be employed in such locations as battery rooms. Lighting in grid substations of the indoor type also requires special treatment, especially the 400 kV variety which is producing multiple reflection problems for which there does not appear to be a ready solution. Normal commercial practice is usually followed for lighting canteens, offices, stores and workshops.

5.12.2. Emergency Lighting

Emergency lighting, energised from the station battery, is arranged to be switched in automatically on the loss of the normal a.c. supply.

In order to avoid undue drain on the battery, which must at the same time sustain other essential services, only a proportion of the total lighting is so arranged. The circuits will be selected so as to include essential areas such as the control rooms, and to provide enough general illumination to allow the operating staff to move about in safety, with particular regard to galleries and stairways.

The change-over is effected by contactors each of which controls a group of lighting points. The d.c. leads to the contactors are run separately from the a.c. leads, on different routes if possible, as a necessary security measure.

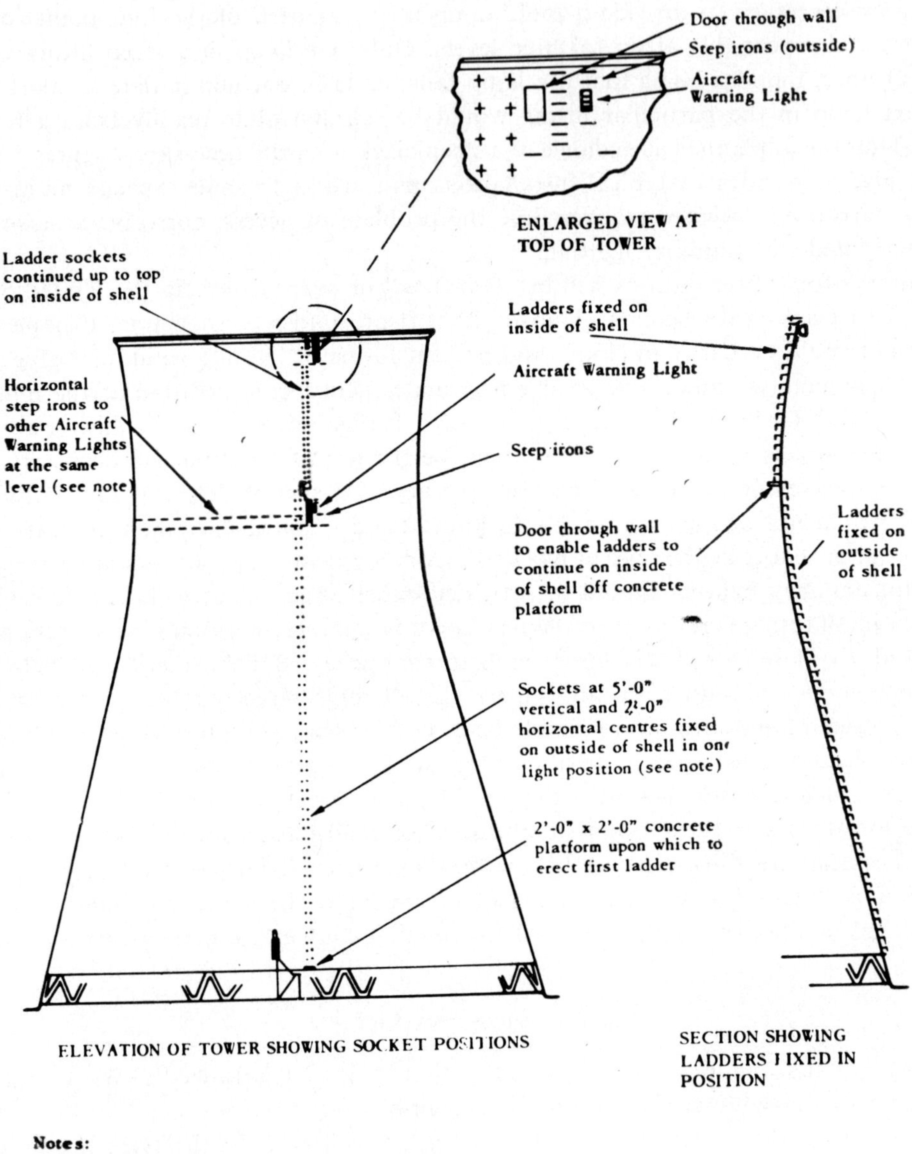

Notes:

Hand-holds and walkways to be provided around the perimeter at the top of concrete chimneys only, irrespective of whether aircraft warning lights are fitted.

Intermediate lines of step irons to be provided on all chimneys exceeding 400 feet in height, irrespective of whether aircraft warning lights are fitted.

FIG. 5.12.1A. Arrangement of ladder sockets and horizontal access step irons on cooling towers.

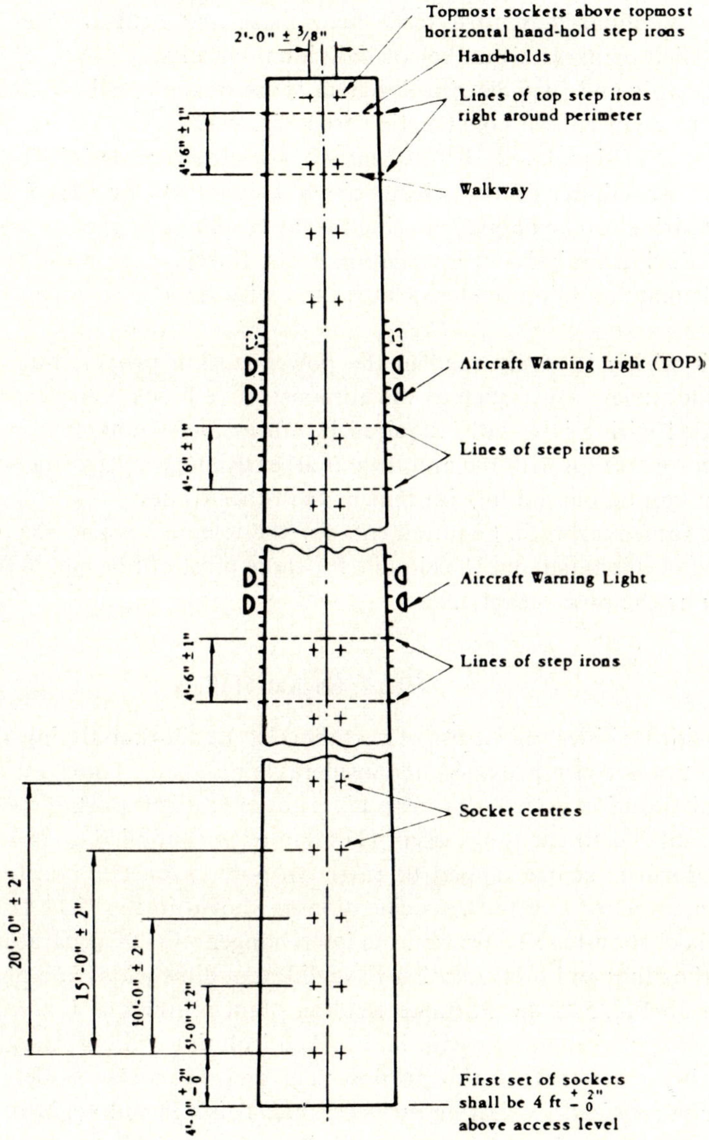

FIG. 5.12.1B. Arrangement of ladder sockets and horizontal access step irons on single-flue chimneys.

5.12.3. **Distribution**

The lighting supplies are derived from the 415 V general services switchboard, which feeds a number of smaller switchboards, each supplying a specific area.

These switchboards in turn feed three-phase and neutral fuse-distribution boards, each of which controls a number of individual circuits.

Switchboards and fuse-distribution boards are of the totally enclosed metal-clad type, and need to be of robust construction (see Fig. 5.12.3).

Paper or PVC-insulated, PVC-sheathed armoured cable will be used for the main feeders, the remainder of the wiring either being PVC or MICS cable. A concentric system of wiring can be employed using the latter, the outer sheath serving as a conductor, thereby reducing capital and installation costs. Special care must be taken, however, to ensure adequate bonding of the sheaths otherwise there is a danger of spurious currents flowing.

Conduit wiring is rarely used in the power station proper, but may be appropriate to "non-industrial" areas such as the administrative block, canteen, etc.

Concealed wiring is usually required in offices and similar accommodation, and this involves co-operation with the architect at an early stage so that the work of the electrical contractor can be phased in with that of the other trades.

For the same reason the requirements for the telephones and electric clocks should be considered at this stage, but the circuits for these must not be run in the same conduits or troughing as the power cables.

5.12.4. **Socket Outlets**

To provide facilities for the use of portable electric tools, soldering irons and handlamps, etc., socket outlets are provided in considerable numbers. Tools are supplied from either a 110 V distribution system or where this is not practicable portable 415/110 V transformers centre tapped to earth are used. Handlamps are supplied at 25 V, via 240/25 V portable transformers centre tapped to earth. In both cases, the transformers are manufactured to a standard C.E.G.B. specification as shown in Fig. 5.12.4A and have integrally mounted socket outlets. To prevent the interchangeability of apparatus for use at differing voltages the plugs and socket outlets have different dimensions and physical arrangements as shown on Fig. 5.12.4B. Portable welding plant requires a 415 V, three phase supply, and 60/100 A socket outlets with interlocked isolating switches are provided adjacent to the main items of plant for this purpose; a.c. and d.c. socket outlets of varying voltages may also be required for testing purposes, etc. It is stressed again that if the correct approved plugs and socket outlets are used, it is only possible to use a plug with its associated socket outlet of the same voltage.

INDUSTRIAL AND OFFICE PATTERNS (INDOOR TYPE)

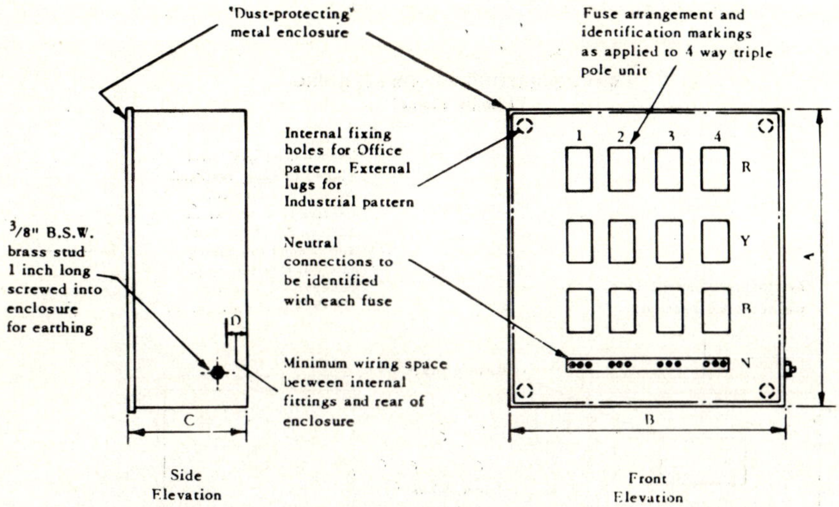

Code		Number of Ways	Rating (Amp)	Number of Poles	Dimensions* (Inches)			
					Max.			Min.
Industrial Pattern	Office Pattern				A	B	C	D*
12/43/029	12/43/162	4				12		
12/43/031	12/43/164	8	15	Single	16½	16		1½
12/43/033	12/43/166	12				22		
12/43/114	12/43/253	4				12	8	
12/43/115	12/43/254	6			24	16		2
12/43/116	12/43/255	8	15			17		
12/43/118	12/43/257	12				22		
12/43/122	12/43/260	4		Triple *		14		
12/43/123	12/43/261	6				18		
12/43/124	12/43/262	8	30		30	20	9½	2½
12/43/126	12/43/264	12				25		
12/43/129	12/43/267	4	60		34½	15		
12/43/131	12/43/269	8				23		

Notes:

1. Dimensions not specified left to manufacturers' discretion.
2. * Dimensions apply only where incoming cables pass behind internal fittings.
3. Provision on front of door for screw attachment of durable identification label, minimum size 4½ x 2 inches.
4. Enclosure adaptable to permit left or right hand door opening.
5. All fuse-boards fitted with neutral bar.
6. Industrial pattern arranged for padlocking, hasp details as follows:-
   Maximum thickness of hasp - 3/8 inch
   Diameter of hole - 1/4 to 3/8 inch
   Maximum thickness of metal between hole and the outside surface of the hasp - 3/8 inch
   Minimum clearance from outside of door to centre line of hole in hasp - 5/16 inch

FIG. 5.12.3. Distribution fuse boards (240/415 V a.c.).

WALL MOUNTING/FLOOR STANDING
(Single Phase)

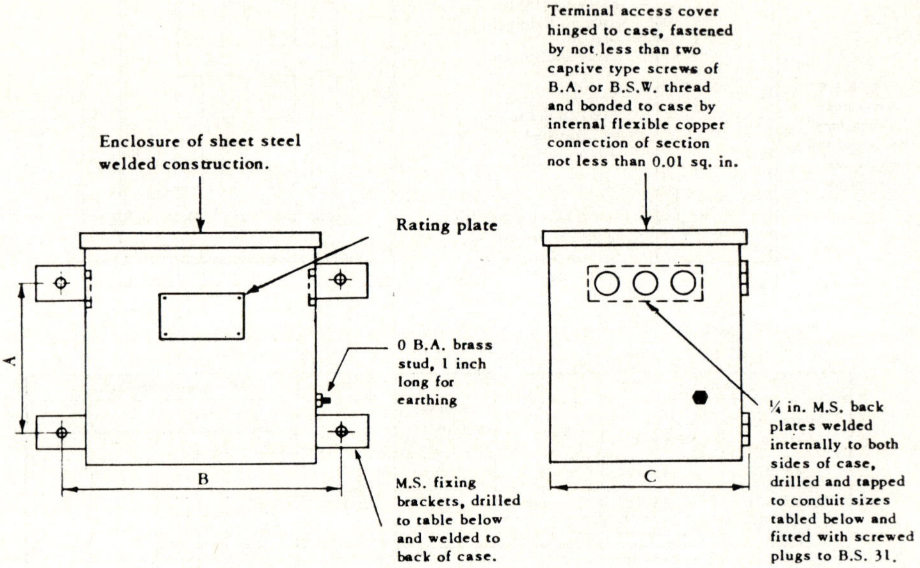

Code No.	Type	Rated Output VA	Rated Primary Voltage Range	Rated Secondary Voltage	Number and Size of Conduit Entries		Dimensions Inches			Diameter of fixing holes Inches
					Primary	Secondary	A	B	•C	
12 58 136	1	630	220-230/ 240-250	25	1 x ¾ in.	2 x ¾ in.	9½	13	9	11/32
12 58 146	2	1,600				2 x 1 in.				13/32
12/58/186	3	630	410-420/ 430-440	110		2 x ¾ in.				11/32
12 58 196	4	1,600				2 x 1 in.				13/32
12/58/206	5	2,500			1 x 1 in.	2 x 1 in.	12	16	11	
12/58 216	6	4,000				3 x 1 in.				17/32

FIG. 5.12.4A. Safety isolating transformers (240/25 : 415/110 V).

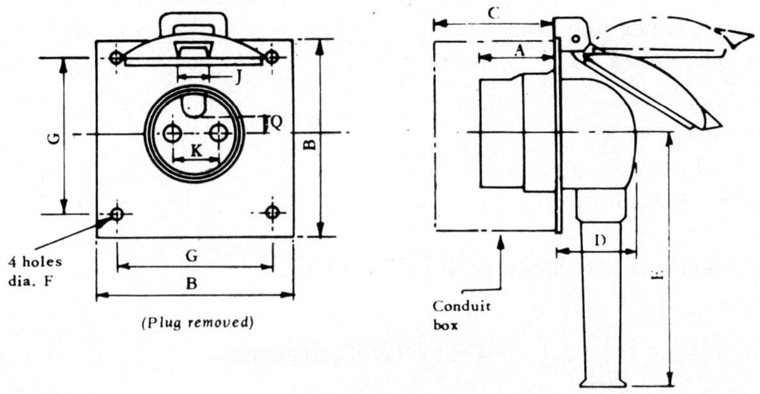

(Plug removed)

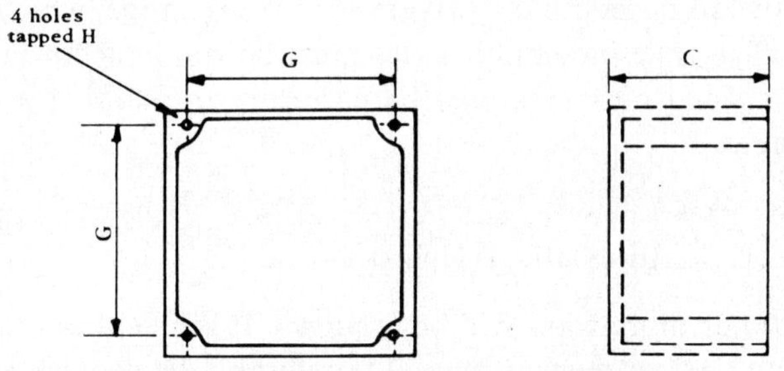

CAST IRON CONDUIT BOX

(For use with above Socket Outlets)

All dimensions in inches

Item	Code No.	Duty		A	B	C		D	E	F	G	H	J	K	Q
		Volts	Amps			Flush	Surface								
Plug Socket Flush Box Surface Box	12/33/111 12/33/191 12/33/241 12/33/253	25	5	$3/4$	3	$1\,3/16$	$1\,1/4$	1	$5\,1/8$	$3/16$	$2\,1/4$	4BA	.375	.625	.281
Plug Socket Flush Box Surface Box	12/33/121 12/33/201 12/33/245 12/33/257	110	15	$1\,19/64$	$3\,3/8$	2	2	$1\,1/4$	$5\,1/4$	$7/32$	$2\,11/16$	2BA	.438	.750	.313

FIG. 5.12.4B. Plugs and socket outlets.

5.13. TELECOMMUNICATIONS

5.13.1. General

Included under this heading are:

Post Office telephones.
Automatic telephones.
Operational telephones.
Staff location system.
Grid control communications.
Fire alarm.
General instructions associated with grid control.

5.13.2. Post Office Telephones

These are sufficiently numerous to necessitate a private branch exchange, which is often combined with the reception office. The incoming cables must be run in accordance with the G.P.O. requirements, and special measures may be necessary to avoid risk of interference due to earth fault currents.

5.13.3. Automatic Telephones

An automatic exchange of 200 or more ways will be required. It is essential that this be accommodated in a room sealed against dust, and should preferably be provided with air-conditioning.

Trunk cables of 25 or 50 pairs connect the P.A.X. with terminal boxes located in convenient situations, whence lines are run to the individual instruments. It is important that telephone cables, whether G.P.O. or private, are not run in the same conduit or trunking as other cables. However, when armoured cables are used, it is not necessary to adhere to this requirement.

5.13.4. Operational Telephones

These are direct wire instruments, and are used for operational instructions between specific points such as central and plant control rooms (if these are not combined) and between locations on the coal conveyor system. Loudspeakers can be used for this application, with the necessary "talk back" facilities incorporated.

5.13.5. Staff Location System

The great extent of the modern power station makes the provision of a staff location system very essential. There are now two different systems in successful use:

"Public address" by loudspeaker.
"V.H.F." personal receiver system.

The public address system needs a very large number of loudspeakers to cover the site. In noisy situations, the use of "sound columns" comprising a number of loudspeakers on one frame have done much to overcome the background noise problem. However, this system can be used to transmit time or fire alarm signals (in conjunction with a high-frequency oscillator) and there is no limit to the number of people who can be called.

The amplifier cubicles for a large installation occupy a considerable amount of space and must be accommodated in a clean and dry situation, such as an equipment or telecommunications room.

The personal receiver system employs a V.H.F. transmitter which radiates signals picked up by a small receiver carried in the pocket. Each receiver is tuned to a different frequency which is radiated when the appropriate call signal is sent out, so that only the individual concerned receives the signal.

The staff location system can be integrated with the automatic (P.A.X.) telephone system, so that the person called can by dialling a predetermined number be put directly into communication with the caller.

5.13.6. **Grid Control Communications**

The communications network for system operation consists, basically, of rented P.O. telephone lines radiating from the grid control centre to each of the power stations in the area.

The facilities provided include direct telephone connection between the grid control engineers and the power station central control room and other main points, such as the switching station. Indication of circuit breaker positions and overload alarms are automatically transmitted to the grid control centre and feeder flow meter readings of MW and MVAR and total generated output are also continuously transmitted.

On the control desk at the power station is mounted an instructor panel. On this panel are displayed the telegraphed instructions from system operation of the required generation together with the amount of generation actually being done. There are also a number of routine telegraph signals usually used for load shedding but whose functions can be changed to suit requirements. The main equipment at the power station is of the telephone type and is mounted in cubicles in the telecommunications room. Provision is commonly made for the supervisory control of small unattended supply points from the power station control room.

A V.H.F. transmitter is also installed with the microphone built into the control desk to enable radio communication to be maintained with the grid control centre in the event of breakdown of the telephone lines. This receiver together with that used for the staff location system can be accommodated in the equipment or telecommunications room, the associated aerials being mounted in a location to achieve maximum effect.

5.13.7. Fire Alarm

This is a statutory requirement, and is met by the installation of a number of suitable sirens or loudspeakers so as to be audible at all parts of the site. As has been mentioned in Section 5.13.5 this can also be incorporated with the public address system. An initiating push button is usually located in the central control room.

Sometimes a direct telephone line is run to the nearest fire station, thus avoiding the necessity of using the ordinary G.P.O. telephone system.

5.13.8. Smoke Detection

Suitable detectors are available which can be installed in situations such as cable tunnels, equipment rooms, etc., to give warning in the event of a concentration of smoke in these areas. An operator is thus able, in many cases, to take the necessary corrective action before a fire occurs.

5.14. ELECTRIC CLOCKS

A precision pendulum clock of the "synchronome" type is required to generate pulses for the metering equipment, and for driving the time-difference instrument in the control room. The clock must be solidly mounted, and preferably situated well away from running plant, so as to be as little subject to vibration as possible.

A second such clock can be used for driving the gatehouse time recorders and the dial clocks throughout the station, although in the latter case ordinary 240 V a.c. synchronous clocks can be used with economy.

There are available, however, sophisticated impulse-driven clock schemes which automatically correct themselves every hour, but the capital cost is high and more associated cabling is required.

Where two clocks are provided, changeover switching is incorporated so that each clock can take over the duties of the other if required.

5.15. SITE ELECTRICAL SUPPLIES

To construct and instal the plant in a 2000 MW power station involves a large number of mechanical and electrical main contractors; the work may be spread over six or more years employing at its peak up to 2500 men. Electricity will be required for a variety of uses such as:

(1) Welding.
(2) Heavy erection plant.
(3) Temporary pumps.
(4) Heating and lighting of contractors' huts.
(5) Heating and lighting of C.E.G.B. site offices.
(6) Heating and lighting of C.E.G.B. site canteens.
(7) Street and other security lighting as necessary.

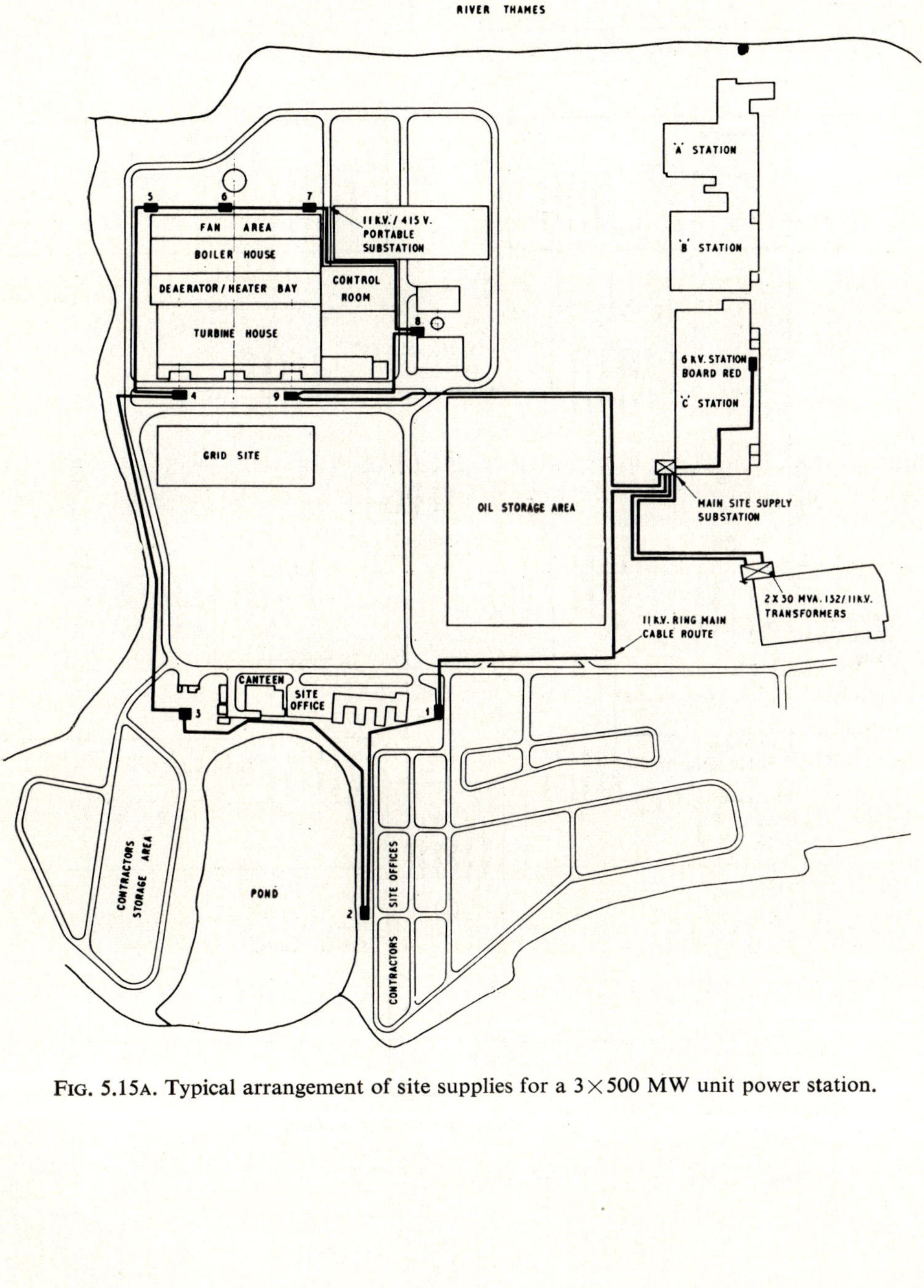

FIG. 5.15A. Typical arrangement of site supplies for a 3×500 MW unit power station.

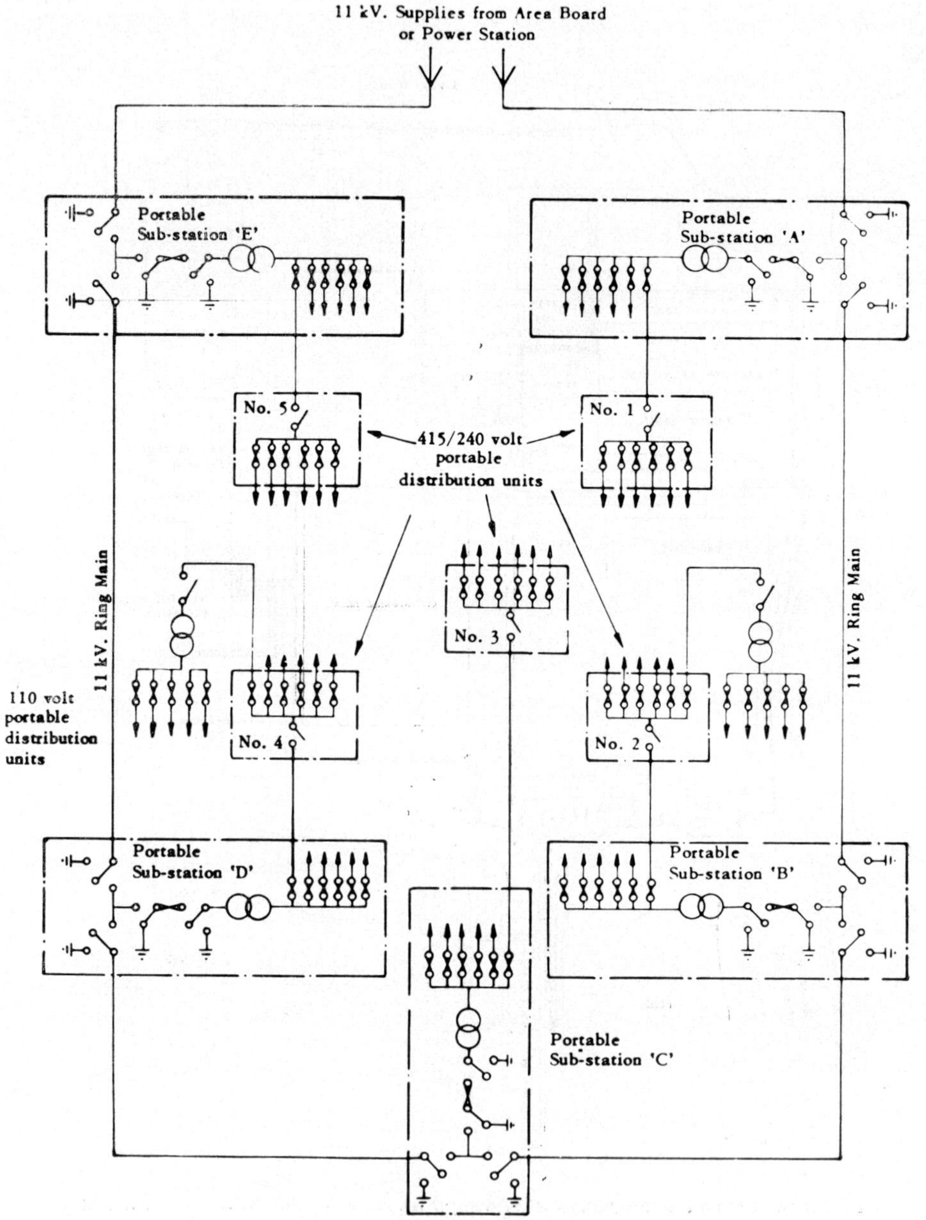

FIG. 5.15B. Typical site supply scheme.

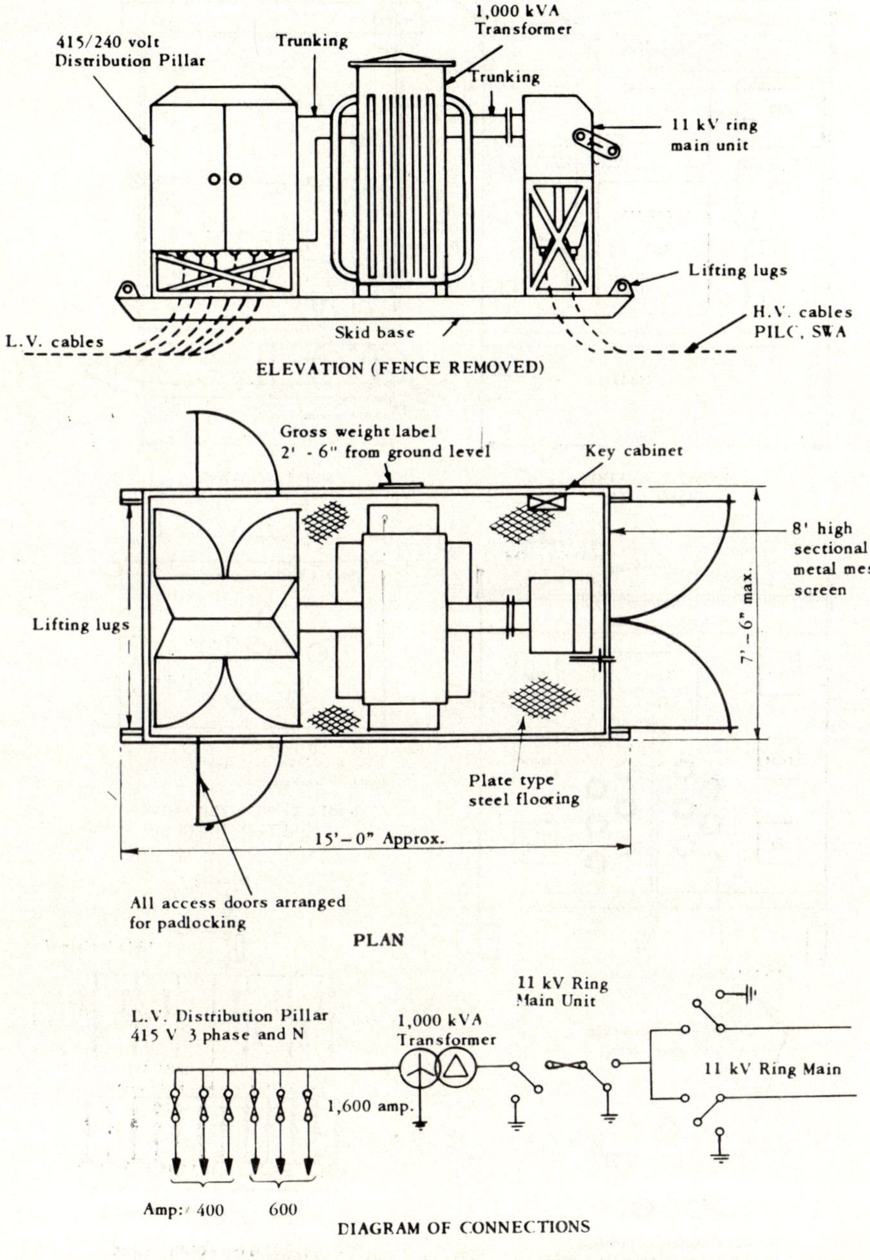

FIG. 5.15C. Portable substation 11,000/415/240 V.

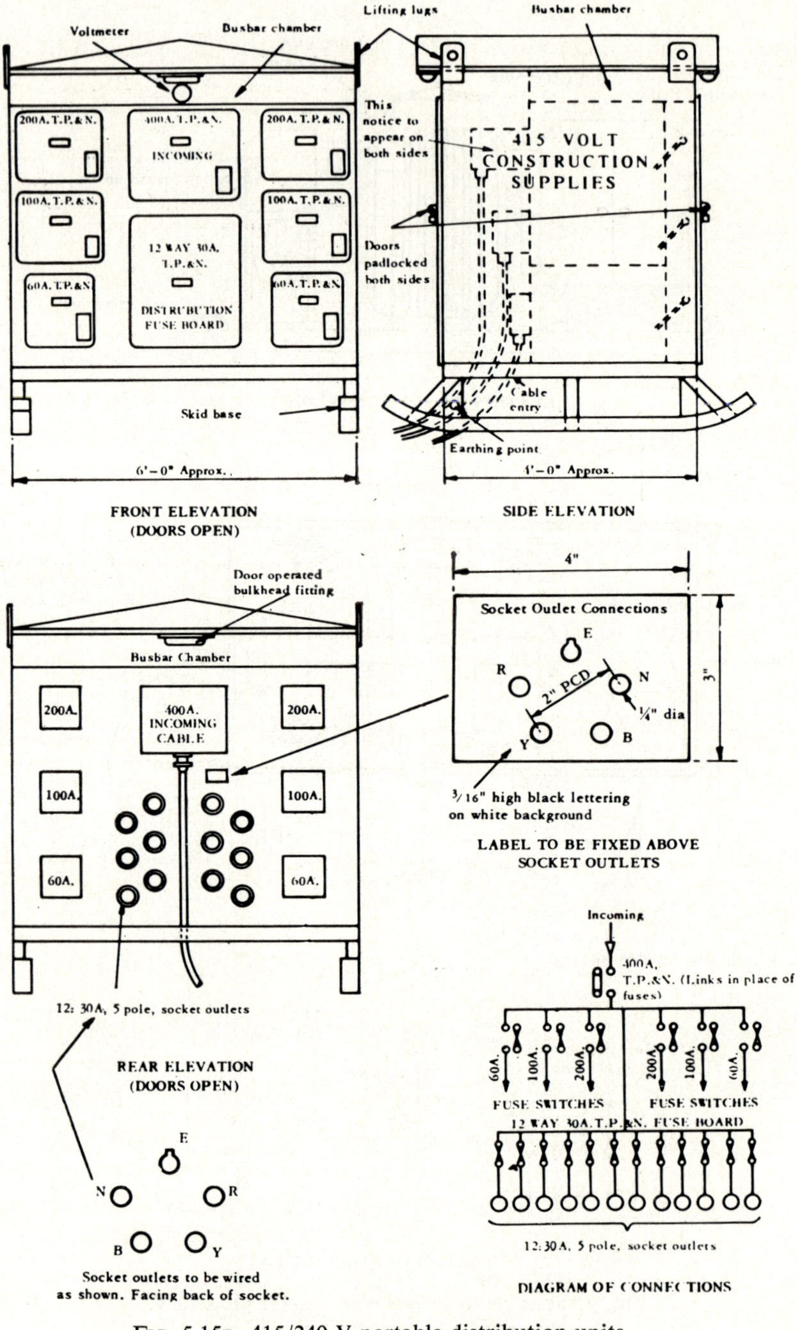

FIG. 5.15D. 415/240 V portable distribution units.

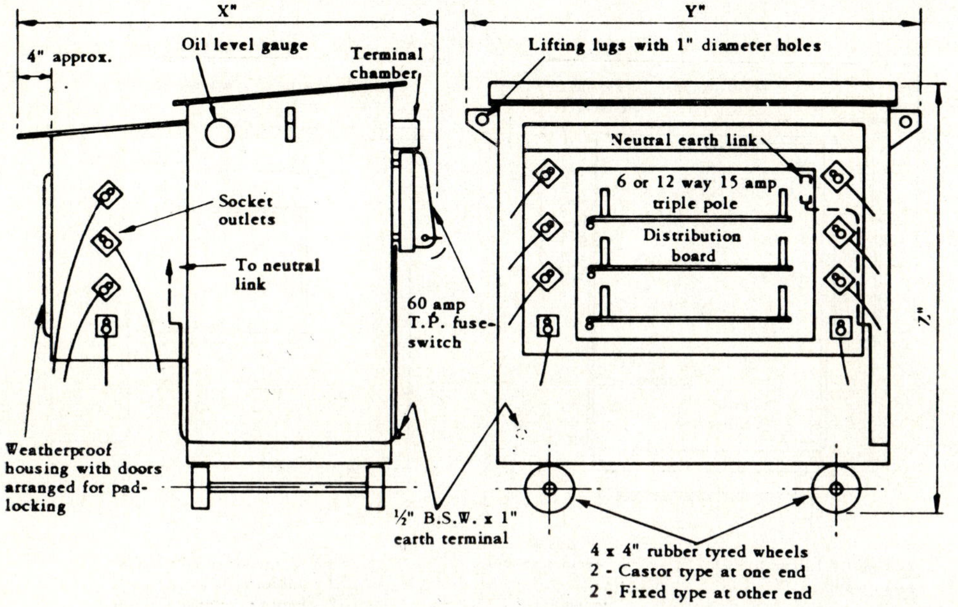

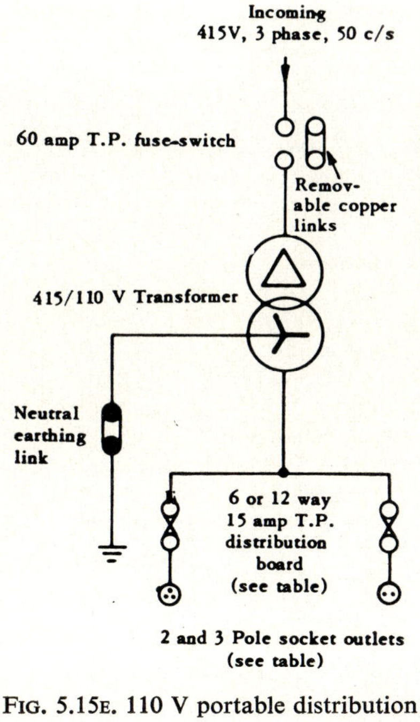

Table

Rating kVA	Distribution Board	No. of Sockets		Approx. Dimensions		
		2 Pole	3 Pole	X	Y	Z
5	6 way	6	2	2' - 6"	2' - 6"	2' - 9"
10	12 way	12	4	3' - 0"	3' - 0"	2' - 9"

FIG. 5.15E. 110 V portable distribution units — 5 and 10 kVA, typical arrangement.

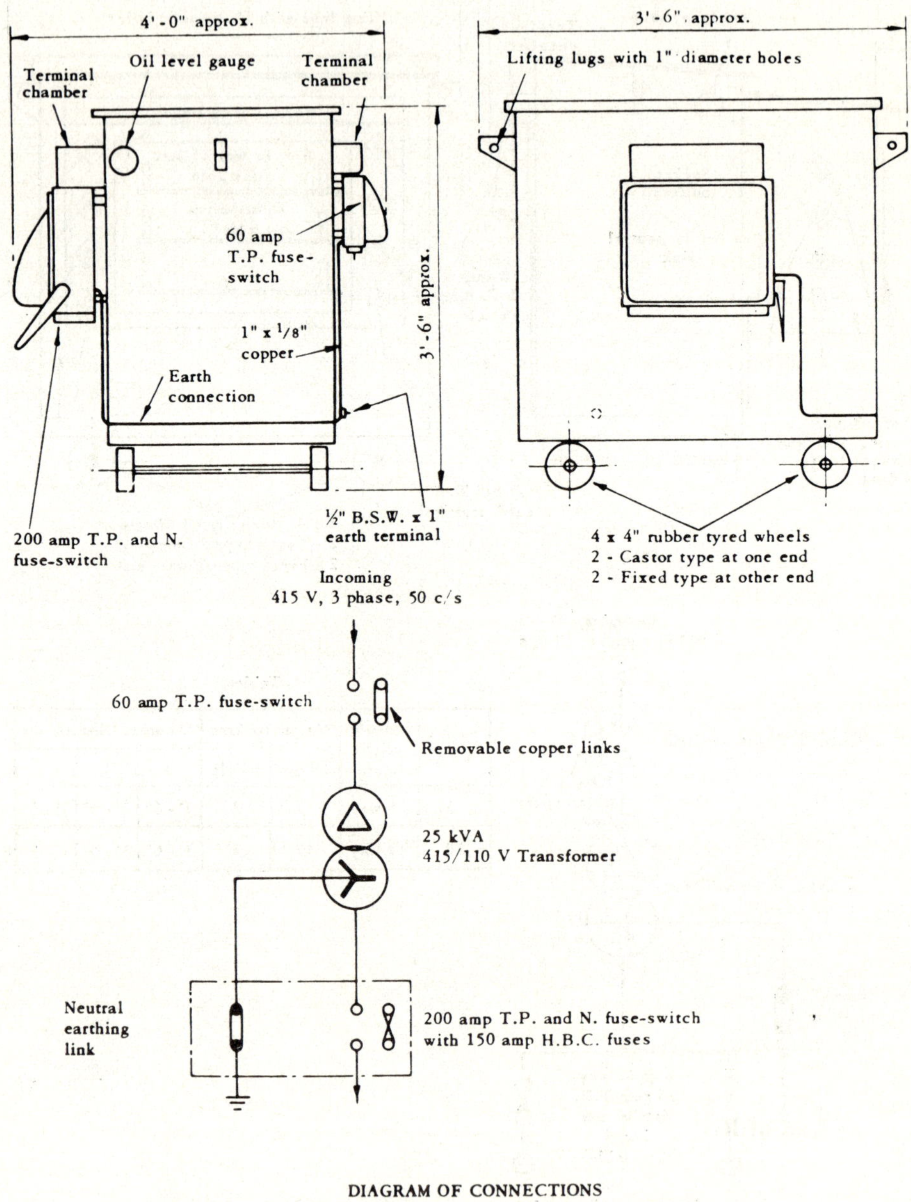

FIG. 5.15F. 110 V portable distribution unit–25 kVA, typical arrangement.

The above loads may have a maximum demand as high as 10 MW and in the general interest of efficiency and safety it is necessary to provide co-ordinated electrical distribution system covering the whole site, and the Board usually undertakes the responsibility of providing this. An 11 kV ring main is established around the perimeter of the construction area and is fed from the Area Board or power station as available. Up to 10×1 MVA capacity portable 11 kV/415/240 V substations are then located at suitable positions around the perimeter as required. Figure 5.15A shows a typical arrangement for a 3×500 MW unit station; Fig. 5.15B shows the schematic arrangement of a typical installation; and Fig. 5.15C the layout of a 1 MVA portable substation. In order to provide electricity nearer the actual working points, 415 V feeders are run in from the perimeter to 415/240 V portable distribution units as shown on Fig. 5.15D. For supplying power to portable tools and hand lamps the Board insists on the use of a 110 V a.c. system derived from a single- or three-phase transformer centre tapped to earth. The Board will usually provide suitable transformers complete with distribution boards and socket outlets or sw/fuses as shown on Figs. 5.15E and F. Electricity is usually provided free to contractors, subject to certain conditions, and a relevant clause to this effect is incorporated in the various specifications. When the construction has been completed, the substations can then be moved with the minimum of effort and installed on another site.

APPENDIX. C.E.G.B. Standards, etc.

List of Relevant C.E.G.B. Standards

1233/1	25, 110 and 240 V Socket Outlets.
1243/1	Standard for Distribution Fuse-boards.
1217/1	Standard for Cable Tray.
1258/1	Standard for Safety Isolating Transformers Portable and Wall Mounting/Floor Standing.
1385/1	Access Fittings for Chimneys and Cooling Towers.
3522/1	Standard for Power Station Auxiliary Transformers.
4388/1	Standard for Lightning and Static Protection of Structures.
4388/2	Standard for Earthing Electrodes.
4499/2	Standard for Electric Motors.
9963/1	Temporary Electrical Supplies.

List of Relevant C.E.G.B. Transmission Plant Design Standards

T.P.S. 3/55	Site Area for Land Acquisition 132 kV Substations.
T.P.S. 3/31	Site Area for Land Acquisition 275 kV Substations.
T.P.S. 3/22	Site Area for Land Acquisition 400 kV Substations.
T.P.S. 3/38	Standard 132 kV Substations and Their Equipment.
T.P.S. 3/20, T.P.S. 3/25	Standard 275 kV Substations and Their Equipment.

T.P.S. 3/23	Standard 400 kV Substations and Their Equipment.
T.P.S. 3/9	Safety Interlocking Principles for Grid Substations.
T.P.S. 3/13	Standards for 415 V Auxiliary Supplies to Grid Substations including the Provision of Diesel-driven Auxiliary Standby Plant.
T.P.S. 3/6	Oil Plant Arrangements for Outdoor Grid Substations.
T.P.S. 4/7	Transformer Noise Enclosures.

List of Relevant C.E.G.B. Generation Design Memoranda

GDM. 28	Design Memorandum for Gas Turbine Generating Plant on New Power Stations.
GDM. 60	Design Memorandum for Siting of Auxiliary Switchboards.
GDM. 61	Design Memorandum for Pulverised Fuel Power Station's Electrical System for Supplying Electrical Equipment on a 4×500 MW unit Station.
GDM. 305	Design Memorandum for Transformer Compound Drainage.

EXERCISES

1. A. What differences in layout might be expected if the grid switching station is located:

(a) Adjacent to the power station.

(b) Two miles distant.

B. What are the main advantages and disadvantages which a 400 kV indoor substation might be expected to show when compared with an outdoor type?

2. With the increasing size of power stations the auxiliary supply voltage has risen from 3·3 kV to 6·6 kV and 11 kV. Why is this?

3. In laying out the switchrooms for the 3·3 kV and 11 kV auxiliary switchgear, what special provisions are made for the safety of personnel?

4. A. Discuss the principal factors which govern the layout of the main generator connections for a large set.

B. What are the principal requirements to be met in the layout of the generator transformer?

5. To what extent are PVC-insulated cables used in power station work? Discuss their advantages and limitations.

6. What is the purpose of the station earthing system and what particular requirements must be met?

7. What fire risks are usually covered by fixed fire protection in power stations, and what systems are employed? What points must be considered in laying out the system?

8. What is the modern tendency with regard to the control of the generating units, and in which way may it be expected to develop?

CONTENTS OF VOLUMES 1-8

Volume 6. Instrumentation, Controls and Testing

1. Boiler and Turbine Instrumentation
2. Electrical Instrumentation and Metering
3. Plant Testing
4. Control Systems
5. Automation, Data Loggers and Computers

Volume 7. Operation and Efficiency

1. Commissioning and Operation of Boilers and Turbo-generators
2. Electrical Performance and Operation of Turbo-generators and System Operation
3. Safety and Maintenance
4. Practical Aspects of Maintenance
5. Plant Optimisation

Volume 8. Nuclear Power Generation

1. Nuclear Physics
2. Nuclear Power Reactors
3. British Nuclear Power Plant
4. Nuclear Power Station Operation
5. Nuclear Techniques